Studies in Surface Science and Catalysis 58

INTRODUCTION TO ZEOLITE SCIENCE AND PRACTICE

Studies in Surface Science and Catalysis

Advisory Editors: B. Delmon and J.T. Yates

Vol. 58

INTRODUCTION TO ZEOLITE SCIENCE AND PRACTICE

Editors

H. VAN BEKKUM

Laboratory of Organic Chemistry, Applied Chemistry Department, Delft University of Technology, Julianalaan 136, 2628 BL Delft, The Netherlands

E.M. FLANIGEN

UOP, Tarrytown Technical Center, Research and Molecular Sieve Technology, Tarrytown, NY 10591, U.S.A.

J.C. JANSEN

Laboratory of Organic Chemistry, Applied Chemistry Department, Delft University of Technology, Julianalaan 136, 2628 BL Delft, The Netherlands

ELSEVIER   **Amsterdam — Oxford — New York — Tokyo   1991**

ELSEVIER SCIENCE PUBLISHERS B.V.
Sara Burgerhartstraat 25
P.O. Box 211, 1000 AE Amsterdam, The Netherlands

Distributors for the United States and Canada:

ELSEVIER SCIENCE PUBLISHING COMPANY INC.
655, Avenue of the Americas
New York, NY 10010, U.S.A.

Library of Congress Cataloging-in-Publication Data

Introduction to zeolite science and practice / H. van Bekkum, E.M.
  Flanigen, J.C. Jansen (editors).
      p.   cm. -- (Studies in surface science and catalysis ; 58)
    Includes bibliographical references and index.
    ISBN 0-444-88969-8
    1. Zeolites.   I. Bekkum, Herman van. II. Flanigen, E.M.
  III. Jansen, J. C. IV. Series.
  TP159.M6I58 1991
  660'.2995--dc20   90-24298
                                                                 CIP

ISBN 0-444-88969-8 (Vol. 58)

This book is printed on acid-free paper.

Printed and bound by CPI Antony Rowe Ltd, Eastbourne
Transferred to digital printing 2006

Dedicated to Richard M. Barrer and Donald W. Breck,

two great scientists and pioneers in the field of zeolites and related materials.

Contents

Chapter 6. Clays from two to three dimensions
R.A. Schoonheydt

Chapter 7. Techniques of zeolite characterization
J.H.C. van Hooff and J.W. Roelofsen

Chapter 8. Solid state NMR spectroscopy applied to zeolites
G. Engelhardt

Chapter 9. Introduction to zeolite theory and modelling
R.A. van Santen, D.P. de Bruyn, C.J.J. den Ouden and B. Smit

Preface

Zeolite science and technology are still in the stage of rapid growth. This is testified
to

- the exponential increase of papers and patent applications in the field, reflecting
  the work of many zeolite scientists, both industrial as well as academic;

- the increasing number of industrial zeolite-based processes and their present rapid
  expansion into organic chemicals manufacture, following important applications of
  zeolites in hydrocarbon processing;

- the recent progress in matters of zeolite accessibility range, matrix behaviour,
  lattice components and satellite structures. In fact, the zeolite horizon seems even
  wider now than some ten years ago;

- the recognition that zeolites, by virtue of their stability and regenerability, are
  perfectly suited for incorporation into a new generation of processes in which clean
  technology and environmental friendliness are major issues.

Some sixteen years have passed since the monograph "Zeolite Molecular Sieves;
Structure, Chemistry and Use" by the late Donald W. Breck was published. The
wealth of information and insights offered by this book, together with its high quality,
guarantee its further intensive use for many years to come. In addition, the student or
professional working in the field of zeolites needs efficient access to the latest
achievements and developments. Some excellent books which deal with parts of the
zeolite field are those by R.M. Barrer, P.A. Jacobs and J.A. Martens, R. Szostak, and
A. Dyer.

The present book was conceived during the Summer School on Zeolites, held in
1989 at Zeist, The Netherlands, on the occasion of the 8th International Conference
on Zeolites which was attended by some ninety - mainly young - persons from many
countries. This Pre-Conference School was held in an informal atmosphere and
generated much enthusiasm among teachers and participants. The programme of
lectures delivered by experts in the zeolite field seemed perfectly suited to serve as a
basis for an introductory book on zeolites and related materials, covering theory and
practice.

All authors expanded their Summer School lectures into a review of the sector or
technique dealt with, starting at an elementary level but giving access to the latest
results and insights and referring to recent publications and reviews.

The Editors are convinced that the present text will not only be useful to students
and workers entering the field. More experienced workers in the field will find a
substantial coverage of the zeolite spectrum, including the latest views on zeolite
structure, characterization and applications. It is particularly the combination of
science *and* technology which will make the book a source of useful information for
those chemists and chemical engineers with an interest in zeolites.

We would like to express our gratitude to the authors who, despite their heavy work-loads, have been so co-operative. In particular we would like to acknowledge Mrs. Mieke van der Kooij-van Leeuwen for typing several chapters of this book into such a pleasing format. She proved to be a perfect coordinator between authors, the editors and publishers in the preparation of the book and in organizing the Pre-Conference School.

Herman (H.) van Bekkum

       Edith (E.M.) Flanigen

              Koos (J.C.) Jansen

Delft/Tarrytown/Delft, September 1990

List of contributors

H. van Bekkum
Laboratory of Organic Chemistry, Delft University of Technology, Julianalaan 136,
2628 BL Delft, The Netherlands

D.P. de Bruyn,
Shell Research B.V., P.O. Box 3003, 1003 AA Amsterdam, The Netherlands

G. Engelhardt
Faculty of Chemistry, University of Konstanz, D-7750 Konstanz, Germany

E.M. Flanigen
UOP Research and Development, Tarrytown Technical Center, Tarrytown, New York
10591, U.S.A.

W.F. Hölderich
BASF A.G., Ammoniaklaboratorium, D-6700 Ludwigshafen, Germany

J.H.C. van Hooff
Laboratory for Inorganic Chemistry and Catalysis, Eindhoven University of
Technology, P.O. Box 513, 5600 MB Eindhoven, The Netherlands

P.A. Jacobs
Department of Surface Science, Catholic University of Leuven, Kardinaal Mercierlaan
92, B-3030 Leuven, Belgium

J.C. Jansen
Laboratory of Organic Chemistry, Delft University of Technology, Julianalaan 136,
2628 BL Delft, The Netherlands

H.G. Karge
Fritz-Haber-Institut der Max-Planck-Gesellschaft, Faradayweg 4-6, 1000 Berlin 33,
Germany

H. van Koningsveld
Laboratories of Organic Chemistry and Applied Physics, Delft University of
Technology, Lorentzweg 1, 2628 CJ Delft, The Netherlands

H.W. Kouwenhoven
ETH Zürich, Technisch-Chemisches Laboratorium, Universitätstrasse 6, CH-8092
Zürich, Switzerland

B. de Kroes
Akzo Chemicals B.V., Research Centre Amsterdam, P.O. Box 15, 1000 AA
Amsterdam, The Netherlands

J.A. Martens
Department of Surface Science, Catholic University of Leuven, Kardinaal Mercierlaan 92, B-3030 Leuven, Belgium

I.E. Maxwell
Shell Research B.V., P.O. Box 3003, 1003 AA Amsterdam, The Netherlands

L. Moscou
Akzo Chemicals, Research Centre Amsterdam, P.O. Box 15, 1000 AA Amsterdam, The Netherlands

C.J.J. den Ouden
Shell Research B.V., P.O. Box 3003, 1003 AA Amsterdam, The Netherlands

M.F.M. Post
Shell Research B.V., P.O. Box 3003, 1003 AA Amsterdam, The Netherlands

J.W. Roelofsen
Akzo Chemicals B.V., Research Centre Amsterdam, P.O. Box 15, 1000 AA Amsterdam, The Netherlands

R.A. van Santen
Laboratory of Inorganic Chemistry and Catalysis, Eindhoven University of Technology, P.O. Box 513, 5600 MB Eindhoven, The Netherlands

R.A. Schoonheydt
Laboratory of Surface Chemistry, Catholic University of Leuven, Kardinaal Mercierlaan 92, B-3030 Leuven, Belgium

B. Smit
Shell Research B.V., P.O. Box 3003, 1003 AA Amsterdam, The Netherlands

W.H.J. Stork
Shell Research B.V., P.O. Box 3003, 1003 AA Amsterdam, The Netherlands

R. Szostak
Zeolite Research Program, Georgia Tech Research Institute, Georgia Institute of Technology, Atlanta, Georgia 30332, U.S.A.

R.P. Townsend
Unilever Research, Port Sunlight Laboratory, Bebington, Wirral, Merseyside L63 3JW, U.K.

S.T. Wilson
UOP Research and Development, Tarrytown Technical Center, Tarrytown, New York 10591, U.S.A.

Chapter 1

THE ZEOLITE SCENE

L. Moscou

Akzo Chemicals, Research Centre Amsterdam,
P.O. Box 15, 1000 AA Amsterdam (The Netherlands)

INTRODUCTION

The present Volume "Introduction into Zeolite Science and Practice" originates from the material that was presented to participants of the "Zeolite Summer School". This school was held in Zeist, Holland prior to and in conjunction with the 8th International Zeolite Conference in Amsterdam.

The material presented during this three days tutorial covered principles, interpretations, recent views and trends in zeolite science and technology. In addition, demonstrations on zeolite synthesis, characterization and applications were given and some experimental training of participants was involved as well.

Professor Herman van Bekkum and Dr. Koos Jansen who where the organizers of this summer school and Dr. Edith Flanigen and many other zeolite experts contributed to the tutorial material and presentations. The wide-spread and overwhelming interest in this summer school prompted the organizers to publish this material. The complete set of presentations have been reworked and extended to form the basis for this book.

The broad academic and industrial interest in zeolites stems from the unique combination of properties of these materials. Many of these properties will be handled in detail by the various authors in their chapter of this book. This preface is an attempt to overview the Zeolite Scene such as the special zeolite character, the various zeolite applications and the market size, their economic impact and the growth of zeolite R&D during the last two decades.

WHAT MAKES ZEOLITES SO SPECIAL ?

The question, what makes zeolites so special when compared to

other crystalline inorganic oxide-materials, can be answered by
mentioning a combination of properties: the <u>microporous</u>
<u>character</u> with <u>uniform pore dimensions</u>, allowing certain
<u>hydrocarbon molecules to enter the crystals</u> while rejecting
others based on too large a molecular size, the <u>ion-exchange</u>
<u>properties</u> which allow to perform all sorts of ion exchange
reactions, the ability to develop <u>internal acidity</u> which makes
the zeolites interesting materials for catalyzing organic
reactions and the high <u>thermal stability</u> of the zeolites.
    We have seen how this "Zeolite Material Science" field has
broadened and deepened by a number of interrelated and
simultaneously occurring developments such as:

* diversification of the composition outside
  aluminosilicates, particularly the crystalline
  aluminophosphates and Ga, Fe, B and Ti containing
  structures.
* growing number of zeolite framework structures.
* increasing insight in crystallization mechanisms.
* development of post-crystallization modification
  procedures.
* development of powerful characterization techniques.

    which makes us again face the question: "what makes zeolites
so special"?
We can add now the following broadened characteristics:

<u>All metal-oxygen tetrahedra exposed to the surface</u>
    Let us first consider non-porous inorganic oxide solids: only
a small fraction of the metal-oxygen atoms in the bulk is
exposed. Surface atoms are accessible only. The surface can be
increased by particle size reduction, but if this is
accomplished by milling etc, it is not a well-controlled
structural modification. Moreover, to reach near 100% exposure,
particle size should be reduced to the nanometer scale.
    Quite different is the situation with zeolites. Via the
entire internal micropore surface all atoms are accessible and
moreover, the surface is not formed by breakage of bonds but at
the pore surface the local coordination of the atoms is unchanged (ref. 1). This is the case because the zeolite structure
is formed by tetrahedra and so the pores are formed by corner

connected tetrahedra. The exception is here the outer surface
of the zeolite crystals where the structure is terminated.
Different surface properties can be expected at the outer
crystal surface compared to the internal surface. The terminated
outer surface of the zeolite crystal is only small compared to
its high, "structure dictated" internal surface area.

Accessibility for modifications

Because all metal-oxygen tetrahedra are exposed to the
internal zeolite surface, they are in principle all accessible
if the pore dimensions allow so. This makes that zeolites are
ideal for all sorts of modifications.
The three main types of modifications include:
1) exchange of charge-compensating cations
2) replacement of Si and Al in the zeolite framework
3) introduction of metal particles

Examples of reactions where countercharge ions are involved
are (reversible) ion-exchange reactions and mostly irreversible
chemical reactions such as reactions of zeolite protons with
silanes as to change pore size or to influence acidity.
Metal atoms involved modifications are often applied in
industrial processes for making zeolite catalysts.
Especially in large pore (Y) zeolites real molecular engineering
is done on large scale to rearrange Si and Al atoms in the
zeolite framework as to reach optimal fitness for use by
fine-tuning the structural and acidic properties. These
framework-rearrangements are known under the names of
substitution (Al$\leftrightarrow$Si), extraction (Al-removal), and insertion
reactions (Si-insertion) and more generally: ultrastabilisation.
Also other metal atoms than Al and Si can be introduced in the
framework.

Stoichiometry of the structure

As each aluminium atom induces one negative charge in the
framework and thus requires one positive counter ion, the
zeolite can be seen as a real stoichiometric inorganic polymer.
Its properties such as ion-echange capacity, proton content,
acidity etc. are quantitatively related to its chemical
composition.
This stoichiometry in combination with the complete

accessibility for smaller molecules and the crystalline and
homogeneous nature of the zeolites, appears to be ideal for
developing and calibrating characterization techniques such as
IR spectroscopy, NMR spectroscopy, NH_3-adsorption and
-desorption and many more.

Acid site tailoring

The fine-tuning of the zeolite structure by all sorts of
modification reactions as described above, can tailor the acid
site properties such as the average site density and the
so-called number of next nearest neighbours (NNN). The latter
property is important in catalytic cracking reactions as it
appears to induce site-selectivity; the balance between
monomolecular and bimolecular reactions can be carefully
regulated by the NNN (refs. 2,3). This is of extreme importance
in the field of gasoline octane-boosting cracking catalysts.

Diversification in zeolite compositions and structures

The ability of aluminosilicate zeolite structures to appear
in compositions where part or all of the metal atoms have been
replaced by other elements and the possibilities to crystallize
in new structures, seem to make the number of compositions and
structures nearly unlimited. In general, new elements in the
frameworks will certainly change the (acidic) properties while
new crystallographic structures seem extremely interesting in
shape- and size -selective catalysis. The new ultralarge pore
zeolites as recently synthesized are attracting a lot of
attention by the catalytic community

Versatile potential for catalytic applications

As we will see in the next section, catalytic zeolite
applications have greatly materialized in the hydrocarbon
processing field (petroleum/petrochemicals). Application of
zeolites in the synthesis of organic fine chemicals is still
underdeveloped; it was suggested that one of the reasons is that
the average synthetic organic chemist is not acquainted with
zeolites and their potentials. These potentials have been
demonstrated for a variety of organic reactions such as
isomerization, aromatic substitution, oxidation and cyclization,
showing the versatile character of zeolites in organic synthesis
reactions (refs. 4,5).

ZEOLITE APPLICATIONS

The four main area's in which zeolites are applied are:

Adsorbants/desiccants/separation processes

This include applications as drying agents, in gas purification, and in important separation processes like n-paraffins from branched paraffins, p-xylene from its isomers etc.

Catalysts

The main industrial catalytic applications are in three area's:
- Petroleum Refining
- Synfuels Production
- Petrochemicals Production

In the Petroleum Refining area, the catalytic cracking is by far the largest application (Y, ZSM-5) followed by hydrocracking catalysts (Y, Mordenite). Other applications are in hydro-isomerisation and dewaxing, using mostly mordenite and ZSM-5. The well-known Methanol-To-Gasoline Process for Synfuel production applies ZSM-5. Also important petrochemical processes such as ethylbenzene by alkylation of benzene, xylene isomerisation, toluene disproportionation run over ZSM-5 containing catalysts.

Detergents

Tonnage wise the use of zeolites in detergent formulations is a large market with even far larger potential. Nearly exclusively zeolite A is serving here as a sequestering agent to substitute phosphates.

Miscellaneous

Either synthetic or natural zeolites are used in a number of applications such as:
- waste water treatment
- nuclear effluent treatment
- animal feed supplements
- soil improvement

ZEOLITE MARKET VOLUME

Our estimate for the zeolite usage in USA, W-Europe and Japan in 1988 is a total of appr. 550.000 metric tons (refs. 6-8).

Fig. 1 shows the dominance of zeolite A in this market as 67% of all synthetic zeolites produced is A-type zeolite for detergent builders (Fig. 2). Zeolite A is an example of a commodity chemical with high production volumes and standard specifications. For catalytic applications more than 98% of the zeolite market is in fluid catalytic cracking and hydrocracking catalysts. The Y-zeolites used are typical specialty products with different complex structures, relatively low or medium production volumes and many novel formulations. Market prices are much higher than for commodity zeolites. This is also the case for a smaller market segment: the ZSM-5 class of zeolites.

The size of the natural zeolite market is negligible in Western-Europe. In Japan considerable amounts are used in soil improvement and other applications to an estimated total of 5000 tons/year. USA consumption is 10000 tons/year for a variety of applications.

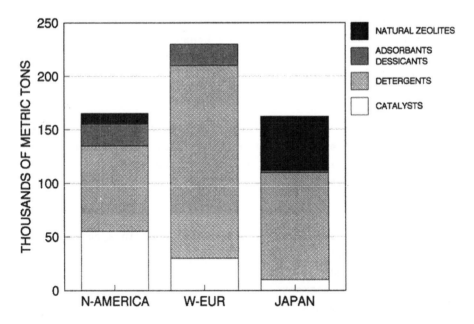

Fig. 1. Diagram showing the estimated zeolite consumption in 1988.

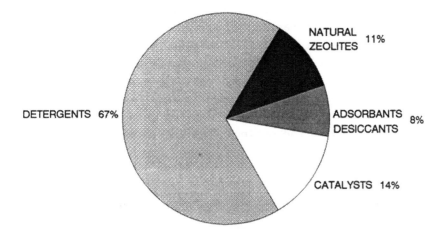

Fig. 2. Diagram showing the zeolite consumption as volume distribution over application types (N-America, W-Europe and Japan in 1988).

ECONOMIC IMPACT

For a so diverse group of zeolite materials and their wide-spread use, it is difficult to give a detailed overview on the economic impact of each specific application. Therefore only a few general observations: the high ion exchange capacity and the specificity in certain ion-exchange reactions makes zeolites economically interesting species. Where small amounts of poison can spoil huge amounts of water, so can small amounts of zeolites clean up large water volumes. Zeolites have high potentials to protect eco-systems, from waste water and gas treatment to water softeners in detergent builders replacing the undesired polyphosphate.

In the hydrocarbon processing field, catalytic cracking is an example where zeolites have had large impact on both crude oil economy and on catcrackerunit-economy. Zeolite application made it possible to increase strongly the catalyst activity and this allowed a further exploitation of the FCC unit's riser technology. At the same time the cracking reactions became much more selective in favour of desired products such as gasoline (Table I). At a worldwide use of more than 8.10^6 tons of oil/day even one percent gasoline yield increase represents an extremely high value. In practice, the accumulated gasoline yield improvements as induced by zeolite technology during the last two decades, has reached levels far over 10%, which means a worldwide added product value of 15-20 billion US Dollars per

year. Even more impressive is that application of zeolite
catalysts in the FCC units has led to enormous savings in crude
oil use to reach the desired gasoline outputs. In other words:
more economic use of our crude oil resources was made possible
by zeolite catalysts. In recent years, not only gasoline yield,
but additionally gasoline quality (octane number) has become
important and the present tailoring of zeolite acid-site
properties is directed to the optimum combination of gasoline
yield with octane quality.

TABLE 1
Evolution in FCC-catalyst quality for VGO-cracking from 1960-1989

Year	1960	1970	1975	1980	1989
Cat-type	Amorphous	Amorphous RE-Y	HD-matrix RE-Y	HD-matrix US-Y	HD-matrix Impr.-Y
Activity	53	60-70	70-80	60-70	70-80
Octane	89	87	84	87	88
Conv.	←		65%		→
Coke	7.8	6.2	4.5	3.0	3.0
Gasoline	40.0	43.5	46.5	44.5	47.0
Gas	17.2	16.0	14.0	17.5	15.0

ZEOLITE R&D
Zeolite publications

The growth of zeolite R&D is well illustrated by the steady
incline of the number of yearly zeolite publications during the
last two decades (Fig. 3). This is based on referenced
titles and key words in Chemical Abstracts (ref. 9). It is
estimated that approximately 2/3 of all the zeolite publications
are "caught" this way. Thus true numbers should be 50% higher
than shown in Fig. 3. From the so estimated 3700 zeolite
publications in 1988, about 1000 are patents.

Although the number or publications is not the only indication for R&D intensity and density, comparison of publications
frequency per country or region can show general trends. Fig. 4
indicates the domination in zeolite publications by North-

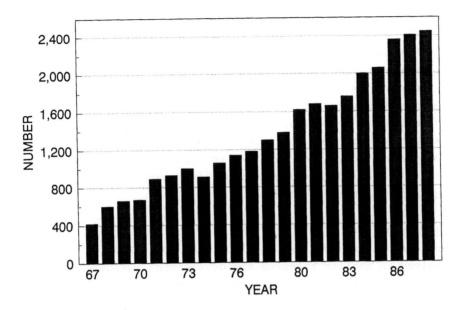

Fig. 3 Number of CA-referenced zeolite publications per year.

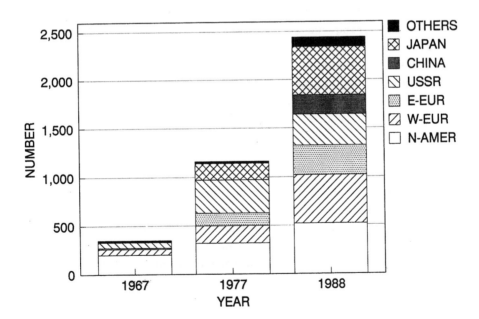

Fig. 4. Distribution of zeolite publications over geographical areas in 1967, 1977 and 1988.

America in 1967. In the following decade, the USSR had strongly
increased their number of publications and so did Japan and
Europe.

In 1988 a more balanced situation appears between these
regions, where China has strongly grown during the last decade.
This information is not very detailed but indicates trends: it
shows that nowadays zeolite R&D is rather well-spread over the
various regions in N-America, Asia and Europe. Two geographic
regions that has not been mentioned so far, because their
relatively small activities compared to the majors, are India an
Australia + N.Zealand, showing strongly increased zeolite
activities during the last decade.

Zeolite scientists

Next question of interest is: "how many zeolite scientists
are involved in zeolite research and development worldwide?"
From the numbers of zeolite publications and estimated zeolite
professionals in Holland and Belgium and the worldwide number of
zeolite publications, we calculated the worldwide number of
zeolite professionals in R&D to be around 15000 (ultimo 1988).
This number does not include the numerous engineers involved in
zeolite manufacturing. We should keep in mind that this simple
calculation is based on the assumption that the percentage of
zeolite work that leads to a publication in the low countries is
about the worldwide average. To compensate for that uncertainty
it seems justified to estimate the true number between 10.000 an
20.000.

ZEOLITE NETWORKS

Several national and international organizations are devoted
to zeolite activities. One of the oldest national ones is the
British Zeolite Association, well-known for their informal
meetings on a yearly basis. The Japanese Zeolite Association was
founded in connection with the 7th International Zeolite
Conference in Tokyo 1986. It is active in distributing zeolite
reference materials and joint characterization studies. In 1988
the German Zeolite Association was founded. Other national
zeolite groups exist in Belgium and France. In the USA, the
American Society for Testing and Materials (ASTM) dedicated a
task group to standardization of Zeolite Test Methods as part of

committee D-32 on Catalysts. The following zeolite catalyst test methods have been and are being standardized:

Issued Standard Testmethods for FCC Zeolite Catalysts
- Unit Cell Dimension of a Faujasite-Type zeolite
- Zeolite area of a catalyst
- Zeolite content by X-ray diffraction
- Catalyst acidity by Ammonia Chemisorption
- Steam deactivation of fresh FCC catalyst
- Catalytic activity by Micro Activity Test (MAT)

Zeolite Standard Testmethods in preparation
- Adsorption capacity
- Crystallinity by X-ray diffraction
- Calcium Exchange Capacity
- Particle size

Finally, two International Zeolite Organizations exist: the ICNZ (the International Committee on Natural Zeolites) and the IZA (International Zeolite Association). They jointly organize the 3-yearly International Zeolite Conferences and sponsor and promote other activities, including publications. Well-known is IZA's Structure Commission which published the "Atlas of Zeolite Structure Types" (ref. 10).
It was also the IZA that sponsored this first zeolite summer school in the Netherlands.

CONCLUSION
No better way to conclude with the first sentence that was written by Donald Breck in 1973 in his famous book: "Zeolite Molecular Sieves". This sentence reads:

> "Rarely in our Technological Society does the discovery of
> a new class of inorganic materials result in such a wide
> scientific interest and kaleidoscopic development of
> applications as has happened with the Zeolite Molecular
> Sieves"

To-day this statement seems more true than ever before!

12

REFERENCES

1   R.A. van Santen, Chem. Magazine, Dec. 1988, pp. 760-761.
2   L.A. Pine, P.J. Maher and W.A. Wachter, J. Catal., 85 (1984)
    pp. 466-476.
3   E. van Broekhoven and H. Wijngaards, in H.J. Lovink (Ed.),
    Akzo Catalysts Symposium '88, Akzo Chemicals Amersfoort, 198
    section F-8.
4   H. van Bekkum and H.W. Kouwenhoven, in M. Guisnet,
    J. Barrault, C. Bouchoule, D. Duprez, C. Montassier and
    G. Pérot (Eds.), Stud. Surf. Sci. Catal., No 41, Elsevier,
    Amsterdam, 1988, pp. 45-59.
5   W. Hölderich, M. Hesse and F. Näumann, Angew. Chem., 100
    (1988), pp. 232-251.
6   N.Y. Chen and T.F. Degnan, Chem. Eng. Progress, 84 (1988),
    pp. 32-41.
7   M. Smart, V. von Schuller-Goetzburg and M. Tashiro, Chemical
    Economics Handbook, SRI International, Zeolites Product
    Review, August 1988.
8   J. Griffiths, Industrial Minerals, Jan. 1987, pp. 19-33.
9   N.G. Bader, Internal Akzo Report.
10 W.M. Meier and D.H. Olson, Atlas of Zeolite Structure Types,
    published on behalf of the Structure Commission of the
    International Zeolite Association, Butterworths, 1987.

Chapter 2

ZEOLITES AND MOLECULAR SIEVES
AN HISTORICAL PERSPECTIVE

Edith M. Flanigen, UOP Research and Development, Tarrytown Technical Center

Tarrytown, New York 10591, U.S.A.

ABSTRACT
     The history of zeolites and molecular sieves is reviewed from the discovery
of the first zeolite mineral in 1756 through the explosion in new molecular sieve
structures and compositions in the 1980's.
     R. M. Barrer's early pioneering work in adsorption and synthesis began the
era of synthetic zeolites. The discovery of the commercially significant
synthetic zeolites A, X and Y by R. M. Milton and D. W. Breck in the late 1940's
to early 1950's led to their introduction by Union Carbide Corporation as a new
class of industrial adsorbents in 1954, and in 1959 as hydrocarbon conversion
catalysts. Today they are used widely throughout the petroleum refining and
chemical process industries as selective adsorbents, catalysts and ion
exchangers, and represent an estimated quarter of a billion dollar industry.
     The last four decades have seen a chronological progression in molecular
sieve materials from the aluminosilicate zeolites to the microporous silica
polymorphs to the microporous aluminophosphate-based polymorphs and
metallosilicate compositions.

1.   EARLY HISTORY

     The history of zeolites began in 1756 when the Swedish mineralogist
Cronstedt discovered the first zeolite mineral, stilbite (ref. 1). He recognized
zeolites as a new class of minerals consisting of hydrated aluminosilicates of
the alkali and alkaline earths. Because the crystals exhibited intumescence when
heated in a blowpipe flame, Cronstedt called the mineral a "zeolite" derived from
two Greek words, "zeo" and "lithos" meaning "to boil" and "a stone". In 1777
Fontana described the phenomenon of adsorption on charcoal (ref. 2). In 1840
Damour observed that crystals of zeolites could be reversibly dehydrated with no
apparent change in their transparency or morphology (ref. 3). Schafhäutle
reported the hydrothermal synthesis of quartz in 1845 by heating a "gel" silica
with water in an autoclave (ref. 4). Way and Thompson (1850) clarified the
nature of ion exchange in soils (ref. 5). Eichhorn in 1858 showed the
reversibility of ion exchange on zeolite minerals (ref. 6). St. Claire Deville
reported the first hydrothermal synthesis of a zeolite, levynite, in 1862
(ref. 7). In 1896 Friedel developed the idea that the structure of dehydrated
zeolites consists of open spongy frameworks after observing that various liquids
such as alcohol, benzene, and chloroform were occluded by dehydrated zeolites
(ref. 8). Grandjean in 1909 observed that dehydrated chabazite adsorbs ammonia,
air, hydrogen and other molecules (ref. 9), and in 1925 Weigel and Steinhoff

reported the first molecular sieve effect (ref. 10). They noted that dehydrated chabazite crystals rapidly adsorbed water, methyl alcohol, ethyl alcohol and formic acid but essentially excluded acetone, ether or benzene. In 1927 Leonard described the first use of x-ray diffraction for identification in mineral synthesis (ref. 11). The first structures of zeolites were determined in 1930 by Taylor and Pauling (ref. 12, 13). In 1932 McBain established the term "molecular sieve" to define porous solid materials that act as sieves on a molecular scale (ref. 14).

Thus, by the mid-1930's the literature described the ion exchange, adsorption, molecular sieve and structural properties of zeolite minerals as well as a number of reported syntheses of zeolites. The latter early synthetic work remains unsubstantiated because of incomplete characterization and the difficulty of experimental reproducibility.

Barrer began his pioneering work in zeolite adsorption and synthesis in the mid-1930's to 1940's. He presented the first classification of the then known zeolites based on molecular size considerations in 1945 (ref. 15) and in 1948 reported the first definitive synthesis of zeolites including the synthetic analogue of the zeolite mineral mordenite (ref. 16).

2. INDUSTRIAL HISTORY
2.1. Synthetic Zeolites

Barrer's work in the mid to late 1940's inspired Milton of the Linde Division of Union Carbide Corporation to initiate studies in zeolite synthesis in search of new approaches for separation and purification of air. Between 1949 and 1954 R. M. Milton and co-worker D. W. Breck discovered a number of commercially significant zeolites, types A, X and Y. In 1954 Union Carbide commercialized synthetic zeolites as a new class of industrial materials for separation and purification. The earliest applications were the drying of refrigerant gas and natural gas. In 1955 T. B. Reed and D. W. Breck reported the structure of the synthetic zeolite A. In 1959 Union Carbide marketed the "ISOSIV" process for normal-isoparaffin separation, representing the first major bulk separation process using true molecular sieving selectivity. Also in 1959 a zeolite Y-based catalyst was marketed by Carbide as an isomerization catalyst (ref. 17).

In 1962 Mobil Oil introduced the use of synthetic zeolite X as a cracking catalyst. In 1969 Grace described the first modification chemistry based on steaming zeolite Y to form an "ultrastable" Y. In 1967-1969 Mobil Oil reported the synthesis of the high silica zeolites beta and ZSM-5. In 1974 Henkel introduced zeolite A in detergents as a replacement for the environmentally suspect phosphates. By 1977 industry-wide 22,000 tons of zeolite Y were in use in catalytic cracking.   In 1977 Union Carbide introduced zeolites for ion-exchange separations.

2.2. Natural Zeolites

For 200 years following their discovery by Cronstedt, zeolite minerals (or natural zeolites) were considered to occur typically as minor constituents in vugs or cavities in basaltic and volcanic rock. Such occurrences precluded their being obtained in mineable quantities for commercial use. From the late 50's to 1962 major geologic discoveries revealed the widespread occurrence of a number of natural zeolites in sedimentary deposits throughout the Western United States. The discoveries resulted from the use of x-ray diffraction to examine very fine-grained (1-5μm) sedimentary rock.   Some zeolites occur in large near monomineralic deposits suitable for mining. Those that have been commercialized for adsorbent applications include chabazite, erionite, mordenite and clinoptilolite (ref. 18).

Japan is the largest user of natural zeolites (see Cpt. 1 in this volume by Moscou). Mordenite and clinoptilolite are used in adsorbent applications including air separation and in drying and purification (ref. 19).   Natural zeolites have also found use in bulk applications as fillers in paper, in pozzolanic cements and concrete, in fertilizer and soil conditioners and as dietary supplements in animal husbandry. The latter bulk uses represent the major volume applications for natural zeolites.

3.   THE 80'S

3.1. Materials

In the 1980's there has been extensive work carried out on the synthesis and applications of ZSM-5 and a proliferating number of other members of the high silica zeolite family.   In 1982 microporous crystalline aluminophosphate molecular sieves were described by Wilson et al. (ref. 20) at Union Carbide, and additional members of the aluminophosphate-based molecular sieve family, e.g., SAPO, MeAPO, MeAPSO, ElAPO and ElAPSO, subsequently disclosed by 1986 (ref. 21). Considerable effort in synthesizing metallosilicate molecular sieves was reported where the metals iron, gallium, titanium, germanium and others were incorporated during synthesis into silica or aluminosilicate frameworks, typically with the ZSM-5 (MFI) topology (ref. 22).   Additional crystalline microporous silica molecular sieves and related clathrasil structures were reported.

The 80's saw major developments in secondary synthesis and modification chemistry of zeolites. Silicon-enriched frameworks of over a dozen zeolites were described using methods of: thermochemical modification (prolonged steaming); mild aqueous ammonium fluorosilicate chemistry; and by high temperature treatment with silicon tetrachloride and low temperature treatment with fluorine gas.

Similiarly, framework metal substitution using mild aqueous ammonium fluorometallate chemistry was reported to incorporate iron, titanium, chromium and tin into zeolite frameworks by secondary synthesis techniques. A review of modification and secondary synthesis chemistry is given in Cpt. 6 in this volume by Szostak.

Overall, the 80's can be described as representing an explosion in new compositions and structures of molecular sieves.

3.2. Characterization Techniques

Application of state-of-the-art sophisticated characterization and structural techniques to molecular sieve materials in the 1980's gave a major advance in our understanding of the structure and chemistry of molecular sieves. These include: the systematic development of hypothetical frameworks (ref. 23); the application of computational chemistry, computer modeling and ab initio calculations to molecular sieve structures; the application of solid state NMR and high resolution electron microscopy techniques; ^{129}Xe NMR characterization; and the use of high energy radiation (synchrotron and intense pulsed neutron) to solve zeolite structures.

3.3. Applications

Applications of zeolites and molecular sieves in the 1980's showed a growth in petroleum refining applications with emphasis on resid cracking and octane enhancement. ZSM-5 was commercialized as an octane enhancement additive in fluid catalytic cracking where Si-enriched Y zeolites served as the major catalytic component in high octane FCC catalysts. The use of zeolite catalysts in the production of organic (fine) chemicals appeared as a major new direction. Zeolites in detergents as a replacement for phosphates became the single largest volume use for synthetic zeolites worldwide (see Moscou this volume, Cpt. 1). Zeolite ion exchange products were used extensively in nuclear waste cleanup at Three Mile Island. New applications emerged for zeolite powders in two potentially major areas, odor removal and as plastic additives.

In adsorption and separation applications the 80's saw a major growth in the use of pressure swing adsorption for the production of oxygen, nitrogen and hydrogen. Processes for the purification of gasoline oxygenate additives were introduced.

3.4. New Directions

An exciting new scientific direction emerged in the 80's for exploring molecular sieves as advanced solid state materials. A recent review by Ozin et al. (ref. 24) speculate "that zeolites (molecular sieves) as microporous molecular electronic materials with nanometer dimension window, channel and cavity architecture represent a 'new frontier' of solid state chemistry with great opportunities for innovative research and development". The applications described or envisioned include: molecular electronics, "quantum" dots/chains, zeolite electrodes, batteries, non-linear optical materials, and chemical sensors.

4. REVIEW OF MOLECULAR SIEVE AND ZEOLITE FUNDAMENTALS

4.1. Molecular Sieves

Molecular sieves are porous solids with pores of the size of molecular dimensions, 0.3-2.0nm in diameter. Examples include zeolites, carbons, glasses and oxides. Some are crystalline with a uniform pore size delineated by their crystal structure, e.g., zeolites. Others are amorphous, e.g., carbon molecular sieves. Most current commercial molecular sieves are zeolites.

The pore size distribution for representative commercial adsorbent types (Fig. 1) contrasts the singular pore size of zeolites 5A and X or Y, with the broader distribution and larger mean pore diameter of the amorphous activated carbon and alumina adsorbents. The carbon molecular sieve has a mean pore diameter intermediate between that of 5A and X or Y zeolite but has a significantly broadened pore size distribution.

4.2. Nomenclature

There is no systematic nomenclature developed for molecular sieve materials. The discoverer of the synthetic species based on a characteristic x-ray powder diffraction pattern and chemical composition assigns trivial symbols. The early synthetic materials discovered by Milton, Breck and coworkers

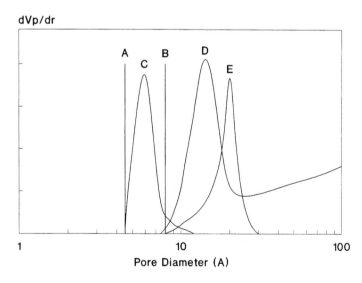

Figure 1. Pore size distribution of representative adsorbents. A) 5A (CaA) zeolites; B) X or Y zeolites; C) carbon molecular sieve; D) activated carbon; and E) porous alumina.

at Union Carbide used the arabic alphabet, e.g., zeolites A, B, X, Y, L. The use
of the Greek alphabet was initiated by Mobil and Union Carbide with the zeolites
alpha, beta, and omega. Many of the synthetic zeolites which have the structural
topology of mineral zeolite species were assigned the name of the mineral, for
example, synthetic mordenite, chabazite, erionite and offretite.

The molecular sieve literature is replete with acronyms: ZSM-5, -11 . . .,
ZK-4 (Mobil); EU-1, FU-1, NU-1 (ICI); LZ-210 and AlPO, SAPO, MeAPO, etc.
(Union Carbide, UOP); and ECR-1 (Exxon). The one publication on nomenclature by
IUPAC in 1979 (ref. 25) is limited to the then known zeolite type materials.

The IZA Atlas of Zeolite Structure Types (1st Ed. 1978; 2nd Ed. 1987)
published by the IZA Structure Commission assigns a three letter code to be used
for a known framework topology irrespective of composition. Illustrative codes
are LTA for Linde zeolite A, FAU for molecular sieves with a faujasite topology,
e.g., zeolites X and Y, MOR for the mordenite topology, MFI for the ZSM-5 and
silicalite topologies, and AFI for the aluminophosphate $AlPO_4$-5 topology. The
acceptance of a newly determined structure of a zeolite or molecular sieve for
inclusion in the official Atlas is reviewed and must be accepted by the IZA
structure commission. The IZA structure commission was given the authority in
1986 at the Tokyo conference to approve and/or assign the three-letter structure
code for new framework topologies.

4.3. Zeolites

Zeolites are crystalline aluminosilicates of group IA and group IIA
elements such as sodium, potassium, magnesium, and calcium (ref. 26). Chemically,
they are represented by the empirical formula:

$$M_{2/n}O \cdot Al_2O_3 \cdot ySiO_2 \cdot wH_2O$$

where y is 2 to 10, n is the cation valence, and w represents the water contained
in the voids of the zeolite. Structurally, zeolites are complex, crystalline
inorganic polymers based on an infinitely extending three-dimensional,
four-connected framework of AlO_4 and SiO_4 tetrahedra linked to each other by the
sharing of oxygen ions. Each AlO_4 tetrahedron in the framework bears a net
negative charge which is balanced by a cation. The framework structure contains
channels or interconnected voids that are occupied by the cations and water
molecules. The cations are mobile and ordinarily undergo ion exchange. The
water may be removed reversibly, generally by the application of heat, which
leaves intact a crystalline host structure permeated by the micropores and voids
which may amount to 50% of the crystals by volume.

The structural formula of a zeolite is based on the crystallographic unit
cell, the smallest unit of structure, represented by:

$$M_{x/n}[(AlO_2)_x(SiO_2)_y] \cdot wH_2O$$

where n is the valence of cation M, w is the number of water molecules per unit cell, x and y are the total number of tetrahedra per unit cell, and y/x usually has values of 1-5. In the case of the high silica zeolites y/x is 10 to 100.

There are two types of structures: one provides an internal pore system comprised of interconnected cage-like voids; the second provides a system of uniform channels which, in some instances, are one-dimensional channel systems. The preferred type has two- or three-dimensional channels to provide rapid intracrystalline diffusion in adsorption and catalytic applications.

In most zeolite structures the primary structural units, the AlO_4 or SiO_4 tetrahedra, are assembled into secondary building units which may be simple polyhedra such as cubes, hexagonal prisms, or octahedra. The final structure framework consists of assemblages of the secondary units. (See Cpt. 3 in this volume by Van Koningsveld)

More than 50 novel, distinct framework structures of zeolites are known. They exhibit pore sizes from 0.3-0.8 nm, and pore volumes from about 0.10 to 0.35 cc/g. Typical zeolite pore sizes using oxygen-packing models are shown in Figure 2. They include small pore zeolites with eight-ring pores with free

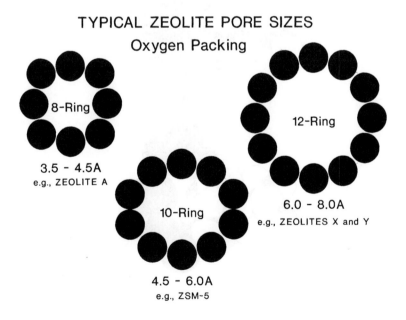

Figure 2. Typical zeolite pore sizes illustrated with oxygen packing model.

20

diameters of 0.30-0.45 nm, e.g., zeolite A; medium pore zeolites formed by a ten ring, 0.45-0.60 nm in free diameter, e.g., ZSM-5; and large pore zeolites with 12-ring pores, ~0.8 nm, e.g., zeolites X and Y. The molecular sieve effect is illustrated in Figure 3 for calcium A zeolite with an eight-ring pore of oxygens. Normal octane (top) readily accesses the internal void through the pore whereas isooctane (bottom) is larger than the pore and is totally excluded.

The zeolite framework should be viewed as somewhat flexible, with the size and shape of the framework and pore responding to changes in temperature and guest species. For example, ZSM-5 with sorbed neopentine has a 0.62nm near-circular pore, but with substituted aromatics as the guest species the pore assumes an elliptical shape, 0.45 to 0.70 μm in diameter.

Some of the more important zeolite types most of which have been used in commercial applications include the zeolite minerals mordenite, chabazite, erionite and clinoptilolite, and the synthetic zeolites type A, X, Y, L, omega, "Zeolon" mordenite, ZSM-5, and zeolites F and W.

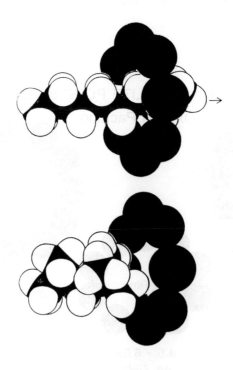

Figure 3. Illustration of molecular sieve effect. Straight chain molecule of normal octane (top) passes through eight ring aperature of 5A (CaA) zeolite; branched molecule of iso-octane (bottom) cannot.

5. HISTORY OF MOLECULAR SIEVE MATERIALS

The theme and research on molecular sieve materials over the last thirty-five year period has been a quest for new structures and compositions. The major discoveries and advances in molecular sieve materials during that period are summarized in Table 1.

TABLE 1
Evolution of Molecular Sieve Materials

Late 40's to Early 50's	Low Si/Al Ratio Zeolites
Mid to Late 60's	High Si/Al Ratio Zeolites
Early 70's	SiO_2 Molecular Sieves
Late 70's	$AlPO_4$ Molecular Sieves
Late 70's to Early 80's	SAPO and MeAPO Molecular Sieves
Late 70's	Metallo- silicates, aluminosilicates
Early to Mid 80's	$AlPO_4$-based Molecular Sieves

The history of commercially significant molecular sieve materials from 1954 to 1979 was reviewed by this author in 1980 (ref. 27). Highlights from that review and the subsequent history are presented here. The reader is referred to Cpt. 3 in this volume by Van Koningsveld for the structures of the materials.

5.1. Aluminosilicate Zeolites and Silica Molecular Sieves

The evolution of aluminosilicate zeolites is summarized in Table 2 based

TABLE 2
The evolution of molecular sieve materials[a]

"Low" Si/Al Zeolites (1 to 1.5):
    A, X

"Intermediate" Si/Al Zeolites (~2 to 5):
    a.   Natural Zeolites:
        erionite, chabazite, clinoptilolite, mordenite
    b.   Synthetic Zeolites:
        Y, L, large pore mordenite, omega

"High" Si/Al Zeolites (~10 to 100):
    a.   By thermochemical framework modification:
        highly siliceous variants of Y, mordenite, erionite
    b.   By direct synthesis:
        ZSM-5

Silica Molecular Sieves:
        silicalite

[a]Adapted from ref. 27

on increasing framework Si/Al composition. The four somewhat arbitrary categories are: 1) "low", 2) "intermediate", 3) "high" silica zeolites, and 4) "silica" molecular sieves.

The transition in properties accompanying the increase in the framework Si/Al is summarized in Table 3. The latter are generalized and should only be viewed as trends. The thermal stability increases from about $700^{\circ}C$ in the low silica zeolites to $1300^{\circ}C$ in the silica molecular sieves. The surface selectivity which is highly hydrophilic in the low silica zeolites is hydrophobic in the high silica zeolites and the silica molecular sieves. The acidity tends to increase in strength with increasing Si/Al ratio. As the Si/Al ratio increases, the cation concentration and ion exchange capacity (proportional to aluminum content) decreases. The structures of the low silica zeolites are predominantly formed with 4, 6, and 8 rings of tetrahedra. In the intermediate silica zeolites we see the onset of 5-rings in mordenite and omega zeolite. In the high silica zeolite structures and the silica molecular sieves we find a predominance of 5-rings of tetrahedra.

TABLE 3
The transition in properties[a]

Transition in:

Si/Al, from 1 to ∞
Stability, from $\leq700^{\circ}C$ to $\sim1300^{\circ}C$
Surface selectivity, from hydrophilic to hydrophobic
"Acidity", increasing strength
Cation concentration, decreasing
Structure, from 4, 6, and 8-rings to 5-rings

---

[a]From ref. 27

The low silica zeolites represented by zeolites A and X are aluminum-saturated, have the highest cation concentration, and give optimum adsorption properties in terms of capacity, pore size and three-dimensional channel systems. They represent highly heterogeneous surfaces with a strongly hydrophilic surface selectivity. The intermediate Si/Al zeolites (Si/Al of 2-5) consist of the natural zeolites erionite, chabazite, clinoptilolite and mordenite, and the synthetic zeolites Y, mordenite, omega and L. These materials are still hydrophilic in this Si/Al range.

The high silica zeolites with Si/Al of 10-100 can be generated by either thermochemical framework modification of hydrophilic zeolites or by direct synthesis. In the modification route stabilized, siliceous variants of Y, mordenite, erionite, and over a half-dozen other zeolites have been prepared by steaming and acid extraction. These materials are reported to be hydrophobic and organophilic and represent a pore size range from 0.4-0.8nm. A very large number of high silica zeolites prepared by direct synthesis have now been reported, including beta, ZSM-5, -11, -12, -21, -34, NU-1 and FU-1, and ferrisilicate and borosilicate analogs of the aluminosilicate structures. Typical of the reported silica molecular sieves are silicalite, fluoride silicalite, silicalite-2 and TEA-silicate. ZSM-5 and silicalite have achieved commercial significance.

The difference in surface selectivity between the hydrophobic silica molecular sieve, silicalite, and the highly hydrophilic zeolite NaX can be observed by comparing their equilibrium adsorption isotherms for water, oxygen and n-hexane (Fig. 4). The hydrophilic NaX pore fills at low partial pressures with all three adsorbates giving a typical Langmuir-type isotherm shape. On silicalite n-hexane and oxygen show a similar typical molecular sieve pore

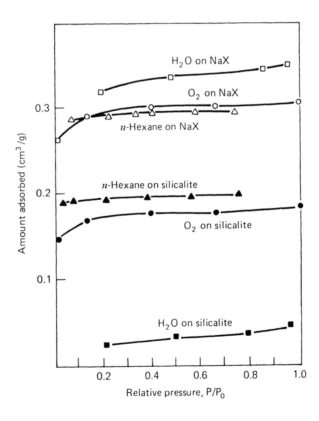

Figure 4. Comparison of adsorption equilibrium isotherms for water, oxygen and n-hexane on NaX zeolite and silicalite. Water and n-hexane at ambient and oxygen at -183°C.

filling at low partial pressures, but illustrative of the hydrophobic characteristic of silicalite, only a small amount of water (approximately ~0.05 cc/g) is adsorbed even at a relative partial pressure near 1. It should be noted that the pore volume of silicalite is substantially smaller than that of NaX.

In summary, if we compare the properties of the low and intermediate zeolites with those of the high silica zeolites and silica molecular sieves, we find that their resulting properties allow the low and intermediate zeolites to remove water from organics and to carry out separations and catalysis on dry streams. In contrast, the hydrophobic high silica zeolites and silica molecular sieves can remove and recover organics from water streams and carry out separations and catalysis in the presence of water.

5.2. Aluminophosphate-Based Molecular Sieves.

In 1982 a major discovery of a new class of aluminophosphate molecular sieves was reported by Wilson et al. (ref. 20). By 1986 some thirteen elements were reported to be incorporated into the aluminophosphate frameworks, Li, Be, B, Mg, Si, Ti, Mn, Fe, Co, Zn, Ga, Ge and As (ref. 21). These new generations of molecular sieve materials designated $AlPO_4$-based molecular sieves comprise more than two dozen structures and two hundred compositions.

5.2.1. Structures

The over two dozen structures of $AlPO_4$-based molecular sieves reported to date include zeolite topological analogues and a large number of novel structures. The major structures are shown in Table 4. They include fifteen novel structures as well as seven structures with framework topologies related to those found in the zeolites, chabazite (34, 44, 47), erionite (17), gismondine (43), levynite (35), Linde Type A (42), faujasite (37), and sodalite (20). Also shown is the pore size and saturation water pore volume for each structure type. The structures include very large pore (1.25nm), large pore (0.7-0.8nm), intermediate pore (~0.6nm), small pore (~0.4nm), and very small pore (~0.3nm) materials. Saturation water pore volumes vary from 0.16 to 0.35 cc/g comparable to the pore volume range observed in zeolites.

The novel structures which have been determined include types 5, 11, 14, 16, 22, 33, 39, 46, 50 and 52. The $AlPO_4$-based structures have been reviewed by Bennett et al. (ref. 28) and are described in this volume in Cpt. 3 by Van Koningsveld.

5.2.2. $AlPO_4$

The new family of aluminophosphate materials ($AlPO_4$-n) includes the first very large pore material, VPI-5 reported by Davis et al. (ref. 29). The VPI-5 structure is outlined by a unidimensional channel of an 18-membered ring with a free pore diameter of 1.25 nm. There is one 12-membered ring structure, $ALPO_4$-5 with a pore size of 0.7-0.8nm; several medium pore structures outlined

TABLE 4
Typical structures in $AlPO_4$-based molecular sieves[a]

Species	Structure Type	Pore Size, nm	Sat'n H_2O Pore Vol. cc/g	Species	Structure Type	Pore Size, nm	Sat'n H_2O Pore Vol. cc/g
Very Large Pore				Small Pore			
VPI-5[b]	Novel,detm.	1.25	0.35	14	Novel,detm.	0.4	0.19
				17	Erionite	0.43	0.28
Large Pore				18	Novel	0.43	0.35
5	Novel,detm.	0.8	0.31	26	Novel	0.43	0.23
36	Novel	0.8	0.31	33	Novel	0.4	0.23
37	Faujasite	0.8	0.35	34	Chabazite	0.43	0.3
40	Novel	0.7	0.33	35	Levynite	0.43	0.3
46	Novel,detm.	0.7	0.28	39	Novel	0.4	0.23
				42	Linde Type A	0.43	0.3
Intermediate Pore				43	Gismondine	0.43	0.3
11	Novel,detm.	0.6	0.16	44	Chabazite-like	0.43	0.34
31	Novel	0.65	0.17	47	Chabazite-like	0.43	0.3
41	Novel	0.6	0.22				
				Very Small Pore			
				16	Novel	0.3	0.3
				20	Sodalite	0.3	0.24
				25	Novel	0.3	0.17
				28	Novel	0.3	0.21

[a]Adapted from ref. 21. [b]From Davis et al., ref. 29.

by 10-membered rings or elliptical 12-rings with pore diameters of 0.6-0.65nm, for example, $AlPO_4$-11, -31 and -41; and small pore size materials such as $AlPO_4$-17 (ERI), with 8-membered ring pores and pore sizes of 0.35-0.45nm in diameter.

The product composition expressed in terms of oxide ratios is $xR \cdot Al_2O_3 \cdot 1.0 \pm 0.2P_2O_5 \cdot yH_2O$, where R is an amine or quaternary ammonium ion. The $AlPO_4$ molecular sieve as synthesized must be calcined at 400 to 600°C to remove the R and water yielding a microporous aluminophosphate molecular sieve.

The characteristics of aluminophosphate molecular sieves include a univariant framework composition with Al/P = 1, a high degree of structural diversity, a wide range of pore sizes and volumes exceeding the pore sizes known previously in zeolite molecular sieves with the VPI-5 18-membered ring material. They are neutral frameworks and therefore have nil ion-exchange capacity. Their surface selectivity is mildly hydrophilic. They exhibit excellent thermal and hydrothermal stability, up to 1000°C (thermal) and 600°C (steam).

5.2.3. Silicoaluminophosphates (SAPO)

The next family of new molecular sieves are the silicoaluminophosphates (SAPO). Sixteen microporous structures have been reported to date, eight of which were never before observed in zeolites. The SAPO family includes a silicon analogue of the 18-ring VPI-5, Si-VPI-5 (ref. 29), a number of large pore 12-ring structures including the important SAPO-37 (FAU), medium pore structures with pore sizes from 0.6-0.65nm, and small pore structures with pore sizes of 0.35-0.45nm including SAPO-34 (CHA). The SAPO's exhibit both structural and compositional diversity.

The SAPO anhydrous composition is $0-0.3R(Si_xAl_yP_z)O_2$ where x, y and z are the mole fraction of the respective framework elements. The mole fraction of silicon, x, typically varies from 0.04 to 0.20 depending on synthesis conditions and structure type. Martens et al. have reported compositions with the SAPO-5 structure with x up to 0.8 (ref. 30). Van Nordstrand et al. have reported the synthesis of a pure silica analogue of the SAPO-5 structure, SSZ-24 (ref. 31).

The introduction of silicon into hypothetical phosphorus sites produces negatively charged frameworks with cation exchange properties and weak to mild acidic catalytic properties. Again, as in the case of the aluminophosphate molecular sieves they exhibit excellent thermal and hydrothermal stability.

5.2.4. Metal Aluminophosphates (MeAPO)

In the metal aluminophosphate (MeAPO) family the framework composition contains metal, aluminum and phosphorus. The metal (Me) species include the divalent forms of Co, Fe, Mg, Mn and Zn, and trivalent Fe. As in the case of SAPO, the MeAPO's exhibit both structural diversity and even more extensive compositional variation. Seventeen microporous structures have been reported, eleven of these never before observed in zeolites. Structure types crystallized in the MeAPO family include framework topologies related to the zeolites, e.g., 34 (CHA) and 35 (LEV), and to the $AlPO_4$'s, e.g.,- 5 and -11, as well as novel structures, e.g., -36 (0.8nm pore) and -39 (0.4nm pore). The MeAPO's represent the first demonstrated incorporation of divalent elements into microporous frameworks.

The spectrum of adsorption pore sizes and pore volumes and the hydrophilic surface selectivity of the MeAPO's are similar to those described for the SAPO's. The observed catalytic properties vary from weakly to strongly acidic and are both metal and structure dependent. The thermal and hydrothermal stability of the MeAPO materials is somewhat less than that of the $AlPO_4$ and SAPO molecular sieves.

The MeAPO molecular sieves exhibit a wide range of compositions within the general formula $0 - 0.3R(Me_xAl_yP_z)O_2$. The value of x, the mole fraction of Me, typically varies from 0.01 to 0.25. Using the same mechanistic concepts described for SAPO, the MeAPO's can be considered as hypothetical $AlPO_4$ frameworks that have undergone substitution. In the MeAPO's the metal appears to substitute exclusively for aluminum resulting in a negative (Me^{2+}) or neutral (Me^{3+}) framework charge. Like SAPO, the negatively charged MeAPO frameworks posses ion-exchange properties and Bronsted acid sites.

5.2.5. Other Compositions

The MeAPSO family further extends the structural diversity and compositional variation found in the SAPO and MeAPO molecular sieves. These quaternary frameworks have Me, Al, P, and Si as framework species. The MeAPSO structure types include framework topologies observed in the binary $AlPO_4$ and ternary (SAPO and MeAPO) compositional systems and the novel structure -46 with a 0.7nm pore. The structure of -46 has been determined (ref. 32).

Quinary and senary framework compositions have been synthesized containing aluminum, phosphorus and silicon, with additional combinations of divalent (Me) metals.

In the ElAPO and ElAPSO compositions the additional elements Li, Be, B, Ga, Ge, As, and Ti have been incorporated into the $AlPO_4$ framework.

Figure 5 shows the compositional relationships among the aluminophosphate-based families of molecular sieves.

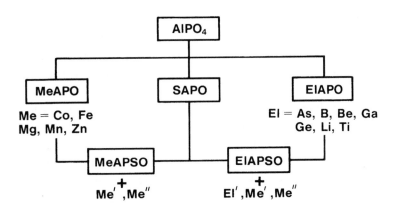

Figure 5. Schematic representation of the relationships in the aluminophosphate-based molecular sieves.

5.3. Metallosilicate Molecular Sieves

A large number of metallosilicate molecular sieves have been reported particularly in the patent literature. Those claimed include silicates containing incorporated tetrahedral iron, boron, chromium, arsenic, gallium and germanium. Most of the work has been reported with structures of the MFI type. Others include metallosilicate analogues of ZSM-11, -12, THETA-1, ZSM-34 and beta. In only a limited number of the reported metal incorporations has sufficient characterization been presented to establish proof of incorporation. To date, only B, Ga, Fe and Ti have been sufficiently characterized to confirm structural incorporation. The metallosilicate molecular sieves are reviewed in detail in Reference 22.

In summary, the progression of molecular sieves from the first zeolite mineral discovered in 1756 to the present has seen a chronological progression from the aluminosilicate zeolites to the microporous silica polymorphs to the microporous aluminophosphate-based polymorphs and metallosilicate compositions.

5.4. Other Framework Compositions

Crystalline microporous frameworks have been reported with compositions of: beryllosilicate, lovdarite (ref. 22); beryllophosphate (ref. 33); aluminoborate (ref. 34); aluminoarsenate (ref. 35); galloarsenate (ref. 36); gallophosphate (ref. 28); antimonosilicate (ref. 37); and germanosilicate (ref. 38).

Harvey et al. (ref. 33) reported the synthesis of alkali beryllophosphate molecular sieves with the RHO, GIS, EDI, and ANA structure topologies, and a novel structure, BPH. Simultaneously, the first metal beryllophosphate mineral species were reported, tiptopite, with the cancrinite topology by Peacor et al. (ref. 39), and pahasapaite, with the RHO topology by Rouse et al. (ref. 40).

6. HISTORY OF SYNTHESIS

A brief description of the history of the synthetic methods developed for the molecular sieves previously described is given here. The reader is referred to Chapter 4 in this volume by Jansen and Wilson for a detailed discussion of synthesis.

The method developed by Milton in the late 1940's involves the hydrothermal crystallization of reactive alkali metal aluminosilicate gels at low temperatures and pressures. The synthesis mechanism is generally described as involving solution mediated crystallization of the amorphous gel, and concepts of cation templating. In the early alkali aluminosilicate synthesis of the low silica zeolites, it has been proposed that the hydrated alkali cation "templates" or stabilizes the formation of the zeolite structural sub-units. A schematic for the early zeolite synthesis is shown in Figure 6a. Alkali hydroxide, reactive forms of alumina and silica, and H_2O are combined to form a gel. Crystallization of the gel to zeolite occurs at a temperature near 100°C.

The addition of quaternary ammonium cations to alkali aluminosilicate gels occurred in the early 1960's first to produce intermediate silica zeolites, e.g., omega and N-A or ZK-4, and subsequently led to the discovery of the high silica zeolites and silica molecular sieves. A schematic for the synthesis of the siliceous zeolites is shown in Figure 6b. The synthesis of high silica zeolites and silica molecular sieves involves synthetic chemistry similar to the initial low silica zeolite method with two important differences: the addition of the quaternary ammonium cation, and a crystallization temperature higher than 100°,

a) Early Zeolite Synthesis

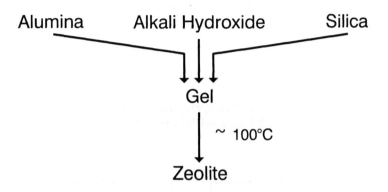

Alumina   Alkali Hydroxide   Silica

Gel

~ 100°C

Zeolite

Hydrothermal Crystallization of Reactive Alkali Aluminosilicate
Gels at Low Temperature and Pressure

b) Siliceous Zeolite Synthesis

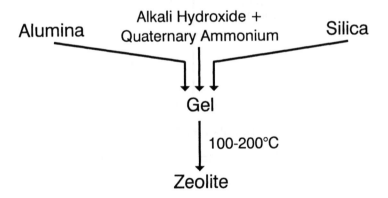

Alumina   Alkali Hydroxide +
Quaternary Ammonium   Silica

Gel

100-200°C

Zeolite

Figure 6. Schematic representation of synthesis method for a) early zeolites;
and b) siliceous zeolites.

30

typically 125 to 200°C. The pH in both the low silica and the siliceous synthesis is highly basic with pH's in the region of 10 to 14.

A schematic of the synthesis method developed in the late 70's for the AlPO$_4$-based molecular sieves is shown in Figure 7. A reactive source of alumina is combined with phosphoric acid and an amine or quaternary ammonium template added to form a reactive gel. Silica or a metal salt is added optionally. The reactive gel is heated to 100 to 200°C for 4 to 48 hours to crystallize the molecular sieve product.

TYPICAL SYNTHESIS OF AlPO$_4$-BASED MOLECULAR SIEVES

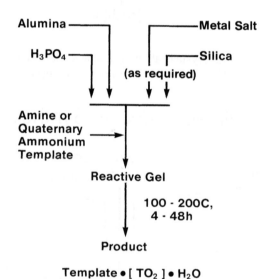

Figure 7. Schematic representation of the synthesis method for AlPO$_4$-based molecular sieves.

There are important differences in the AlPO$_4$-based synthesis compared to the aluminosilicate and silica systems: the amine or quaternary ammonium ion is frequently the only template species (no alkali metal); and the initial pH in the gels is typically mildly acidic to mildly basic.

The history of the cation templates used in molecular sieve synthesis is summarized in Table 5. Over the period of thirty-five years of molecular sieve synthesis a number of concepts have been developed by various workers in cation-structure specificity. These are variously described as structure-directing, the templating effect, clathration phenomena, stereospecific replication, and nucleation effects.

TABLE 5
History of cation templates

Low Si/Al Zeolites	Alkali Metal Cations
	Na^+ A,X,Y
	K^+ L
High Silica Zeolites	Alkali + Quaternary Ammonium
Silica Molecular Sieves	Na^++TPA ZSM-5, Silicalite
Aluminophosphate-based	Amines and Quaternary
Molecular Sieves	Ammonium Cations

---

7. THE FUTURE

7.1. Materials

As noted in Section 3.1 the 80's have seen an explosive increase in discovery of new compositions and structural topologies. Based on the very high activity in this area in the last two to three years, we can expect a continuation of the proliferation of new compositions and structures of molecular sieves. At the Amsterdam meeting, Bedard et al. (ref. 41) reported the crystallization of a large number of structures and compositions of microporous metal sulfides extending the crystalline microporous materials from metal oxides to metal sulfides. Further advances can also be expected in novel compositions derived from modification and secondary synthesis chemistry.

When we consider the very large number of structures and compositions now reported in the molecular sieve area and compare that with the number of commercial molecular sieves, what is the probability of future commercialization of a new material? There are many factors affecting achievement of commercial status: the market need, the market size, the costs of development and marketing, and the cost and degree of difficulty in manufacturing. As a result, it is likely based on historical experience that no more than a few of the prolific number of new molecular sieve materials of the 80's will achieve commercial status in the 1990's.

7.2. Applications

Molecular sieve adsorbents will continue to be used in the now-practiced separation and purification applications throughout the chemical process industry. New directions in the 90's include environmental and biopharmaceutical applications which have only recently received attention.

Future trends in catalysis in the 90's include a continuing accelerated discovery of new catalytic materials; an expanded use in petroleum refining particularly in the area of high octane gasoline and in the development of reformulated gasoline; commercial development in conversion of alternate resources to motor fuels and base chemicals; and as routes to organic chemical intermediate or end products.

The large application of zeolites as ion exchangers in detergents will continue to show growth during the 1990's particularly in Europe and the Far East. The other applications of zeolites as ion exchangers in the nuclear industry, in radioactive waste storage and cleanup, and metals removal and recovery will probably remain a relatively small fraction of the worldwide market for molecular sieve materials.

Among the new application areas the use of molecular sieves as functional powders, for odor removal and as plastic additives, could become large volume application areas.

8. HISTORY OF INTERNATIONAL CONFERENCES AND ORGANIZATIONS

In 1957 the first informal molecular sieve conference was held at Pennsylvania State University in the U.S.A. In 1967 the first of a series of international molecular sieve conferences chaired by Professor R.M. Barrer was held in London. Subsequently, international meetings have been held every three years (with one exception at Chicago) - in 1970 at Worcester, 1973 at Zurich, 1977 at Chicago, 1980 at Naples, 1983 in Reno, 1986 in Tokyo, and this year 1989 in Amsterdam.

An international molecular sieve organization was first formed in 1970 in conjunction with the Worcester Conference, and called the International Molecular Sieve Conference (IMSC). Its responsibility was to continue the organizational implementation of future international molecular sieve conferences on a regular basis. In 1977 at the Chicago meeting the name of the organization was changed to the International Zeolite Association (IZA) and its scope and purpose expanded to "promote and encourage all aspects of the science and technology of zeolitic materials", as well as organizing "International Zeolite Conferences" on a regular basis. The term zeolite in the new organization "is to be understood in its broadest sense to include both natural and synthetic zeolitic as well as molecular sieves and other materials having related properties and/or structures" (ref. 42). International Zeolite Association regional affiliates established include the British Zeolite Association (BZA) in 1980; the Japan Association of Zeolites, (JAZ) in 1986; and regional zeolite associations in France, Germany and Belgium in 1988.

In addition to these organizations the IZA has several established Commissions. The first is the Structure Commission formed in 1977 which has published two editions of the Atlas of Zeolite Structure Types (1978, 1987). In 1986 commissions were organized in the area of catalysis and synthesis. An adsorption and ion exchange commission is under study. In 1988 the Consortium for Theoretical Frameworks was organized by J. V. Smith at the University of Chicago.

9.   HISTORICAL EPILOGUE

Key factors in the growth of molecular sieve science and technology include the pioneering work of Barrer in molecular sieve separations and synthesis, the key discoveries of Milton and Breck and associates at Union Carbide, the rapid commercialization of the new synthetic zeolites and their applications by Union Carbide (1949-1954); the major development at Union Carbide in adsorption process design and engineering technology; major discoveries in hydrocarbon conversion catalysts at Union Carbide, Exxon, Mobil Oil, Shell and other industrial laboratories; the discovery and commercialization of sedimentary zeolite mineral deposits in the United States in the 60's, and last, but not least, the dedication and contribution of so many high quality scientists and engineers.

REFERENCES

1   A.F. Cronstedt, Akad. Handl. Stockholm, 18 (1756) 120-130.
2   F. Fontana, Memorie Mat. Fis. Soc. ital. Sci., 1 (1777) 679.
3   A. Damour, Ann. Mines, 17 (1840) 191.
4   G.W. Morey and E. Ingerson, Econ. Geol., 32 (1937) 607.
5   H.S. Thompson, J. Roy. Agr. Soc. Engl., 11 (1850) 68;
     J.T. Way, ibid. p. 313.
6   H. Eichhorn, Poggendorf Ann. Phys. Chem., 105 (1858) 126.
7   H. de St. Claire-Deville, Compt. Rend., 54 (1862) 324.
8   G. Friedel, Bull. Soc. Franc. Mineral. Cristallogr., 19 (1896) 94-118.
9   F. Grandjean, Compt. Rend., 149 (1909) 866-68.
10   O. Weigel and E. Steinhoff, Z. Kristallogr., 61 (1925) 125-54.
11   R.J. Leonard, Econ. Geol., 22 (1927) 18-43.
12   W.H. Taylor, Z. Kristallogr., 74 (1930) 1.
13   L. Pauling, Proc. Nat. Acad. Sci., 16 (1930) 453; Z. Kristallogr., 74 (1930) 213.
14   J.W. McBain, The Sorption of Gases and Vapors by Solids, Ch. 5, Rutledge and Sons, London, 1932.
15   R.M. Barrer, J. Soc. Chem. Ind., 64 (1945) 130.
16   R.M. Barrer, J. Chem. Soc., (1948) 2158.
17   R.M. Milton, in M.L. Occelli and H.E. Robson (Eds.), Zeolite Synthesis, ACS Sympos. Ser. 398, American Chemical Society, Washington, D.C., 1989, pp. 1-10.

34

18   F.A. Mumpton, in D. Olson and A. Bisio (Eds.), Proc. 6th Intl. Zeolite
      Conf., Reno, USA, July 10-15, 1983, Butterworths, Guilford, Surrey, UK,
      1984, pp. 68-82.
19   K. Torii, in L.B. Sand and F.A. Mumpton (Eds.), Natural Zeolites,
      Occurrence Properties, Use, Pergamon Press, New York, 1978, pp. 441-50.
20   S.T. Wilson, B.M. Lok, C.A. Messina, T.R. Cannan and E.M. Flanigen, J. Am.
      Chem. Soc., 104 (1982) 1146-47.
21   E.M. Flanigen, B.M. Lok, R.L. Patton and S.T. Wilson, in Y. Murakami,
      A. Ijima and J.W. Ward (Eds.), New Developments in Zeolite Science and
      Technology, Proc. 7th Intl. Zeolite Conf., Tokyo, Aug. 17-22, 1986,
      Kodansha Ltd., Tokyo and Elsevier, Amsterdam, 1986, pp. 103-12.
22   R. Szostak, Molecular Sieves, Principles of Synthesis and Identification,
      Van Norstrand Reinhold, New York, 1989, pp. 205-77.
23   J.V. Smith, in P.A. Jacobs and R.A. van Santen (Eds.), Zeolites: Facts,
      Figures, Future, Stud. Surf. Sci. Catal. 49A, Elsevier, Amsterdam, 1989,
      pp. 29-47.
24   G.A. Ozin, A. Kuperman and A. Stein, Angew. Chem. Int. Ed. Engl., 28
      (1989) 359-76.
25   R.M. Barrer, Pure and Appl. Chem., 51 (1979) 1091-1100.
26   D.W. Breck, Zeolite Molecular Sieves, Structure, Chemistry and Use, John
      Wiley & Sons, Inc., New York, 1974; reprinted by Krieger, Malabar,
      Florida, 1984.
27   E.M. Flanigen, in L.V.C. Rees (Ed.), Proc. 5th Intl. Conf. on Zeolites,
      Naples, Italy, June 2-6, 1980, pp. 760-80.
28   J.M. Bennett, W.J. Dytrych, J.J. Pluth, J.W. Richardson and J.V. Smith,
      Zeolites, 6 (1986) 349-60.
29   M.E. Davis, C. Saldarriaga, C. Montes, J. Garces and C. Crowder, Nature
      (London), 331 (1988) 698-9.
30   J.A. Martens, M. Mertens, P.J. Grobet and P.A. Jacobs, in P.J. Grobet,
      W.J. Mortier, E.F. Vansant and G. Schulz-Ekloff (Eds.), Innovation Zeolite
      Mater. Sci., Stud. Surf. Sci. Catal. 37, Elsevier, Amsterdam, 1988, pp.
      97-105.
31   R.A. Van Nordstrand, D.S. Santilli and S.I. Zones, in W.H. Flank and T.E.
      Whyte, Jr. (Eds.), Perspect. Mol. Sieve Sci., ACS Symp. Ser. 368, 1988,
      pp. 236-45.
32   J.M. Bennett and B.K. Marcus, in P.J. Grobet, W.J. Mortier, E.F. Vansant
      and G. Schulz-Ekloff (Eds.), Innovation Zeolite Mater. Sci., Stud. Surf.
      Sci. Catal. 37, Elsevier, Amsterdam, 1988, pp. 269-79.
33   G. Harvey and W.M. Meier, in P.A. Jacobs and R.A. van Santen (Eds.),
      Zeolites: Facts, Figures, Future, Stud. Surf. Sci. Catal. 49A, Elsevier,
      Amsterdam, 1989, pp. 411-20.
34   J. Wang, S. Feng and R. Xu, in P.A. Jacobs and R.A. van Santen (Eds.),
      Zeolites: Facts, Figures, Future, Stud. Surf. Sci. Catal. 49A, Elsevier,
      Amsterdam, 1989, pp. 143-50.
35   G. Yang, L. Li, J. Chen and R. Xu, J. Chem. Soc., Chem. Commun., (1989)
      948.
36   J. Chen and R. Xu, J. Solid State Chem., 80 (1989) 149-51.
37   Y. Yamagishi, S. Namba and T. Vashima, in P.A. Jacobs and R.A. van Santen
      (Eds.), Zeolites: Facts, Figures, Future, Stud. Surf. Sci. Catal. 49A,
      Elsevier, Amsterdam, 1989, pp. 459-67.
38   Z. Gabelica and J.L. Guth, in P.A. Jacobs and R.A. van Santen (Eds.),
      Zeolites, Facts, Figures, Future, Stud. Surf. Sci. Catal. 49A, Elsevier,
      Amsterdam, 1989, pp. 421-30.
39   D.R. Peacor, R.C. Rouse and J.-H. Ahn, Am. Mineral. 72 (1987) 816-20.
40   R.C. Rouse, D.R. Peacor and S. Merlino, Am. Mineral. 74 (1989) 1195-202.
41   R.L. Bedard, S.T. Wilson, L.D. Vail, J.M. Bennett and E.M. Flanigen, in
      P.A. Jacobs and R.A. van Santen (Eds.), Zeolites: Facts, Figures, Future,
      Stud. Surf. Sci. Catal. 49A, Elsevier, Amsterdam, 1989, pp. 375-87.
42   International Zeolite Association, Newsletter 1, Aug. 22, 1977.

Chapter 3

STRUCTURAL SUBUNITS IN SILICATE AND PHOSPHATE STRUCTURES

H. VAN KONINGSVELD

Laboratories of Organic Chemistry and Applied Physics, Delft University of Technology, Lorentzweg 1, 2628 CJ DELFT, The Netherlands

SUMMARY
   Similarities within fifty structure types are described. A survey is given of several kinds of single, double and triple chains of T atoms observed in many structure types. Within the existing structure types four sub-groups are considered. The first group contains structure types which can be built by connecting double chains. The second group, the "pentasil-family", can be characterized by the presence of triple chains. The third group has (pseudo) hexagonal symmetry. Many structure types in this group can arise from different close-packing arrangements of identical puckered layers. The last group of structure types may originate from the double 4-ring or its derivatives.
   A simplified description of the structure types reveals an interrelationship between structural subunits through simple hydrolysis and condensation processes. A great majority of the structure types can be envisaged to develop from condensation of 4-rings, 6-rings or chains of T atoms. In some structure types an additional single T atom is needed to form the observed structural subunits.
   A special type of the double crankshaft chain is only observed in several synthetic frameworks containing Al and P.
   Unit-cell dimensions very often provide information on the chain type present.

1. INTRODUCTION

The "Atlas of Zeolite Structure Types" (ref. 1) contains 64 topologically distinct tetrahedral TO_4 frameworks, where T may be Si, Al, P, Ga, B, Be etc. The compiled structure types do not depend on composition, distribution of the various T atoms, cell dimensions or symmetry. Zeolites are so diverse that secondary building units (SBU; Fig. 1) are needed in describing and classifying their topologies. The non-chiral SBU are invented on the assumption that each zeolite framework can be generated from one type of SBU only (ref. 2). Actually, many structures can be described by more than one type of SBU. Sometimes, two types of SBU are needed to make up the framework.

Various other ways, sometimes involving structual subunits (SSU) of greater complexity than the SBU from Fig. 1, have been developed to describe the framework topologies. The search for new SSU is useful: any subunit is a potential candidate for synthesis because it might be a precursor from which the zeolite grows.

36

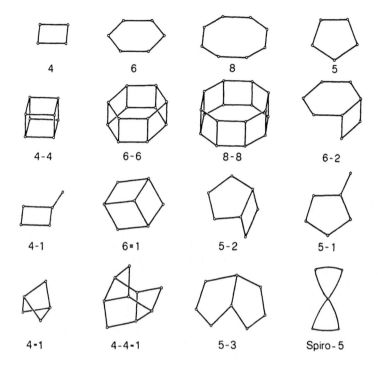

Fig. 1. Secondary Building Units identifiable in zeolite frameworks. T atoms are at cross-section and termination of lines. O atoms (not shown) are located about midway between T atoms. (Reproduced from ref. 1. with permission)

In "Structural Chemistry of Silicates" (ref. 3), Liebau classifies the silicates by (branched) chains and -rings using several parameters to describe the linkedness, branchedness, dimensionality, multiplicity and periodicity of the anions, chains or rings.

Smith (ref. 4) uses an edge, rings, polyhedra, coplanar and noncoplanar chains as SSU and investigates how the subunits can be modified. A systematic way is searched for in which the SSU can be linked together in new ways leading to an extension of the list of theoretical frameworks.

Gottardi et al. (ref. 5) describe the topologies of the majority of the natural zeolites by using as SSU 4- and 6-rings (single and double), chains and puckered layers, sometimes with "handles".

Inspection of existing structure types may give clues for choosing targets for syntheses from the list of theoretical frameworks. Study of the analogies between existing structure types may lead to ideas on the relationship between framework geometry, i.e. the precursors to crystallization, and synthesis conditions. It may even provide synthesis conditions for new materials with tailored pore geometry.

This contribution describes the similarities observed in some fifty structure types containing at least 8-ring apertures. Clathrasils, where the windows are too small to let pass encaged species or their decomposition products (ref. 3), and zeolites with interrupted frameworks are not considered.

About ten of the fifty structure types describe natural zeolites only: no synthetic counterparts have (yet) been reported. Approximately twenty structure types represent synthetic framework topologies only. Another twenty structure types have natural as well as synthetic analogues. They are referenced throughout this paper by their mnemonic codes (ref. 1) with a black dot, (CODE)•, an open circle, (CODE)⁰, and without a superscript, (CODE), respectively.

A tentative description will show that all structure types considered can be built using SSU originating from condensation of chains and rings (4- and 6-ring) or from hydrolysis and degradation of the double 4-ring. This simplification of the framework structures reveals an interrelationship between SSU through simple condensation processes.

Section 2 describes some basic concepts. In Section 3 the SSU observed in the structure types are summarized. Section 4 presents subgroups of types using the basic concepts and SSU. Finally, in Section 5 some conclusions are drawn.

2. SOME BASIC CONCEPTS

2.1 3-connected 2-dimensional nets

In 1954 Wells wrote his papers (refs. 6, 7) on the geometrical basis of crystal chemistry. After his book "Structural Inorganic Chemistry" (ref. 8) the monograph "Three-dimensional Nets and Polyhedra" (ref. 9) appeared summarizing the invented 4-connected 3-dimensional (4-conn. 3D) nets. Smith and collaborators gave a systematic enumeration of 4-conn. 3D nets in a series of papers the last of which appeared recently (ref. 10).

The 4-conn. 3D nets can be derived from the simpler planar 3-conn. nets. The present "state of the art" shows a list of 66 3-conn. 2D nets, of which 8 have not yet been found in the topological analysis of the 230 3D nets (ref. 11). It is appropriate here only to summarize those nets found in the majority of the existing structure types. More nets will be discussed in the next paragraph.

Fig. 2 shows five 3-conn. 2D nets which are more or less frequently observed in zeolites. Three of the listed nets (1, 4 and 5) were already presented as early as 1954 by Wells (ref. 6). The 3-conn. 2D nets represent networks of polyhedra. A node of the network is described by its surrounding polygons, the number of equal polygons being recorded by a right superscript. The network is

38

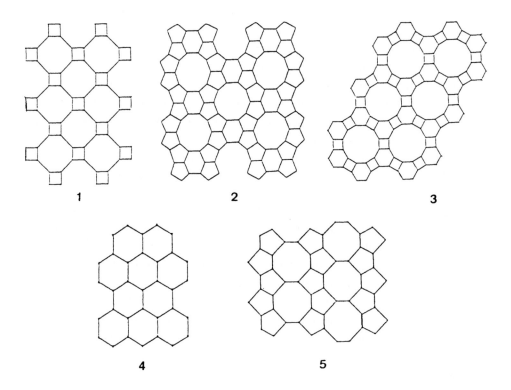

Fig. 2. Some examples of 3-connected 2-dimensional nets observed in zeolites. The nets are defined by their Schläfli symbols (see text). 1): 4.8^2-net; 2): $(5^2.6)_1(5^2.10)_3(5.6.10)_2$-net (or: the "pentasil"-net); 3): 4.6.12-net; 4): 6^3-net; 5): $(5^2.8)_2(5.8^2)_1$-net.

described by listing the different nodes with their relative frequency as right subscript (ref. 3): the 4.8^2 net (Fig. 2; 1) has one kind of nodes surrounded by one quadrangle and two octagons.

In Sections 4.1 and 4.2 the 4.8^2 and $(5^2.6)_1(5^2.10)_3(5.6.10)_2$ (Fig. 2; 2) 3-conn. 2D nets, respectively, are converted into 4-conn. 3D nets of existing structure types by conversion of some or all of the edges into chains.

Sometimes it is convenient to consider the 3-conn. 2D nets as non-planar layer-like SSU. Each node then represents one tetrahedron with an indication of its directionality, i.e. whether it is linked upwards or downwards to the neighbouring layers. In Section 4.3 the 4.6.12- and 6^3 layer-like SSU (Fig. 2; 3 and 4) are converted into 4-conn. 3D structure types by "layer-stacking" and "cylinder-stacking", respectively.

2.2 The sigma transformation

Not all the existing structure types can be directly derived from the
3-conn. 2D nets just described. More nets can easily be constructed using the
concept of the σ transformation. It is a recipe which "expands the single
tetrahedron TO_4, or a tetrahedrally connected structure by imaginary fission of
T atoms lying on specified planes or surfaces running through the structure,
and creation of new oxygen bridges connecting the pair of T atoms resulting
from the fission" (ref. 12, 13). Some examples of the T expansion are depicted
in Fig. 3: an edge is transformed into a 4-ring (a), the 6^3 net changes into
the $(6.4.8)_2(6.8^2)_1$ net (b) and, by three successive transformations, the
sodalite (or β) cage is expanded to the α cage (c). The inverse operation is
called σ contraction: the just mentioned $(6.4.8.)_2(6.8^2)_1$ net can also be

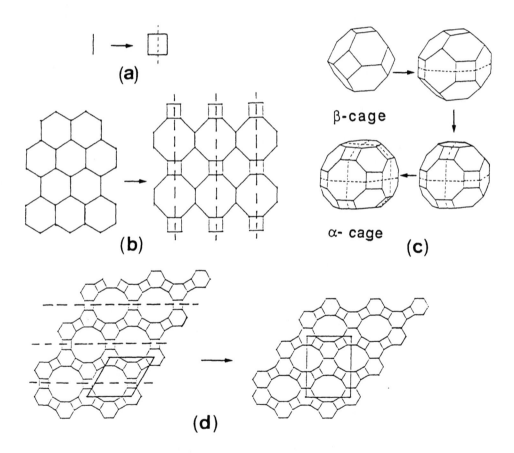

Fig. 3. The sigma transformation. Three examples of σ expansion (a, b and c)
and one example of contraction (d) are shown.

obtained by contraction of the 4.8^2 net (Fig. 2) and the (pseudo) hexagonal 4.6.12 net transforms to the orthorhombic $(4.6.10)_4(6.6.10)_1$ net (Fig. 3 (d)). Yet another σ transformation, using a 6-ring chair as the fission surface, has been described recently by Bennett et al. (ref. 14). Sigma transformations of the "pentasil" net (Fig. 2, nr. 2) will be dealt with separately in Section 4.2.

3. STRUCTURAL SUB UNITS (SSU)

3.1 Single and double chains

Frameworks of many zeolites can readily be constructed from several kinds of chains. The chain types given in Fig. 4 are three of the fundamental chains described by Liebau (ref. 3): the unbranched zweier-, dreier- and vierer chain with identity periods of $\approx$ n * 2.5 Å, where n = 2, 3 and 4, respectively. In mineralogy, the unbranched single- and double zweier chains are equivalent to the wollastonite and epididymite chain. The unbranched double vierer chain is called the feldspar chain. In this contribution the single and double chains are referenced by zigzag (Z and ZZ), saw (S and SS), and crankshaft (C, CC-1, and CC-2), respectively (ref. 4). The double chains can be considered as condensation products of single chains (Color Plate 1 and Fig. 4). The CC-2 chain can be obtained from two C chains by shifting one C chain over half the identity period with respect to the other. The CC-2 chain is more dense and has a significantly smaller identity period than CC-1. The projection of the double chains along the chain axis is (in idealized form) a square or a rectangle.

The double chains from Fig. 4 can as well be described as buckled ladders of 4-rings. These ladders might arise from condensation of single 4-rings. The rings can condense in different ways depending upon which oxygen atoms are involved in linking the 4-rings into ladders. The bridging oxygen atom can be above or below the 4-ring plane. The adjacent ring atoms are said to have upwards (U) or downwards (D) directionality. The resulting 4-ring ladders are also illustrated in Color Plate 1. The ZZ, SS and CC-1 chains can come into being through condensation of UUDD rings. The CC-2 chain has UDUD rings.

There is at least one example illustrating that CC-1 can be changed into CC-2 by heat treatment. When APC^0 is heated above 200 ^{0}C, the CC-1 chain irreversibly transforms into the CC-2 chain of APD^0 (ref. 15). Fig. 5 shows how the solid state reaction might proceed. One of the possible hydrolysis products of CC-1 is given in the Figure between brackets. The CC-2 chain can readily be obtained from this intermediate chain of single connected 4-rings by changing the directionality of some of its T atoms followed by condensation. The assumption of the existence of such an intermediate is not completely

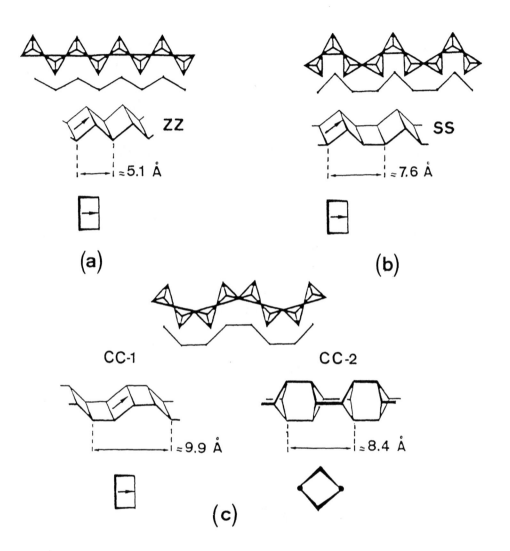

Fig. 4. Some chains in zeolites. (a): zigzag chain, (b): saw chain, (c): crankshaft chain. The top of each drawing shows a single chain constructed from TO_4-building units followed by its symbolic representation by T-T linkages only. Next, the double chain is depicted together with its projection along the chain axis. The arrow gives the down-chain direction (ref. 5) and the black dots indicate downward linkages.

hypothetical. In YUG* a similar single connected 4-ring chain has been observed (see Fig. 5). This chain is called the BS (bifurcated square) chain by Smith (ref. 4). Like the suggested intermediate, the BS chain might stem from a partly hydrolyzed CC-1 chain or from partly condensed 4-rings.

42

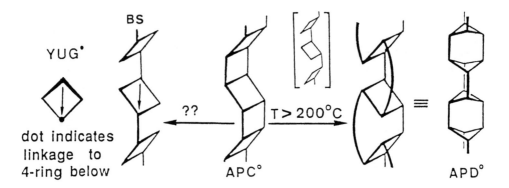

Fig. 5. Irreversible transition of CC-1 into CC-2. The BS chain in YUG• is similar to the suggested intermediate. At the extreme left the down-chain projection symbol in YUG⁰ is added.

A second structure in which the same solid state transition may occur is ATF⁰. There are indications (ref. 16) that, upon calcination, the CC-1 chain in ATF⁰ is transformed to the CC-2 chain. The reported unit-cell dimension along the chain axis of 8.4 Å (ref. 1) indeed strongly points to the presence of the CC-2 chain. In that case the published structure type in ref. 1 is in error.

Yet another chain, the fibrous (FI) chain (Fig. 6) has UDUD rings. Single T-atoms connect the 4-rings into the FI chain. The NAT, EDI and THO• structure types are obtained by various linkages of FI chains (see Section 4.1).

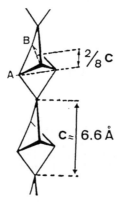

Fig. 6. The FI chain, observed in fibrous zeolites, may be obtained from condensation of single UDUD-rings and single T atoms. Possible linkages to neighbouring chains are at levels A and B.

No natural or synthetic zeolites have been observed so far based on chains with UUUD rings.

The double 4-ring with UUUU (and DDDD) rings is discussed separately in the last paragraph of this Section.

3.2 Triple chains

Triple chains can be created by condensation of single chains, possibly via double chains. There is only one way to connect three single zigzag chains into a triple chain. The connections in the triple saw chain can be made in three different ways. Figure 7 shows the two chain types with their possible connections and the projection symbols used (ref. 17). The triple zigzag and saw chains are also shown in Color Plate 2 together with their projection along the chain axis. The ZZZ chain is the only triple chain, which can also be built from the 5-1 SBU (Color Plate 2). The ZZZ and SSS-1 triple chains are frequently observed in zeolites with a "pentasil" net (see Section 2.2). The SSS-2 chain has not been observed yet. Very recently the structure of ZSM-57 has been reported (ref. 18). It is the only example so far of a zeolite containing, besides the SSS-1 subunit, the SSS-3 subunit.

Crankshaft chains can condense in different ways. The resulting triple chains are shown in Color Plate 3. None of these chains, except the "side by side" chain, has been observed in any material yet. The "side by side" triple chain has recently been observed in VPI[0] (ref. 19). See, however, the alternative description of VPI[0]-5 in Section 4.3.

3.3 Layer-like structural subunits

In this paragraph some finite and infinite 3-conn. 2D nets, which may act as puckered layer-like SSU (see Section 2.1), are discussed. In Figure 8 (parts of) the 4.8^2 and 4.6.12 net are presented. The arrows indicate the downwards direction in the 4-rings.

The finite part of the 4.8^2 net in Figure 8(a) can arise from association of four 4-rings with "handles" or, equally well, from four 6-rings around a four-fold operator axis (Color Plate 4). Two of these entities can condense to form the α cage. The α cage is a SSU for LTA[0], RHO and KFI[0]. Another cage, in which one entity is rotated over 60[0] around an axis perpendicular to the 8-ring before condensation starts, has not yet been observed in zeolites. Other finite parts of the 4.8^2 net, constituing of four 4-rings around a 4-fold axis (equal to the entity without "handles" shown in Fig. 8(a)) can give the γ cage (ref. 3, p. 157). Linked α and γ cages form the SSU for PAU•.

Finite parts of the 4.6.12 net (Figure 8(b)), which may develop from association of three 4-rings, can give the sodalite (or β) cage and the

44

TRIPLE CHAIN :

PROJECTION ALONG
THE CHAIN AXIS :

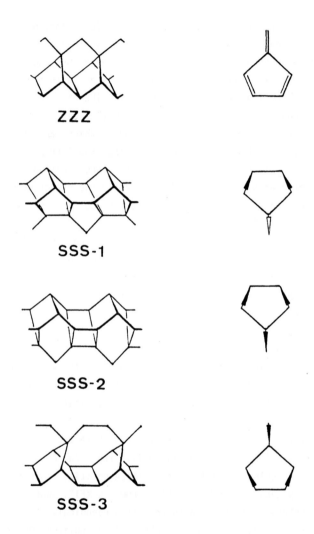

ZZZ

SSS-1

SSS-2

SSS-3

Fig. 7. Triple chains obtained from zigzag and saw chains together with the symbolic representation of their chain axis projection. The open arrow indicates a shift of the S chain over half the chain period.

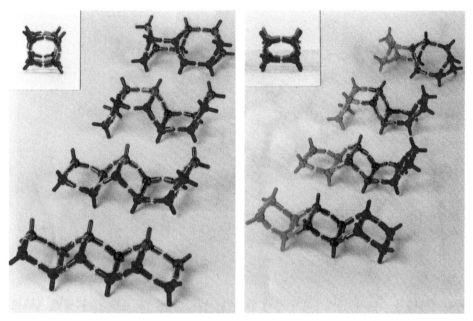

Color Plate 1. Double chains shown as condensation products of single chains (left) and of single 4-rings (right). From bottom to top: double zigzag (ZZ), -saw (SS) and -crankshaft (CC-1 and CC-2) chains. The inset gives a down-chain projection.

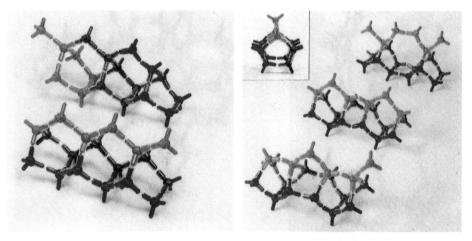

Color Plate 2. The triple zigzag chain (left) and triple saw chains (right) shown as condensation products of single chains. The zigzag triple chain can also be built using the 5-1 SBU. The down-chain projection of all chains is presented in the inset.

46

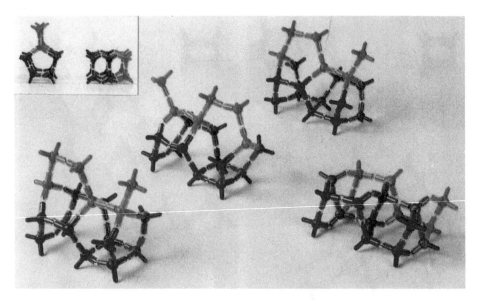

Color Plate 3. The triple crankshaft chains. The "side by side" triple chain (at the extreme right) is the only triple chain of the crankshaft type observed up till now. The inset gives the two down-chain projections.

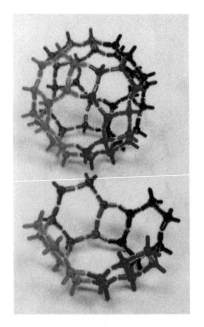

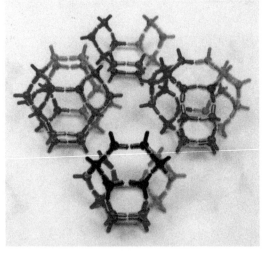

Color Plate 4. Finite part of the 4.8^2 puckered layer-like SSU shown as condensation product of four 6-rings around a 4-fold axis (bottom left); two of these entities can give the α cage (top left). Finite part of the 4.6.12 SSU shown as condensation product of three 4-rings around a 3-fold axis (top right); two of these entities can give the SOD (or β) and GMEL cages (middle right). Condensation with a single 6-ring can produce the CAN (or ϵ) cage (bottom right).

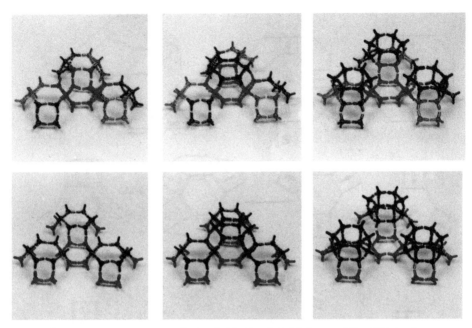

Color Plate 5. Presentation of the infinite 4.6.12 layer-like SSU in two ways: as condensation product of the entity shown at the top in the previous Plate (top left) and as condensation product of single 6-rings (bottom left). Further condensation with 6-rings (black) produces two new SSU with CAN cages (middle) or D6R (right).

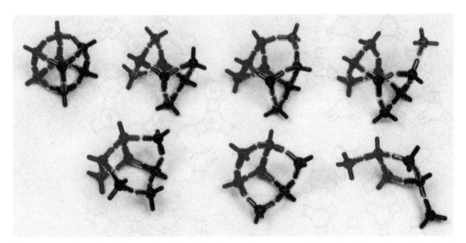

Color Plate 6. The D4R and SSU's which can be interrelated through relatively simple hydrolysis and condensation processes.

48

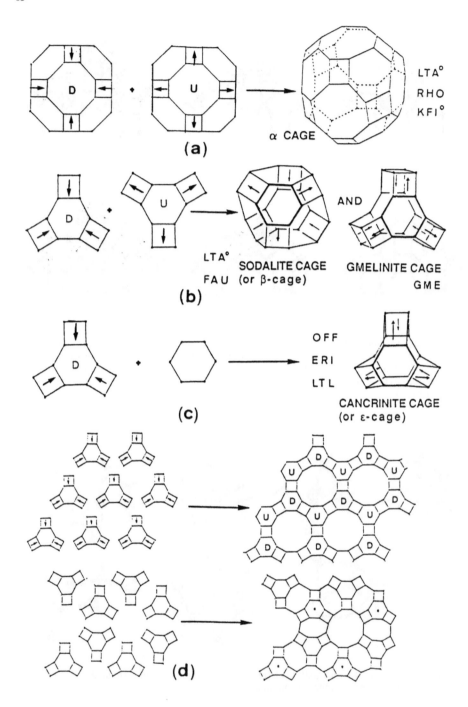

Fig. 8. Possible condensation of parts of puckered layer-like SSU to cages (a, b and c) and to infinite layers (d).

gmelinite cage (Color Plate 4). The cages differ by the rotational orientation of the two constituent entities with respect to each other. The directing role of kations and template molecules used in the synthesis may be crucial in this respect. The β cage and GME cage are SSU for LTA[0] and FAU, and GME, respectively.

The cancrinite (or ϵ) cage can come about when the same finite part of the 4.6.12 net condenses with a 6-ring (Figure 8(c); Color Plate 4) or from condensation of three 6-rings around a three-fold operator axis (the three "boats" in Figure 8(c)). The CAN cage is a SSU for OFF, ERI and LTL.

The structure types given so far are those which can be built from <u>complete</u>, individual cages. SOD, CAN, EAB[0] and MAZ are not quoted here because in these structure types the cages share faces. There is, however, another approach which covers in one common description all the (pseudo) hexagonal structure types.

In Figure 8(d) two infinite layer-like SSU are given which can originate from condensation of the finite SSU just discussed. The infinite 4.6.12 SSU can also arise from condensation of 6-rings (Color Plate 5). The structure types CAN, GMEL, SOD, CHA and the very recently reported APO-52[0] (ref. 20) can <u>all</u> be described by different close-packing of identical 4.6.12 layers (see Section 4.3). Color Plate 5 shows two other layer-like SSU. They may arise from two different ways in which the infinite 4.6.12 SSU and single 6-rings condense. The resulting hexagonal layers contain cancrinite cages and double 6-rings (D6R), respectively. Close-packing of these layers produce OFF and ERI and EAB[0] and LEV, respectively.

Analogous condensation of single 6-rings with the second SSU in Figure 8(d) give SSU (not shown) from which LTL and MAZ can be built.

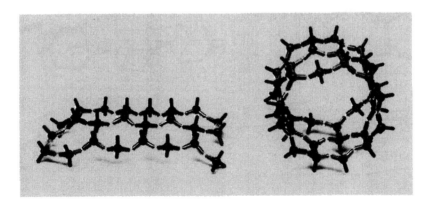

Fig. 9. Transformation of (part of) the 6^3-net into a cylindric SSU. The cylinder wall consists of fused 6-rings only.

Finally, some hexagonal structures are based on the 6^3 net. Figure 9 (left-hand side) shows the 6^3 SSU in which all nodes have U (upwards) directionality. Such plane 6^3 SSU never lead to a 3D framework structure. A double layer, well known from the clay minerals, is formed instead. However, different sized parts of such 6^3 SSU can spontaneously curl into cylinders with different radii (Figure 9 (right-hand side)). The cylinders can be stacked to form 3D frameworks of AEL°, AFI° and VPI-5°.

In Section 4.3 existing structure types are generated using the SSU described in this paragraph.

3.4 Chains from (modified) double 4-rings (D4R)

The D4R and silicate anions which can be derived from the D4R are given in Color Plate 6. Each of the seven units can be linked to produce chains presented in Figure 10. The chains are depicted following the sequence of the SSU shown in Color Plate 6 from left to right and from top to bottom. Simple

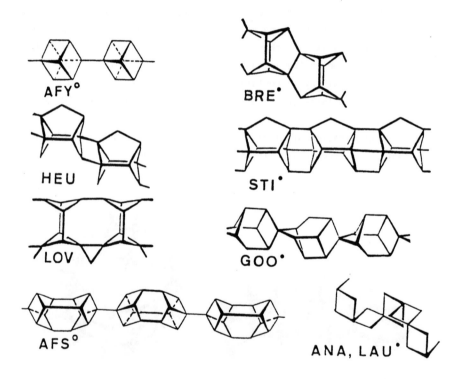

Fig. 10. Chains generated from the SSU shown in Color Plate 6.

symmetry operations link these chains into the 3D structure types indicated in the same Figure and shown in Section 4.4.

The chain in AFY^0 is formed by the D4R. In BRE• the constituent SSU is the D4R in which two opposing bonds are hydrolysed; the unit can also be described as a segment of the CC-2 chain. The next unit depicted in Color Plate 6 is formed when a single T-atom is added to the BRE• unit. Two different ways of linking the resultant unit produce the HEU and STI• chain. When the HEU (or STI•) unit is hydrolysed the LOV unit is obtained with a single T-atom at a terminal position. In GOO• two other bonds of the D4R are hydrolysed. Finally, the D4R has lost one T atom and two T atoms in the AFS^0, and ANA and LAU• units, respectively.

The structure types are shown, with the chain axis parallel to the plane of the paper in Section 4.4.

4. DESCRIPTION OF THE STRUCTURE TYPES (Ref. 21)

This Section, which mainly contains Photographs and Tables, shows how 3-conn. 2D nets (discussed in Section 2.1 and 2.2) are converted into 4-conn. 3D nets of existing structure types.

In Section 4.1 and 4.2 the conversion is realized by transformation of some or all edges of the 2D net into chains described in Section 3.1 and 3.2, respectively.

Section 4.3 presents the construction of the 3D structure types by different stacking of the SSU described in Section 3.3.

Finally, in Section 4.4 the structure types built from chains obtained from the (modified) double 4-ring are shown.

4.1. Structure types with 4.8^2 or $(6.4.8)_2(6.8^2)_1$-net

4-Conn. 3D nets of 12 known structure types can be obtained by converting the edges of the 4.8^2 net into different chain types. Photo 1 summarizes the structure types. The Photo shows the frameworks seen along the chain axis, the mnemonic codes, the type-, connection- and downwards direction of the chains. GIS has a 4.8^2 net in two projections. ATF^0, as well as ATT^0 in a second projection, exhibits the $(6.4.8)_2(6.8^2)_1$ net. This net can be obtained by σ contraction of the 4.8^2 net (Section 2.2). The projection symbols used in describing the fibrous zeolites (EDI, THO• and NAT) indicate at what levels the chains are linked (see Fig. 6). The numbers give the height of the chain in fractions of eight along the chain axis. Table 1 presents the unit-cell dimensions along the chain axis. From the Table, it can be concluded that the unit-cell dimensions can provide information on the chain type present. There

52

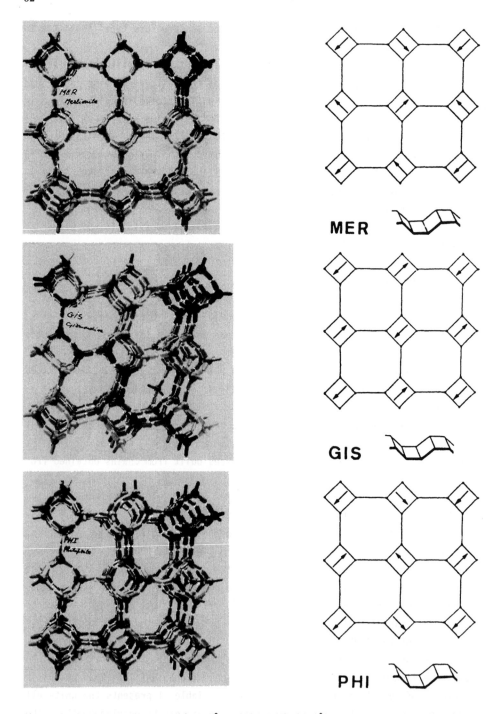

Photo 1. Structure types with 4.8^2 and $(6.4.8)_2(6.8^2)_1$ connectivity. The black rectangle indicates a chain projection. The chain type is given after the mnemonic code. The projection symbols used are defined in Figs. 4, 5 and 6.

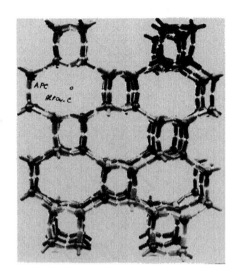

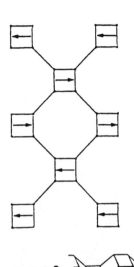

APC°

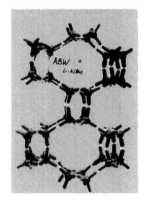

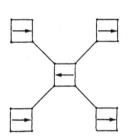

ABW°

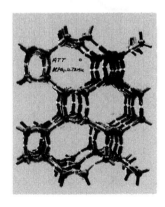

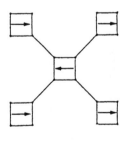

ATT°

Photo 1 - Continued.

54

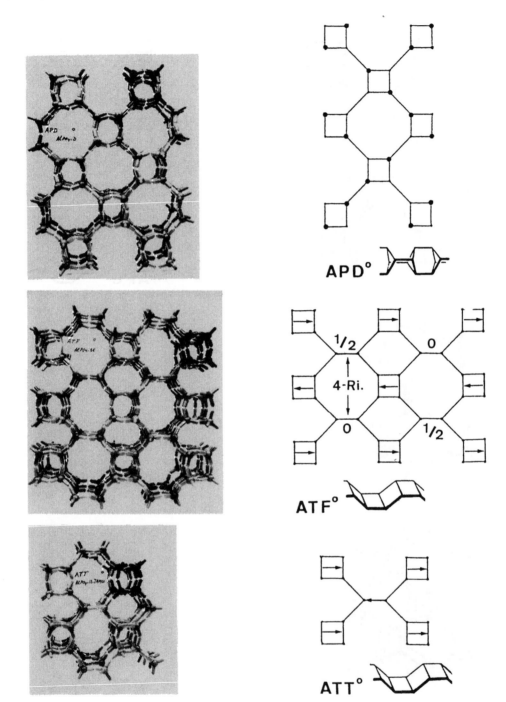

Photo 1 -
Continued.

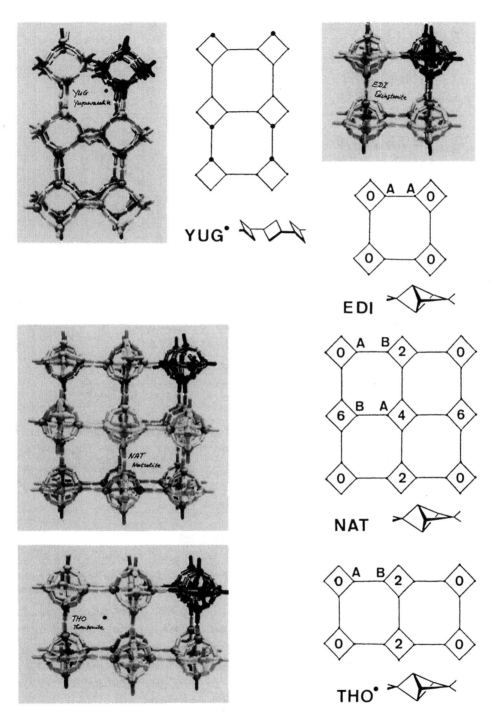

Photo 1 - Final page.

TABLE 1.

Unit cell dimension (Å) along chain axis and chain type in structures ⌐ (4.8²) or $(6.4.8)_2(6.8^2)_1$ connectivity.

Code	Axis (Å)	Chain type	Code	Axis (Å)	Chain type
MER	10.0		ABW°	5.0	
GIS	9.8		ATT°	7.3	
*GIS	10.0				
PHI	9.9		EDI	6.5	
APC°	8.9		THO•	6.6	
•ATT°	10.3		NAT	6.6	
•ATF°	8.4				
YUG•	10.0		▲APD°	8.6	

* GIS has a 4.8²-net in two projections
• $(6.4.8)_2\ (6.8.^2)_1$ - connectivity
▲ Obtained from APC above 200° C

are no natural zeolites containing the CC-2 chain. Only the synthetic zeo⌐ APD°, and perhaps ATF° (ref. 16), contains this type of chain. The struc⌐ types in this group have 8-ring apertures only. In a tentative description ⌐ could say that all structure types can come into being by condensatioⁿ single 4-rings with different directionality.

4.2. Structure types with a (modified) pentasil net

The number of 3-conn. 2D nets underlying the 11 known structure types of "pentasil"-family is much larger than in the previous paragraph. In Fig. 11 $(5^2.6)_1(5^2.10)_3(5.6.10)_2$ net is shown. This "pentasil" net is observed in T(FER, MFI°, MEL° and very recently in ZSM-57°(ref. 18). Sigma expansion of pentasil net in two different ways produce nets of MOR and MTW°, respectivᵉ The new net of MTT° is created when certain σ planes in TON° (denoted blacⁱ Photo 2) act systematically as mirror planes on a sub unit-cell level. Finaⁱ σ expansion of the $(5^2.8)_2(5.8^2)_1$ net of BIK (and DAC• and EPI• in projection) gives the net of DAC• in a second projection.

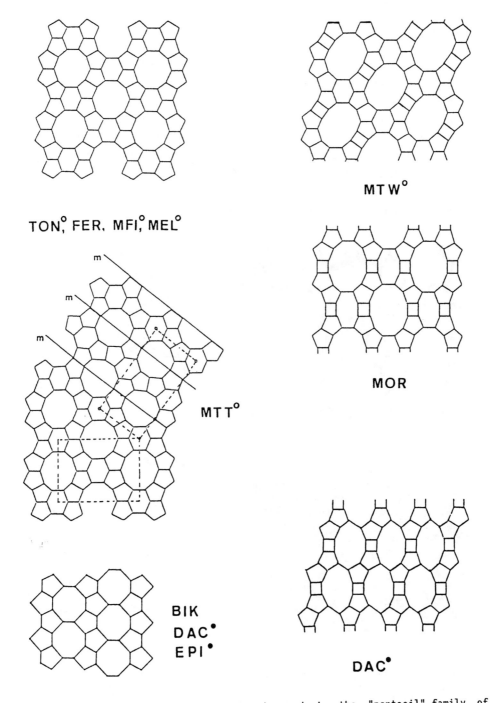

Fig. 11. Some of the 3-conn. 2D nets observed in the "pentasil"-family of zeolites.

58

Photo 2 shows the 11 structure types in a sequence based on the nets just discussed. The Photo gives the structure types seen along the (triple) chain axis together with the mnemonic codes. The 3D nets are given with the projection symbols indicating the chain types. In several structures 4-rings appear. The height of these rings is shown in fractions of the chain axis. The chain type in MFI[0] and MEL[0] is rather complicated (see Table 2) and has therefore not been indicated by projection symbols. These two structures are frequently described using the pentasil chain, shown in black in Photo 2. DAC[•] and EPI[•] show the same projection along the zigzag chain. A second projection, with the chains parallel to the plane of the paper, reveals the differences between the two structure types.

The "pentasil" structure types can arise from condensation of single chains.

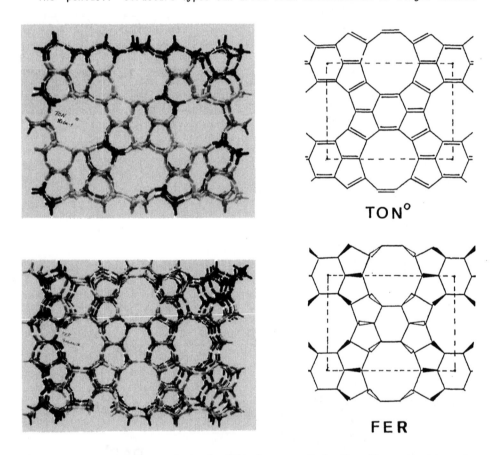

Photo 2. Structure types of the (modified) pentasil family. The projection of a triple chain is indicated in black. The projection symbols used are defined in Figure 7. In MFI[0] and MEL[0] the pentasil chain is shown in black.

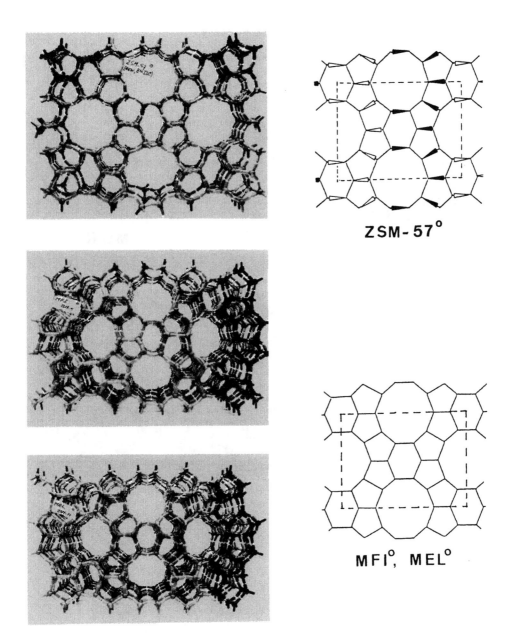

ZSM-57°

MFI°, MEL°

Photo 2 - Continued.

60

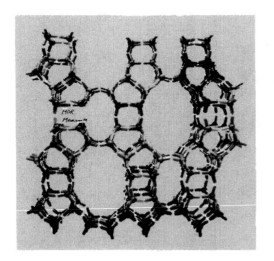

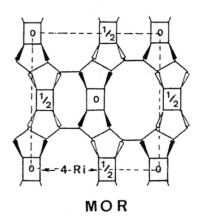

MOR

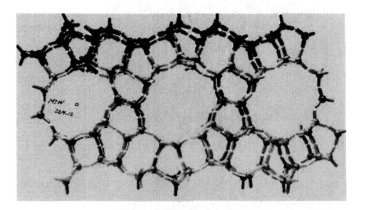

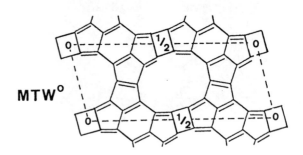

MTW°

Photo 2 - Continued.

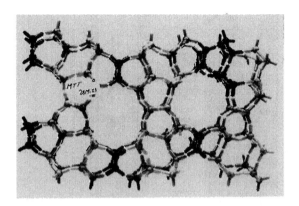

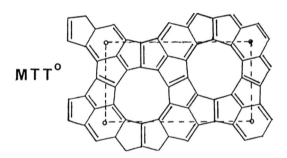

MTT^{o}

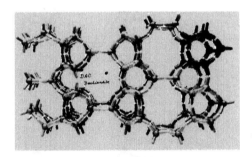

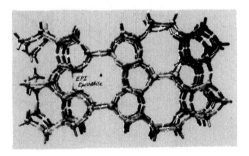

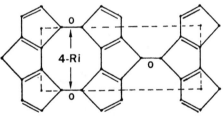

DAC$^{•}$, EPI$^{•}$

Photo 2 - Continued.

62

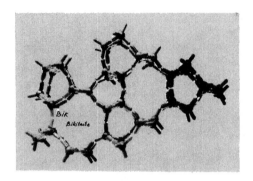

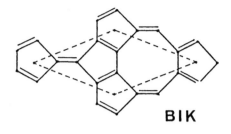

BIK

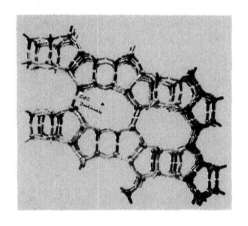

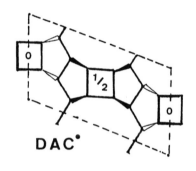

DAC°

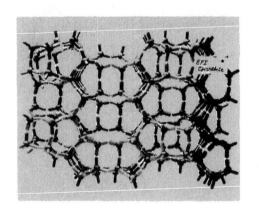

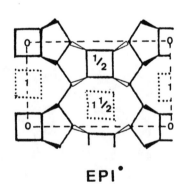

EPI°

Photo 2 - Final page.

TABLE 2.
Unit-cell dimension (Å) along chain axis and chain type in structures with a (modified) pentasil projection along the chain axis.

Code	Chain Axis (Å)	Chain type
BIK	5.0	
TON°	5.0	
MTW°	5.0	
MTT°	5.0	
ZSM-57°	7.5	
FER	7.5	
MOR	7.5	
DAC•	7.5	
EPI•	7.5*	
DAC•	10.3	
EPI•	10.2	
MFI°	19.9	
MEL°	20.1	

$* = a \sin \beta$

Single chains condense, ultimately, into triple chains observed in all the frameworks. However, the existence of complete triple chains as SSU is less probable because in most of the structure types the triple chains have one single chain in common. Planes containing these "common" single chains very often behave as fault planes (ref. 1): layer-like segments of the structure can be stacked in more than one way. When these "faults" appear systematically on a sub unit-cell level a new structure type is generated (compare TON° and MTT°). Currently, we are searching for other SSU in this group of zeolites from which the frameworks can be built by simple symmetry operations.

Table 2 summarizes the unit-cell dimensions of the structure types along the chain axis. The Table again illustrates that unit-cell dimensions can provide information on the chain type present. Neither natural nor synthetic zeolites exhibit the C chain; an exception might be (part of) the chain in MFI° and MEL°. 10-Ring apertures are frequently observed in these structure types.

4.3 Structure types based on (parts of) layer-like SSU

Parts of the 4.8^2 and 4.6.12 SSU may develop by association of single 6- and single 4-rings. These entities can condense to the α, β or γ cages or to infinite 4.6.12 subunits (see Section 3.3).

Linkage of <u>complete</u> α cages via 4-, 6-, or 8-ring faces (resulting in D4R, D6R and D8R) gives the structure types LTA⁰, KFI⁰ and RHO, respectively. When complete β cages are linked via 4- or 6- ring faces the structure type LTA⁰ or FAU arise. LTA⁰ can thus orginate from condensation of both cage types. The four structure types, shown in Photo 3, can also develop from condensation of D4R, D6R or D8R. LTA⁰, KFI⁰ and RHO have 8-ring pores and FAU exhibits 12-ring apertures. The rather complicated framework of PAU● with 8-ring apertures is not shown. PAU● can be obtained by linking α and γ cages through D8R.

The three non-planar infinite 4.6.12 subunits (Color Plate 5) are given in a stylized way in Fig. 12. The numbers 1, 2 and 3 represent possible shifts of the origin along the long diagonal in the (pseudo) hexagonal SSU. For the present description 1 is chosen as the origin. Identical layers can be stacked together in different ways: the origin of successive layers can be above 1, 2 or 3 of the first layer. The shift of successive layers may be accompanied by a 60⁰ rotation of the layer with respect to the first one. Type of (hydrated) cation and template molecule present may play an important role in directing the close-packing sequence. Some stacking arrangements are shown in Fig. 13.

CAN, GMEL, SOD, CHA and the very recently reported APO-52⁰(ref. 20) can be obtained by different stacking arrangements of identical 4.6.12 SSU depicted in Fig. 12(a). The thickness of the layer, in the framework, equals the zigzag repeat distance of 5.0 Å. Although SOD has only 6-ring apertures, it has been included here because it nicely meets the close-packing description when transformed from the cubic to the (pseudo) hexagonal lattice. The packing arrangements and the lengths of the hexagonal axes are given in Table 3.

OFF and ERI, and EAB⁰ and LEV can be built by stacking of the layers shown in Fig. 12(b) and (c), respectively. The SSU, given in Fig. 8(d) (bottom-right), may condense with 6-rings to a SSU containing CAN cages from which LTL can be built. A more complicated condensation results in the SSU found in MAZ. The thickness of the layers in these 6 structure types equals the saw repeat of 7.5 Å. Relevant data are summarized in Table 3. All the structure types have 8- and/or 12-ring apertures.

Finally, three additional structure types are given in this paragraph. They show, in projection, the same hexagonal 4.6.12 net or a closely related net. It is, however, impossible to describe these frameworks by stacking of the layer-like SSU depicted in Fig. 12. The structure types AFI⁰, AEL⁰ and VPI⁰-5 can be formed by stacking of cylinders with 6^3 connectivity (Fig. 9). The cylinder has

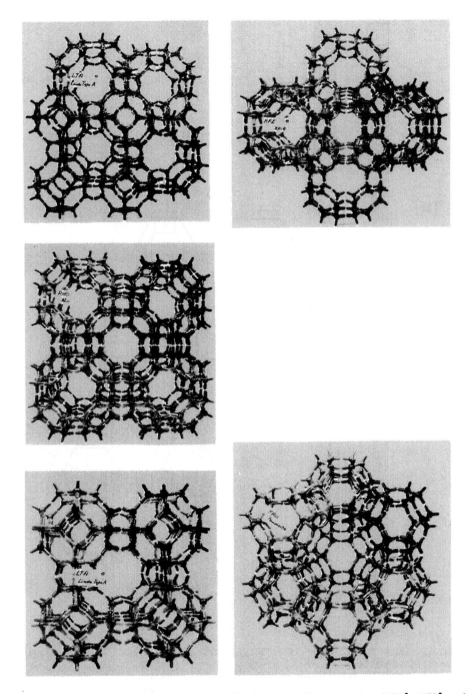

Photo 3. Structure types obtained by linking complete α cages (LTA⁰, KFI⁰ and RHO) or β cages (LTA⁰ and FAU). In LTA⁰ (top left corner) both cage-types can be recognized.

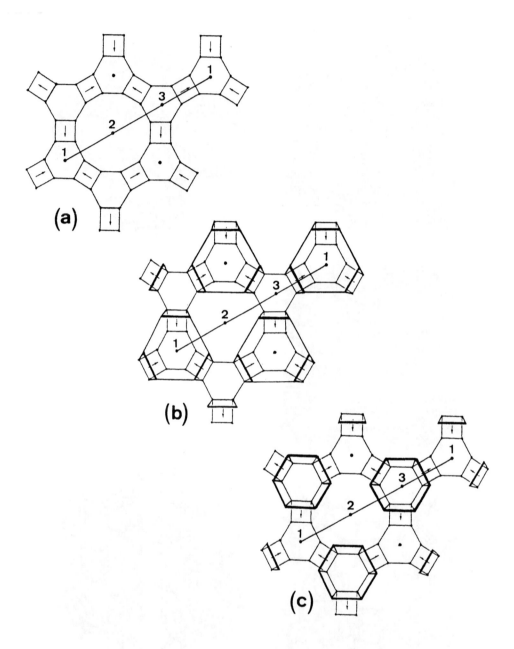

Fig. 12. Three 4.6.12 layer-like SSU. Compare Color Plate 5 for a spatial impression of site 1.
(a) The simple puckered layer in CAN, GMEL, SOD, CHA and APO-52.
(b) The puckered layer in OFF and ERI, with CAN cages. Obtained from (a) by condensation of single 6-rings on top of site 1.
(c) The layer in EAB^0 and LEV, with double 6-rings. Obtained from (a) by condensation of single 6-rings on top of site 3.

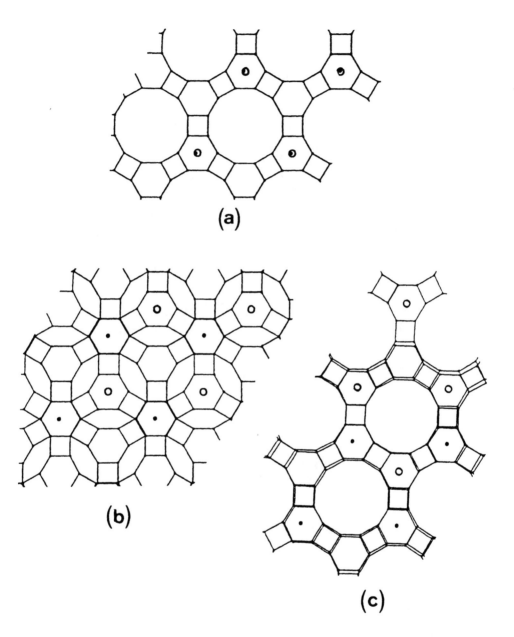

Fig. 13. Three examples of stacking two identical simple puckered 4.6.12 layers. Origins in the two layers are defined by ● and o, respectively.
(a) The 11-stacking: the origin of the second layer coincides with the origin of the first layer.
(b) The 12-stacking: the origin of the second layer is shifted over 1/3 of the long diagonal to position 2.
(c) The 13*-stacking: the shift of the second layer over 2/3 of the long diagonal is accompanied by a 60⁰ rotation of the layer (symbolized by a *).

TABLE 3.

Dimension (Å) along the (pseudo) hexagonal axis in structures with different stacking of identical hexagonal puckered layers or cylinders of 6-rings.

Code	c(Å)*	Equal to:	Packing
CAN	5.1	1x	1 1
GMEL	10.0	2x	1 3*1
SOD	15.4	3x	1 2 3 1
CHA	15.1	3x	1 3 2 1
APO-52	28.9	6x	1 3 2 1 *2*3*1
OFF	7.6	1x	1 1
ERI	15.1	2x	1 1*1 ▲
EAB°	15.2	2x	1 1*1
LEV	23.0	3x	1 2 3 1
MAZ	7.6	1x	1 1 ▲
LTL	7.5	1x	1 1
VPI°	8.1	1x	
AFI°	8.4	1x	cylindrical
AEL°	8.4!	1x	6-Ri sheets

* Accompanied by a 60° rotation with respect to the first layer
⁺ The other axis is ≈ 13.0 Å in all cases except for MAZ, LTL and VPI where a ≈ 18.5 Å
! Orthorhombic
▲ Differ by different SSU

a different radius in each structure type. The stacking introduces crankshaft chains with the characteristic 8.5 Å repeat distance (Table 3) The AFI° framework can therefore as well originate from hexagonal linkage of CC-2 chains resulting in a (6x2)-ring aperture. In AEL°, with a σ contracted net (Fig. 3(d)) and orthorhombic symmetry, in addition to the CC-2 chain, a single C chain is needed in order to generate the observed 10-ring aperture. Linkage of CC-2 chains around a 5-fold operator axis does not lead to a 4-conn. 3D framework. AEL° is the only structure type with a 10-ring pore. In VPI°-5 hexagonally connected "side by side" CCC chains (Color Plate 3) are observed leading to (6x3)-ring apertures. All 14 structure types, seen along the (pseudo) hexagonal axis, are illustrated in Photo 4.

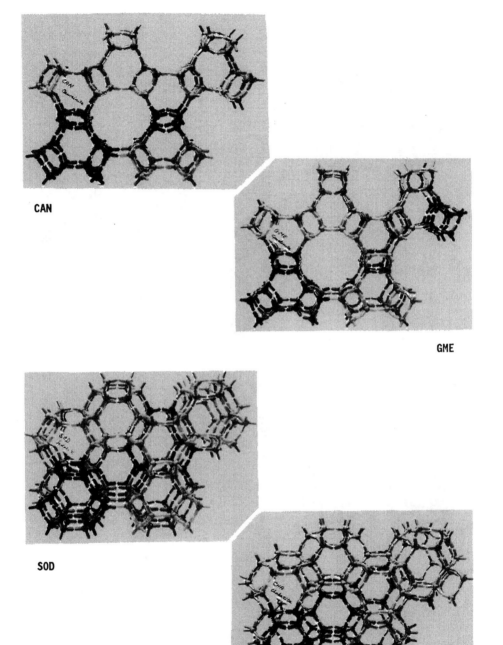

CAN

GME

SOD

CHA

Photo 4. Structure types with (pseudo) hexagonal symmetry. The surrounding of site 1 in each layer and the hexagonal chain is depicted in black. In AFI⁰, AEL⁰ and VPI⁰-5 the 6-ring cylinder wall is indicated in black as well.

70

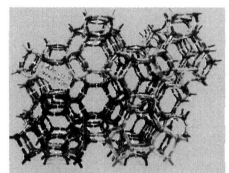

APO-52°

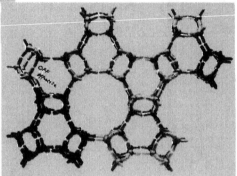

OFF

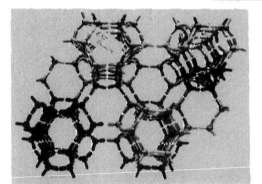

ERI

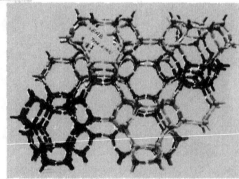

EAB

Photo 4 - Continued.

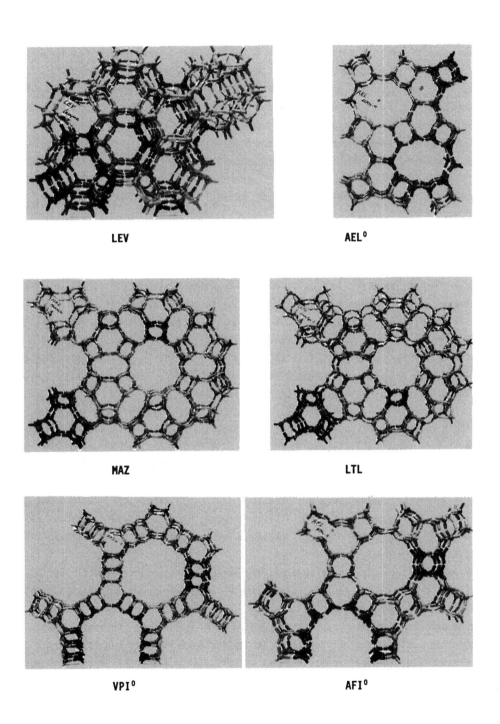

LEV   AEL⁰

MAZ   LTL

VPI⁰   AFI⁰

Photo 4 - Final page.

4.4 Structure types containing the (modified) D4R

This paragraph presents in Photo 5 the 9 structure types which may originate from the D4R or its derivatives. The structure types show diverse maximum pore sizes: (highly distorted) 8-rings in ANA, BRE° and GOO°, 10-rings in LAU°, STI° and HEU and 12-rings in the hexagonal structures AFS⁰ and AFY⁰. The framework structure of LOV is unique: it contains 3-rings and has 9-ring pores. LOV is the only structure type with these geometrical aspects observed so far. It is, however, not necessary to introduce the new spiro-5 SBU (ref. 1) to describe its structure. It is one of the rarely occurring structure types with Be in the framework. In ref. 22 the structure of metal sulfide-based microporous solids are discussed. Their frameworks seem to contain 3-rings quite often.

5. CONCLUSIONS

The 12 structure types with (modified) 4.8^2 connectivity have 8-ring apertures only. 10-Ring pores are frequently observed in the 11 structure types of the "pentasil"-family. The 14 (pseudo) hexagonal structure types nearly all exhibit 8- and/or 12-ring apertures; the exceptions are AEL⁰ (10-ring) and VPI⁰-5 (18-ring). The apertures in the 9 structure types which can be envisaged to develop from the D4R or its derivatives, as well as in the 5 structure types built from α (and γ) or β cages contain 8, 10 or 12 T atoms. LOV is the only structure type with 3-rings, 9-ring apertures and Be as T atom.

There are no (known) natural zeolites with the CC-2 chain. Only synthetic AlPO's contain this type of chain. This might indicate that, on the long term, the CC-2 chain is less stable than the CC-1 chain. The transformation of the CC-1 into the CC-2 chain observed at elevated temperature in APC⁰ and very probably in ATF⁰ might confirm the difference in stability.

The unit-cell dimensions of a zeolite can provide information on the modular units of its framework: the 5.0, 7.5, 10.0 and 8.4 Å repeats are characteristic for the (single, double or triple) zigzag, saw, crankshaft-1 or crankshaft-2 chains, respectively.

The structure types described can all arise from condensation of 4-rings, 6-rings and/or chains. In some cases single T atoms are needed to form the SSU. The presumed directing role of cations and template molecules in such a condensation process is still unknown. A comparison between synthesis conditions of the synthetic zeolites and hydrothermal growth conditions of the corresponding natural zeolites seems useful in this respect. The ultimate goal is to get more insight into the relation between synthesis conditions and silicate anions present during the nucleation and subsequent growth of a zeolite. The search for "elegant" SSU may give precursors from which the

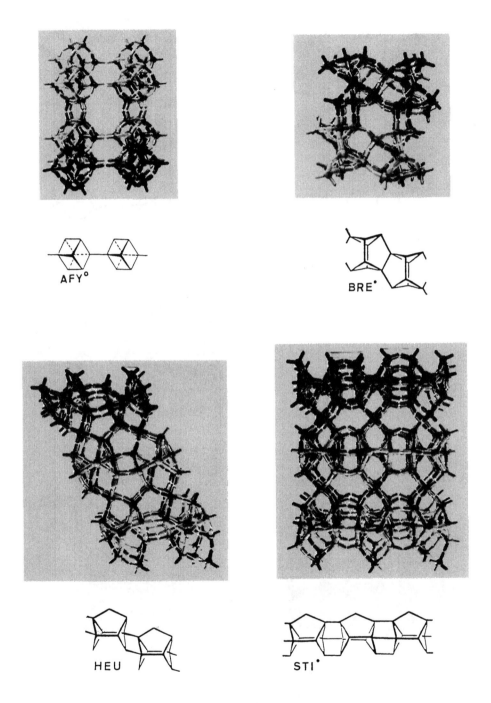

Photo 5. Structure types which can be inferred from the D4R or its derivatives. The chains, parallel to the plane of the paper, are shown in black.

74

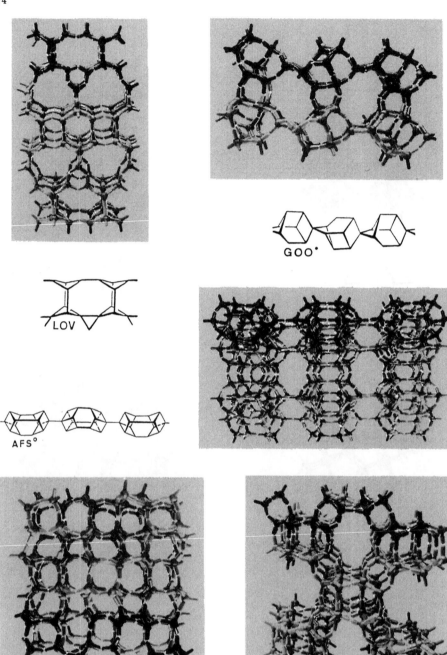

GOO°

LOV

AFS°

Photo 5 - Final page.   ANA, LAU°

zeolites grow. Their actual existence has still to be proven. What has been written in this chapter on the formation of zeolites is therefore merely tentative. This simplified description of the structure types reveals an interrelationship between SSU through simple hydrolysis and condensation processes.

REFERENCES

1 W.M. Meier and D.H. Olson, Atlas of Zeolite Structure Types, 2nd rev. edn., Butterworths, London, 1987.
2 W.M. Meier in: Molecular Sieves, Soc. Chem. Ind., London (1968), pp. 12-16.
3 F. Liebau, Structural Chemistry of Silicates, Springer Verlag, Berlin, Heidelberg, 1985.
4 J.V. Smith, Chem. Rev., 88 (1988) 149-182.
5 G. Gottardi and E. Galli, Natural Zeolites, Springer Verlag, Berlin, Heidelberg, 1985.
6 A.F. Wells, Acta Cryst., 7 (1954) 535-544.
7 A.F. Wells, Acta Cryst., 7 (1954) 545-554.
8 A.F. Wells, Structural Inorganic Chemistry, 4th edn., Oxford University Press, London, 1975.
9 A.F. Wells, Three-dimensional Nets and Polyhedra, Wiley-Interscience, New York, 1977.
10 F.C. Hawthorne and J.V. Smith, Z. Kristallogr., 183 (1988) 213-231.
11 J.V. Smith, "Towards a comprehensive mathematical theory for the topology and geometry of microporous materials", in: P.A. Jacobs and R.A. van Santen (Eds.), Zeolites: Facts, Figures, Future. Proc. 8th Int. Zeolite Conf., Amsterdam, July 10-14, 1989, Elsevier, Amsterdam, 1989, pp. 29-47.
12 R.M. Barrer, Zeolites and Clay Minerals as Sorbents and Molecular Sieves, Academic, London, 1978, pp. 63-65.
13 R.M. Barrer, Hydrothermal Chemistry in Zeolites, Academic, London, New York, 1982, pp. 11-17.
14 J.M. Bennett and B.K. Marcus, "The crystal structures of several metal aluminophosphate molecular sieves ", in: P.J. Grobet et al. (Eds.), Proc. Int. Symp. on Innovation in Zeolite Materials Science, Belgium, 1987, Elsevier, Amsterdam, 1987, pp. 269-279.
15 E.B. Keller, Thesis, Zürich, 1987.
16 S.T. Wilson, private communication.
17 R. Gramlich-Meier, Z. Kristallogr., 177 (1986) 237-245.
18 J.L. Schlenker, J.B. Higgins and E.W. Valyocsik, in Recent Research Reports presented during the 8th Int. Zeolite Conf., Amsterdam, July 10-14, 1989, pp. 287-288.
19 M.E. Davis, C. Saldarriaga, C. Montes, J. Garces and C. Crowder, Zeolites, 8 (1988) 362-366.
20 J.M. Bennett, R.M. Kirchner and S.T. Wilson, "Synthesis and idealized topology of $AlPO_4$-52, a new member of the ABC six-ring family", in: P.A. Jacobs and R.A. van Santen (Eds.), Zeolites: Facts, Figures, Future. Proc. 8th Int. Zeolite Conf., Amsterdam, July 10-14, 1989, Elsevier, Amsterdam, 1989, pp. 731-739.
21 All structure types, shown in this paper, were built after consulting the original literature cited in ref. 1.

76

22 R.L. Bedard, S.T. Wilson, L.D. Vail, J.M. Bennett, E.M. Flanigen, "The next
    generation: synthesis, characterization, and structure of metal sulfide-
    based microporous solids", in: P.A. Jacobs and R.A. van Santen (Eds.),
    Zeolites: Facts, Figures, Future, Proc. 8th Int. Zeolite Conf., Amsterdam,
    July 10-14, 1989, Elsevier, Amsterdam, 1989, pp. 375-387.
23 J.J. Pluth, J.V. Smith, D.G. Howard and R.W. Tschernich, in Recent Research
    Reports presented during the 8th Int. Zeolite Conf., Amsterdam, July
    10-14, 1989, pp. 111-112.

Note added in Proof

Recently the structure of boggsite, BOG [•] (Photo 6), was presented (ref. 23). It
is the first example of a structure with a three-dimensional channel system
with both 10- and 12-rings. Using our previous description, BOG [•] belongs to the
structure types containing the modified D4R (Color Plate 6). The constituing
SSU is the D4R in which two opposing bonds are hydrolysed, like in BRE [•] (Photo
5). Simple symmetry operations generate layers. These layers are connected by
double pentasil chains. The chains called pp-chains by Smith (ref. 4), are
different from the pentasil chains mentioned on p. 24. The pp-chain can
originate from hydrolysis of doubly connected D4R's (see Photo 7). Further
hydrolysis of the pp-chain can give a double chain which is also recognizable
in MFI[o] and MEL[o].

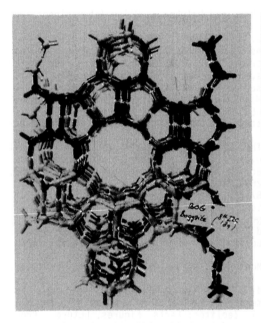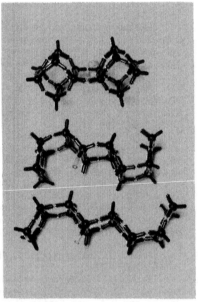

Photo 6   Photo 7

Chapter 4

THE PREPARATION OF MOLECULAR SIEVES

A. Synthesis of zeolites   J.C. Jansen
B. Synthesis of AlPO$_4$-based molecular sieves   S.T. Wilson

A. Synthesis of zeolites

J.C. Jansen

University of Technology Delft, Laboratory of Organic Chemistry,
Julianalaan 136, 2628 BL Delft, The Netherlands

I. INTRODUCTION

a. General

Nature provided mankind with zeolites (ref. 1). Massive zeolite deposits have been discovered at many places in the world (ref. 2). The occurrence of natural zeolites can be assigned to certain geological environments or hydrological systems (refs. 3,4).

Natural zeolites generally form by reaction of mineralizing aqueous solutions with solid aluminosilicates. The main synthesis parameters are: (i) the composition of the host rock and interstitial solutions; pH ~ 10, (ii) the time; thousands of years and (iii) the temperature; often < 100 ^{0}C.

The first systematic studies on zeolite synthesis could thus be guided by the geological and mineralogical findings of the natural species (ref. 5). From 1946 on many additional zeolite types without a natural counterpart have been synthesized (ref. 6). The evolution in the preparation of one of the most studied zeolites is illustrated in Figure 1 by the number of papers and patents on the material denoted as ZSM-5 (ref. 7 and Section XII of this chapter).

NUMBER OF REPORTS

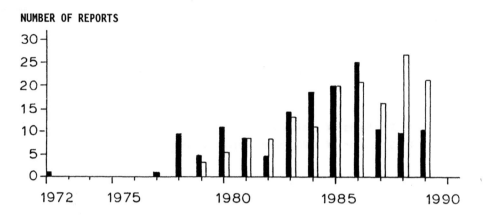

Fig. 1. The annual number of papers (□) and patents (■) on the preparation of zeolite ZSM-5 since the first publication in 1972.

Throughout the last four decades molecular sieves were mainly prepared by precipitation/crystallization of an aqueous mixture of reagents at 6 < pH < 14 and temperatures between 100-200 $^{\circ}$C. As shown in Scheme 1 a relatively large effort is needed on the optimal preparation procedure of the reactant mixture,

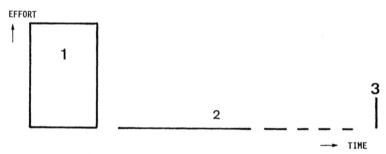

Scheme 1. The effort for the preparatory (1), the reaction (2) and the isolation (3) versus time.

whereafter the hydrothermal reaction process (2) runs autoclaved for a few days or weeks without manual intervention. Isolation (3) of the crystalline material is a simple final step in the synthesis procedure.

The zeolite synthesis field is not only extended and refined by useful data of modern zeolite characterization and application techniques but also by interfacing areas of physical, chemical and mathematical science, see Scheme 2.

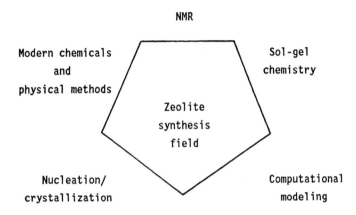

Scheme 2. Areas of chemical, physical and mathematical science interfacing the zeolite preparation field.

Studies in the sol-gel chemistry and NMR analysis area have contributed substantially to the knowledge of the hydrothermal reaction process.

Papers and reviews regarding subjects within the different areas which are mentioned in Scheme 2 and which are of interest for zeolite synthesis are given in Table 1.

Besides the annually new zeolite preparations the extensive exploratory efforts of "zeolite scientists" in the last decade has resulted in the synthesis of porous materials like the $AlPO_4$-group (part B of this chapter), the metal-sulfides (ref. 30) and the clathrasils (ref. 31). Accordingly, zeolite synthesis appears to remain a promising area for future research.

The crystallinity of different synthesis products is well illustrated in Plate 1. The morphology and forms of the crystals give a first indication of the type of zeolite present and the purity of the product.

Table 1. Examples of subjects from areas of physical, chemical and mathematical science which delivered contributions to the knowledge of the zeolite synthesis process together with references.

Area	Subject	Reference
Sol-gel chemistry	Hydrolysis and condensation of silicates	8,9
	The sol-gel process	10
NMR	Structure of (alumino)silicate-clusters in solution	11,12
Computational modeling	Lattice energy calculation	13
	Local interactions in lattice	14
Modern chemical and physical methods	Alkoxides as reagents	15
	Fluorides as reagents and mineralizing agents	16
	Gravity - reduced	17
	- elevated	18
	CVD (chemical vapour deposition)	19
	Microwave	20
Nucleation/crystal-lization theory-practise	Mathematical analyses of zeolite Crystallization. A review	21
	Are the general laws of crystal growth applicable to zeolite synthesis	22
Zeolite Characterization Application	ZSM-5/-11 intergrowth Catalysis	23
	- The catalytic site activity	24,25,26
	- The catalytic properties and the crystal size	27

b. This chapter

In this part of the chapter the preparation of two subgroups of the micro-porous tectosilicates (see Chapter 3) i.e. the aluminosilicates and silicates, both including the clathrasils, will be presented. The division between aluminosilicates and silicates is often discussed on Al-poor rather than Al-free level (ref. 30).

The aluminosilicates, starting from Si/Al ratio 1 up to e.g. Si/Al ratio of 10000, do reveal the presence of Al in synthesis, in characterization as well as in application, see Fig. 2 (ref. 31).

The Al-poor zeolites show no, at least no detectible, Al-dependent behaviour and are therefore, together with the Al-free materials, denoted as silicates.

The presence of aluminium, the guest-host interaction and the nucleation and crystallization all contribute to the synthesis events which are chronologically described in Sections II to VII of this chapter.

Section VIII is focussed on the reaction parameters.

In Sections IX and X the silicates and clathrasils are presented.

Examples of research syntheses performed with certain procedures and/or mixture compositions are listed in Section XI.

Sections XII and XIII contain literature sources on zeolite preparations and the references, respectively.

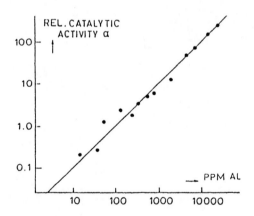

Fig. 2. The relative catalytic activity of H-ZSM-5 versus Al content on ppm scale (ref. 31).

82

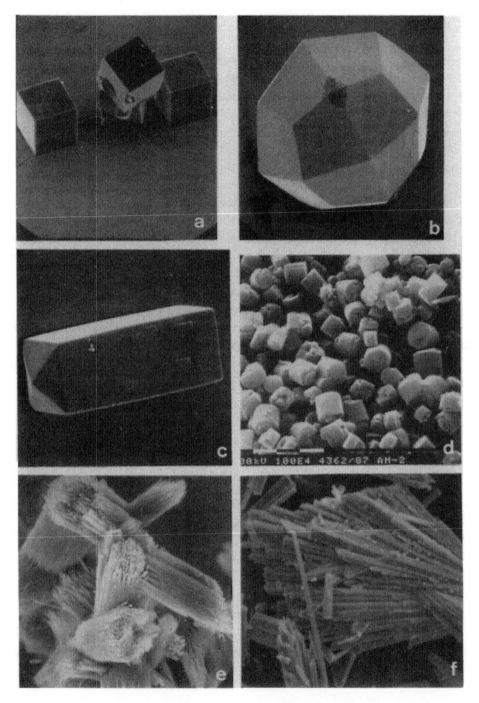

Plate 1. The crystalline nature of zeolites. a) Single crystals of zeolite A
and b) and c) of analcime and of natrolite, respectively. d) A
batch of zeolite L, e) typical needle aggregates of zeolite
mordenite and f) of Nu-10.

II. PREPARATORY

a. Reactants

The chemical sources which are in principle needed for zeolite syntheses are given in Table 2.

Table 2. Chemical sources and their function in zeolite synthesis.

Sources	Function(s)
SiO_2	Primary building unit(s) of the framework
AlO_2^-	Origin of framework charge
OH^-	Mineralizer, guest molecule
Alkali cation, template	Counterion of framework charge, guest molecule
Water	Solvent, guest molecule

Within each type of source a variety of chemicals (ref. 32), has been used as the differences in physical nature and chemical impurities strongly influence the zeolite synthesis kinetics (ref. 34) and sometimes the properties as catalysts (refs. 24-27).

Data on the specifications of regularly used chemical sources are given in the following survey.

- SiO_2-sources

Recent synthesis papers of the Proceedings of the International Zeolite Conferences (refs. 34-37) and of other zeolite conferences (refs. 38-40) reveal that for laboratory scale particular Si-sources are often used, see Table 3a.

Depending upon the particular synthesis a certain Si-source might favour a specific crystallization. For instance, the Aerosil 200 product can be readily dissolved compared to the Optipur and Gold label material because of the difference in particle size, see Figure 3. As the rate of dissolution can influence the rate of nucleation and crystallization (ref. 41) the product formation can be affected. At the same time the Al and other potentially Si replacing impurities are more than 10000 times higher in the Aerosil 200 product compared to the Optipur and Gold label materials.

The influence of impurities can change the crystal form (ref. 42) and the chemical properties (refs. 24-26).

Table 3. Specifications and the references of recent, regular used sources, and high-pure, [*]Si-, and [*]Al-sources.

Si-source (a) Al-source (b)	Specifications		Reference
	Phys.	Chem. impurities (ppm)	manufacturer

(a)

Silicon compounds

$Si(OCH_3)_4$ Tetramethylorthosilicate (TMOS)	liquid	Na,Ca < .5	Merck
$Si(OC_2H_5)_4$ Tetraethylorthosilicate (TEOS)	liquid	Al,Pt < .2	
$Na_2SiO_3.9H_2O$ (or $Na_2H_2SiO_4.8H_2O$) Na_2O 11%, SiO_2 29% Water glass		Al < 200 Fe < 120 Ti < 60 heavy metals < 50	"N" Philidelphia Quartz Co.

Colloidal silica

Ludox-AS-40 SiO_2 40 wt % NH_4^+ (counterion) Ludox-HS-40 SiO_2 40 wt % Na^+ (counterion)	sol	Al < 500 Zr Fe < 50 Ti B < 10	DuPont de Nemours

Fumed silica

Aerosil-200 CAB-O-SIL M-5	D_p ~ 12 nm	Al < 10 Fe < .6 Ti < 10	Degussa BDH

* Silica

Optipur Gold label	D_p ~ 200 μm D_p ~ 800 μm	Al < .001 Fe < 0.01	Merck Aldrich

Table 3, continued

(b)

- NaAlO$_2$			Riedel de Hahn
Na$_2$O 54%		Fe < 4	Carlo Erba
Sodium aluminate			BDH Ltd.
- AlOOH			
Pseudo-boehmite	$D_p \sim mm$		
Al$_2$O$_3$ 70%		Fe < 4	Vista
H$_2$O 30%		Ti < 40	
Catapal-B			
- Al(OH)$_3$			
Gibbsite		Fe < 3	Merck
- Al(NO$_3$)$_3$.9H$_2$O			
- Al$_2$O$_3$	$D_p \sim nm$		
* Aluminiumoxide (Ultrex)		Fe < 0.01	Baker

Therefore, a careful choice of the reactants is needed. The high grade Si-alkoxides of which even double alkoxides like -Si-O-Al- are available (ref. 43) do not have the above discussed disadvantages, except for the rate of hydrolysis of the alkoxide groups.

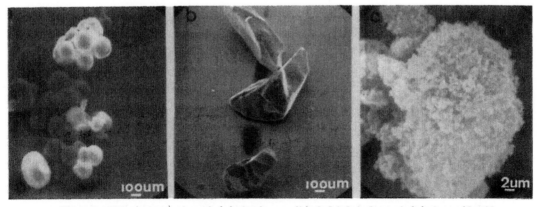

Fig. 3. SEM photographs of (a) Optipur, (b) Gold label, and (c) Aerosil 200.

- AlO_2^- source

Often used Al-sources, collected from the same references as given for the Si-sources, are listed in Table 3b together with the main chemical impurity. Though the very pure Al_2O_3 product consists of small particles it is not easily dissolved.

- Alkali cation/template

The inorganic cations in the zeolite synthesis are mainly alkaline or ammonium ions. The organic cations/templates used may be devided in charged and neutral molecules containing functional atoms or groups. The large number of organic molecules used in zeolite synthesis is extensively listed in several publications (refs. 44, 45) together with the specific zeolite product formed. To illustrate in general the variation in organic template molecules some of the more common templates are listed in Table 4.

- OH^-

Most zeolite syntheses are performed under basic conditions using OH^- as a mineralizing agent. A second agent is F^- (refs. 16 and 46) of which the different nature compared to OH^- will be discussed in the section on reaction parameters.

Both anions are the counterion of the inorganic or organic cations used for the syntheses. Depending upon the quality of the mineralizing agent impurities such as Al^{3+} and Fe^{3+} are present at ppm unit scale.

- The overall reactant mixture

In general the chemical behaviour of impurities like Fe^{3+} and Ti^{4+} are of minor importance compared to Al^{3+} in high Si/Al zeolites in the heterogeneous catalysis when based on Bronsted activity. However, in the case of an all silica zeolite, or modified zeolites like B-ZSM-5, Fe-ZSM-5 and Ga-ZSM-5 traces of Al^{3+} from reactants as given in Table 3 may play an unexpected dominant role in the Bronsted activity (refs. 24-26). Extensive information on this point is given in Chapter 5 on the modification of zeolites. Another example of the influence of impurities from reactants is K^+. The crystallization time can be retarded by factors when K^+ is present in the syntheses of e.g. zeolite Na-A or Na-ZSM-5 (refs. 47, 48).

Impurities like trivalent metal ions sometimes change physical conditions in the reaction mixture indicated by the crystal form or morphology (ref. 42).

Table 4. Type of organic templates, functional atoms/groups and references.

Organic template	Functional atom/group	Ref.	Organic template	Functional atom/group	Ref
amine	$-N\langle$	49	penta-erythritol	$\begin{array}{c} C\text{-OH} \\ \mid \\ HO\text{-}C\text{-}C\text{-}C\text{-OH} \\ \mid \\ C\text{-OH} \end{array}$	65
	$-N\overset{\frown}{}C_n$ n=4,5	50			
	$N\langle\rangle$	51	amine + alcohol	$\rangle N\text{-}(C_n\text{-OH})_x$ n=2,3 x=1–3	66
di-amine	$\rangle N\text{-}C_n\text{-}N\langle$ 3≤n≤9	52			
	$N\langle\rangle N$	53	ammonium + alcohol	$-\overset{+}{N}\text{-}C_n\text{-OH}$ n=2	67
ammonium	$-\overset{+}{N}-$	54			
	$\rangle\overset{+}{N}\overset{\frown}{}C_n$ n=4,5	55	acetal	(6-membered ring with O, O, O)	68
	$-\overset{+}{N}\langle\rangle$	56	amine + ether	$-N\langle\rangle O$	69
	$C_n\overset{\frown}{}\overset{+}{N}\overset{\frown}{}C_n$ n=4,5	57	N-oxide + ammonium	$^-O\overset{+}{-}N\langle\rangle N^\pm$	70
di-ammonium	$-\overset{+}{N}\text{-}C_n\text{-}\overset{+}{N}-$ 3≤n≤9	58	phosphonium	$-\overset{+}{P}-$	71
	$^\pm\overset{+}{N}\langle\rangle N^\pm$	59			
tri-ammonium	(triaryl/tri-N⁺ structure)	60		(tetraphenylphosphonium type structure)	72
amine + ammonium	$N\langle\rangle N^\pm$	61			
alcohol	$C_n\text{-OH}$ n=1–6	62			
di-ol	$HO\text{-}C_n\text{-OH}$ n=2–6	63	di-phosphonium	$-\overset{+}{P}\text{-}C_6\text{-}\overset{+}{P}-$	73
tri-ol	$\begin{array}{c} HO \quad OH \\ \mid \quad\; \mid \\ C\text{-}C\text{-}C \\ \mid \\ OH \end{array}$	64			

88

b. The reaction vessel and hydrothermal conditions

Depending upon the reaction temperature chosen, mainly between 60 °C and 300 °C, reaction vessels can vary as shown in Table 5. Various Teflon inserted autoclaves, see Fig. 4a, together with relatively low priced (stainless steel) and high priced (reenforced polyetheretherketon) autoclaves are shown in Fig. 4b and 4c.

To follow the course of the events taking place in the synthesis mixture, a lookthrough autoclave, see Fig. 4d, can be used. It was concluded, using such experimental conditions, that nucleation and crystallization of zeolite ZSM-5 occurred *on* and *in*, respectively, gelspheres of about 2 mm (ref. 42).

Another possibility to monitor synthesis events *in situ* is the application of IR internal reflectance using a crystal embedded in an autoclave, as shown in Fig. 5 (refs. 74, 75).

Table 5. Regular used lab-scale reaction vessels, the typical impurities and temperature range.

Reaction vessel	Volume	Impurity	Temperature
Plastic bottle	< 1 l	Zn^{2+}	< 100 °C
Stainless steel autoclave	< 5 l	Fe^{3+}, Cr^{3+}	>> 200 °C
Stainless steel + teflon lining	< 2 l	nuclei of preceding synthesis	< 200 °C
Quartz autoclave	< 5 ml	Si	< 200 °C

The autoclaves must be filled between 30 and 70 vol % in the case of an aqueous reaction mixture between 100 and 200 °C to maintain a liquid phase (ref. 76).

Cleaning of the reaction vessels can be considered in some cases, e.g. the teflon-lined autoclaves. As memory effects caused by nuclei of preceding synthesis in cavities of the teflon wall can be encountered in subsequent experiments it is important to clean the vessel with either HF and water at room temperature or NaOH and water at the reaction temperature.

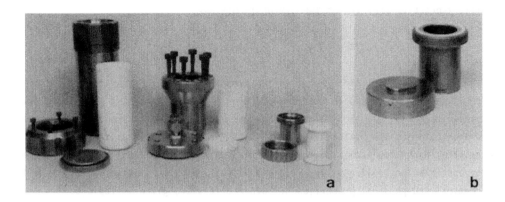
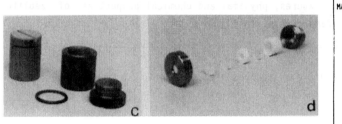
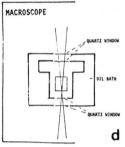

Fig. 4. Different autoclaves for laboratory use. a) Teflon lined autoclaves up to 1000 ml, b) stainless steel autoclave of 25 ml, c) "Arlon" (polyetheretherketon) reenforced with carbon fiber or glass fiber) autoclaves and d) stainless steel look through autoclave with quartz windows and Teflon inserts in exploded view together with a schematic drawing of the experimental set up.

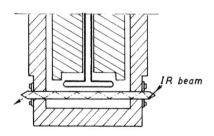

Fig. 5. Parr mini autoclave. IR internal reflection via a crystal embedded in the autoclave makes monitoring of zeolite synthesis events possible (ref. 74).

90

III. ZEOLITE PRODUCT VERSUS THE SYNTHESIS MIXTURE

a. Two synthesis examples

The fine tuning and differences in the preparation of each zeolite type is too complex to be discussed in this introduction on the synthesis of zeolites. In specific parts of the chapter, however, is chosen for a more detailed presentation of the synthesis of two substantial different zeolite types, i.e. zeolite Na-A and zeolite TPA-ZSM-5. The two zeolites present roughly all the groups in which zeolite types are divided (ref. 77). The synthesis mixtures and chemical and physical properties of both zeolites are given in Table 6.

Table 6. The synthesis mixtures, physical and chemical properties of zeolites Na-A and TPA-ZSM-5 (refs. 78-80).

Na-A		TPA-ZSM-5
	- An example of synthesis mixtures -	
	(molar oxyde ratio)	
1	SiO_2	1
.5	Al_2O_3	< .14
1	Na_2O	.16
	TPA_2O	.3
17	H_2O	49
< 100	t (^0C)	> 150
	- Physical and chemical properties -	
3D, holes connected via windows	pore arrangements	2D, intersecting channels
1.28	density (g/cm^3)	1.77
.37	pore volume (cm^3/g)	.18
Na^+, H_2O	lattice stabilization	TPA^+
1	Si/Al	≥ 12
low	Bronsted activity	high
hydrophylic	affinity	hydrophobic

b. Zeolite product versus synthesis mixture

The most simple zeolite product composition can be given by the overall Si/Al ratio and the cation type/content.

More often the unit cell composition of the zeolite crystal is expressed, e.g.

Na-A: $Na_{12}[Al_{12}Si_{12}O_{48}].27\ H_2O$

TPA-ZSM-5: $4\ TPA[Al_nSi_{96-n}O_{192}]H_2O$
$n^* \leq 4$

* At higher loadings than 4 Al/uc, TPA^+ is replaced by the smaller cation Na^+ (ref. 81).

The zeolite reaction mixture is often formulated in the molar oxide ratio of the reactants, e.g. $SiO_2:Al_2O_3:Na_2O:(TPA_2O):H_2O$. The ratios of H_2O/SiO_2, OH^-/SiO_2, SiO_2/Al_2O_3 and (Na_2O/TPA_2O) then give an impression of the concentration, solubility and the expected zeolite types, respectively (ref. 82). Correlation between the synthesis mixture and the product can be obtained from ternary composition diagrams (see Fig. 6a,b) (refs. 83-86), or from graphs of crystallization fields of zeolite types as a function of reactant ratios, see Fig. 6c and Section XI.b.3.

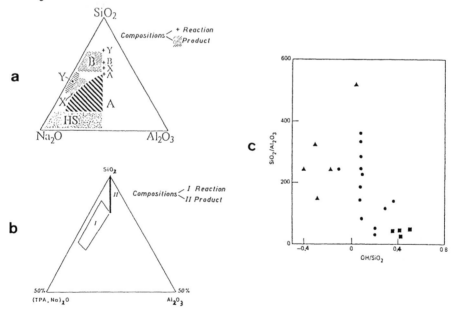

Fig. 6. Zeolite product versus the synthesis mixture. a) and b), ternary compositon diagrams with an inorganic and organic cation/template, respectively. c) Crystallization fields, indicating (●) ZSM-5, (■) ZSM-35, and (▲) ZSM-39 (ref. 87).

92

The product fields at certain P,T, depicted in Fig. 6, are obtained from experimental data which are not always expected from a thermodynamic point of view. As the inevitably heterogeneous synthesis mixture contains micro-domains with different reactant ratios, kinetic parameters might induce other product phases than those derived from the ternary synthesis composition diagram. Because particularly the nucleation is kinetically determined it is thus of interest to understand the different factors, e.g. type of Si-source, cation, Al-source, additives and physical parameters, influencing the kinetic stage of nucleation. The influence of these factors can be recognized in the subsequent events of the zeolite preparation which are given in Table 7 and discussed hereafter in detail.

Table 7. The subsequent events present in the course of the zeolite preparation.

Temperature	Subsequent events
Low (< 60 °C)	Reactant solutions
	Reactant mixture - gel formation
Low → high	Gel rearrangement
(< 60 °C → < 200 °C)	Dissolution of gel
	Dissociation of silicate
High (< 200 °C)	Pre-nucleation phase
	Nucleation
	Crystallization
Low (< 60 °C)	Isolation

IV. THE LOW TEMPERATURE REACTION MIXTURE

a. Introduction

The reaction mixture events occurring at low temperature (< 60 °C) will be discussed for two reasons.

i) Reaction mixtures are prepared at low temperature. Drastic chemical and physical changes take place then.

ii) Substantial knowledge about the zeolite reaction mixture at low temperature has been obtained using characterization methods such as the molybdate method (ref. 88), the paper chromatography method (ref. 89), the tri-methylsilylation method (ref. 90), IR- and laser-Raman spectroscopy (ref. 91), single crystal structure analysis (refs. 92, 93) and the NMR technique (ref. 94).

Mostly, starting reaction mixtures typically consist of a gel phase and a liquid phase which means that nucleation is initiated at high temperature by the presence of a residual gel phase, though there are a few exceptions (refs. 91, 95). The (alumino)silicate gel phase consists of either a homogeneous dispersed phase of branched chains of sol particles, see Fig. 7[I], or a more separated solid phase of an ordered aggregate of sol particles (like opals), see Fig. 7[II](refs. 96, 97).

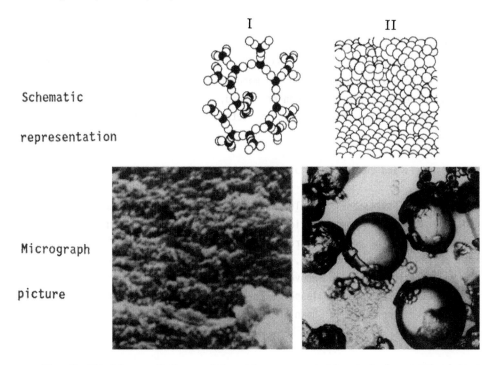

Fig. 7. Alkaline gel forms. Schematic representation and micrograph picture of I) a dispersed low density gel (ref. 96b) of branched chains of sol particles and II) a separated high density gel form resulting in spheres consisting of an ordered aggregate of sol particles (like opals).

94

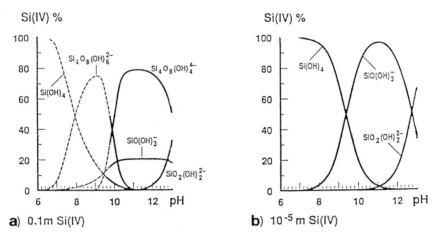

Fig. 8. Silicate distribution versus pH at a) high and b) low silicate
concentration (ref. 98).

The pH of the liquid phase, in the case of OH⁻ as the mineralizing agent, lies
generally between 8-12. As depicted in Fig. 8 the most abundant form(s) of Si-
species at relatively high pH are the monomeric ions, whereas at lower pH value
monomeric neutral Si-species can be formed, when the Si-concentration is low.
At high concentration, however, cyclic tetramers are most abundant species
(ref. 98).

b. Hydrolysis and condensation of silicate

Monomers and oligomers in solution are in equilibrium with the gel phase. At
this ambient stage of the reaction mixture monomeric silica species can be
released from the gel via hydrolysis reactions and are present in solution as
e.g. $Si(OH)_3O^-$ and $Si(OH)_2O_2^{2-}$. The dissolution of the gel is promoted by the
OH⁻-coordination of silicon above four, thus weakening the other siloxane bonds
to the gel network. This nucleophilic mechanism is presented to occur via a
S_N2-Si transition state as shown in Scheme 3a (ref. 9).

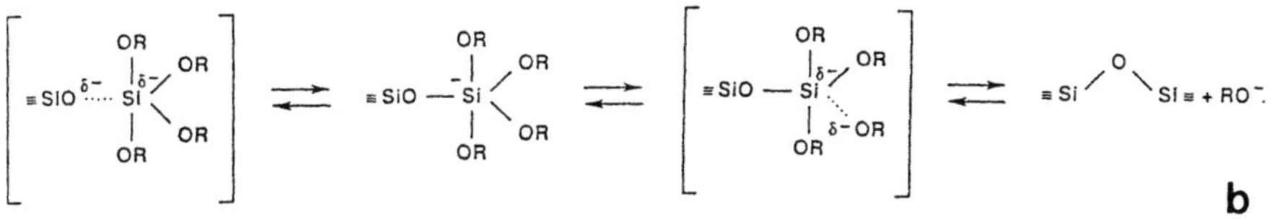

b

Scheme 3. a) Hydrolysis and b) condensation mechanism of silicate species at room temperature.

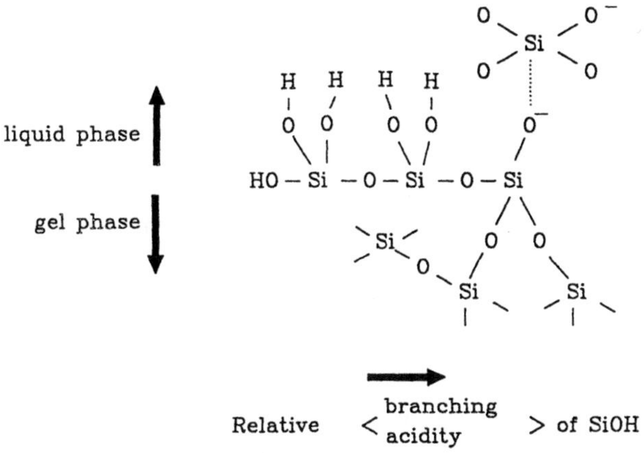

Scheme 3c. Growth site in the gel phase for monomers from solution.

The mechanism of the condensation reactions in aqueous systems at high pH involves the attack of a nucleophilic deprotonated silanol group on a monomeric neutral species as represented in Scheme 3b (ref. 9).

The acidity of the silanol group depends on the number and type of substituents on the silicon-atom. The more silicon substituents are present, the more acidic the OH-groups of the central silicon atom. As shown in Scheme 3c, at high pH the most favourable polymerization is the reaction between large most highly branched species and the monomer silica species.

At more neutral pH, hydrolysis and condensation of clusters, containing specific bonding configurations, see Fig. 9, indicate that inversion in the pentacoordinate state of Si, illustrated in Scheme 3, is not essential (ref. 99).

96

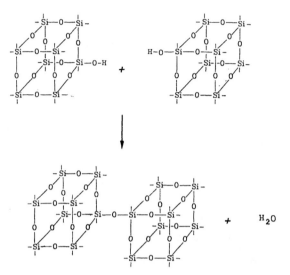

Fig. 9. Condensation of octamers,   with retention of the configuration.

The pentacoordinate silicon intermediate state is, however, confirmed crystallographically (ref. 100).

Condensation of the monomers lead, as the pH of the zeolite synthesis mixture is above the isoelectric point of silica (ref. 101) to ramified clusters. Such clusters can be reorganized into fewer larger particles with a corresponding reduction in surface energy, according to the Ostwald ripening principle. The structural evolution of a growing cluster is schematically given in Fig. 10.

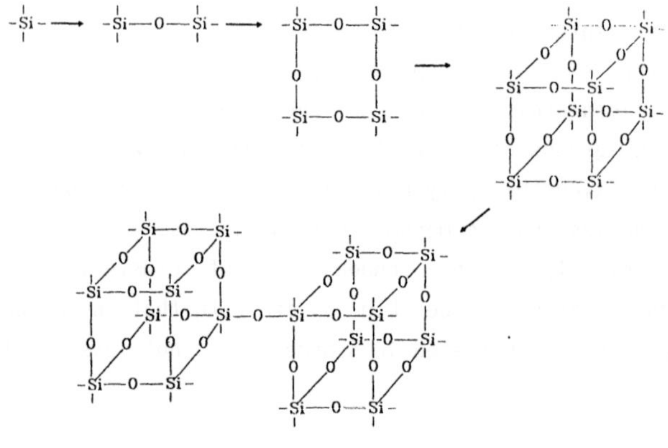

Fig. 10. Structural evolution of silicate clusters.

c. Evidence for silicate clusters

In the course of the gel dissolution the monomers form dimers, according to ^{29}Si-NMR studies (ref. 94), via condensation reactions whereafter trimers and tetramers, cyclic trimers and tetramers and higher order rings are observed as condensation products, see Fig. 11.

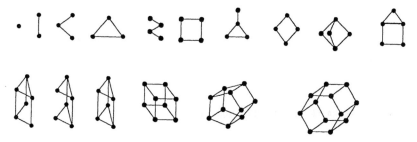

Fig. 11. Numerous oligomers characterized in solution at low temperature by ^{29}Si-NMR (ref. 101).

Evidence for the existence of e.g. double four rings resulted from the single crystal structure analysis of so-called pseudo-A, a material, *not a zeolite*, crystallized at ambient temperature from a mixture of SiO_2, TBAOH and H_2O, see Fig. 12 (ref. 93).

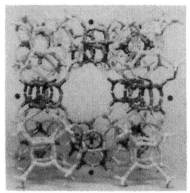

Fig. 12. Model of a part of the framework of pseudo-A; the double four ring units are indicated by asterisks (*).

The silicate species identified in the liquid phase by a.o. NMR, SAXS (ref. 102) and IR, (ref. 103) are products in a simple reaction mixture of SiO_2, NaOH and H_2O at room temperature.

98

The interaction of alkali-ions in such systems is not clear. It is often suggested (ref. 104) that the ordered hydration sphere of a.o. Na^+ stabilizes silicate species. Recent NMR results indicate that interaction between cations and silicate species (ref. 105) do occur.

An organic cation/template added as ingredient(s) to the simple reaction mixture shows in typical experiments according to NMR measurements interaction with the gel and silicate species, respectively (refs. 106-108). However, the highly complicated set of interactions and fast changing equilibria, due to the increased number of type of species after addition of template and/or Al^{3+} has not been unravelled yet.

V. THE TEMPERATURE RAISE OF THE REACTION MIXTURE

Temperature raise, from < 60 ^{0}C up to < 200 ^{0}C, can be performed in several ways as shown in Fig. 13 for one type of autoclave and reaction mixture. The different heating rates are achieved in static systems.

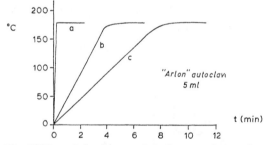

Fig. 13. Different heating rates for one type of autoclave achieved by (a) microwave, (b) hot sand bath and (c) hot air oven.

The size of the autoclave, the viscosity of the reaction mixture and the way of agitating e.g. static, tumbling or turbostirring are factors modulating the temperature raise of the reaction mixture.

During the temperature raise of the reaction mixture from ambient to reaction conditions primary events are:
- Accelerated dissolution of the gel into monomeric silicate species.
- Dissociation of silicate oligomers in solution and increase of monomers as measured by NMR up to ~ 100 ^{0}C (refs. 109-112). As shown in Fig. 14 a model study with NMR on trimethylsilylated silicate confirms (ref. 109) a shift of the silicate anion equilibrium from relative high-molecular, mainly double four rings, to low-molecular weight, monomers and dimers.

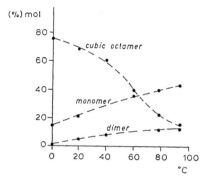

Fig. 14. Main changes in composition (% mol) of trimethylsilylated silicate solution versus temperature.

- Higher concentration and mobility of monomeric silicate- and eventually aluminate species.
- Association of primary building units.
- Possible nucleation and crystallization of unwanted (metastable) phases.

Some secondary events are:
- The start of the degradation of quaternary ammonium ions, which can be substantial in a ZSM-5 synthesis (ref. 42) as depicted in Fig. 15.
- Start of the drop in pH caused by the Hoffman degradation.

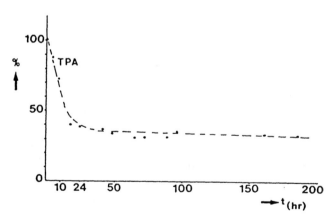

Fig. 15. Degradation of tetrapropylammonium versus time.

VI. THE HIGH TEMPERATURE REACTION PROCESS

a. Introduction

The main event occurring in the synthesis mixture at the reaction temperature
is the formation of zeolites from amorphous material. Chemical reaction
processes accelerated by the high temperature lead to:

  i) further reorganization of the low temperature synthesis mixture;

 ii) whereafter primary (homogeneous or heterogeneous) and secondary (seed
     crystals (ref. 113)) nucleation;

iii) and finally, precipitation (based on reactions) as a form of
     crystallization.

b. Nucleation

At the high temperature of the reaction mixture the zeolite crystallization is
expected after an induction period in which the nucleation occurs. During the
induction period the gel and species in solution (aforementioned in the low
temperature section) rearrange from a continuous changing phase of monomers and
clusters, e.g. polysilicates and aluminosilicates. These clusters form and
disappear through inhomogeneities in the synthesis mixture via condensation and
hydrolysis processes. The continuous dissolution of the gel phase increases,
however, the amount of clusters and the possibility of further association of
the clusters and cations. In the course of this process particles become
stable. Nuclei of certain dimensions, e.g. ~ 10 Å for zeolite Na-A (ref. 114)
and ~ 20 Å for zeolite ZSM-5 (ref. 115), are formed and crystallization starts.

c. Crystallization

The lines along which ideas on zeolite crystal formation are developed, either
based on bulk and macroscopic observations or on molecular mechanistic scale
are described in this paragraph. Four cases of nucleation and crystallization
are schematically presented in Table 8. (a) Zeolite crystallizations, which
might occur in clear synthesis solutions, or, more often, in heterogeneous
reaction mixtures where (b) highly dispersed or (c) dense gel forms are
present, see also Fig. 7. In some occasions (d) metastable solid phases undergo
transformation during synthesis. Homogeneous nucleation whereafter crystallization has been observed in (a) clear solution experiments (refs. 91, 92).

Table 8. Four cases of crystal growth environment and schematic representation of nucleation and crystallization.

Crystal growth environment	Nucleation (·) Crystallization (→)
(a) Clear solution	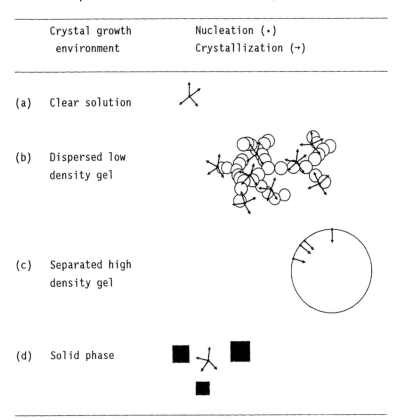
(b) Dispersed low density gel	
(c) Separated high density gel	
(d) Solid phase	

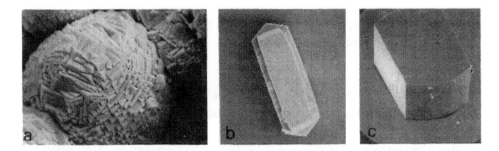

Fig. 16. a) Powder and b) a twinned elongated prismatic crystal of ZSM-5 from a dispersed gel phase and c) a cubic form of ZSM-5 from a dense gel phase.

Nucleation (heterogeneous) occurs at the liquid-gel interface in the dispersed gel-solution mixtures (b) (ref. 108). The forms of the crystallization products in the case of a dispersed gel phase are shown for ZSM-5 in Fig. 16a,b. Similarly to the clear synthesis solutions, the driving force for crystallization is equal in all directions as shown in Table 8a,b. In the case of a dense gel phase present in the synthesis mixture, see Table 8c, crystallization proceeds *into* the gel (ref. 42) as shown schematically in Fig. 17. Deviating crystal forms compared to crystal forms from dispersed gel systems are then observed, as shown in Fig. 16c.

Generally, the typical form and morphology of a zeolite crystal reveals not only information on the type of the zeolite formed but also on the crystal growth history, as shown above.

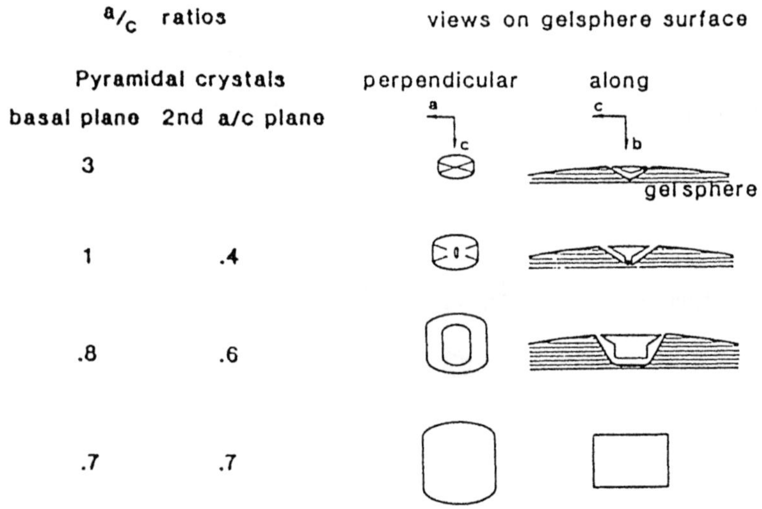

Fig. 17. Average a/c ratios of developing crystals and schematic drawing of growth process in the gel spheres.

As a liquid phase is continuously present between the dissolving dense gel phase and the growing crystal, the crystallization is, however, still solvent mediated.

When a metastable solid phase, e.g. a zeolite, is present in the synthesis mixture, a transformation into a more stable phase is possible, according to the Ostwald rule of successive transformations (ref. 116).

The nucleation and crystallization of the new phase, illustrated in Table 8d, occurs in the supersaturated solution generated by the dissolution of the former phase (ref. 117).

In the last three cases of Table 8 dynamic equilibria between successive steps of dissolution, ion transportation and precipitation, can be recognized (ref. 118). Especially, the precipitation/crystallization step, i.e. the type of crystal building units and the way of crystal growth on molecular level, has been subject to many studies.

d. Crystal building units

At least three types of crystal building units have been suggested which are described below.

d.1. The primary building unit

Arguments that primary building units, i.e. tetrahedral monomeric species, can be involved in the crystallization are:

   i) The general view from crystal growth theories that crystals are formed via primary building units (ref. 119);
  ii) The general view in sol/gel chemistry (refs. 8, 10) that the most favoured condensation reaction occurs between a monomeric and polymeric species. In terms of the zeolite crystallization: between a primary building unit and a crystal surface; see Section IVb;
 iii) At raising temperatures (up till 100 $^{\circ}$C) the concentration of monomers increases (ref. 109) at the expense of clusters. Though in situ measurements (till 200 $^{\circ}$C) are not actually performed, the above experimental results might indicate that at reaction temperatures mainly monomers are present;
  iv) Studies on the crystallization of zeolite have shown that the growth of a zeolite occurs by a surface reaction of monomeric anions (ref. 120).

d.2. A typical cluster as building unit

As shown in Chapter 3 of this book secondary building units (SBU's) are relatively low ($\leq$ 16-Si-tetrahedra) polymer units. SBU's were introduced several decades ago (ref. 121) and used since to present structural (ref. 6) and further physical features of the zeolites. At the same time SBU's acting as non-chiral independent units can generate a certain zeolite structure. It is, however, though the SBU's show sometimes a superficial resemblance to silicate

104

anions, not likely that SBU's are the building blocks of the growing crystal (ref. 122). On the other hand, the building of the porous and different zeolite frameworks with monomers condensating in the right topology seems less favourable compared to a typical cluster building unit (ref. 123). From this point of view suggestions are raised about a typical or common cluster building unit for all zeolite structures.

d.3. The cation templating theory

Organic as well as inorganic cations show structure directing, i.e. water-ordering, properties. Typical examples are given in a review of single crystal structure analysis of organic water clathrated cations (ref. 124). The water molecules comprising a tetrahedral network in the first layer around the cation might be partly replaced by silicate and aluminate anionic tetrahedra. The clathrated cations might serve this way as crystal building units.

An example of such a templating/clathrating role is the formation of sodalite with tetramethylammonium (TMA$^+$) cations (ref. 125).

The high temperature events, discussed above, are summarized in Scheme 4.

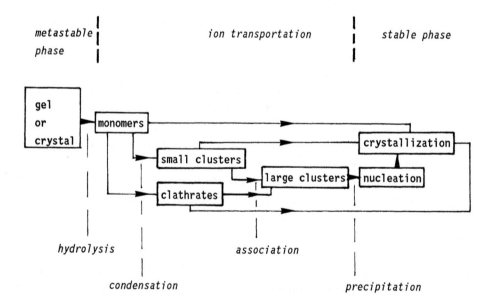

Scheme 4. Representation of successive steps in the evolution of zeolite crystallization.

e. Nucleation-crystallization kinetics

Nucleation and crystallization events are generally illustrated on characteristic S-shaped crystallization curves (ref. 126). The yield (wt % of crystalline material), often determined by indirect methods, plotted against time gives an impression of the nucleation and crystallization time and certain reaction temperatures. More accurate information on the crystallization kinetics can be provided when, based on crystal size and size distribution, the linear crystal growth rate and the rate of nucleation can be determined. Of the studies (ref. 127) on zeolite crystallization, one contribution (ref. 128) reporting on a method to collect kinetic data is briefly described here.

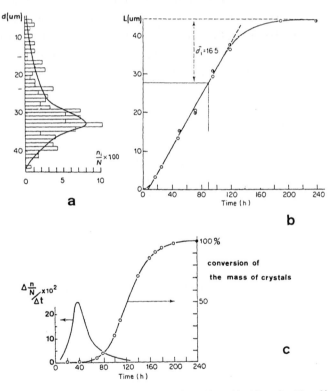

Fig. 19. a) Histogram of the crystal size distribution in the final product, b) diameter of the largest crystals of different unfinished crystallization runs versus time, resulting in the crystal growth rate graph and c) (i) the nucleation rate (number of crystals of each unfinished crystallization run versus time) together with (ii) the yield curve.

A number of identical synthesis experiments, but differing in total synthesis time, were performed. The average diameter of the largest crystals which could be collected from the various products was measured. In the case of zeolite Na-X it was found that in a plot of crystal size versus time the linear crystal growth rate (.5 ΔL/Δt) was constant, irrespective of the crystal size, even until near exhaustion of the crystal building units, see Fig. 19b.

The nucleation time can be determined now for any crystal in the final product of this Na-X crystallization, knowing the growth rate. For instance, a crystal of 16.5 μm nucleated at t ~ 90 h. Together with the particle size distribution curve, Fig. 19a, the rate of nucleation was found, see Fig. 19c.

The nucleation rate curve and the yield curve calculated from both the growth rate and particle size distribution curve, indicate that as soon as the crystallization starts the chemical nutrients are consumed for crystal growth. The formation of fresh nuclei is from then on largely suppressed. In conclusion, it can be said that zeolite synthesis, resulting in a good crystalline product can deliver accurate information on nucleation and crystallization kinetics.

f. Energy of activation

Though zeolitic material can be prepared at low temperature (20-60 °C) most nucleation and crystallization processes are performed at temperatures between 60 and 250 °C. The choice of the reaction temperature is governed by the energy of activation required for the zeolite crystallization.

Table 9 shows the energy of activation (E_a) as a function of the Si/Al ratio.

Table 9. E_a's of different zeolite framework types and Si/Al ratios.

Guest molecule	Framework	Si/Al	E_a (kcal/mol)	Ref. 129
Na^+; H_2O	Y	1.5	11.8	
		1.8	12.3	a
		2.2	14.1	
		2.5	15.6	
TPA^+	MFI	30	7	b
TPA^+	MFI	∞	11	
Na^+; H_2O	MFI	80	18	c

It appears that the E_a's are not related to diffusion of crystal building units in solution (E_a (diff.) < 5 kcal mol^{-1}) but to condensation reactions between the crystal surface and crystal building unit. As shown in Table 9 the E_a of Na-X changes as a function of the Si/Al ratio which indicates that the more silicious the zeolite, the larger the E_a. Generally, this trend is also observed between different zeolites, although the contribution to E_a of cations and templates, as shown for ZSM-5, can be substantial.

VII. ISOLATION OF THE ZEOLITE PRODUCT

Products of zeolite preparations can be composed of either one pure zeolitic phase, a mixture of zeolitic phases or a mixture of a zeolitic phase and e.g. quartz, cristobalite or gel phase. Mostly the product is isolated by decantation/centrifugation or filtration.

If the product consists of crystals with a uniform crystal form which is recognized as characteristic for the expected product, the zeolite can be separated by decanting the mother liquor followed by washing with water.

If there is, however, e.g. some gel phase present, this may be either co-precipitated as a separate phase or adsorbed on the crystals. Careful dissolution of the gel phase with e.g. a dilute basic OH^- solution at slightly elevated temperature is strongly advisable prior to the isolation of the zeolite. Especially in the case of adsorbed gel on the crystal surface elemental analysis (AAS, ICP or EMPA) is required to control the Si/Al ratio of the crystals before and after the washing procedure (ref. 130). The final step in the zeolite preparation is the drying or calcination procedure after which the zeolite void volume is free for different modification and/or application.

VIII. REACTION PARAMETERS

a. Introduction

The type of reactants, the way the reactant mixture is made, the pH and the temperature typically affect the crystallization kinetics and product formation.

108

Furthermore the pre-treatment of the reaction mixture, the addition of crystal growth inhibitors, the reaction mixture temperature trajectory and the use of seeds have an influence on the zeolite preparation.

Some aspects of the type of the above mentioned factors are discussed in the following paragraphs. Illustrations are mainly given on the zeolite A and ZSM-5 formation.

b. The Si-source

As mentioned in Section II of this chapter the different types of the Si-sources contain impurities which may affect zeolite crystallization. Another parameter, the specific surface area of these sources, can result in different nucleation and crystallization times as shown for zeolite A in Fig. 20a (ref. 47). The shorter induction and crystallization times lead to more and smaller crystals, see Fig. 20b.

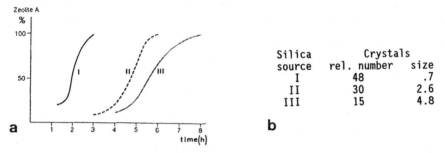

Fig. 20. a) The yield of zeolite A versus time of different silica sources.
b) The specific surface areas of the silica sources (I > II > III) result in different amounts and sizes of crystals.

c. The type of template

Many types of template are regularly used (see e.g. Section II of this chapter). The surprising performance of certain templates on stabilizing the type of zeolite framework formed is illustrated in Table 10. One type of template can be used to crystallize various zeolites whereas the same type of zeolite may be crystallized while using different templates.

Table 10. Single and mixture of templates/cations in the preparation of different zeolite types.

Single template	Zeolite	Ref.	Mixture of template/cation		Zeolite	Ref.
TMA$^+$ ⟨	Sodalite	131	TMA$^+$, Na$^+$		A, X, Y	135
	Gismondine	132			L (+ K$^+$)	136
					Sodalite, P, S and R	136
					ZSM-6 and ZSM-47	137
					Omega	138
TPA$^+$ ⟩ Na$^+$	ZSM-5	133 134	TEA EDA Ethanolamine Propanolamine Alcohol Glycerol Morpholine Hexanediol TPA	, Na$^+$	ZSM-5	139

The role of the single template/cation in stabilizing subunits of different zeolite types is not unravelled yet.

A common factor, however, appears to be the size of a certain free void diameter in the structures of sodalite and gismondine, 6.8 Å and 7.0 Å, respectively, and the diameter of ~ 6.7 Å of the template TMA$^+$, see Fig. 21.

Fig. 21. Models of a) the sodalite and b) the gismondine void and the void filler/template/cation TMA$^+$.

Although TPA$^+$ and Na$^+$ are rather different templates/cations a common factor might be the stabilization of voids (either intersection of channels or channel windows), see Fig. 22.

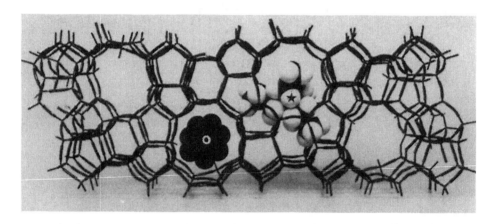

Fig. 22. View along straight channels of wire model of ZSM-5 with either TPA$^+$ (*) or hydrated Na$^+$ (0) on intersections of channels and channel windows, respectively.

Charged templates compensate negative framework charges, due to isomorphous substitution of Si^{4+} by Al^{3+}. A range of Si/Al ratios is possible, see Scheme 5. If, however, the number of charged templates required for charge compensation cannot be accommodated for dimensional reasons the zeolite combines charged templates with e.g. Na^+. This way, still various Si/Al ratios for one zeolite type are possible as shown in Scheme 5 for zeolite ZSM-5. Sodalite can be prepared with two different Si/Al ratios.

ZSM-5	Si/Al	Sodalite	Si/Al
TPA^+	23 - <10000	TMA^+	5
TPA^+/Na^+	23 - 11		
Na^+	11	Na^+	1

Scheme 5. Different Si/Al ratios for ZSM-5 and sodalite.

d. The reactant mixture

The way reactant mixtures are made, e.g. the addition sequence of the reactants, the stirring and gel aging can result in method-dependent factors influencing nucleation. As shown in Fig. 23 crystals of zeolite A started growing in a zeolite X synthesis mixture whereafter zeolite X crystals started growing on and over the zeolite A crystal (refs. 140, 141).

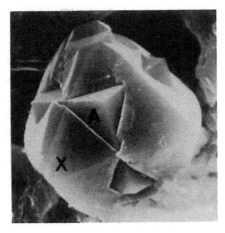

Fig. 23. Overgrowth of zeolite X onto zeolite A.

Though the thermodynamic variables were correctly chosen to prepare zeolite X, synthesis mixtures of zeolite A and X, given below, do have comparable elements and apparently local kinetic factors initiated the synthesis of zeolite A.

	$Na_2SiO_3.9H_2O$	$NaAlO_2$	Triethanolamine	H_2O	Ref. (142)
zeolite A:	.4	.1	.7	28	(molair)
zeolite X:	.4	.05	.7	28	

e. The pH

e.1. Introduction

The pH and the solubility of reactants in the synthesis mixture are governed by the presence of OH^- or F^-.

An advantage of F^- compared to OH^- is the higher solubility of e.g. Fe^{III} and Ti^{IV} and the condensation capability for e.g. Ge^{IV}. A too high concentration of F^-, however, prevents the polycondensation mechanisms. A compromise between solubility of certain elements and inhibition of zeolite framework formation leads to F^- synthesis mixtures which are less supersaturated than OH^- media. Hence, only a few zeolite types are obtained, until now (ref. 16).

e.2. OH^-

Raising pH of synthesis mixtures using OH^-, mainly influences the crystallization of a certain zeolite in a positive way within the synthesis field. As depicted in Fig. 24a and b for zeolite A and zeolite ZSM-5, respectively, increasing the pH shows an increase in the crystallization rate. The OH^- is a strong mineralizing agent for bringing reactants into solution. The higher the pH and thus the concentration of dissolved reactants the more the rate of crystal growth of zeolites is enhanced (refs. 47, 143).

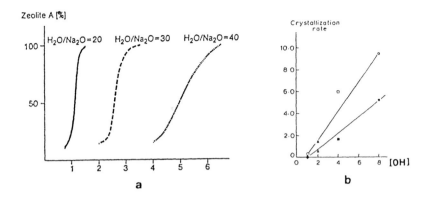

Fig. 24. The influence of alkalinity on a) zeolite A and b) ZSM-5 crystallization.

e.3. F⁻

After the first publications on synthesis with F⁻ (ref. 46) extensive studies have been undertaken to investigate the effect of F⁻ and possibilities in zeolite synthesis (refs. 16, 46). Replacing OH⁻ by F⁻ with e.g. NH_4HF, NH_4F and BF_3 the pH values of the synthesis mixture lies generally between 3 and 10. A typical synthesis formulation is given in Section XII of this chapter. The zeolites obtained so far by this route are silica-rich materials of which the structure types are:

ZSM-5

Ferrierite

Theta-1

and ZSM-23

f. The temperature

It has been shown for many zeolites that raising synthesis temperatures within a certain zeolite synthesis field increases the crystal growth rate (refs. 47, 144, 145). As shown in Fig. 25a and b for zeolite A and zeolite ZSM-5, increasing temperature influences the crystal growth rate whereas in the case of zeolite A the crystal size does not change significantly compared to substantial variations in the ZSM-5 product.

114

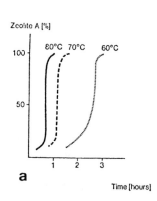

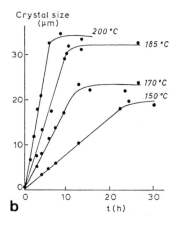

Fig. 25. Influence of temperature on the crystallization of a) zeolite A and
b) zeolite ZSM-5.

IX. ALL SILICA MOLECULAR SIEVES

a. Introduction

Two preparation routes can be followed to obtain all silica molecular sieves:
i) A direct synthesis to crystallize molecular sieves with a SiO_2 composition
   and well known zeolite topologies.
ii) A secondary synthesis. After the direct synthesis of a zeolite a de-
    alumination procedure, e.g. steaming (ref. 146), ammonium silicon
    hexafluoride (ref. 147) or silicon tetrachloride (ref. 148) can lead to an
    all silica molecular sieve.

b. Synthesis

Though the neutral all silica molecular sieves do formally not need to be
stabilized with cations the silica structures usually contains the cations used
in the synthesis. For example, tetrapropylammonium for silicalite-1,
tetrabutylammonium for silicalite-2, and tetraethylammonium for silica-ZSM-12.
Recently, amines, di-amines (ref. 149) and poly-amines (ref. 150) have been
used as templates.

Table 11 contains a list of all silica molecular sieves with two examples of
synthesis recipes and references.

Table 11. All silica zeolites, recipes and references.

Product	Ref.
Silica-ZSM-48	150

Recipe: 14.6 g of triethylenetetramine is dissolved in 18 ml H_2O whereafter the solution is stirred into dry 1.2 g SiO_2. The smooth dispersion is then autoclaved between 120-180 $^\circ$C for 28-105 days, respectively.

Silica-ferrierite	151

Recipe: 0.75 g $Si(OC_3)_4$ is added to a solution of 1.2 g ethyldiamine (EDA) in 10 ml H_2O. After adding 2 ml 1 M aqueous boric acid the solution is sealed in a silica tube and heated at ~ 170 $^\circ$C for 56 days.

Silicalite-1 (Silica-ZSM-5)	152
Silicalite-2 (Silica-ZSM-11)	153
Silica-ZSM-22	149

c. Remark

The main property of the silica molecular sieves is the strong hydrophobic character of the pores. The preferential uptake of e.g. traces of organic compounds (ref. 152) from water, which is not accommodated (ref. 154), in silicalite-1 is a good example.

X. CLATHRASILS or actually silicates and zeolite molecular sieves?

a. Introduction

The name "clathrasil" has been introduced for a subclass of porous tectosilicates different from zeolites. The windows of the framework, connecting the cages, are too small to let guest species, stabilized during the synthesis, pass.

116

This characteristic of a clathrate together with the all silica composition is considered as specific for the members of the clathrasils (ref. 29).

There are, however, exceptions. The recently synthesized decadodecasil-3R (DD-3R) (ref. 155) contains windows of eight-rings of oxygen, indicating that diffusion of small molecules through the porous structure is possible after calcination. This structure can therefore be considered to form an interface between the clathrasils and the silica molecular sieves.

A modified type of DD-3R denoted as Sigma-1 (ref. 156) can, however, be seen as a link between clathrasils and zeolites, because some Si-framework sites are isomorphously substituted by Al. Finally, there is a novel tectosilicate, Sigma-2 (ref. 157). The recently solved structure of which two different polyhedra, see Fig. 26, have not been found before, reveals eight-rings of oxygen and cages with a free diameter of 75 nm. Sigma-2 has been prepared in the silicalite as well as in the zeolite form and can thus be considered as an intermediate between clathrasils, zeolites and silicates.

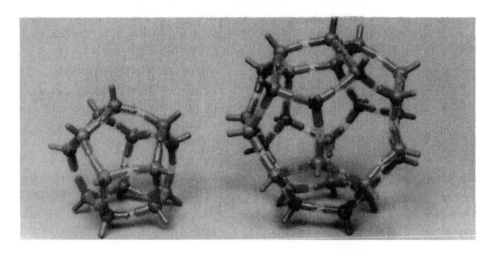

Fig. 26. The nonahedral and eikosahedral cages of Sigma-2.

b. Experimental

The clathrasils can be synthesized generally from .5 M silica, prepared by hydrolyzing an alkoxysilane, e.g. $Si(OCH_3)_4$, in solutions containing an amine as guest template molecule. The syntheses are mainly carried out between 160-240 °C. The clathrasils, together with detailed synthesis data and products expected, are given in Table 12.

Table 12. Clathrasils with references to synthesis prescriptions and two
examples of synthesis recipes.

Product	Ref.
Melanophlogite	158
Dodecasil 3C	159
Dodecasil 1H	160
Silica-sodalite	161
Sigma-1	156
DD-3R (silica)	155

Recipe: 0.75 g $Si(OCH_3)_4$ is added to a
solution of 1.2 g ethylenediamine (EDA)
in 10 ml H_2O. After adding 350 mg 1-amino-
adamantane the solution is sealed in a
silica tube and heated at 170 °C for
~ 70 days.

Sigma-2 (silica and aluminosilicate)   157

Synthesis example: The molar oxyde ratio
of the synthesis system is:

Na_2O	3	
AN	20	(1-adamantanamine)
Al_2O_3	(0.6)	(Al-wire)
SiO_2	60	(colloidal silica)
H_2O	2400	

The system was crystallized at 180 °C and
continuously stirred for a few days.

c. Remark

The templating role of some of the guest molecules is illustrated in Fig. 26.
Polyhedra of different clathrasils are filled with a guest molecule. As only
4-, 5- or 6-ring faces are present in most of the polyhedra it looks like the
crystal building units have formed around the guest molecule as this molecule
is too large to pass through one of the rings. The clathrate formation might
therefore be obtained and based on single building units in solution and/or at
the growing crystal surface (ref. 162).

118

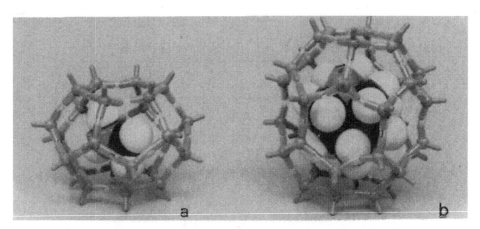

Fig. 27. Orientation of various guest molecules in clathrasils (ref. 163). a) H_3CNH_2 in $[5^{12}6^2]$ and b) adamantylamine in $[5^{12}6^8]$ of melanophlogite and dodecasil 1H, respectively.

XI. EXAMPLES OF SYSTEMATIC RESEARCH in the field of molecular sieves preparation to reach various objectives

a. Introduction

A main thrust of research is:
- to synthesize new molecular sieves
- further optimization of recipes
- to gain knowledge on the essential functions of reactants, e.g. structure directing role of cation/template
- to prepare relative large single crystals for fundamental studies

The list can be longer, however, the examples given below illustrate generally the purpose and variety in the research of molecular sieves preparation.

b. Research examples

Objective	Parameter(s)
1) Preparation of zeolites	Non-aqueous solvents
2) Preparation of zeolites	F^- as mineralizing agent
3) Investigation of crystallization fields with pyrrolidine as template	$Na_2O-Al_2O_3-SiO_2-H_2O$ system was varied

b. Research examples (continued)

Objective   Parameter(s)

4a) Investigation of template-zeolite   Systematic variation of
    interaction   template
4b) Directing role of template in   Use of bis-quaternary ammonium
    the crystallization   compounds

5) Large single crystals   Knowledge on nucleation/
                                               crystallization

6) Morphology and form of zeolite   Change of [SiO$_2$], template,
    products   cation or additives

b.1. The use of non-aqueous solvents

In contrast to the rich crop of zeolite types synthesized in aqueous systems
the results in non-aqueous solvents are poor (refs. 164, 165). Solvents used,
of which the choice was a.o. based on boiling point (100-200 ^{o}C) and relative
permittivity (10-45) (water: 78), are given in Table 13.

Table 13. Zeolite products formed.

Solvents	Na$^+$	K$^+$	Li$^+$	Ca^{++}
Glycol	HS	-	-	-
Glycerol	HS	-	-	-
DMSO	HS	-	-	-
Sulfolane	HS	-	-	-
C$_6$C$_7$ alcohol	HS	-	-	-
Ethanol	HS	-	-	-

HS: Hydroxysodalite.

120

Generally mixtures within the following molar oxyde ratio were used:

MeO	1-20
Al_2O_3	1
SiO_2	1-100
Solvent	5-350
and MeO/SiO_2	0.1-10

As shown in the Table zeolite products could only be obtained in the case of Na^+. The use of other inorganic and organic cations was not successful.

As the boiling point is a less critical factor than high relative permittivity (reduces the Coulomb force between ions and polar compounds thus enhancing dissolution) other non-aqueous solvents for zeolite crystallization which might be subject to zeolite synthesis tests are given below.

Non-aqueous solvent	E_r
formic acid	57
formamide	84
hydrogen peroxide	93
hydrocyanic acid	95

b.2. The use of F^- (ref. 166)

The compositional ratios of the reaction mixtures used were:

	Al or B/Si	F/Si	Templ./Si	H_2O/Si
in molar ratio	0-0.5	0.05-6	0.05-6	4-500

with the pH of the mixtures between 1.5-10. The reaction mixtures were heated at 60-250 °C and autoclaved for a few hours to a few months. After isolation the products were washed with water and dried.

A typical example of a "F^-" synthesis of ZSM-5 is given below:
- Reaction mixture composition:

36 g	Ammonium aluminiumsilicate (Si/Al ~ 7; NH_4/Al ~ 1)		
18.5 g	NH_4F	pH	= 7
33.2 g	TPABr	t	= 172 °C
180 g	H_2O	time	= 11 days

- Product
  Unit cell composition:
  1.8 NH_4^+ + 4.1 TPA^+ $[Al_{2.9}Si_{93.1}O_{192}]$
  Crystal size
  30 x 12 μm

Advantages and differences using F^- instead of OH^- as concluded so far:
- Low pH compared to OH^-
- Incorporation in the framework of elements sparingly soluble in alkaline medium, e.g. Fe^{III}
- Synthesis without alkaline cations
- New possibility to directly incorporate cations as NH_4^+ and divalent cations such as Co^{2+} as well
- Good stability of usual templates such as TAA^+ in this medium
- Highly crystalline materials

b.3. Pyrrolidine as template (ref. 87)

The crystallization of zeolites in the system $Na_2O-Al_2O_3-SiO_2-H_2O$ + pyrrolidine as a template was studied. The reaction mixture compositions used are given in Table 14 in molar oxyde ratio.

Table 14. Reaction mixture ranges in molar oxyde ratios.

Na_2O	0.05-0.5	
Al_2O_3	0.002-0.05	
SiO_2	1	+ ~ 0.7 pyrrolidine
H_2SO_4	0-0.4	
H_2O	20-80	

Two procedures were used:
  I. To a stirred aluminium sulphate solution, calculated amounts of sodium silicate, sulphuric acid and pyrrolidine were added dropwise.
 II. Calculated amounts of aluminium nitrate, colloidal silica and pyrrolidine were added to a stirred sodium hydroxide solution.

122

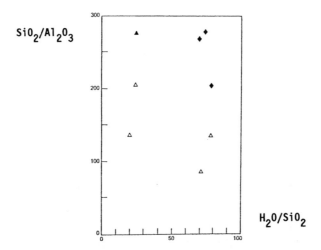

Fig. 28. Crystallization fields of product SiO_2/Al_2O_3 versus H_2O/SiO_2 of ZSM-39 (▲), ZSM-48 (◆) and KZ-1 (△).

The follow-up of both procedures was to autoclave the reaction mixture for 7-40 h at 423-435 K with stirring. After isolation the product was washed with water and dried.

The results of the experiments are given in Fig. 6 for procedure A whereas the results of the experiments with procedure B are given in Fig. 28.

The main conclusion of the study is that pure ZSM-5, ZSM-35, ZSM-39, ZSM-48 and KZ-1 can be crystallized with pyrrolidine in the aforementioned synthesis system. No common factor, based on the use of pyrrolidine, could be recognized in the various zeolite products.

b.4.a. *The use of bis-quaternary ammonium compounds in molecular sieves synthesis (ref. 58)*

The objective in this study was the systematic variation of template in the synthesis. An example of the synthesis is given below together with the products formed, see Table 15.

The general formula of this bis-quaternary template (T) is:

$$T = [(CH_3)_3N(CH_2)_xN(CH_3)_3]^{2+}$$

Table 15. Synthesis mixtures and product formation with bis-quat as template.

Synthesis conditions molar oxyde ratio		x	Product formation
SiO$_2$	60		
Al$_2$O$_3$	1		_Zeolite phases_
Na$_2$O	10	3	ZSM-39
TBr$_2$	10	4	ZSM-12
H$_2$O	3000	5,6	EU-1
		7,8	ZSM-23
			Silica phases
		3	EU-4
(without Al$_2$O$_3$)			
		4,9	EU-2

The reaction conditions were 180 °C, three days and crash-cooling after the syntehsis was terminated.

b.4.b. Another example (ref. 167)

Systematic variation of the chain length of the template (T) given below in the general formula

$$T = \quad \begin{matrix} H \\ H \end{matrix} {>} N-(CH_2)_x-N {<} \begin{matrix} H \\ H \end{matrix}$$

resulted in the products, given in Table 16.

Table 16. Zeolite formations, obtained with bis-quat.

x	Zeolite phase
2-5	Ferrierite
	ZSM-5
5-6	ZSM-5
7-10	ZSM-11

The full synthesis description is given in ref. 167.

b.5. The synthesis of relatively large single crystals of molecular sieves

b.5.1. Introduction

Pertaining to e.g. the viscosity of the synthesis mixture several systems, clear solution, diluted gel and dense gel phase have been investigated.

b.5.2. Crystallization of ZSM-22 from a clear solution (ref. 149)

In a typical experiment tetramethoxysilane was hydrolyzed in 3 M diethylamine (DEA) according to the following reactions:

$$Si(OCH_3)_4 + 2 H_2O \xrightarrow{\hspace{1.5cm}} SiO_2 + 4 CH_3OH \xrightarrow{DEA} 2 SiO_2(C_2H_5)_2NH$$
$$180 \ ^oC$$
$$100 \ days$$

Single crystals of silica-ZSM-22 of 45 x 100 x 225 μm were isolated and used for structure determination.

b.5.3. Synthesis of elongated prismatic ZSM-5 crystals

The objective of this study was to obtain large single crystals of ZSM-5. Systems using Na^+-TPA^+, Li^+-TPA^+ and NH_4^+-TPA^+ were investigated applying a reaction mixture given in molar oxyde ratio for e.g. NH_4^+-TPA^+:

TPA_2O	4	
$(NH_4)_2O$	123	T = 453 K
Al_2O_3	1	t = 7 days
SiO_2	59	
H_2O	2280	

Products

Alkaline-free, homogeneous elongated prismatic single crystals of ZSM-5 of 350 μm in length at maximum (ref. 168).

b.5.4. Synthesis of cubic shaped single crystals of ZSM-5 (ref. 169)

The synthesis of this type of crystals developed recently (ref. 169) was subject of a study on the crystal growth history (ref. 42) of this type of crystals. The objective was: to pinpoint the driving forces which change the ZSM-5 crystal form from elongated prismatic into cubic. The crystal growth history study revealed that the cubic crystal growth occurred in a dense gel phase.

Perfect single crystals up to 500 μm of zeolite (ZSM-5) and all silica (silicalite-1) molecular sieve type, see Fig. 16c, could be obtained using the following molar oxyde ratio:

	ZSM-5	Silicalite-1
SiO_2	12	12
Al_2O_3	1	
Na_2O	44	44
TPA_2O	44	44
H_2O	2000	2000

After 5 days at 180 ^{0}C crystals could be isolated and selected from the product.

b.5.5. Synthesis of single crystals of zeolite A and X (ref. 142)

Single crystals of zeolite A and X up to 100-500 μm in size could be obtained using the following procedures.

Procedure for zeolite A:
Solution I: 100 g $Na_2SiO_3.9H_2O$ in 350 ml H_2O + 50 ml TEA
Solution II: 80 g $NaAlO_2$ in 350 ml H_2O + 50 ml TEA
Both solutions are filtered with milipore filters, whereafter solution II is added to solution I with stirring. The crystallization is performed at 75-85 ^{0}C for 2-3 weeks, without stirring.

Procedure for zeolite X:
Identical to the procedure for zeolite A, only 40 g of $NaAlO_2$ is used in solution II now. The crystallization time is 3-5 weeks.

126

Remark

Careful filtering of the starting solutions substantially reduces the amount of heterogeneous nuclei such as dust and foreign particles in the starting chemicals. The lower the number of nuclei, the larger the crystals.

b.6. Morphology and form of mordenite and ZSM-5

The morphology and/or form of zeolite crystals appear generally to be influenced by:
- [SiO_2]
- Guest molecule type
- Cation (ref. 171)
- Crystal growth inhibitors
A frequently observed crystal form of mordenite is the needle form (with pore channel system parallel to needle direction), see Fig. 29a.

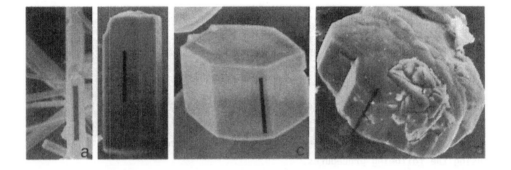

Fig. 29. Different forms of mordenite. The needle form a), the intermediate forms b) and c) and the disk form d). The pore direction is indicated by a bar (ref. 85).

As shown in Fig. 29b, c and d, completely different crystal forms of mordenite can be prepared. According to the synthesis system used (ref. 85) the main influence in the shape of the crystals seems to be the [SiO_2]. The higher the [SiO_2], i.e. the more the crystallization occurs in a dense gel, the more the elongated form is reduced and changed into a disk form.

The increase in pore entries and decrease in pore length going from needle to disk form is evident and may be of interest in catalysis (ref. 170).

The elongated prismatic form is the most frequently found crystal form of ZSM-5. Changing the [SiO$_2$] can change the crystal form as shown in Fig. 30a and b for relatively low and high [SiO$_2$] concentrations, respectively.

Fig. 30. The elongated prismatic crystal form (a) and the cubic crystal form (b) of zeolite ZSM-5.

Changing the template type, i.e. replacing TPA$^+$ for the divalent bi-quaternary ammonium ion, hexapropyl-1,6-hexanediammonium, resulted in different crystal forms for low as high [SiO$_2$] as well, see Fig. 31a and b.

Fig. 31. The modification of the crystal form at low (a) and at high (b) [SiO$_2$] of zeolite ZSM-5 prepared with biquat as template.

In the case of an additive (inhibitor) like boric acid an enrichment of crystal faces in the c-direction was observed as shown in Fig. 32.

128

Fig. 32. Additional crystal face {001} compared to regular elongated prismatic form of zeolite ZSM-5.

XII. LITERATURE SOURCES PERTAINING ZEOLITE PREPARATION ASPECTS

Though most of the literature sources are given in Section XIII, a more extended list of sources is given below for reasons of clarity and ease.

- Chemical Abstracts

A literature search in the Chemical Abstracts (CA) can be successful when Controlled Vocabulary Index Terms (CVIT's) are used. As CVIT's after 1976 are not only assigned to words in the title and the abstract, but also throughout the text of the paper (open literature or patent) the search will be thoroughly. The choice of CVIT's must be correct. In the case the word "synthesis" is used instead of "preparation" the main part of the search "hits" will pertain to reactions with the aid of zeolites whereas the preparation of zeolites is then difficult to extract.

- Proceedings of International Zeolite Conferences (IZC)

1. "Molecular Sieves", Soc. Chem. Ind., London, 1968; Proceedings of the 1st IZC, London, U.K., 1967.
2. "Molecular Sieves I and II", Adv. Chem. Ser., 101 and 102, ACS, Washington, D.C., 1971; Proceedings of the 2nd IZC, Worcester, Mass., U.S.A., 1970.
3. "Molecular Sieves", Adv. Chem. Ser., 121, ACS, Washington, D.C., 1973; W.M. Meier and J.B. Uytterhoeven, Eds., Proceedings of the 3rd IZC, Zürich, Switzerland, 1973.

4. "Molecular Sieves-II", ACS Symp. Ser., 40, ACS, Washington, D.C., 1977; J.R. Katzer, Ed., Proceedings of thre 4th IZC, Chicago, Ill., U.S.A., 1977.

5. "Proceedings of the 5th International Conference on Zeolites", Heyden, London, Philadelphia, Rheine, 1980; L.V.C. Rees, Ed., Proceedings of the 5th IZC, Naples, Italy, 1980.

6. "Proceedings of the 6th International Conference on Zeolites", Butterworths, Guildford, 1984; D. Olson and A. Bisio, Eds., Proceedings of the 6th IZC, Reno, Nev., U.S.A., 1983.

7. New Developments in Zeolites Science and Technology", Kodansha, Tokyo, Elsevier, Amsterdam, Oxford, New York, Tokyo, 1986, Stud. Surf. Sci. Catal., 28; Y. Murakami, A. Iijima and J.W. Ward, Eds., Proceedings of the 7th IZC, Tokyo, Japan, 1986.

8. "Zeolites: Facts, Figures, Future", Elsevier, Amsterdam, Oxford, New York, Tokyo, 1989, Stud. Surf. Sci. Catal., 49; P.A. Jacobs and R.A. van Santen, Eds., Proceedings of the 8th IZC, Amsterdam, Netherlands, 1989.

- *Synthesis part in recent international conferences*

"Zeolites, Synthesis, Structure, Technology and Application", Elsevier, Amsterdam, Oxford, New York, Tokyo, 1985, Stud. Surf. Sci. Catal., 24; B. Drzaj, S. Hocevar and S. Pejovnik, Eds.
"Innovation in Zeolite Materials Science", Elsevier, Amsterdam, Oxford, New York, Tokyo, 1988, Stud. Surf. Sci. Catal., 37; P.J. Grobet, W.J. Mortier, E.F. Vansant and G. Schulz-Ekloff, Eds.
"Zeolite Synthesis", ACS Symp. Ser., 398, ACS, Washington, D.C., 1989; M.L. Occelli and H.E. Robson, Eds.

- *Journal*

Zeolites, L.V.C. Rees and R. von Ballmoos, Eds., Publishers, Butterworth, Heinemann, Stoneham, MA, U.S.A.

- *Books*

"Zeolite Molecular Sieves", Structure, Chemistry and Use, John Wiley & Sons, New York, London, Sydney, Toronto, 1974; D.W. Breck.
"Hydrothermal Chemistry of Zeolites", Academic Press, London, New York, 1982; R.M. Barrer FRS.

130

"Synthesis of High-Silica Alumiosilicate Zeolites", Elsevier, Amsterdam, Oxford, New York, Tokyo, 1987, Stud. Surf. Sci. Catal., 33; P.A. Jacobs and J.A. Martens, Eds.

"Molecular Sieves, Principles of Synthesis and Identification", Van Norstrand Reinhold, New York, 1989; R. Szostak.

"An Introduction to Zeolite Molecular Sieves", John Wiley and Sons, Chichester, 1988, A. Dyer.

ACKNOWLEDGMENT. I like to thank Dr. H. Kouwenhoven for reading the manuscript.

XIII. REFERENCES

1 G. Gottardi and E. Galli, Minerals and Rocks, Natural Zeolites, Springer-Verlag, Berlin, 1985.
2 L.B. Sand and F.A. Mumpton, Natural Zeolites, Occurrence, Properties and Use, Pergamon Press, Oxford, 1978.
3 R.L. Hay, Geologic Occurrence of Zeolites, in: L.B. Sand and F.A. Mumpton (Eds.), Natural Zeolites, Pergamon, Oxford, 1978, pp. 135-143.
4 A. Iijima, Geology of Natural Zeolites and Zeolitic Rocks, in: L. Rees (Ed.), Proc. 5th Int. Conf. on Zeolites, Naples, Italy, June 2-6, 1980, Heyden, London, 1980, pp. 103-118.
5 R.M. Barrer, Synthesis of Molecular Sieve Zeolites, in: Molecular Sieves, London, England, Soc. Chem. Ind., London, 1968, pp. 39-46.
6 W.M. Meier and D.H. Olson, Atlas of Zeolite Structure Types, 2nd edn., Butterworths, London, 1987.
7 C.A.G. Konings, Delft University of Technology, Central Library. The Library search was performed using Controlled Vocabulary Index Terms of the Chemical Abstracts.
8 C.J. Brinker, D.E. Clark and D.R. Ulrich (Eds.), Symp. Proc. Mat. Res. Soc., Vol. 32, Better Ceramics through Chemistry, Albuquerque, U.S.A., February, 1984, Elsevier, New York, 1984.
9 C.J. Brinker, J. Non-Crystalline Solids, 100, 1988, 31-50.
10 C.J. Brinker, D.E. Clark and D.R. Ulrich (Eds.), Symp. Proc. Mat. Res. Soc., Vol. 73, Better Ceramics through Chemistry II, Palo Alto, U.S.A., April 15-19, 1986, Mat. Res. Soc., Pittsburgh, 1986.
11 S.D. Kinrade and T.W. Swaddle, Inorg. Chem., 27, 1988, 4253-4259.
12 A.V. McCormick, A.T. Bell and C.J. Radke, J. Phys. Chem., 93, 1989, 1741-1744.
13 R.A. van Santen, G. Ooms, C.J.J. den Ouden, B.W. van Beest and M.F.M. Post, Computational Studies of Zeolite Framework Stability, in: M.L. Occelli and H.E. Robson (Eds.), Zeolite Synthesis, ACS Symp. Ser. 398, ACS, Washington, DC, 1989, pp. 617-633.
14 A.G. Pelmenshschikov, G.M. Zhidomirov and K.I. Zamaraev, Quantum-chemical Interpretation of Intrazeolite Chemistry Phenomena, in: P.A. Jacobs and R.A. van Santen (Eds.), Stud. Surf. Sci. Catal., 49B, Elsevier, Amsterdam, 1989, pp. 741-752.
15 D.C. Bradley, Chem. Rev., 89, 1989, 1317-1322.
16 J.L. Guth, H. Kessler, J.M. Higel, J.M. Lamblin, J. Patarin, A. Seive, J.M. Chezeau and R. Wey, Zeolite Synthesis in the Presence of Fluoride Ions, in: M.L. Occelli and H.E. Robson (Eds.), Zeolite Synthesis, ACS Symp. Ser., 398, ACS, Washington, DC, 1989, pp. 176-195.
17 L.B. Sand, A. Sacco, Jr., R.W. Thomson and A.G. Dixon, Zeolites, 7, 1989, 387-392.

18 D.T. Hayhurst, P.J. Melling, Wha Jung Kim and W. Bibbey, Effect of Gravity on Silicalite Crystallization, in: M.L. Occelli and H.E. Robson (Eds.), Zeolite Synthesis, ACS Symp. Ser., 398, ACS, Washington, DC, 1989, pp. 233-243.

19 M. Niwa and Y. Murakami, J. Phys. Chem. Solids, 50, 1989, 487-496.

20 Chu Pochen, F.G. Dwyer and V.J. Clarke, Crystallization Method Employing Microwave Radiation, E.P. 0358827.

21 R.W. Thomson and A. Dyer, Zeolites, 5, 1985, 202-210.

22 F. Di Renzo, F. Fajula, F. Figueras, S. Nicolas and T. des Courieres, Are the General Laws of Crystal Growth Applicable to Zeolite Synthesis?, in: P.A. Jacobs and R.A. van Santen (Eds.), Stud. Surf. Sci. Catal., 49A, Elsevier, Amsterdam, 1989, pp. 119-132.

23 L.Y. Hou and L.B. Sand, Determinations of Boundary Conditions of Crystallization of ZSM-5/ZSM-11 in one System, in: D. Olson and A. Bisio (Eds.), Proc. 6th Int. Conf. on Zeolites, Reno, U.S.A., July 10-15, 1983, Butterworths, London, 1984, pp. 887-893.

24 E.G. Derouane, L. Baltusis, R.M. Dessau and K.D. Schmitt, Quantitation and Modification of Catalytic Sites in ZSM-5, in: B. Imelik (Ed.), Catalysis by Acids and Bases, Elsevier, Amsterdam, 1985, pp. 135-146.

25 C.T-W. Chu, G.H. Kuehl, R.M. Lago and C.D. Chang, J. of Catal., 93, 1985, 451-458.

26 M.F.M. Post, T. Huizinga, C.A. Emeis, J.M. Nanne and W.H.J. Stork, An Infrared and Catalytic Study of Isomorphous Substitution in Pentasil Zeolites, in: H.G. Karge and J. Weitkamp (Eds.), Stud. Surf. Sci. Catal., 46, Proc. of an Int. Symp., Sept. 4-8, 1988, Würzburg, F.R.G., Elsevier, Amsterdam, 1988, pp. 363-375.

27 W.O. Haag, R.M. Lago and P.B. Weisz, Faraday Discuss. Chem. Soc., 72, 1981, 317-330.

28 R.L. Bedard, S.T. Wilson, L.D. Vail, J.M. Bennett and E.M. Flanigen, The Next Generation, Synthesis, Characterization, and Structure of Metal Sulfide-based Microporous Solids, in: P.A. Jacobs and R.A. van Santen (Eds.), Stud. Surf. Sci. Catal., 49A, Elsevier, Amsterdam, 1989, pp. 375-387.

29 F. Liebau, H. Gies, R.P. Gunawardane and B. Marler, Zeolites, 6, 1986, 373-377.

30 L.V.C. Rees, Nature, 296, 1982, 491-492.

31 P.B. Weisz and J.N. Miale, J. Catal., 4, 1965, 527-529.

32 R.M. Barrer, Hydrothermal Chemistry of Zeolites, Academic Press, London, 1982, p. 106.

33 Ref. 32, p. 105.

34 L. Rees (Ed.), Proc. 5th Int. Conf. on Zeolites, Naples, Italy, June 2-6, 1980, Heyden, London, 1980.

35 D. Olson and A. Bisio (Eds.), Proc. 6th Int. Conf. on Zeolites, Reno, U.S.A., July 10-15, 1983, Butterworths, London, 1984.

36 Y. Murakami, A. Iijima and J.W. Ward (Eds.), New Developments in Zeolite Science and Technology, Proc. 7th Int. Conf. on Zeolites, Tokyo, Japan, August 17-22, 1986, Kodansha, Tokyo, and Elsevier, Amsterdam, 1986.

37 P.A. Jacobs and R.A. van Santen (Eds.), Stud. Surf. Sci. Catal., 49, Zeolites, Facts, Figures, Future, Proc. 8th Int. Conf. on Zeolites, Amsterdam, The Netherlands, July 10-14, 1989, Elsevier, Amsterdam, 1989.

38 B. Drzaj, S. Hocevar and S. Pejovnik (Eds.), Stud. Surf. Sci. Catal., 24, Zeolites, Synthesis, Structure, Technology and Application, Portoroz, Yugoslavia, September 3-8, 1984, Elsevier, Amsterdam, 1985, Synthesis Part.

39 P.J. Grobet, W.J. Mortier, E.F. Vansant and G. Schulz-Ekloff (Eds.), Stud. Surf. Sci. Catal., 37, Innovation in Zeolite Materials Science, Nieuwpoort, Belgium, September 13-17, 1987, Elsevier, Amsterdam, 1988, Synthesis Part.

132

40 M.L. Occelli and H.E. Robson (Eds.), Zeolite Synthesis, ACS Symp. Ser., 398, Los Angeles, U.S.A., September 25-30, 1988, ACS, Washington, DC, 1989.

41 P.A. Jacobs and J.A. Martens, Stud. Surf. Sci. Catal., 33, Synthesis of High-silica Aluminosilicate Zeolites, Elsevier, Amsterdam, 1987, p. 71.

42 Ref. 40, Crystal Growth Regulation and Morphology of Zeolite Single Crystals of the MFI Type, pp. 257-273.

43 J.C. Pouxviel and J.P. Boilot, J. Mat. Sci., 24, 1989, 321-327.

44 B.M. Lok, T.R. Cannan and C.A. Messina, Zeolites, 3, 1983, 282-291.

45 E. Moretti, S. Contessa and M. Padovan, La Chimica e l'Industria, 67, 1985, 21-34.

46 (a) E.M. Flanigen and R.L. Patton, US Pat. 4073865, 1978. (b) J.L. Guth, H. Kessler, M. Bourgogne, R. Wey and G. Szabo, Fr. Pat. 2567868, 1986.

47 W. Meise and F.E. Schwochow, Kinetic Studies on the Formation of Zeolite A, in: W.M. Meier and J.B. Uytterhoeven (Eds.), Molecular Sieves, Proc. 3rd Int. Conf. on Zeolites, ACS Symp. Ser., 121, Zürich, Switzerland, September 3-7, 1973, ACS, Washington, DC, 1973, pp. 169-178.

48 J.B. Nagy, P. Bodart, H. Collette, J. El Hage-Al Asswad, Z. Gabelica, R. Aiello, A. Nastro and C. Pellegrino, Zeolites, 8, 1988, 209-220.

49 (a) M.R. Rubin, E.J. Rosinski and C.J. Plank, US Pat. 4151189, 1979. (b) G.F. Dwyer and P. Chu, Eur. Pat. 11362, 1980.

50 J.M. Nanne, M.F.M. Post and W.H. Stork, NL Pat. 4251499, 1981.

51 G.T. Kerr, US Pat. 3459676, 1969.

52 (a) D.A. Hickson, BE Pat. 8886833, 1981. (b) L.D. Rollman, US Pat. 4296083, 1981. (c) L. Marosi, M. Schwarzmann and J. Stabenow, Eur. Pat. 49386, 1982. (d). L.D. Rollman and E.W. Valyocsik, US Pat. 4139600, 1979. (e) L.D. Rollman and E.W. Valyocsik, US Pat. 4108881, 1978. (f) L.D. Rollman and E.W. Valyocsik, Eur. Pat. 15132, 1980.

53 M.K. Rubin and E.J. Rosinski, US Pat. 4331643, 1982.

54 R.J. Argauer and G.R. Landolt, US Pat. 3702886, 1972.

55 R. Le Van Mas, O. Pilati, E. Moretti, R. Covini and F. Genoni, US Pat. 4366135, 1982.

56 A.J. Tompsett and T.V. Whitham, Eur. Pat. 107457, 1984.

57 W. Sieber and W.M. Meier, Helv. Chim. Acta, 57, 1974, 1533.

58 J.L. Casci, Bis-quaternary Ammonium Compounds as Templates in the Crystallization of Zeolites and Silica Molecular Sieves, in: Y. Murakami, A. Iijima and J.W. Ward (Eds.), Proc. 7th Int. Conf. on Zeolites, Tokyo, Japan, August 17-22, 1986, Kodansha, Elsevier, Tokyo, Amsterdam, 1986, pp. 215-222.

59 (a) G.T. Kerr, Science, 140, 1963, 1412. (b) G.T. Kerr, J. Inorg. Chem., 5, 1966, 1539.

60 J. Ciric, US Pat. 3950496, 1976.

61 G.T. Kerr, US Pat. 3459676, 1969.

62 C.J. Plank, E.J. Rosinski and M.K. Rubin, US Pat. 4175114, 1979; US Pat. 4199556, 1980.

63 J.L. Casci, B.M. Lowe and T.V. Whittam, Eur. Pat. 42225, 1981.

64 M. Taramasso, G. Perego and B. Notari, Bel. Pat. 887897, 1981.

65 T.V. Whittam, Eur. Pat. 0054386, 1982.

66 E. Moretti, M. Padovan, M. Solari, C. Marano and R. Covini, Germ. Pat. 3301798, 1983.

67 M.K. Rubin, C.J. Plank and E.J. Rosinski, US Pat. 4021447, 1977.

68 J. Keysper, C.J.J. den Ouden and M.F.M. Post, Synthesis of High-silica Sodalite from Aqueous Systems, in: P.A. Jacobs and R.A. van Santen, Stud. Surf. Sci. Catal., 49A, Zeolites, Facts, Figures, Future, Elsevier, Amsterdam, 1989, pp. 237-247.

69 Idenitsu Kosan Co., Ltd. Jpn., JP Pat. 8207816, 1982.

70 C.A. Audeh and E.W. Valyocsik, US Pat. 4285922, 1981.

71 P. Chu, US Pat. 3709979, 1973.

72 See ref. 71 and ref. 41, p. 148

73 M. Baacke and P. Kleinschmit, Eur. Pat. 91537, 1983.

74 W.R. Moser, J.E. Cnossen, A.W. Wang and S.A. Krouse, J. Catal., 95, 1985, 21-32.
75 W.R. Moser, C.C. Chiang and J.E. Cnossen, Advances in Materials Characterization, Plenum, New York, 1985.
76 R.A. Laudise, Chemical & Engineering News, 1987, 30-43.
77 Ref. 32, p. 18.
78 D.W. Breck, Zeolite Molecular Sieves, John Wiley & Sons, New York, USA, 1974.
79 W.O. Haag, R.M. Lago and P.B. Weisz, Nature, 309, 1984, 589-591.
80 F.J. van der Gaag, Thesis Delft, The Netherlands, 1987.
81 G. Debras, A. Gourgue, J.B. Nagy and G. de Clippeleier, Zeolites, 6, 1986, 161-168.
82 L.D. Rollmann, Synthesis of Zeolites, An Overview, in F.R. Ribeiro, A.E. Rodrigues, L.D. Rollmann and C. Naccache, Proceedings of the NATO ASI on Zeolites, Science and Technology, Alcabideche, Portugal, May 1-12, 1983, Martinus Nijhoff Publ., The Hague, The Netherlands, 1984, pp. 109-126.
83 Ref. 78, p. 270.
84 A. Erden and L.B. Sand, J. Catal., 60, 1979, 241-256.
85 P. Bodart, J.B. Nagy, E.G. Derouane and Z. Gabelica, Study of Mordenite Crystallization, in: P.A. Jacobs (Ed.), Structure and Reactivity of Modified Zeolites, Elsevier, Amsterdam, 1984, pp. 125-132.
86 Z. Gabelica, N. Dewaele, L. Maistriau, J.B. Nagy and E.G. Derouane, Direct Parameters in the Synthesis of Zeolites ZSM-20 and Beta, in: M.L. Occelli and H.E. Robson (Eds.), Zeolite Synthesis, ACS Symp. Ser., 398, ACS, Washington, DC, 1989, pp. 518-543.
87 K. Suzuki, Y. Kiyozumi, S. Shin, K. Fujisawa, H. Watanabe, K. Saito and K. Noguchi, Zeolites, 6, 1986, 290-298.
88 E. Thilo, W. Wieker and H. Stade, Z. Anorg. Allg. Chem., 340, 1965, 261-276.
89 W. Wieker and D. Hoebbel, Z. Anorg. Allg. Chem., 366, 1969, 139-151.
90 L.S.D. Glasser and E.E. Lachowski, J. Chem. Soc. Chem. Comm., 1980, 973-974.
91 J.L. Guth, P. Caullet, P. Jacques and R. Wey, Bull. Soc. Chim. Fr., 3-4, 1980, 121-126.
92 D. Hoebbel, G. Engelhardt, A. Samoson, K. Ujszasky and Yu. I. Smolin, Z. Anorg. Allg. Chem., 552, 1987, 236-420.
93 G. Bissert and F. Liebau, Zeitschrift für Kristallographie, 179, 1987, 357-371.
94 See Chapter 8 and references cited therein.
95 S. Ueda, N. Kageama and M. Koizumi, Crystallization of Zeolite Y from Solution Phase, in: D. Olson and A. Bisio (Eds.), Proc. 6th Int. Conf. on Zeolites, Reno, USA, July 10-15, 1983, Butterworths, London, 1984, pp. 905-913.
96 R.K. Iler, The Chemistry of Silica, Wiley, New York, 1979.
97 J.V. Sanders, Journal de Physique, C3, 1985, 1-8.
98 C.F. Baes and R.E. Mesmer, The Hydrolysis of Cations, Wiley, New York, USA, 1976.
99 W.G. Klemperer, V.V. Mainz and D.M. Milar, A Molecular Building-block Approach to the Synthesis of Ceramic Materials, in: C.J. Brinker, D.E. Clark and D.R. Ulrich (Eds.), Better Ceramics through Chemistry II, Materials Research Society, Pittsburgh, USA, 1986, pp. 3-13.
100 F. Liebau, Inorg. Chim. Acta, 89, 1984, 1-7.
101 A.V. McCormick and A.T. Bell, Catal. Rev.-Sci. Eng., 31 (1 & 2), 1989, 97-127.
102 D.W. Schaefer and K.D. Keefer, Structure of Soluble Silicates, in: C.J. Brinker, D.E. Clark and D.R. Ulrich (Eds.), Symp. Proc. Mat. Res. Soc., 32, Better Ceramics through Chemistry, Albuquerque, USA, February 1984, Elsevier, New York, 1984.
103 G. Boxhoorn, O. Sudmeijer and P.H.G. van Kasteren, J. Chem. Soc. Chem. Commun., 1983, 1416-1418.

134

104 Z. Gabelica, E.G. Derouane and N. Blom, Factors Affecting the Synthesis of Pentasil Zeolites, in: T.E. Whyte, Jr., A. Dalla Betta, E.G. Derouane and R.T.K. Baker, Catalytic Materials, Relationship between Structure and Reactivity, ACS Symp. Ser., 248, ACS, 1984, pp. 219-236.

105 A.V. McCormick, A.T. Bell and C.J. Radke, The Influence of Alkali Metal Hydroxides on Silica Condensation Rates, in: C.J. Brinker, D.E. Clark and D.R. Ulrich (Eds.), Better Ceramics Through Chemistry III, 1988, Mat. Res. Soc., Pittsburgh, 1988.

106 Z. Gabelica, J.B. Nagy, P. Bodart, N. Dewaele and A. Nastro, Zeolites, 7, 1987, 67-72.

107 Q. Chen, J.B. Nagy, J. Fraissard, J. El Hage-Al Asswad, Z. Gabelica, E.G. Derouane, R. Aiello, F. Crea, G. Giordano and A. Nastro, in: Proc. NATO Workshop, April 24-27, 1989, Dourdan, France.

108 J.B. Nagy, P. Bodart, H. Collette, C. Fernandez, Z. Gabelica, A. Nastro and R. Aiello, J. Chem. Soc. Faraday Trans. 1, 85 (9), 1989, 2749-2769.

109 E.J.J. Groenen, A.G.T.G. Kortbeek, M. Mackay and O. Sudmeyer, Zeolites, 6, 1986, 403-411.

110 G. Engelhardt and D.Z. Hoebbel, Chem., 23, 1983, 33.

111 F. Schlenkrich, E. Beil, O. Rademacher and H. Scheler, Z. Anorg. Allg. Chem., 519, 1984, 41.

112 O. Rademacher, O. Ziemans and H. Scheler, Z. Anorg. Allg. Chem., 519, 1984, 165.

113 R.D. Edelman, D.V. Kudalkar, T. Ong, J. Warzywoda and R.W. Thomson, Zeolites, 9, 1989, 496-502.

114 R.W. Thompson and A. Dyer, Zeolites, 5, 1985, 292-301.

115 See ref. 41, Chapter 1.

116 See ref. 32, p. 174.

117 B. Subotic, D. Skrtic and I. Smit, J. of Crystal Growth, 50, 1980, 498-508.

118 P. Bodart, J.B. Nagy, Z. Gabelica and E.G. Derouane, J. Chim. Phys., 83, 1986, 777-790.

119 Private communcation with Profs. P. Bennema and G.M. van Rosmalen.

120 E. Grujic, B. Subotic and L.J.A. Deprotovic, Transformation of zeolite A into hydroxysodalite III, in P.A. Jacobs and R.A. van Santen (Eds.), Stud. Surf. Sci. Catal., 49A, Amsterdam, 1989, pp. 261-270.

121 W.M. Meier, in: "Molecular Sieves", London, England, Soc. Chem. Ind., London, 1968, p. 10.

122 The similarity between the structures of silicate species in solution and the SBU's is too small, e.g. the open five-, six- and eight-membered ring systems of the SBU's are unknown in aqueous solution.

123 Ref. 32, p. 122 and 153.

124 G.A. Jeffrey, Hydrate Inclusion Compounds, in: J.L. Atwood, J.E.D. Davies and D.D. MacNicol (Eds.), Inclusion Compounds, Vol. 1, Structural Aspects of Inclusion Compounds formed by Inorganic and Organometallic Host Lattices, Academic Press, London, 1984, pp. 135-190.

125 T.C.W. Mate, J. Chem. Phys., 43, 1965, 2799.

126 a) J. Ciric, J. Colloid Interface Sci., 28, 1968, 315-323. b) E.F. Freund, J. Cryst. Growth, 34, 1976, 11-15.

127 a) R.B. Borade, A.J. Chandvadkan, S.B. Kulkarniu and P. Ratnasamy, Indian J. of Techn., 21, 1983, 358-362. b) R. Aiello and R.M. Barrer, J. Chem. Soc., A, 1970, 1470.

128 S.P. Zdhanov and N.N. Samlevich, Nucleation and Crystal Growth of Zeolites, in: L.V.C. Rees (Ed.), Proc. of the 5th Int. Conf. on Zeolites, Naples, Italy, June 2-6, 1980, Heyden, London, 1980, pp. 75-84.

129 a) H. Kacirek and H. Lechert, J. Phys. Chem., 80, 1976, 1291. b) K.-J. Chao, T.C. Tasi, M.-S. Chen and I. Wang, J. Chem. Soc. Faraday I, 77, 1981, 465. c) E. Narita, K. Sato, N. Yatabe and T. Okabe, Ind. Eng. Chem. Prod. Res. Dev., 24, 1985, 507-512.

130 R. von Ballmoos, Thesis Zürich, 1981.

131 C. Baerlocher and W.M. Meier, Helv. Chim. Acta, 52, 1969, 1853-1860.

132 C. Baerlocher and W.M. Meier, Helv. Chim. Acta, 53, 1970, 1285-1293.

133 a) H. Nakamoto and H. Takahasi, Chem. Lett., 1981, 1739-1742, b) F. Crea, J.B. Nagy, A. Nastro, G. Giordano and R. Aiello, Thermochimica Acta, 135, 1988, 553-357. c) D.T. Hayhurst, A. Nastro, R. Aiello, F. Crea and G. Giordano, Zeolites, 8, 1988, 416-422.

134 a) See ref. 129c. b) F.-Y. Dai, M. Suzuki, M. Takahashi and Y. Sato, in: Y. Murakami, A. Iijima and J.W. Ward (Eds.), Proc. 7th Int. Conf. on Zeolites, Tokyo, Japan, Aug. 17-22, 1986, Kodansha, Tokyo, and Elsevier, Amsterdam, 1986, pp. 223-230. c) V.P. Shiralkar and A. Clearfield, 9, 1989, 363-370.

135 a) R.M. Barrer and P.J. Denny, J. Chem. Soc., 1961, 971. b) R.M. Barrer, P.D. Denny and E.M. Flanigen, US Pat. 3306922, 1967.

136 R. Aiello and R.M. Barrer, J. Chem. Soc. A, 1970, 1470.

137 E.M. Flanigen and E.B. Kellberg, US Pat. 4241036, 1968.

138 G.T. Kokotailo and S. Sawruk, US Pat. 4187283, 1980.

139 An extensive list of organic templates is given in Table 5 of refs. 44, 45.

140 A. Gutze, J. Kornatowski, H. Neels, W. Schmitz and G. Finger, Cryst. Res. & Technol., 20, 1985, 151-158.

141 E. de Vos Burchart, J.C. Jansen and H. van Bekkum, Zeolites, 9, 1989, 423-435.

142 J.F. Charnell, J. Cryst. Growth, 8, 1971, 291.

143 D.T. Hayhurst, A. Nastro, R. Aiello, F. Crea and G. Giordano, Zeolites, 8, 1989, 416-423.

144 Ref. 32, p. 145.

145 N.N. Feoktistova, S.P. Zhdanov, W. Lutz and M. Bülow, Zeolites, 9, 1989, 136-139.

146 C.A. Fyfe, G.C. Gobbi, G.J. Kennedy, J.D. Graham, R.S. Ozubho, W.A. Murphy, A. Bothner-By, J. Dadok and A.S. Chesnick, Zeolites, 5, 1985, 179-183.

147 G.W. Skeels and D.W. Breck, Zeolite Chemistry V, in: D. Olson and A. Bisio (Eds.), Proc. 6th Int. Conf. on Zeolites, Reno, USA, July 10-15, 1983, Butterworths, London, 1989, pp. 87-96.

148 H.K. Beyer and I. Belenijkaja, A New Method for the Dealumination of Faujasite-type Zeolites, in: B. Imelik, C. Naccache, Y. Ben Taarit, J.C. Vedrine, G. Coudurier and H. Praliaud (Eds.), Stud. Surf. Sci. Catal., 5, Elsevier, Amsterdam, 1980, pp. 203-210.

149 B. Marler, Zeolites, 7, 1987, 393-397.

150 R.P. Gunawardane, H. Gies and B. Marler, Zeolites, 8, 1988, 127-131.

151 H. Gies and R.P. Gunawardane, Zeolites, 7, 1987, 442-445.

152 E.M. Flanigen, J.M. Bennett, R.W. Grose, J.P. Cohen, R.L. Patton, R.M. Kirchner and J.V. Smith, Nature, 271, 1987, 512-516.

153 D.M. Bibby, N.B. Inlestone and L.P. Aldridge, Nature, 280, 1979, 664-665.

154 D.H. Olson, W.O. Haag and R.M. Lago, J. Catal., 61, 1980, 390-396.

155 H. Gies, Zeitschrift für Kristallographie, 175, 1986, 93-104.

156 A. Stewart, D.W. Johnson and M.D. Shannon, Synthesis and Characterization of Crystalline Aluminosilicate Sigma-1, in: P. Grobet, W.J. Mortier, E.F. Vansant and G. Schulz-Ekloff (Eds.), Stud. Surf. Sci. Catal., Proc. Int. Symp., September 13-17, 1987, Nieuwpoort, Belgium, Elsevier, Amsterdam, 1988, pp. 57-64.

157 A. Stewart, Zeolites, 9, 1989, 140-145.

158 H. Gies, Z. Kristallogr., 164, 1983, 247-257.

159 H. Gies, Z. Kristallogr., 167, 1984, 73-82.

160 H. Gerke and H. Gies, Z. Kristallogr., 166, 1984, 11-22.

161 D.M. Bibby and M.P. Dale, Nature, 317, 1985, 157-158.

162 R.M. Barrer, Porous Crystals: A Perspective, in: Y. Murakami, A. Iijima and J.W. Ward (Eds.), Proc. 7th Int. Conf. on Zeolites, Tokyo, Japan, August 17-22, 1986, Kodansha, Elsevier, Tokyo, Amsterdam, 1986, pp. 3-11.

163 F. Liebau, Structural Chemistry of Silicates, Springer-Verlag, Berlin, New York, Tokyo, 1985, p. 243.

136

164 W.A. van Erp, H.W. Kouwenhoven and J.M. Nanne, Zeolites, 7, 1987, 286-288.
165 Xu Wenyang, Li Jianquan, Li Wengyuan, Zhang Huiming and Liang Bingchang, Zeolites, 9, 1989, 468-473.
166 J.L. Guth, H. Kessler and R. Wey, New Route to Pentasil-type Zeolites using a non Alkaline Medium in the Presence of Fluoride Ions, in: Y. Murakami, A. Iijima and J.W. Ward (Eds.), New Developments in Zeolite Science and Technology, Proc. 7th Int. Conf. on Zeolites, Tokyo, Japan, Aug. 17-22, 1986, Kodansha, Tokyo and Elsevier, Amsterdam, 1986, pp. 121-128.
167 E.W. Valyocsik and L.D. Rollmann, Zeolites, 5, 1985, 123-125.
168 U. Müller and K.K. Unger, Zeolites, 8, 1988, 154-156.
169 H. Lermer, M. Draeger, J. Steffen and K.K. Unger, Zeolites, 5, 1985, 131-134.
170 C.W.R. Engelen, unpublished results.
171 D.E.W. Vaughan, Secondary Cation Effects on Sodium and Potassium Zeolite Synthesis at Si/Al_2 = 9, in: M.M.J. Tracy, J.M. Thomas and J.M. White (Eds.), Mat. Res. Soc. Symp. Proc., Microstructure and Properties of Catalysts, Vol. 111, Nov. 30-Dec. 3, 1987, Boston, M.R.S., Pittsburgh, U.S.A., 1988, pp. 89-100.

B. SYNTHESIS OF AlPO$_4$-BASED MOLECULAR SIEVES

S.T. WILSON

UOP, Research and Development, Tarrytown Technical Center, Tarrytown, NY, 10591, USA

ABSTRACT

The synthesis of AlPO$_4$-based molecular sieves has been achieved using hydrothermal synthesis techniques. Reaction mixtures containing Al, P, and one or more of 13 additional elements have been crystallized in the presence of organic amines or quaternary ammonium cations. These organic templating agents direct structure formation and are usually occluded in the voids of the crystalline microporous product. In addition to templating, other variables affecting synthesis are raw materials, composition of the reaction mixture, method of combining the reactants, crystallization time and temperature, and reaction pH.

INTRODUCTION

Since 1982, with the first report of molecular sieves with aluminophosphate lattices (the AlPO$_4$ family), the compositional and structural diversity of AlPO$_4$-based molecular sieves has continued to grow. There are currently more than 27 different structures and at least 13 elements in addition to Al and P have been incorporated into AlPO$_4$-based frameworks. [refs. 1-5]

HISTORICAL BACKGROUND

The original choice of aluminophosphate chemistry for preparing a new family of molecular sieves was based in part on the extensive hydrothermal synthesis literature and the structural similarities of AlPO$_4$ to SiO$_2$. Dense phase AlPO$_4$ exists in all the structural polymorphs of SiO$_2$ - quartz, cristobalite, and tridymite. Each polymorph contains equimolar amounts of Al and P in tetrahedral coordination with oxygen. Hydrothermal synthesis of the quartz form of AlPO$_4$, called berlinite, has been described in numerous papers. This synthesis is usually carried out under quite acidic conditions (in the presence of excess H$_3$PO$_4$) and at temperatures greater than 150°C.

Various hydrates of AlPO$_4$ have been prepared hydrothermally under milder conditions, usually near 100°C, but still acidic. These include synthetic forms of AlPO$_4 \cdot$ 2 H$_2$O (metavariscite and variscite) and six hydrates (designated H1 through H6) first prepared and characterized by F. d'Yvoire [ref. 6]. These hydrates were all reported to transform into dense phase $AlPO_4$ on heating (usually at temperatures significantly less than 600°C).

The first aluminophosphate molecular sieve synthesis attempts were patterned after the well established preparation of SiO_2 molecular sieves under hydrothermal conditions using organic, quaternary ammonium ions as structure-directing (or templating) agents. When quaternary ammonium cations were employed in aluminophosphate media under hydrothermal conditions, the first $AlPO_4$ molecular sieves were prepared. The preferred pH range for synthesis is mildly acidic to mildly basic, in contrast to the pH ranges usually employed in the synthesis of dense phase $AlPO_4$ (more acidic) or SiO_2 molecular sieves (more basic). The products typically contain Al in tetrahedral coordination.

$AlPO_4$-BASED MOLECULAR SIEVES
Acronyms and Compositions [ref. 3]

The first family of aluminophosphate-based molecular sieves, designated $AlPO_4$-n, contained Al and P in lattice T-sites. Subsequent efforts to incorporate other elements were successful and appropriate acronyms were introduced for each new family (Table 1). A total of 15 elements, with oxidation states ranging from +1 to +5, have been incorporated into $AlPO_4$-based molecular sieve frameworks.

TABLE 1. Compositional Acronyms for $AlPO_4$-Based Materials

Acronym	Framework T-Atoms	(Me or El T-Atoms)
$AlPO_4$	Al, P	
SAPO	Si, Al, P	
MeAPO	Me, Al, P	(Co, Fe, Mg, Mn, Zn)
MeAPSO	Me, Al, P, Si	
ElAPO	El, Al, P	(As, B, Be, Ga, Ge, Li, Ti)
ElAPSO	El, Al, P, Si	

Based on a considerable amount of structural and spectroscopic evidence, as well as elemental analysis, the substitution patterns of many of these elements have been elucidated. The SAPO, MeAPO, and other structures can best be described using a hypothetical $AlPO_4$ lattice with alternating Al and P sites as a basis. In SAPO's the element Si is incorporated into P sites alone or into both Al and P sites. Divalent and trivalent elements substitute into Al sites, minimizing any charge imbalance.

Structure Types

Within each compositional family a wide variety of structure-types have been observed and most have been determined by one or more of the methods: 1) X-ray diffraction, 2) neutron diffraction, 3) model-building or 4) analogy with existing frameworks. Each structure-type with a known or probable tetrahedral net has been designated with a number (Table 2). The same number is used for a given structure-type regardless of framework composition, e.g., $AlPO_4$-5 and SAPO-5 have the same structure-type. Although structures first prepared by UOP workers have been named following this procedure, other researchers sometimes invent their own nomenclature. For example, the very large pore $AlPO_4$ molecular sieve VPI-5 [ref. 7], the zeolite ZSM-5, and $AlPO_4$-5 all have distinctly different structures.

TABLE 2. Pore Sizes in $AlPO_4$-Based Structure-Types

Pore sizes	T-atoms in largest ring	Structure-Types
Very Large	18	VPI-5
Large	12	5, 36, 37, 40, 46, 50
Medium	10	11, 31, 41
Small	8	14, 17, 18, 22, 26, 33 34, 35, 39, 42, 43, 44 47, 52
Very Small	6	16, 20

To date, at least 27 structure-types have been identified. The largest class is small pore (8R), containing over half of the known structures.

SYNTHESIS
Raw Materials

As in zeolite synthesis, the choice of raw materials can dramatically affect synthesis. In the synthesis of $AlPO_4$-based molecular sieves, sources of Al and P and optionally, Si, Me, or

El, are combined with an amine or quaternary ammonium templating agent to produce a reaction mixture that typically has a starting pH between 3 and 10.

The preferred and most utilized source of phosphate is orthophosphoric acid, an essentially monomeric form of PO_4. Compared to the hydroxides of Si and Al, the "hydroxide" of P is distinctly acidic, lowering the initial pH of the reaction mixture. Reaction pH tends to increase, however, as the phosphoric acid is consumed. The ester or "alkoxide" of phosphoric acid, triethylphosphate, has been used sparingly as a phosphate source. It is less reactive and tends to hydrolyze more slowly than the alkoxides of Si or Al, but does offer a means of manipulating reaction mixture acidity independent of the templating agent. The typical, water soluble, inorganic phosphate salts are undesirable since they introduce high concentrations of alkali metal or ammonium which interfere with nucleation or reduce the stability of the final product.

The most studied sources of alumina are the pseudoboehmite form and the alkoxide, both of which introduce no extraneous cations. Of the two, pseudoboehmite apparently reacts less during the initial preparation of the reaction mixture, giving mixtures with lower initial pH. The alkoxides, such as aluminum isopropoxide or sec-butoxide, hydrolyze rapidly to give a very reactive form of Al which reacts quickly with phosphoric acid. This behavior gives higher initial pH and probably produces different precursor species. The use of alkoxides may introduce alcohol into the reaction mixture, but there are means of removing it prior to crystallization.

Other hydroxides of aluminum have been used much less frequently. Gibbsite, for example, is less reactive than pseudoboehmite, and is more prone to dense phase formation. The use of aluminum salts, such as aluminum sulfate or sodium aluminate, introduces large quantities of extraneous cations or anions into the reaction mixture, which may interfere with pH control and nucleation. The soluble form of Al, aluminum chlorhydrol, has not been well studied but initial results are positive.

Crystalline forms of aluminum phosphate, such as metavariscite or $AlPO_4 \cdot 1.5 \ H_2O$ (H3 of d'Yvoire), can serve as the source of both Al and P. Tapp et al. have reported the preparation of

metavarscite in situ and then converted it into $AlPO_4$-11 [ref. 8].
Commercially available monoaluminum phosphate solution
(Al/P = 0.33) has also been used successfully.

In the preparation of SAPO's all the typical forms of silica
have been successfully employed - colloidal, precipitated, fumed,
and alkoxide. There is no clear preference among the various
forms.

The metal Me or element El can be added to the reaction mixture
as a water soluble salt (such as the acetate or sulfate) or the
metal or metal oxide can be dissolved in phosphoric acid, forming
a phosphate salt in situ.

In addition to the various oxide constituents that will form the
molecular sieve framework, the reaction mixture contains a
templating (or structure-directing) agent in the form of an amine
or quaternary ammonium salt. Over 100 templates have been
reported, some examples of which are shown in Table 3. Some of
these templating agents direct the formation of only one or two
structures (e.g. TMAOH) under a variety of conditions while others
(e.g. nPr_2NH) will nucleate numerous structures depending on
synthesis variables such as time, temperature, and template
concentration.

TABLE 3. Typical Templates for Selected $AlPO_4$-Based Structures

Structure	Template	IZA Code
5	Tripropylamine (Pr_3N)	AFI
11	Di-n-propylamine (nPr_2NH)	AEL
14	Isopropylamine	
16	Quinuclidine	AST
17	Quinuclidine	ERI
18	Tetraethylammonium hydroxide (TEAOH)	
20	Tetramethylammonium hydroxide (TMAOH)	SOD
31	nPr_2NH	
33	TMAOH	ATT
34	TEAOH	CHA
35	Quinuclidine	LEV
36	Pr_3N	
37	Tetrapropylammonium hydroxide and TMAOH	FAU
39	nPr_2NH	
41	nPr_2NH	
46	nPr_2NH	AFS
52	Pr_3N + TEAOH	AFT
VPI-5	nPr_2NH	VFI

The final component of the reaction mixture is the solvent. Although water is the most commonly used, appreciable quantities of alcohol can also be present if alkoxides are used as raw materials. These alcohols serve as cosolvents and may be water miscible. Von Ballmoos and Derouane [ref. 9] have used water-immiscible n-hexanol in the two phase synthesis of SAPO molecular sieves. In their initial reaction mixtures the silica source (tetraethylorthosilicate) is dissolved in the hexanol and the Al, P, and template sources are initially present in the aqueous phase. Crystallization occurs under conditions of vigorous agitation.

Guth, Kessler, and coworkers have expanded the synthesis conditions normally used to prepare zeolites by using fluoride-containing media [ref. 10]. More recently, they reported the preparation of AlPO$_4$ and SAPO molecular sieves from such media. The fluoride, usually added in the form of aqueous HF, seems to behave as a mineralizer and causes the growth of larger crystals.

Reaction Mixture Composition

Typical reaction mixture compositions for AlPO$_4$, SAPO, and MeAPO are shown below. In each case the template R is present in excess of the product requirement and serves in part to control reaction pH:

$$AlPO_4 \quad - 1.0 \; R \cdot Al_2O_3 \cdot P_2O_5 \cdot 40 \; H_2O \tag{1}$$
$$SAPO \quad - 1.0 \; R \cdot 0.6 \; SiO_2 \cdot Al_2O_3 \cdot P_2O_5 \cdot 40 \; H_2O \tag{2}$$
$$MeAPO \quad - 1.0 \; R \cdot 0.4 \; MeO \cdot 0.8 \; Al_2O_3 \cdot P_2O_5 \cdot 40 \; H_2O \tag{3}$$

For example, two synthesis mixtures illustrate the broad range of reactant concentrations that can produce AlPO$_4$-5 [refs. 2, 11]:

$$1.5 \; Pr_3N \cdot Al_2O_3 \cdot 1.0 \; P_2O_5 \cdot 40 \; H_2O \tag{4}$$
Pseudoboehmite, 150°C, 8 hours

$$3.5 \; Et_3N \cdot Al_2O_3 \cdot 1.7 \; P_2O_5 \cdot 300 \; H_2O \tag{5}$$
Aqueous aluminum hydroxide, 175°C, 15 hours

In both cases the product is pure AlPO$_4$-5 with equimolar Al and P. In reaction 5 the template and the phosphate are considerably in excess of the product requirement, the extra template serving to counteract the pH depressing effects of the extra phosphoric acid. To improve the chances of preparing a pure product, the Al concentration is usually chosen to be limiting, since it is the least soluble. These two templates are but two of the more than 25 templates that make the 5 structure-type.

The addition of Si to an aluminophosphate reaction mixture produces a SAPO on crystallization. Of all the elements incorporated into aluminophosphate-based frameworks, Si shows the widest compositional latitude, since Si can substitute for both Al and P. Two reaction mixtures that produce SAPO-5 are shown below:

$$1.5 \text{ triethanolamine} \cdot 0.6 \text{ SiO}_2 \cdot \text{Al}_2\text{O}_3 \cdot \text{P}_2\text{O}_5 \cdot 45 \text{ H}_2\text{O} \qquad (6)$$
pseudoboehmite, Cab-o-sil™ EH-9, 200°C, 24 hours, stirred

$$1.0 \text{ cyclohexylamine} \cdot 1.7 \text{ SiO}_2 \cdot \text{Al}_2\text{O}_3 \cdot \text{P}_2\text{O}_5 \cdot 87 \text{ H}_2\text{O} \qquad (7)$$
aluminum isopropoxide, Ludox™ AS, 200°C, 50 hours, tumbled

In reaction (7) [ref. 12], the product SAPO-5 has a higher framework Si content. In many cases the product Si content can be increased by increasing the SiO_2 content of the reaction mixture, but other synthesis factors, such as template, template concentration, and crystallization temperature contribute to the success of this strategy. Further increasing the SiO_2 in reaction mixture (7) to 20 produced a different structure, SAPO-44, under otherwise identical conditions.

In the preparation of MeAPO's the Al concentration of the reaction mixture is typically decreased as the Me concentration is increased, to avoid unreacted alumina. Figures 1 and 2 illustrate the response of MAPO (Me = Mg) synthesis to Mg concentration and crystallization temperature using TEAOH as template [ref. 13].
At 100°C MAPO-34 is the major product for x = 0.2-0.4 but crystallographically pure product is observed over a relatively narrow range near x = 0.25. Crystallization at both higher and lower Mg concentrations produces progressively more impurities, $AlPO_4$ hydrate structure-types (H3, metavariscite, variscite) or unidentified Mg-rich structures. At 150°C MAPO-5 is the major product over a broad range of Mg concentrations but it is never observed free of impurities like MAPO-34.

Combining the Reactants

A variety of methods for preparing the reaction mixture have been described in the literature. Some of these are described

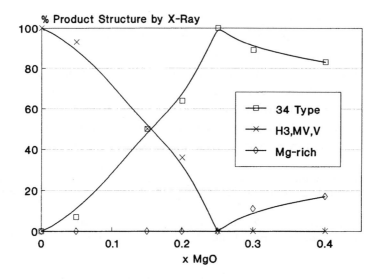

Fig. 1. Product distribution vs. initial Mg concentration in MAPO-34 synthesis with TEAOH (100°C, 48 hours). [1.0 TEAOH • x MgO • 0.8 Al_2O_3 • P_2O_5 • 40 H_2O • 2x HOAc], magnesium acetate, pseudoboehmite

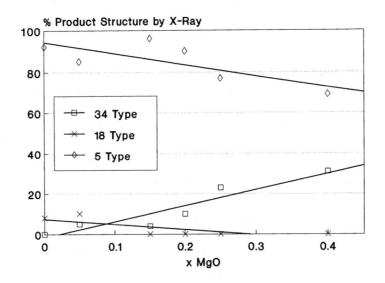

Fig. 2. Product distribution vs. initial Mg concentration in MAPO-34 synthesis with TEAOH (150°C, 72 hours). [1.0 TEAOH • x MgO • 0.8 Al_2O_3 • P_2O_5 • 40 H_2O • 2x HOAc], magnesium acetate, pseudoboehmite

below, using the abbreviations R=template, H=free water, P=phosphoric acid, Al=alumina source, Si=Si source, Me=metal salt:

AlPO$_4$ MeAPO

    1 - (P+H)+Al+R   5 - (P+H)+Al+(Me+H)+R
    2 - (P+H+R)+Al   6 - (P+H+Me)+Al+R

SAPO   7 - (Me+H)+Al+(P+H)+R

    3 - (P+H)+Al+Si+R
    4 - (H+P+R)+Si+Al

Although the effects of the order of addition are not well understood, it is clear that the mix order can affect mix rheology and pH. For example, combining the template R with the phosphoric acid before the addition of Al, reduces the initial reaction rate of the phosphate with the alumina and usually produces a more acidic pH and a less viscous mixture.

Some workers have promoted the benefits of aging a reaction mixture prior to crystallization [e.g., SAPO-37, ref. 14]. In some cases product purity or morphology may be affected by such a procedure, but none of the syntheses reported so far require such a pretreatment for crystallization to occur and the general desirability of an aging step has not been demonstrated.

Reaction pH

During crystallization the pH of a reaction mixture typically varies within the range of 3 to 10 and usually moves toward neutral pH. The initial pH can be manipulated primarily by varying the amounts of the phosphoric acid and the template (i.e., the strongest acid and base in the mixture), and to a lesser extent, by changing the reactivity of the alumina, aging the mixture, or changing the mix order. When the initial pH is in the low range, it tends to rise as the phosphoric acid is consumed. At initial pH values less than 3.0, dense phases are more likely. Crystallization above pH 10 can be more problematic because the product tends to become more soluble at higher pH and many of the divalent metals form insoluble hydroxides and become less reactive.

The Al and Si concentrations are usually chosen to be limiting since unreacted forms are often solid over the reaction pH range and may contaminate the product. Al and Si are typically less soluble and the P, template, and Me are typically more soluble within the crystallization pH range. Even higher relative concentrations of template and phosphoric acid can produce reaction media which are initially clear solutions. Some structures have

146

been crystallized from such clear solutions [refs. 15-16]. Kessler
and coworkers have added appreciable amounts of HF to reaction
mixtures to study the effects on crystallization. Remarkably the
HF has very little effect on the initial pH and the product
structure-types are so far identical to those obtained without HF
present [ref. 10]. UOP workers have added HCl to an AlPO$_4$-5
synthesis mixture and observed a lower initial pH and severe
alteration of the product mix [ref. 2].

Crystallization

The crystallization time varies considerably, based on
temperature, the structure and the composition, and other
variables. Some crystallizations are very fast, e.g. AlPO$_4$-5
crystallizes in less than 4 hours at 150°C. Other crystallizations
require several days, e.g. AlPO$_4$-18 requires 120 hours at 150°C for
full crystallinity. The rate depends on the system, with AlPO$_4$ and
MeAPO synthesis typically faster than SAPO.

Crystallization temperatures are typically 100-200°C. Higher
temperatures may accelerate crystallization or nucleate a different
structure. Most of the reported AlPO$_4$ syntheses have been in the

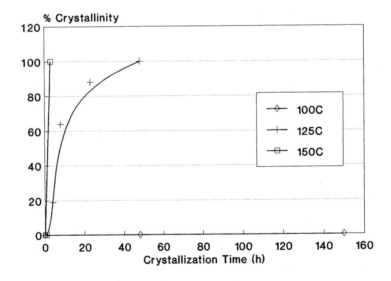

Fig. 3. Rate of crystallization of AlPO$_4$-5 from mix composition:
1.75 Pr$_3$N • 0.25 TEAOH • Al$_2$O$_3$ • P$_2$O$_5$ • 40 H$_2$O [pseudoboehmite
alumina, quiescent, mix method: (H + Al) + P + TEAOH + Pr$_3$N]

range 150-200°C, although AlPO$_4$-20 can be crystallized at 100°C, and AlPO$_4$-5, AlPO$_4$-18, and VPI-5 can be crystallized at 125-130°C. The effects of temperature on crystallization of an AlPO$_4$-5 are shown in figure 3. This mixture shows no evidence of crystallization at 100°C, even after 168 hours. At 125°C, crystallization to AlPO$_4$-5 is slow, reaching full crystallinity after 48 hours. At 150°C full crystallinity is reached within 3 hours.

Most MeAPO syntheses occur at 100-150°C. Interestingly, the MeAPO structure-types which crystallize at 100°C are the zeolite types, 20 (SOD), 34 (CHA), and 35 (LEV). Some reaction mixtures give different structures depending on temperature (e.g. MAPO-5 and MAPO-34, figs. 1 and 2). SAPO and MeAPSO compositions require temperatures of 150-200°C for reasonable rates. For a given structure-type the SAPO composition usually requires a higher temperature than the AlPO$_4$ or MeAPO compositional variants. Most of the reported syntheses of aluminophosphate-based structures have been under quiescent conditions. Stirring may speed up crystallization or allow a lower crystallization temperature. For example, AlPO$_4$-11 was originally synthesized quiescently at 200°C in 24 hours using reaction mixture (8):

$$1.0 \ nPr_2NH \cdot Al_2O_3 \cdot P_2O_5 \cdot 40 \ H_2O \qquad (8)$$

With stirring, this reaction mixture crystallizes to dense phase AlPO$_4$ under otherwise identical conditions, but at 150°C a similar mixture crystallizes to AlPO$_4$-11 after 4 hours. In principle, agitation during crystallization can compensate for heterogeneities in the reaction mixture but it may affect nucleation, changing the structure-type formed.

Work-up

The aluminophosphate-based molecular sieves are typically recovered by filtration or centrifugation. Before a choice of procedure is made, the sedimentation behavior of the product slurry should be observed to determine uniformity, and the product should be examined under the optical microscope. Many products have a characteristic size and morphology, and the presence of multiple morphologies would suggest that some form of fractionation may be necessary to recover a pure product. Excess, non-crystalline silica or alumina typically has a lower density than the $AlPO_4$ or SAPO product.

Both filtration and centrifugation have their advantages and disadvantages. In the centrifugation method, the product is settled by centrifugal force, separated from the supernatant by decanting, and then reslurried in fresh water. The centrifuge/decant/reslurry procedure should be repeated several times for adequate washing. Filtration and subsequent thorough washing offer a better method for removing the last traces of mother liquor from the product. However, filtration of the product reaction mixture without any intervening processing may not always be possible, since non-crystalline impurities can blind filter paper. Filtration alone will tend to recover all the product solids, so an intervening fractionation or sedimentation step is advisable to separate and characterize any phases that differ significantly in density or morphology from the expected product.

The oxide efficiency, or amount of starting oxides recovered as product, typically varies within the range 40-90%. Increasing the template or solvent concentrations tends to lower efficiencies, affecting $AlPO_4$ recovery more than MeAPO or SAPO.

Almost all as-synthesized forms of $AlPO_4$-based structures have occluded organic template. In most cases air calcination gives the easiest and most complete removal of the organic via pyrolysis and combustion. If the organic can undergo a clean Hofmann elimination without subsequent coking, then pyrolysis in vacuum or an inert atmosphere may be effective. Following such calcination or pyrolysis, framework charge is balanced primarily by protons. If the organic is not required to balance framework charge, as in the $AlPO_4$'s, and if the pore system allows diffusion of the template, then desorption of the template intact may occur on mild heating.

In one reported case, hydration of an $AlPO_4$-based molecular sieve after removal of the occluded organic leads to partial or complete loss of crystallinity. Barthomeuf and coworkers found that SAPO-37 following calcination loses crystallinity on exposure to atmospheric moisture at ambient temperature [ref. 17]. If the calcined SAPO-37, however, is exposed to water vapor at temperatures greater than 80°C, it remains stable. In this case the adsorbed water at lower temperatures is "liquid water" and appears more damaging than the water vapor present at higher temperatures.

OTHER ISSUES

Templating has been a frequently discussed phenomenon in zeolite and molecular sieve synthesis, particularly as more organics are used. One of the more recent definitions describes templating as a phenomenon "occurring during either the gelation or the nucleation process whereby the organic molecule organizes the oxide tetrahedra into a particular geometric topology around itself and thus provides the initial building block for a particular structure-type." [ref. 18]

Most of the high silica zeolites and silica molecular sieves, and all of the reported $AlPO_4$-based structures are prepared with an organic amine or quaternary ammonium present, and most are recovered with the occluded organic intact within the framework. Very few zeolite structures which were first synthesized with the aid of an organic template were then subsequently prepared without an organic present. ZSM-5 is one such example. In all the other cases the organic plays a unique and critical role.

The role of the organic in $AlPO_4$-based synthesis includes structure-direction, where charge density, shape and size are important, and pH control, where pK_b and solubility become important. Instances of templating where there is a unique, hand-in-glove fit between organic and lattice are few. One structure often has several effective templates and most templates will direct formation of more than one structure. Templating as understood here should be viewed as the ability of an organic to select or nucleate a small number of structure-types from a much larger number of possible structures.

The template plays at least two additional roles in the product - it stabilizes voids and balances framework charge. By packing the cages and channels the organic can increase the overall thermodynamic stability of the template/lattice composite, so that the metastability of the lattice alone is less critical.

Although the template appears to dominate nucleation, for a given template, other synthesis conditions such as template concentration, crystallization temperature, and Si or Me concentration can exert a secondary effect on structure-direction. The complexity of nucleation is evident in Table 4, wherein conditions are given for preparing six different MAPO structures using a single template, di-n-propylamine [ref. 13].

As observed in zeolite synthesis, the seeding of a reaction mixture with a small amount of a previously crystallized, pure structure-type can also influence nucleation of an $AlPO_4$-based structure.

TABLE 4. Typical MAPO Synthesis Conditions with nPr_2NH

Structure	w	x	y	Time	Temp	Seed
11	1.0	0.17	0.92	24	200	No
31	1.5	0.20	1.00	24	150	Yes
39	1.0	0.40	0.80	24	150	No
41	2.0	0.30	0.85	24	200	Yes
46	2.0	0.30	0.85	168	150	No
50	2.5	0.30	0.85	144	150	No

Reaction mixture: w nPr_2NH · x MgO · y Al_2O_3 · P_2O_5 · 45 H_2O

FINAL COMMENTS

The synthesis of $AlPO_4$-based molecular sieves is no longer in its infancy. Since the first publications in 1982, research activity has accelerated to the point there were more than 100 publications on $AlPO_4$-based molecular sieves in 1989. Many of the same issues encountered in zeolite synthesis are being addressed. With all the powerful analytical techniques available now that were not available to early zeolite workers, such as synchrotron XRD and solid state MAS-NMR, our understanding of these new materials has been enhanced.

REFERENCES

1 S.T. Wilson, B.M. Lok, C.A. Messina, T.R. Cannan and E.M. Flanigen, G.D. Stucky and F.G. Dwyer (Eds.), Intrazeolite Chemistry, Vol. 218, ACS Symposium Series, 1983, American Chemical Society, Washington, DC, USA, pp. 79-106.
2 S.T. Wilson, B.M. Lok, C.A. Messina, and E.M. Flanigen, in D. Olson and A. Bisio (Eds.), Proc. of the 6th Int. Zeolite Conference, Reno, USA, July 10-15, 1983, Butterworth, Guildford, Surrey, UK, 1983, pp.97-109.

3   E.M. Flanigen, B.M. Lok, R.L. Patton and S.T. Wilson, in
    Y. Murakami, A. Ijima and J.W. Ward (Eds.), New Developments in
    Zeolite Science and Technology, Vol. 28, Stud. Surf. Sci.
    Catal., 1986, Elsevier, USA, pp. 103-12.
4   J.M. Bennett, W.J. Dytrych, J.J. Pluth, J.W. Richardson, Jr. and
    J.V. Smith, Zeolites, 6 (1986) 349-60.
5   E.M. Flanigen, R.L. Patton and S.T. Wilson, in P.J. Grobet,
    W.J. Mortier, E.F. Vansant and G. Schulz-Ekloff
    (Eds.),Innovation in Zeolite Materials Science, Vol. 37, Stud.
    Surf. Sci. Catal., 1988, Elsevier, New York, pp. 13-27.
6   F. d'Yvoire, Bull. Soc. chim. France, 1762 (1961).
7   M.E. Davis, C. Montes and J.M. Garces, in M. Occelli and
    H.Robson (Eds.), Zeolite Synthesis, Vol. 398, ACS Symp. Ser.,
    1989, American Chemical Society, Washington, DC, USA, pp.
    291-304.
8   N.J. Tapp, N.B. Milestone and D.M. Bibby, Zeolites, 8 (1988)
    183-8.
9   E.G. Derouane and R. Von Ballmoos, U.S. Patents # 4,647,442 and
    4,673,559 (1987).
10 H. Kessler, in J. Klinowski and P.J. Barrie (Eds.), Recent
    Advances in Zeolite Science, Vol. 52, Stud. Surf. Sci. Catal.,
    1989, Elsevier, New York, pp. 17-37.
11 E. Jahn, D. Mueller, W. Wieker, and J. Richter-Mendau, Zeolites,
    9 (1989) 177-81.
12 J.A. Martens, M. Mertens, P.J. Grobet and P.A. Jacobs, in
    P.J.Grobet, W.J. Mortier, E.F. Vansant and G. Schulz-Ekloff
    (Eds.), Innovation in Zeolite Materials Science, Vol. 37, Stud.
    Surf. Sci. Catal., 1988, Elsevier, New York, pp. 97-105.
13 S.T. Wilson and E.M. Flanigen, in M. Occelli and H. Robson
    (Eds.), Zeolite Synthesis, Vol. 398, ACS Symp. Ser., 1989,
    American Chemical Society, Washington, DC, USA, pp. 329-345.
14 L. Maistriau, N. Dumont, J.B. Nagy, Z. Gabelica and
    E.G. Derouane, Zeolites, 10 (1990) 243-250.
15 D.A. Lesch and R.L. Patton, EP Patent Application 293919 (1988).
16 P. Wenqin, Q. Shilun, K. Quibin, W. Zhiyun and P. Shaoyi, in
    P.A. Jacobs and R.A. van Santen (Eds.), Zeolites: Facts,
    Figures, Future, Vol. 49A, Stud. Surf. Sci. Catal., 1989,
    Elsevier, New York, pp. 281-289.
17 M. Briend, A. Shikholeslami, M.J. Peltre, D. Delafosse,
    D. Barthomeuf, J. Chem. Soc., Dalton Trans., (1989) 1361-2.
18 B.M. Lok, T.R. Cannan and C.A. Messina, Zeolites, 3 (1983)
    282-291.

Chapter 5

Modified Zeolites

R. Szostak
Zeolite Research Program
Georgia Tech Research Institute
Georgia Institute of Technology
Atlanta, Georgia 30332 USA

SUMMARY
    The application of secondary or post-synthesis treatment of
zeolites and molecular sieves has produced modified materials
with the desired properties for their application as catalysts.
Acid leaching, steam stabilization, and vaporous metal halides
as well as low temperature reactions with silicon fluoride salts
all produce zeolites which are similarly deficient in framework
aluminum yet have divergent catalytic properties.

INTRODUCTION

    Post-synthesis modification of a zeolite encompasses a
variety of techniques to further control the acid activity
and/or the shape selectivity of a specific zeolite structure.
There are three major types of modifications which can be
applied to the zeolite: structural modification in which the
framework SiO_2/M_2O_3 (where M = Al or other metal cation such as
Fe, B or Ga) is changed resulting in a change in acid activity.
Reactions with steam would be an example of this type of
modification; the modification of the surface of the crystal to
further select the size of pore opening. This would include the
addition of large organometallic species which cannot adsorb
into the pore system; and internal pore modifications which
block or alter the structural acid sites and/or restrict the
internal pore diameter. This is exemplified by the adsorption
and subsequent decomposition of small metal hydrides in the
zeolite.

    For each method of modification, a number of
characterization techniques have been applied in order to
quantitatively identify the changes brought about within the
material upon treatment. Of all of the techniques utilized,
which includes I.R., N.M.R., X-ray diffraction, electron
microscopic and surface sensitive techniques (ESCA, AUGER,
SIMS), examination of adsorption and catalytic properties
remains to be the most powerful tools to discern the effect of
the modification technique on the properties of the zeolite.

    The term "modified zeolite" has actually been used to
describe a wide range of materials, from those which have simply

been ion exchanged to materials where total structural collapse
has been observed. The effects of modifications of a zeolite on
acid acidity and selectivity properties are shown in Table 1.
Ion exchange in a zeolite to modify the selectivity and acidity
will be discussed elsewhere in this book. Complete structural
collapse of the zeolite framework does not result in a catalyst
of significant utility. This discussion will encompass only the
materials which have been altered in an essentially irreversible
manner, unlike ion exchange, yet maintain a reasonable amount of
structural integrity.

TABLE 1

Summary of the effect various modification methods have on the
acid activity and shape selectivity of the zeolite modified.

	Activity Change	Pore Modification
Ion Exchange	yes	yes
Mineral Acid	yes	sometimes
H_2O (steam)	yes	mesopores formed
EDTA	yes	mesopores formed
$SiCl_4$	yes	mesopores formed
$(NH_4)_2SiF_6$	yes	--
HF	yes	--
Organic Adsorption	yes	yes
Coke	yes	yes
CVD of organometallics	sometimes	yes
"Al Reinsertion"	(yes)	?

EXTRACTION OF ALUMINUM FROM ZEOLITE FRAMEWORKS

Early ion exchange studies of zeolites had shown that not
all of the zeolite materials could be directly ion exchanged
with acid cations due to loss in crystallinity upon exchange in
the acidic media[1]. Some zeolites, however, did not appear to
exhibit such behavior as one early researcher commented, on
studies of the higher silica containing zeolites, "preliminary
results show that this material is a unique cation exchanger,
operating over the entire pH range...without change in the
crystal structure from the parent material" [2]. Though
structural degradation was not occurring in these materials, it

was not realized at the time that structural changes via framework aluminum removal were indeed taking place.

TABLE 2

Preparation methods of aluminum-deficient zeolites [5]. (Reproduced with permission of the American Chemical Society.)

---

1. Dealuminated Y Zeolites

A. Hydrothermal treatment with NH_4Y zeolites
    (formation of ultrastable Y zeolites)

B. Chemical Treatment
    1. Reaction with chelating agents (e.g. EDTA,ACAC)
    2. Reaction with $CrCl_3$ in solution
    3. Reaction with $(NH4)_2SiF_6$ in solution
    4. Reaction with $SiCl_4$ (or other halide) vapor
    5. Reaction with F_2 gas

C. Hydrothermal and chemical treatment

    Reaction of ultrastable Y zeolites with:
    1. Acids (e.g. HCl, HNO_3)
    2. Bases (e.g. NaOH)
    3. Salts (e.g. KF)
    4. Chelating agents (e.g.EDTA, ACAC)

2. Dealuminated mordenite and other high silica zeolites

A. Chemical treatment
    1. Reaction with acids (HCl, HF)
    2. Reaction with $SiCl_4$ vapors
    3. Reaction with $(NH_4)_2SiF_6$
    4. Reaction with F_2 gas
    5. Reactions with chelating agents

B. Hydrothermal and chemical treatment
    1. Steaming and acid leaching
    2. Repeated steaming and acid leaching

Hydrated alumina becomes appreciably soluble at a pH of around 4 but in the presence of hydrated silica the solubility of aluminum is only slightly suppressed. It would be expected that a zeolite would be susceptible to acid attack and subsequent aluminum removal. In most cases, the aluminum in a zeolite is essentially surface aluminum. This is unlike clays which are acid leaching resistant because the aluminum is located between silicate layers and not in contact with the aqueous acid phase. Selective aluminum removal without structure collapse was first described by Barrer with the essentially complete removal of aluminum from clinoptilolite by treatment with strong mineral acid [3]. Mordenite, erionite and zeolite L, all silica rich zeolites, were also shown to be dealuminated by direct leaching with strong mineral acids. Values of 600 were reported for the SiO_2/Al_2O_3 ratios of acid dealuminated mordenite. Mordenite was of special interest to the early workers in this field due to 1) its availability and 2) its potential application as a catalyst [3,4]. Dealumination of mordenite and other higher silica zeolites will be described in detail in a later section. Table 2 summarizes the variety of methods which can result in the dealumination of a zeolite.

CRACKING CATALYST DEVELOPMENT: HISTORICALLY A QUESTION OF STABILITY

Of the zeolites which have been modified by the ever increasing number of methods, zeolite type Y has seen the greatest concentration of effort. With the realization that zeolites had enormous potential as cracking catalysts, specifically the faujasite structure with its three dimensional large pore system, an intense interest in increasing the thermal stability of these materials ensued. It was already known that increasing the SiO_2/Al_2O_3 ratio increased thermal stability. Mordenite, clinoptilolite and erionite, three silica rich zeolite structures exhibited inherently higher thermal stability than the more alumina rich materials such as zeolite Y. However, the faujasite structure was resistant to crystallizing with a framework SiO_2/Al_2O_3 ratio greater than six. Such resistance to the preparation of higher silica content for this structure through synthetic methods limited the desirable thermal stability found in more silica rich zeolites. It was the potential for successful commercial application of this zeolite which closely corresponded to understanding and controlling the factors that affect stability. It is interesting to note that the more subtle changes in the material which occur with dealumination methods developed nearly twenty years ago are still not completely understood today.

Unlike the results obtained for mordenite, the direct dealumination of type Y with mineral acid induced structural collapse, thus development of other techniques to dealuminate zeolite Y to produce a more thremally stable material was necessary. Two of these early and successful methods proposed and demonstrated by Kerr [6-13] included the use of chelating agents such as EDTA (ethylenediamine tetraacetic acid) and thermal treatment of the ammonium ion exchanged form in the presence of steam. A secondary acid leaching and recalcining was found to further stabilize the structure [5]. More recent innovations in the modification of zeolite Y to increase its SiO_2/Al_2O_3 ratio include the use of $SiCl_4$ at high temperatures [14] and mild solution phase modification using ammonium fluorosilicate as a dealumination agent [15].

METHODS FOR DEALUMINATING ZEOLITE Y

The details of the methods developed to remove the framework aluminum from the zeolite Y structure to increase thermal stability are presented here for comparison. It is important to point out that the first two procedures require the presence of sufficient amounts of steam as shown by Kerr [9-11] in his classic study of the importance of a hydrous environment. Using deep bed and shallow bed calcination methods, he obtained products with divergent properties, illustrating his point. The dependence of the degree of dealumination on the partial pressure of steam was further described by Engelhardt, et al., [16,17], who showed that the greater the hydrothermal environment, the more aluminum is removed from the framework.

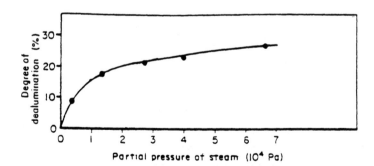

Figure 1: Dependence of the degree of dealumination on the partial pressure of steam in the shallow bed treatment of zeolite Y at 550°C [16]. (Reproduced with permission of Butterworths.)

This relationship between degree of dealumination and the partial pressure of the water vapor is shown in Figure 1.

158

Hydrothermal stabilization involves hydrogen exchange, aluminum extraction, and dehydroxylation in the presence of some water vapor. Contraction of the framework structure provided physical evidence that a change had taken place while the bulk SiO_2/Al_2O_3 ratio may remain unaffected. Other treatments, such as with ethylenediamine tetraacetic acid, EDTA, (acetylacetone, ACAC, has also been used [18,19,105]), $SiCl_4$ and $(NH_4)_2SiF_6$ all result in the "washing" of various amounts of spurious aluminum away from the material during the dealumination process, thus an increase in bulk SiO_2/Al_2O_3 ratio is observed.

Some of the differences in the catalytic behavior between materials dealuminated by the different methods have been attributed to the contribution of dendrital aluminum to catalytic activity, the presence of mesopores and hydroxyl groups at defect sites, all points of discussion and speculation for many years. Thus dealumination is not the only process occurring during these post-synthesis treatments. Migration of silica to fill the voids left by dealumination had also been suggested as a factor in stabilization and such migration is sensitive to the dealumination method employed [20-22]. Intense investigation into the movement of both aluminum and silicon species using a myriad of physical characterization techniques has resulted in numerous proposals of the types of species formed during this process and thier contribution to the catalytic activity of the modified zeolite.

Steam Stabilization (Methods A and B) [23]

Method A: This procedure illustrates the stabilization of the structure following a simple ammonium exchange while the zeolite still contains 10 to 25% of its original alkali-metal cations.

1. A Y-type zeolite is ion exchanged with an ammonium salt solution to reduce the level of sodium to about 10 to 25% of its original value. The remaining sodium is increasingly difficult to remove by conventional exchange because of its location in the small sodalite cages of the zeolite.

2. The partially-exchanged zeolite is washed free of excess salt and stabilized by heating in a static or steam-containing atmosphere to a temperature between 600°C and

825°C, so that the unit cell shrinks by about 1%, or about 0.2 to 0.3Å.

3. The thermal stability of the stabilized structure can be increased still further by removing the major portion of the remaining sodium which has now become readily exchangeable.

Method B: Complete Alkali removal.

1. A Y-type zeolite is ion-exchanged with an ammonium salt solution to reduce the level of the sodium to about 10 to 25% of its original value.

2. The partially exchanged zeolite is washed free of excess salts and heated to a temperature between 200° and 600°C so that the sodium ions have been redistributed but the structure has not undergone significant shrinkage.

3. The major portion of the remaining sodium is removed by again ion exchanging with an ammonium salt solution. The zeolite at this stage is in an extremely precarious metastable state and can easily be made unstable by a variety of conditions not yet fully understood.

4. The low-sodium, metastable zeolite is rapidly heated in a static or steam-containing atmosphere to a temperature between 600° and 800°C to minimize the time during which the hydrogen form might be present. This stabilizing calcination causes a shrinkage of approximately 1 to 1.5% in the unit cell."

Steam Stabilization/Acid Extraction Method

This method is based on the steam treatments described above with the exception that the extralattice aluminum is removed via an acid leaching.

1. NH_4Y is prepared from NaY then calcined at 540°C for two hours in steam.

2. The material is ammonium ion exchanged and further calcined at 815°C for three hours in steam.

3. The resulting material is treated for two hours with HCl(1N).

EDTA Method [7,8,23]

"This illustrates the preparation of an ultrastable structure in which framework aluminum is extracted from a Y type zeolite with a solution of the acid form of EDTA.

1.   Approximately 0.25 to 0.50 mole of H_4EDTA per equivalent of zeolite cation is slowly added to a slurry of the zeolite in water under refluxing conditions, the complete addition requiring at least 18 hours.

2.   The aluminum-deficient zeolite is heated with a purge of inert gas to $800^{\circ}C$ resulting in a contraction of the lattice of about 1%."

SiCl$_4$ Treatment [14,24]

Dealumination takes place in an upflow reactor with a bed depth of about 50 mm and diameter of 30 mm.

1.   Hydrated NaY (or HY) powder was placed in the reactor and the temperature increased $10^{\circ}C$/min to $347^{\circ}K$ and held constant for two hours and purged with nitrogen.

2.   The reactor was cooled to $247^{\circ}C$, still under nitrogen. The N_2 stream was then diverted through a bubbler containing $SiCl_4$ until a constant temperature was obtained (the reaction being exothermic). Further heating of $10^{\circ}C$/min was undertaken until the final exposure temperature was reached. It was then held at that temperature for 40 minutes after which pure N_2 was reintroduced, the tube purged of $SiCl_4$ and the reactor cooled to room temperature.

Ammonium Fluorosilicate Method [25,26]

1.   NaNH$_4$Y zeolite with 80% of the total Na$^+$ exchanged by NH$_4^+$ was introduced in a 0.4 M solution of $(NH_4)_2SiF_6$ at $70^{\circ}C$, the zeolite to $(NH_4)_2SiF_6$ ratio being 3.3 g/g.

2.   After the zeolite was well dispersed, the temperature of the bath was increased to $95^{\circ}C$ and kept at this value during three hours with agitation. Then the product was filtered and thoroughly washed with boiling water. The sample was dried and exchanged twice at $80^{\circ}C$ with an ammonium solution and calcined at $550^{\circ}C$. Further dealumination by this procedure required intermediate calcination. It must be noted that dealumination further

than 25 Al/U.C. provokes an appreciable loss in crystallinity. The unit cell size generally range from 24.60Å to 24.35Å and has been reported to be more stable than other dealuminated Y preparations [27].

Characteristics of Modified Type Y

After post-synthesis modification as described in the previous sections, the gross properties of the modified zeolite are similar to those of the parent zeolite. The most significant difference lies in the greatly enhanced stability and drastically different catalytic properties. For the materials which have been thermally treated in the presence of steam, the chemical composition of the resulting stable structure is almost indistinguishable from that of the unstable ammonium exchanged only zeolites. This is shown in Table 3.

TABLE 3

Composition of zeolite Y, one which is stabilized by steam treatment and one which is not stabilized [23]. (Reproduced with permission of the American Chemical Society.)

	Stabilized	Not Stabilized
SiO_2/Al_2O_3 of starting zeolite Y	5.2	5.2
Chemical Analysis		
Na_2O	0.10%	0.10%
Al_2O_3	21.10%	21.10%
SiO_2	78.77%	78.77%
SiO_2/Al_2O_3	6.35	6.35
Unit cell Å	24.34	24.69
Unit cell of starting zeolite Y, Å	24.65	24.65
Surface area after 2 hours at:		
$700^{\circ}F$	$837 (m^2/g)$	$1008 (m^2/g)$
1500	851	254
1550	793	132
1650	842	18
1700	743	15
1725	678	
1800	542	

162

The indication that a change has occurred is the contraction of the unit cell with a decrease in the unit cell dimensions of the order of 1-1.5% relative to the parent material. The exact amount of such decrease depends on 1) sodium content, 2) degree of aluminum extraction and 3) the calcination conditions. The range of changes observed in the unit cell of type Y with treatment starting with differing SiO_2/Al_2O_3 parent materials is shown in Figure 2. The SiO_2/Al_2O_3 of the ordinate is the framework ratio not representative of the bulk.

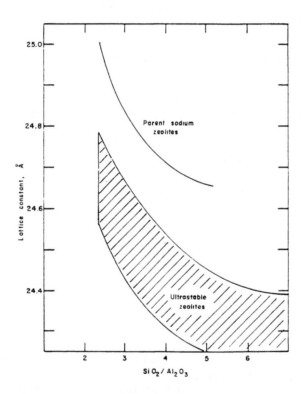

Figure 2: Unit cell contraction in ultrastable zeolites [23]. (Reproduced with permission of the American Chemical Society.)

Breck and Flanigen were the first to show that the number of Al ions in the structural framework of type Y can be determined from the unit-cell constant [29]. Kerr has summarized the relationship in the form [28]:

$$N(Al) = m(a_o-X)$$

where N(Al) equals the framework aluminum ions per unit cell and
a_o is the unit cell parameter in Å. The values for slope and
intercept have been determined by various groups and their
results are provided in Table 4.

TABLE 4
Values for the slope and the intercept used to determine the
amount of framework aluminum in zeolite type Y based on the unit
cell parameter [28]. (Reproduced with permission of
Butterworths.)

	Slope,m	Intercept, X
Breck and Flanigen [29]	115.2	24.191
Kerr et al. [28],		
Dempsey, et al. [30]	112.1	24.222
Sohn et al. [31]	107.1	24.238

In using the relationship between unit cell size and
framework aluminum content it is critical for reproducability to
have all samples equilibrated at constant humidity before the
measurement of the cell size [29,36]. The variations reported
by other groups can generally be attributed to the presence of
defect "holes" in the structure, which are dependent on the
method used to dealuminate the zeolite. The amount of healing
or replacement of aluminum with silicon in the structure is
method dependent. Since the aluminum-depleted defect site is
proposed to be larger than that of a healed (silicon-
substituted) site [34] it is difficult to directly compare
aluminum contents of the different materials on the basis of
unit cell size alone. The presence of such method dependent
defects has been identified through a number of other techniques
[32,33]. Care must therefore be taken in using unit cell data to
exclusively quantitatively determine the amount of framework
aluminum.

Determinination of the unit cell size has been an important
technique utilized at the refineries [35]. In catalytic
cracking, the size of the unit cell has been shown to correlate
well with the RON (research octane number) and MON (motor octane
number), the ratio of olefin/paraffin and the yields of several
of the fractions including C_3 gas, gasoline, light cycle oil,
bottoms and coke. The smaller unit cell volume correlates with

the lower production of coke. In the refinery the unit cell
size is used to predict RON and the dry gas yield.

The unit cell size, however, is only one step in
characterizing the type Y cracking catalysts [37]. In addition
to acid strength and site density which can be changed by
controlling the modification procedure; levels and types of the
nonframework component as well as mesoporosity development
within the crystal also alter the performance of the cracking
catalyst. As discussed above, these cannot be ascertained from
the unit cell measurements. The variability in sites based on
the type of modification employed is proposed to change the
cracking intermediates from the carbocations to radical cation
type species.

EFFECT OF SODIUM CONTENT
Figure 3 shows the important role of sodium in the
contraction of the unit cell of zeolite Y [23]. The presence of
sodium has a two-fold effect on stability. High temperature
stability is limited by its presence, a long recognized fact in
the catalyst art [38]. The presence of greater than 25% sodium
prevents unit cell contraction during calcination of the
partially ammonium exchanged zeolite, as the presence of sodium

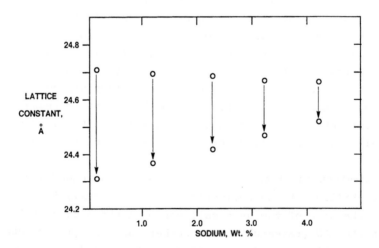

Figure 3: Unit cell contraction of ammonium-exchanged zeolite Y
during stabilization as a function of sodium content [23].
(Reproduced with permission of American Chemical Society.)

inhibits aluminum extraction. Generally speaking, the unit cell contraction is influenced by sodium content, but it is not a direct result of sodium removal. Little contraction occurs with ion exchange alone; in fact, a slight increase in the unit cell is observed when the sodium ions are replaced by ammonium ions.

Extraction of aluminum from the framework occurs when the partially exchanged form undergoes an intermediate calcination and not during the exhaustive exchange process. Low sodium levels are of practical importance as they aid in enhancing vanadium tolerance in the dealuminated zeolite cracking catalysts.

ALUMINUM EXTRACTION: CHARACTERIZATION OF FRAMEWORK AND NONFRAMEWORK COMPONENTS

The temperature and water partial pressure also induce differing degrees of framework aluminum loss as shown in Table 5 [39]. With steam treatment only, the bulk SiO_2/Al_2O_3 ratio remains constant with the only change to indicate framework aluminum removal being the change in the unit cell volume. Unit-cell volume decreases with increasing temperature (severity) of steaming. Crystallinity after treatment was not reported in this study; however, the indication is that, depending on the treatment employed, only 20% of the sample will remain crystalline under the more severe treatments [39]. The bulk SiO_2/Al_2O_3 ratio was altered through subsequent leaching of the sample with acid as can be seen from the data presented in Table 5. Changes are observed in the amount of residual sodium as well as the bulk aluminum content, with little observed change in the unit cell volume, indicating that the role of the acid leaching is mainly to remove the spurious aluminum species not contained within the framework.

The drastic differences in thermal stability between a steam stabilized and an unstabilized type Y zeolite can be seen from the determination of the surface areas of the materials after thermal treatment for two hours at progressively increasing temperature as shown in Table 2. The steam stabilization also induces the formation of a system of mesopores within the structure. Such a system can be observed in the textural changes between an unsteamed and steamed type Y zeolite under transmission electron microscopic investigation [39-41]. High resolution images of the steamed sample show many areas of low contrast which are attributed to the presence of amorphous domains or hole and macro-cavities within the zeolite crystal.

Adsorption studies also indicate presence of a mesoporous network in the materials [42-45]. In addition, NMR has been applied to try to quantify the degree of mesoporosity in

TABLE 5

Treatment of zeolite Y under various conditions and its effect on SiO_2/Al_2O_3 ratio and on the unit cell dimensions [39]. (Reproduced with permission of Elsevier Publishing.)

Samples	Treatment	Na%	SiO_2/Al_2O_3(b)	a_o/Å(a)
NH_4Y	NaY exchanged in NH_4Cl	2.2	4.8	24.70
Deal 6-2	NH_4Y steamed 600°C in 100% H_2O, exchanged with NH_4Cl, steamed 820°C in 60% H_2O	0.27	4.8	24.35
Deal 6-3	Same but final treatment 890°C in 100% H_2O	0.27	4.8	24.32
Deal 6-4	Deal 6-3 steamed 920°C in 100% steam	0.27	4.8	24.22
Deal 9-2	Same as Deal 6-2 but final treatment 890°C, 100% H_2O	0.27	4.8	24.19
Deal 9-2ext	Deal 9-2 treated with HCl (1 N) at 90°C	0.07	48	24.20
Deal 10 ext	Same as Deal 9-2,ext but steaming and extraction repeated two times	0.07	98	24.18
Deal 16 ext	Same as Deal 9-2 ext	0.04	156	24.20
HY	NaY exchanged in NH_4Cl (eight times) outgassed at 300°C in vacuum	1	4.8	

(a) unit cell constant; (b) from chemical analysis

the zeolite [46-48]. The "swiss cheese" model of this dealuminated zeolite is linked to the severity of the high-temperature, hydrothermal treatment of the material, as such mesopore volume is not as evident in type Y samples dealuminated

via the $SiCl_4$ or $(NH_4)_2SiF_6$ routes [49]. A comparison was made
of the difference in the extent of a mesopore system in three
dealuminated samples of Y prepared by three different methods:
steam treatment (high level of mesopores), $SiCl_4$ treatment
(moderate level of mesopores), and treatment with $(NH_4)_2SiF_6$
(some mesoporosity) [43]. However, subsequent reports on
throughly washed and well characterized samples of $(NH_4)_2SiF_6$
treated Y zeolite have indicated no mesoporosity can be observed
after such treatment [106]. Treatment with EDTA is also shown to
induce a mesopore system [50].

27Al NMR examination of type Y dealuminated after the
differing treatments suggests that octahedral Al species may
still be contained within the zeolites dealuminated with steam
as well as chemically stabilized with $(NH_4)_2SiF_6$. This is shown
in Figure 4. Dealumination with EDTA, which extracts the

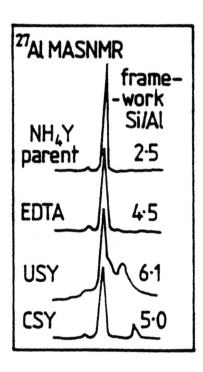

Figure 4: 27Al NMR of Zeolite type Y (NH_4Y parent), EDTA
treated type Y (EDTA), Steam stabilized type Y (USY) and
$(NH_4)_2SiF_6$ treated type Y (CSY). Framework Si/Al provided on
the right side of the figure [50]. (Reproduced with permission
of Elsevier Publishing.)

aluminum from the pores, shows little difference in the 27Al NMR
spectrum from that of the parent material. The differences in
the peak positions for the nonframework aluminum in USY and type
Y treated with $(NH_4)_2SiF_6$ in Figure 5 are attributed to the
presence of fluorine bound to the nonframework aluminum in the
silicon fluoride treated Y. Chemical analysis substantiates the
presence of residual fluorine in the sample examined [50]. Such
residual fluorine was shown to affect the catalytic activity of
this dealuminated zeolite. Butane cracking is increased for
the$(NH_4)_2SiF_6$ treated material relative to the similar material
steam dealuminated [25,106]. The increased activity of ammonium
fluorosilicate treated type Y was attributed to residual
fluorine content [50].

In addition to proposed contributions by the non-framework
aluminum species to the activity of these catalysts, the
difficult question of determining selective extraction of
aluminum (i.e. framework site selectivity) based on the method
of dealumination has also been considered. Examination of the
29Si NMR for type Y dealuminated under various deep and shallow

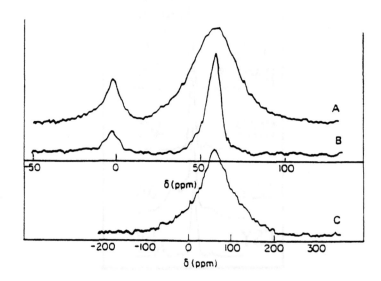

Figure 5: 27Al FT-NMR spectra of an 85 DeNa-Y 400 DB sample at
vl= 70.34 MHz: rehydrated sample without MAS (A) and with MAS at
a rotation frequency of 2.5 kHz (B), dehydrated sample without
MAS (C) [107]. (Reproduced with permission of Butterworths.)

bed conditions reveals that the steaming method employed is not
site specific for the removal of aluminum from the framework.
The ordering of silicon and aluminum in the dealuminated species
appears to be determined only by the final ratio achieved
[16,17]. The question of ordering of Si/Al in the framework
based on the method of dealumination, however, has not yet been
completely resolved [43].

SPECTROSCOPIC TECHNIQUES APPLIED TO UNDERSTANDING DEALUMINATION

NMR has been utilized in examining these dealuminated
materials to a varying degree of success. Trying to apply these
techniques quantitatively is difficult because they are highly
technique and interpretation sensitive. For example,
controversy has arisen with regards to the presence of aluminum
in the zeolite which appears to be invisible to NMR analysis.
Such "invisible aluminum" occurs in dehydrated samples; however,
it was found, with adequately hydrated materials, this is not
the case as they contain "ideal" symmetry for the octahedral
andtetrahedral species. The effect of dehydration and
rehydration can readily be seen in Figure 5. In this figure a
dealuminated
sample of zeolite type Y shows only the presence of tetrahedral
framework aluminum in the dehydrated state. With hydration of
the material, the presence of the nonframework octahedral
component can be observed. Magic angle spinning further
sharpens the lines but has no further effect on the resolution
of the two broad peaks under the experimental conditions
employed in this study. The presence of "NMR invisible"
aluminum has been brought up by several authors [32,51].

Grobet and coworkers [52], in examining $SiCl_4$ dealuminated
type Y using the combination of NMR and IR, conclude that the
$SiCl_4$ treatment still produces hydroxyl nests in the material
which can be healed through silica migration during a subsequent
steaming process. They indicate that the hydroxyl nests in the
aluminum deficient spots contribute to an intensity increase in
the 29Si NMR line at -107 ppm and that the extra lattice
aluminum hydroxyls enhance the 100.6 ppm line. This is contrary
to the interpretation of Englehardt, et al. [53].

Using infrared techniques, Hanke and Moller [33] have
identified combination and overtone bands in the steam treated

zeolite type Y in the near infrared spectra of those materials, and Anderson and Klinowski have observed differences in the hydroxyl group infrared spectra of materials treated with $SiCl_4$ and steamed samples [54]. Zi and Yi [34] have also utilized several techniques including infrared spectroscopy to compare different dealuminated materials. All authors have concluded that differences in the nature of the material, ranging from small to significant, do occur and depend on the method of dealumination. These differences include the number of structural defect sites (holes) relative to silica insertion and the presence of mesoporosity based on method of treatment.

Radial electron distribution studies and x-ray structural studies show the presence of both octahedral and tetrahedral species occupying nonframework sites, including octahedral-octahedral [Al(Oh)-Al(Oh)], octahedral-tetrahedral [Al(Oh)-Al(Td)], and tetrahedral-tetrahedral [Al(Td)-Al(Td)] species, all of which appear similar to the presence of known alumina and hydroxylated alumina fragments [39,55] and confirm the data obtained through NMR techniques [56-58]. The presence of amorphous silica-alumina phases based on infrared spectroscopic data is proposed by Garralon, et al., [59]. Ray and coworkers [60] find that treatment of zeolite Y with $SiCl_4$ results in the formation of several different types of SiOH and detect the presence of amorphous $Si(OSi)_4$ after dealumination with $SiCl_4$ as well as with $OCCl_2$ [61]. Scherzer [5] details the possible extra-framework species which have been suggested based on selected characterization techniques. Many are in agreement with those of Brunner [56] and Mauge [39]. The proposed species are shown in Table 6.

APPLICATION OF SURFACE SPECTROSCOPIC TECHNIQUES

Depth profiling using Fast Atom Bombardment/Secondary Ion Mass Spectroscopy (FABMS/SIMS), for example, comparing EDTA, steam and fluorosilicate treated type Y, provides further evidence that these methods produce different materials. The treatment with the $(NH_4)_2SiF_6$ produces the most homogeneously dispersed material, with SiO_2/Al_2O_3 ratios consistent throughout the crystal. On the other hand, EDTA treatment produces a silica rich surface layer which is expected as the extraction would take place at the first point of contact, the surface of

the crystal. Ultrastabilization with steam shows progressive
enrichment of the surface with aluminum as a function of the
time of the steam treatment [50]. A summary of the effect
different treatments have on the surface composition of
dealuminated faujasite is presented in Table 7.

TABLE 6:
Extra framework species (taken in part from Scherzer [5]).

Cationic	Neutral
Al^{+3}	$AlO(OH)$
AlO^{+}	$Al(OH)_3$
$Al(OH)^{+2}$	Al_2O_3
$Al(OH)_2^{+}$	
$[Al-O-Al]^{+4}$	

$$\underset{O}{\overset{O}{[Al\diagdown Al]}}^{+2}$$

Other

$$\underset{[Al-O-\overset{|}{Al}-O-Al]}{\overset{O}{}}^{+3}$$

Aluminosilicate species
Silicate species

DOES GETTING THERE BY A DIFFERENT ROUTE GET YOU THE SAME PLACE?
     Though a variety of different methods have been developed
to accomplish the same goal, that of stabilizing the zeolite
type Y structure with the removal of framework aluminum, it
appears that each modifies a different aspect of the material.
For example, the higher silica containing materials prepared
from treatment with ammonium fluorosilicate make 2 to 5% more
gasoline, 0.2 to 0.4% more light cycle oil and 1 to 2% less coke
than catalysts made with the steam treated Y catalysts. In the
steam-stabilized Y catalyst, more coke is formed, attributable
to amorphous debris in the pores which is decreased with the
fluorosilicate treated materials [27]. The $SiCl_4$ treatment of
type Y produces differing product distributions in the

TABLE 7

Surface Composition and depth profile of dealuminated zeolites

Treatment	Depth Profile
Steaming	Al rich surface
Acid (mordenite and ZSM-5)	Al deficient surface
Steam + acid (mordenite and ZSM-5)	Al deficient surface
Steam + acid (type Y)	Near uniform distribution
EDTA	Al deficient surface
$SiCl_4$	Al rich surface*
$(NH_4)_2SiF_6$	Near uniform distribution

* Si rich surface for smaller pore zeolites due to surface deposition

conversion of n-decane when compared to the steamed zeolite [62]. It is evident from the catalytic results in both model cracking reactions, as well as in actual use of these catalysts under conditions similar to the industrial cat crackers, that the differences observed between modification techniques determined through physical characterization methods translate to observed changes under catalytic operating conditions. The complex subtleties of these differences, however, are still not completely understood.

AND THEN THERE IS REALITY

It is important to note that any "initial" material used as an FCC catalyst immediately begins a course of change upon entering the cracking unit. "Equilibrium" catalysts in the unit differ from the "initial" materials so carefully characterized using the techniques described in this chapter. The "equilibrium" catalyst is a mixture of many different modified zeolites. As the catalysts in the unit are exposed to very high temperatures in the regenerator, dealumination of the framework continues. Such exposures to the regenerator temperatures are for differing lengths of time, ranging from a few hours to a few months. The acid activities of the "equilibrium" catalyst may therefore span an order of magnitude, from extremely high activity found in recently added fresh catalyst to the other extreme, near zero activity for a catalyst which has been circulating for several months. These "equilibrium" catalysts differ from unit to unit, from feed type to feed type, not only from catalyst to catalyst.

As more methods are developed to prepare highly silicious faujasite structures, it is important to understand that a concomitant decrease in the number of acid sites results in producing low catalytic activity in cracking reactions. In the optimization of cracking catalysts, a compromise must be reached between high thermal stability and enough acid sites to maintain activity. Selectivity towards gasoline and related products of high commercial importance is the ultimate goal, regardless of the method.

MODIFICATION OF HIGH SILICA ZEOLITE STRUCTURES: DIFFERENCES FROM ZEOLITE Y

The application of post-synthesis modifications to already thermally stable zeolites containing a high proportion of silica has been explored in detail over the last 10 years. Increased hydrophobicity, making the material more organophilic and diluting the number of the highly acidic sites within the structure have been the basis for much of the work in this area. High silica zeolites have been structurally modified using the same methods listed in Table 1. Dealumination as a method of changing the SiO_2/Al_2O_3 ratio produces more than just aluminum removal from the framework, as mesopore systems, and silanol defect sites [63,64] are also observed thus producing materials which differ from those prepared via direct synthesis methods.

REACTIONS OF ZEOLITES WITH FLUORIDE

Fluorine treatment is another way of modifying a zeolite; however, it is sensitive to the type of zeolite as well as to the conditions of the treatment. Such variability between zeolites must be taken into account when applying any of these modification techniques. Aluminates and silicates react with fluorine under extremely mild conditions. In a zeolite dealumination will occur with the subsequent formation of $AlF_x(OH)_y$ species, depending on the amount of fluoride and water present when exposed to a fluorine containing species. F_2 [65],

HF [66-69], boron trifluoride [70], aluminum fluoride or silicon fluoride containing species [71] have all been reported to structurally modify zeolites. Structures which have been exposed to fluorine in one of the forms listed above include mordenite, erionite, L, omega, silicalite, and ZSM-5, as well as type Y. The reactions with fluorine compounds induce an increase in the hydrophobicity of the zeolites treated, most notably the higher silica containing ones, due to framework

TABLE 8

Fluorine treatment conditions [65]. (Reproduced with permission of the American Chemical Society.)

	Treatment Conditions					
Zeolite	F_2 Vol%	O_2 Vol%	Duration (min)	Temp (^oC)	Type	F_2 Content (wt %)
LZ-105[a]*	10	2	10	60	Severe	N.A.
LZ-105[b]	5	5	15	60	Severe	2.4[f]
H-Zeolon**	10	2	10	60	Severe	3.5[f]
H-Zeolon	5	0[g]	30	25	Severe	10.0
H-Zeolon	1	0[g]	45	25	Mild	N.A.
Erionite[c]	5	5	5	60	Severe	4.6
Erionite[d]	5	5	5	60	Severe	N.A.
NH_4,TMA-Omega	5	5	5	60	Severe	4.4
NH_4,K-L	5	5	5	60	Severe	4.8
NH_4,Y[c]	2	0[g]	5	60	Mild	N.A.
NH_4,Y[c]	2	0[g]	15	25	Mild	2.2
NH_4,Y[c]	1	0[g]	15	25	Mild	N.A.

a. 20 wt% alumina bonded.
b. Acid washed and 20 wt% alumina bonded.
c. Mild steaming followed by NH_4^+ exchange with hot NH_4Cl solution.
d. Mild steaming and NH_4Cl exchange repeated.
e. F_2 content analyzed after fluorination.
f. F_2 content analyzed after 600^oC calcination and 2 1/2 hr. soxhlet extraction subsequent to fluorination.
g. Samples treated in a flow-through reactor with premixed F_2-N_2 mixture.

* Zeolite in sample is silicalite
** Mordenite
N.A. -- Not analyzed.

aluminum removal. In the reaction with F_2, not all zeolites treated show an increase in thermal stability with treatment. Both zeolites L and omega are reported to loose crystallinity when calcined to temperatures of $600^{\circ}C$ after treatment. The methods used for direct fluorination of selected zeolites are presented in Table 8. The degree of dealumination, as can be seen from this table can be controlled by varying the treatment conditions. An important feature of this method of modification is the persistence of fluorine residue after treatment which, several authors have noted when using various fluorine containing reagents, may not be easily extracted [65,50].

The crystallinity of these materials remains very high based on oxygen adsorption capacity measurements. Significant indication that fluorine treatment modifies the hydrophilic character of the material is shown by water adsorption capacities for silicalite, mordenite, erionite and type Y (previously steam stabilized) which decrease significantly at low P/P_0. In competitive adsorption tests using butanol and water, the mordenite material (Zeolon from Norton Co.) and type Y show the most significant change in properties after F_2 treatment. Untreated mordenite does not exhibit selectivity towards n-butanol removal, increasing to 42% after fluorine treatment. Similar increases are observed for type Y. Generally, cracking activity of this material decreases with fluorination associated with dealumination; however, both mordenite and omega show significant increases in activity after treatment, attributable to the possible presence of active nonframework aluminum fluoride containing species which were not removed during washing or calcination.

ENHANCEMENT IN ACTIVITY IN ZSM-5 WITH STEAM

The modification of the zeolite framework structure to increase thermal stability with decreasing framework aluminum, has led to the observation of an enhancement in activity for the zeolite in catalytic applications. In addition to Kerr and coworkers [12], observing this for type Y, Mirandatos and Barthomeuf [72] found that the activity in mordenite is enhanced after hydrothermal treatment. Lago and coworkers [73] and Haag and Lago [74], Sendoda, et al [75], and Szostak [76] saw similar behavior in the hydrothermal treatment of zeolite ZSM-5. Such

176

enhanced activity may result from interplay between active sites rather than merely a decrease in the concentration of framework aluminum as is recognized in the faujasite material. In zeolite Y, sites of different acidity have been ascribed to the difference in composition or the number of next nearest neighbors which are aluminum ions. Acidity increases as the number of next nearest aluminum neighbors decreases [77,18,19,30,12]. The activity of these zeolites is sensitive to conditions of hydrothermal treatment which include time, temperature and water pressure. The enhancement in activity for the higher silica zeolites only occurs when mild steaming conditions are employed. Under severe steaming conditions, reduced activity is found in direct proportion to the loss of aluminum from the framework sites. The enhancements observed when the zeolite is mildly steamed are not reaction specific, i.e., paraffin (n-butane, n-hexane) cracking, toluene disproportionation and paraffin isomerization reactions all reflect such enhancement in activity of the zeolite ZSM-5 [73,74,76,78].

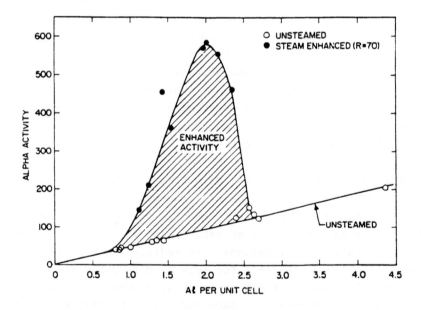

Figure 6: Dependance of alpha activity on the tetrahedral Al content of unsteamed and steam enhanced HZSM-5 [73]. (Reproduced with permission of Elsevier Publishing.)

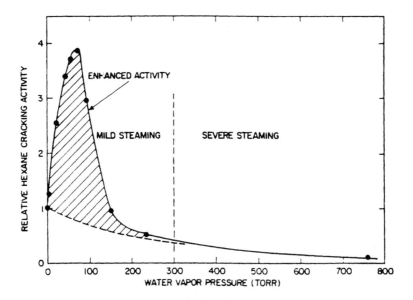

Figure 7: The effect of steaming severity (pressure) on the hexane cracking activity of HZSM-5 (R=70) [73]. (Reproduced with permission of Elsevier Publishing.)

Lago et al. [73] provide a detailed examination of ZSM-5 correlating the degree of enhancement with the SiO_2/Al_2O_3 ratio as well as with the water vapor pressure. These are shown in Figures 6 and 7. The lower the water vapor pressure, the greater the enhancement in activity, with pressures less than 100 torr providing the greatest enhancement in cracking activity towards n-hexane. Such low vapor pressures are indeed easily obtained under normal calcination conditions where high environmental humidities may be encountered or localized steaming environments generated (i.e. decomposition of the organic template to produce steam) suggesting a need for care in is over 300 torr, "severe steaming" conditions are encountered in which only a decrease in activity is observed. A wide range of cracking activities are encountered with the mild steam enhancement of ZSM-5, with increasing activities found for decreasing SiO_2/Al_2O_3. A maximum between activity and the number of aluminum ions per unit cell was determined to be two which is consistent with the proposed site pairs for aluminum in the framework of zeolite ZSM-5. An examination of the cracked product distribution for butane cracked over mildly steamed ZSM-

178

preparing these materials as catalysts. When the vapor pressure
5 shows little difference selectivity based on the envelope of
products from that observed for the parent material [76] which
suggests similar Bronstead activity in both the parent and the
steam enhanced material as shown in Figure 8.

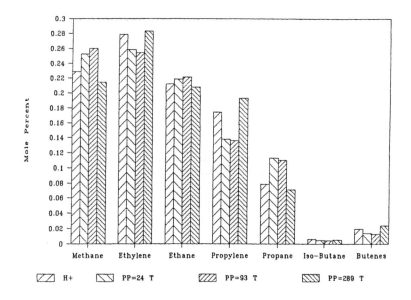

Figure 8: Similarity in the product envelope for the acid
catalyzed cracking of butane over HZSM-5(50) and mildly steamed
HZSM-5(50) at steam partial pressures of 24, 93 and 289 torr.

On the other hand, a detailed NMR (27Al,29Si and 1H) study
of these materials [56] does not appear in agreement with the
proposed aluminum site pairing model proposed by Lago [73]. No
line in the NMR was observed which indicates the presence of
Si(2Al) which makes up 5% of the aluminum atoms in the material.
The data collected using both NMR and catalytic techniques is
shown in Table 9. All of the aluminum species in these
materials were accounted for in the NMR that is no "NMR
invisible" aluminum was determined to be present.

Though numerous types of species have been proposed for the
nonframework aluminum in dealuminated zeolites which include
cationic, anionic as well as neutral species as described in the

previous sections, agglomerate formation of the aluminum species appears to be low in these mildly steamed materials. Rehydration results in the detection of AlOOH and $Al(H_2O)_6^{+3}$ species. Brunner [56] as well as Ione and coworkers [79] indicate that these nonframework species are composed of non-charged, nonhydroxylated and nonstable small complexes and propose, that as electron pair acceptors, they behave as Lewis sites. Due to the low concentration of such species, direct evidence of their presence is difficult to obtain.

TABLE 9

Concentration of aluminum species and hydroxyl groups and results of catalytic activity measurements with dependence on the water vapor pressure p [56]. (Reproduced with permission of Butterworths.)

	Concentration/species per 24 framework tetrahedra								
p torr (1)	AlF (2)	AlNF oct (3)	AlNF tetr (4)	1H (5)	SiOH (6)	SiOAl H (7)	AlOH (8)	ko (9)	B (10)
0	1.3	0.1	--	1.6	0.3	1.3	--	6.0	0.6
14	1.0	0.1	0.4	1.4	0.3	1.1	--	12.8	0.9
50	0.7	0.1	0.7	1.3	0.3	0.7	0.3	16.0	1.0
100	0.5	0.2	0.8	1.2	0.4	0.6	0.2	12.7	1.1
300	0.4	0.2	0.9	1.0	0.3	0.4	0.3	4.1	3.3
700	0.4	0.2	0.9	0.9	0.3	0.4	0.2	3.1	15.5

Column: (1) Water vapor pressure p; (2) concentration of framework aluminum atoms determined by 27Al MAS NMR (60ppm line); (3) and (4) concentration of nonframework aluminum species in an octahedral (0 ppm line) and a tetrahedral (30 ppm line) coordination respectively; (5) total concentration of 1H obtained from the initial value of the free-induction decay; (6),(7) and (8) concentration of SiOH, bridging OH, and AlOH groups, respectively, determined by 1H MAS NMR; (9) reaction rate constant ko in the units of mol*MPa-1g-1h-1; (10) parameter B, characterizing the activity/time stability in the unit 10-3 MPa*g*mol-1. The experimental error is +/- 0.1 in columns (2)-(8), +/- 20% in column (10). "O Torr" indicates a thermal treatment in a water free nitrogen stream. All samples were steamed for 150 min at 540°C.

HYDROTHERMAL ENHANCEMENT OF ACTIVITY IN METALLOSILICATES

As described above, hydrothermal modification of high silica zeolites, depending on conditions of steaming, results in an initial increase in activity followed by a decay in that

180

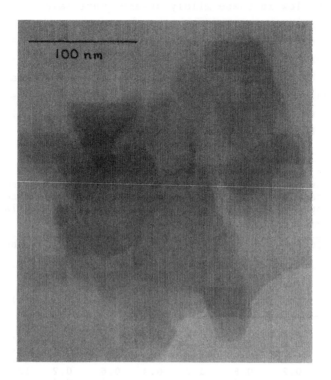

Figure 9: TEM image of a section of iron silicate showing migration of small iron oxide particles to the edges and grain boundaries of the crystal.

activity as increasing amounts of aluminum are removed from the framework. Similarly, demetallation results when a metallosilcate is contacted with steam. Significant differences arise as the metal component, originally in the framework inducing acid activity through the presence of Si-O(H)-M sites within the structure, is dislodged from the framework. Liberation from the framework increases catalytic metal activity. Two metallosilicate systems have been examined in detail, the gallosilicate and the iron silicate analogs of zeolite ZSM-5.

Physical characterization of the materials shows a decline in acidity based on ammonia adsorption with time of steam treatment [80] as well as a decrease in ion exchange capacity. In the case of the iron silicate, a significant change in color is observed with steaming, shown to be due to the formation of

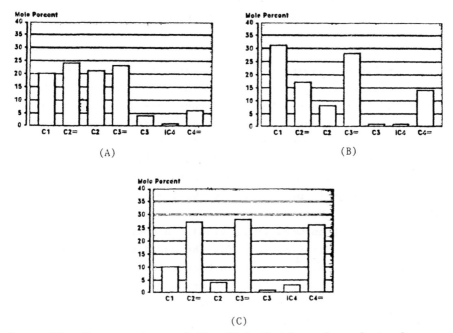

(A)   (B)

(C)

Figure 10: A comparison of the distribution of products from
the cracking of butane over modified metal silicate molecular
sieves with the ZSM-5 structure. (a) HZSM-5(70) (b) H-[Fe]ZSM-
5(79) steamed one hour (c) H-[Ga]ZSM-5(88) steamed one hour

nonframework octahedral Fe-O-Fe species [81]. The iron system
lends itself to analysis with a wide variety of techniques
including Mossbauer, which substantiates the increasing presence
of octahedral nonframework iron in the material with severity
(time) of steaming.

     The migration of iron from framework to nonframework sites
and the subsequent migration to the outside of the crystal and
to grain boundaries could be more easily identified than in the
aluminosilicates due to the greater mass of the iron producing
higher contrast in the HREM. Images of these steamed samples
show the generation of particles in the range of 15Å to 150Å
depending on the conditions of steam treatment, duration and
temperature [82]. Steaming temperatures around 700°C produced
very large agglomerates of iron oxide while milder temperatures
(550°C) induced smaller, more highly dispersed particles to
form. A TEM image of the iron oxide particles in a modified

iron silicate with the ZSM-5 structure is shown in Figure 9. In this sample particles could be observed at the surface of the very small crystallite as well as dispersed throughout the crystal.

The catalytic activity of these materials was increased, with the increase strongly dependent on the reaction employed. In the cracking of butane, a distinct enhancement in catalytic activity was seen for the modified gallosilicate. In addition, the products distributions were significantly altered due to the presence of nonframework gallium(oxide) with the predominant formation of olefins. The modified iron silicate catalyst also resulted in a change in product selectivity when used in the cracking of butane. Methane formation was enhanced due to the presence of the nonframework iron(oxide). A comparison between the cracked product distributions for the modified alumino-, gallo- and iron silicate is shown in Figure 10. Comparisons with free metal oxide showed that the production of specific cracked products directly resulted from the presence of the nonframework metal oxide component.

CVD METHODS APPLIED TO ZEOLITES: $SiCl_4$ TREATMENT

It must be noted that $SiCl_4$, as a reagent, is used in modifying surfaces for coatings in ceramic and semiconductors. Though spectacular results have been obtained in stabilizing and dealuminating zeolite type Y with $SiCl_4$, such treatment of the smaller pore zeolites, such as ZSM-5 have proved to be more selective. The treatment of ZSM-5 with $SiCl_4$ studied by Namba [83] produced materials which were surface enriched in silicon. Surface enrichment in silica has a two-fold benefit in this system, 1) surface acid sites are removed, decreasing the possibility of non-specific acid catalyzed reactions and thus improving the shape selectivity of the material and 2) coking, proposed to form selectively on the surface of the zeolite, can be minimized, thus improving the catalyst life for organic reactions. The improved selectivity of such a modification is shown in the alkylation of 1,2,4-trimethylbenzene with methanol in Table 10.

The treatment of smaller pore zeolites with reactive gases such as $SiCl_4$ borders between surface deposition and dealumination. Hidalgo [84] has examined the modification of

dealumination. Hidalgo [84] has examined the modification of mordenite with $SiCl_4$, $GeCl_4$, $TiCl_4$ and $SnCl_4$ and found both deposition and dealumination to occur. The surface of the zeolite becomes more resistant to the adsorption of hydrocarbons, an indication of pore mouth deposition. An observed decrease is seen in ammonia adsorption, indicating a loss in acidity probably through dealumination by free chloride species liberated after decomposition. Klinowski, et al. [85] used NMR to identify dealumination in mordenite treated with $SiCl_4$ vapor. The results of Hidalgo and Klinowski indicate that $SiCl_4$ provides deposition as well as dealumination. Germanium and titanium chloride enter to only a limited extent and the tin deposits only at the outer surface of the zeolite. A discussion concerning using these techniques to "(re)insert" species into a zeolite structure is found at the end of this chapter.

TABLE 10

Results[a] of the alkylation of 1,2,4-trimethylbenzene with methanol on the parent HZSM-5 and DAI-HZSM-5 [83] (dealuminated with $SiCl_4$) zeolites[b]. (Reproduced with permission of Butterworths.)

Catalyst	Parent	DAI
(Si/Al) bulk	19	24
(Si/Al) surface	18	32
Conversion of		
1,2,4-trimethylbenzene (%)	24.4	23.5
Yield (%):		
Toluene	0.5	0.2
Xylenes	6.3	4.1
1,2,3- and 1,3,5-		
Trimethylbenzenes	2.6	2.6
1,2,4,5-tetramethylbenzene	10.2	16.9
1,2,3,5-tetramethylbenzene	1.8	0.6
1,2,3,4-tetramethylbenzene	0.7	0.3
Pentamethylbenzene	0.1	0.1
Fraction of 1,2,4,5-isomer in		
tetramethylbenzene produced (%)	80	95

a. In the initial stage; process time = 0 = 30 min
b. The reaction conditions are described in Namba [83]

INSERTION OF SILICON INTO $AlPO_4$ STRUCTURES

Since $AlPO_4$ molecular sieves possess neither a net framework charge nor an exchangeable cation sites they have little or no inherent catalytic activity. Post-synthesis modification could induce catalytic activity through introduction of framework charge by the insertion of silicon [86,87]. Theoretically, substitution of silicon could occur for either aluminum or phosphorous in the $AlPO_4$ structure. Thermodynamic calculations on model compounds indicate that substitution may be preferred for the phosphorous and not the aluminum producing a cationic Si-O-P moities in the framework [88]. Such a substitution is contrary to the accepted rules for substitutions of metal into molecular sieves put forth by Flanigen [89]. Ammonium fluorosilicate treatments of $AlPO_4$-5 between 25 and $98^{\circ}C$ shows little consistency with low repeatability for selected incorporation of silicon. The reaction with $SiCl_4$, on the other hand is shown to be more feasible. Reaction temperature between $450^{\circ}C$ and $750^{\circ}C$ were examined. The higher temperatures resulted in a greater loss in crystallinity with nearly 60% of the structure degraded when the $AlPO_4$ was treated at $750^{\circ}C$. The toluene methylation reaction was used to identify the possibility of generation of either acidic or basic sites within the modified aluminophosphate. Little basic activity was observed in these materials suggesting the generation of only Bronstead acid sites. The alkylation reaction did show that the $SiCl_4$ treated $AlPO_4$-5 was intermediate in catalytic activity between the synthesized SAPO-5 and the parent $AlPO_4$-5 with selectivities more similar to the $AlPO_4$-5 starting material. Cumene cracking activities of the $SiCl_4$ modified $AlPO_4$-5 were found to be intermediate between the $AlPO_4$-5 and a synthesized SAPO-5. XPS examination indicated that the $SiCl_4$ modification did not deposit SiO_2 on the surface of the $AlPO_4$-5 crystal thus indicating that the $SiCl_4$ did adsorb into the pores to react with the framework structure itself. In general, the reaction of $SiCl_4$ with the aluminophosphate molecular sieves appears to be limited in its ability to react within reasonable periods of time with the aluminophsophate framework to give only moderate increases in acid activity of the material.

CVD AS A METHOD OF PORE MOUTH RESTRUCTURING

The ability to selectively restrict an adsorbate from contacting acid sites within the pores of the zeolite based on its size is accomplished through selective pore mouth closure. A variety of materials have been utilized to increase the selectivity of one adsorbate over the other. Production of an acid free crystal surface by blocking the surface acid sites is one incentive for modifying the external crystal surface. Shape selectivity in the acid catalyzed conversion thus becomes more pronounced as non-selective surface reactions are minimized. The ability to control the constriction of the pore mouth which also contributes to enhanced shape selectivity in the zeolite by either completely blocking the adsorption of one molecular species over the other or by changing the diffusivity, would be valuable in designing specific catalyst/adsorbent systems [90].

One such means for selective catalyst design is the utilization of chemical vapor deposition (CVD) methods developed for surface and coatings applications. This technique is the deposition of a coating from the vapor phase onto the surface of a substrate. The discussions here concerning CVD only considers the modification of the surface of pore and not the modification of the structure itself. Structural modification is considered under dealumination techniques described earlier in this chapter. An exception is the solid/vapor reaction of the zeolite with $SiCl_4$ which modifies the structure as well as the surface of the crystal, depending on the zeolite. On the other hand, $Si(OMe)_4$, or $Ge(OMe)_4$, reacts only with the surface of a crystal. Experimental difficulties can be encountered in utilizing CVD technique on zeolites. Generally, coatings are deposited on smooth surfaces, not on fine powders. Deposition on powder is fraught with experimental difficulties. Deposition uniformly on the external surface of all of the small zeolites crystals in the bed of the CVD unit must be considered, as many of the species used in CVD are extremely reactive under the conditions employed. Coating occurs only on the top layer of the sample, providing inhomogeneities of the coating. Too slow deposition leads to possible adsorption of the depositing material into the crystal. Fluidized bed techniques are now being developed to further insure homogeneous surface coverage of the deposited material on very fine particles.

All standard zeolite characterization techniques (X-ray, compositional analysis, microscopy, IR and NMR) have been employed to examine the zeolite materials modified through CVD techniques, but the most informative methods are adsorptive and catalytic characterization. For example, the reaction of tetramethylorthosilicate ($Si(OMe)_4$) and also tetramethylorthogermanate ($Ge(OMe)_4$) with crystals of mordenite have been examined by Niwa and coworkers [90,91]. They confirm the precipitation of silicon containing species on the exterior surface of the crystal. In their experiments, deposition was performed in a quartz balance in order to monitor the rate of deposit and the final amount, thereby quantifying their results. Surface-only deposition was confirmed based on the maintenance of acidity within the material. The water adsorption capacity did not decrease significantly, a further indication that diffusion of the $Si(OMe)_4$ into the crystals with subsequent internal blockage was not occurring. On the other hand, the

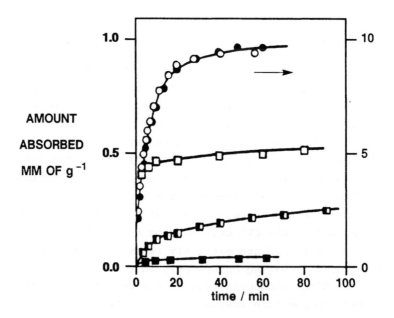

Figure 11: Adsorption of H_2O (o,●) and o-xylene (□ ■ ▫) on HM (O □), SiHM (1.4 wt %) (▫) and SiHM (2.7 wt %) (■) [92]. (Reproduced with permission of the Royal Society of Chemistry.)

shape selective adsorptive properties were extremely sensitive
to the amount of silicon deposited. XPS data indicated the
presence of a 0.6 (1.6 wt % Si(OMe)$_4$) to 8.0Å (4.7 wt %
Si(OMe)$_4$) coating of surface silicon. The 8Å coating completely
suppressed hydrocarbon adsorption.

Mordenite shows little selectivity for the isomers of
xylene as ortho-xylene rapidly adsorbs into the crystal.
Modification of the surface with the silicon species greatly
reduces the diffusivity of the ortho isomer. Increasing amounts
of deposited material further control the diffusivity. This can
be seen in the adsorption profile shown in Figure 11. However,
the rate of diffusion is also suppressed for the very small
critical diameter molecule, n-hexane. The diffusion constants
for hexane, p-xylene and o-xylene are shown in Table 11 at
various degrees of silica deposition. From these diffusion
studies, the size of the opening for mordenite has been
restricted from 6.5 x 7.0Å to that of around 6Å [91,92].

TABLE 11
Diffusion constants on SiHM (in 10^{-14} cm2/s), mordenite which
has been modified with varying amounts of Si(OMe)$_4$ [92].
(Reproduced with permission of the Royal Society of Chemistry.)

adsorbate	hexane	p-xylene	o-xylene
SiHM (1.4 wt%)	--	2.1	0.28
SiHM (2.7 wt%)	3.4	<0.01	<0.01
SiHM (4.8 wt%)	0.18	<0.01	--

Catalytic shape selectivity is also modified by deposition
of the organosilicates. In the hydrocracking of octane isomers,
selectivities are extremely sensitive to the amount of deposited
silica. Silica loadings of 3.2 wt% provide significant cracking
selectivity differences between the isomers, not observed on the
parent mordenite material. A 0.2 wt% increase in deposition
totally suppresses conversion of the branched isomers. This can
be seen in Figure 12.

Niwa and coworkers [93] also examined the ability of the
silica to coat the surface of a metal loaded zeolite to
determine if reactions on external surface metal could be

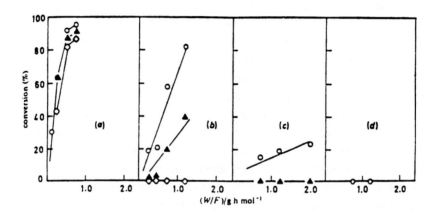

FIGURE 12: Hydrocracking of octane (○), 3-methylheptane (▲) and 2,2,4-trimethylpentane (◇) over (a) PtHM, (b) SiPt HM (3.2 wt %), (c) SiPtHM (3.4 wt%) and (d) SiPtHM (3.7 wt %). Amount of zeolite W divided by the flow rate (F= 0.091 mol/h) is shown on the abscissa [93]. (Reproduced with permission of the Royal Society of Chemistry.)

suppressed. In the samples of platinum loaded on mordenite, the $Si(OMe)_4$ could not adequately coat the platinum particles on the surface of the crystal. Cyclohexane conversion, known to occur over the surface platinum metal sites, was not suppressed with silicon deposition thus indicating some limitation in this technique for suppressing surface metal activity.

The use of the vaporous metal chlorides are well known for depositing metal on surfaces. As discussed in a preceding section, MCl_4 (where M=Si,Ge,Ti and Sn) can be deposited on the surface of a zeolite. The use of metal chlorides is limited however, as the liberated chloride has the potential to induce dealumination of the structure. $MoCl_5$ has also been claimed to modify mordenite [94]. The results reported indicate a significant level of surface deposition occurs on the crystals as nitrogen adsorption is only somewhat suppressed. No studies indicating the extent of change in the acidity or selectivity were presented in that study.

INTERNAL PORE MODIFICATION

The modification of the internal zeolite pore structure through secondary synthesis methods is a way of "engineering" a shape selective zeolite. Besides ion exchange, which will not be discussed in this chapter, preadsorption of polar molecules which stick to the charge sites within the pores of the zeolite have also been used as a way to regulate diffusion and molecular selectivity in the zeolites. Polar molecules such as water and amines all induce a secondary selectivity factor which influences the diffusivity within the structure [95]. Coke, too, can influence diffusivities, with adsorption rates declining when surface coke is deposited. Such transient effects which influence and modify the zeolite surface will not be considered in detail here, as they do not represent "irreversible" change to the material. Coke can be burned off and adsorbed molecules desorbed (or burned off).

Two small inorganic molecules, silane and diborane are extremely reactive towards zeolite hydroxyl groups producing a permanent change to the properties. Silanes, such as SiH_4, are known to react with SiOH and can further hydrolyze by reacting with adjacent silanol groups. H_2 liberation accompanies such changes [96-98]. Boranes behave in a similar manner. Such modifications have been used to encapsulate gases such as krypton and xenon. Temperature is a critical parameter in influencing the degree of this reaction. The effect such modification has on adsorption of krypton and xenon is shown for mordenite in Figure 13 [95]. Selectivity occurs when small amounts of SiH_4 are deposited (i.e. 0.48 mmol).

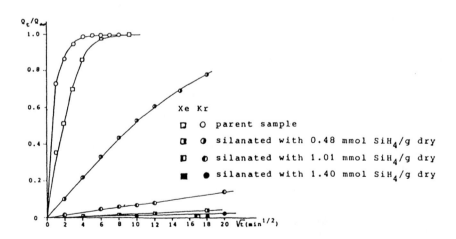

FIGURE 13: Krypton and Xenon sorption at $0^{\circ}C$ [95] on mordenite LP (Reproduced with permission of Elsevier Publishing.)

COMING FULL CIRCLE: PUTTING ALUMINUM BACK INTO THE STRUCTURE

The generation or regeneration of acid sites within a zeolite structure by post-synthesis techniques has been reported by several authors [99-103]. Exchanges between aluminum and other trivalent element such as boron [104] have also been noted. The identification of such insertion is fraught with as much difficulty and controversy as that found for the location of aluminum and identification of active sites in dealuminated materials discussed in the previous sections. The reaction of aluminum halide appears to be nonspecific. Deposition of extralattice aluminum is observed [102] as shown in Table 12. Ammonia desorption studies show a distinct similarity between zeolites synthesized with aluminum in the framework and zeolites which have been treated with $AlCl_3$, suggesting insertion may take place. 29Si NMR studies of the aluminated materials show the presence of a signal at -106 ppm, attributed to the presence of Si-(1Al) in the material. Development of new hydroxyl bands in the IR with treatment with $AlCl_3$ also suggests that such insertion is possible. The reaction with $AlCl_3$ does not occur under mild conditions, however. High temperatures are required. Mechanistically, it has been proposed that such insertions or exchanges between elements are a result of the filling of vacancies left in the framework by reaction with the silanol

TABLE 12

Bulk and framework aluminum concentrations and amounts of NH_3 desorbed [102]. (Reproduced with permission of Elsevier Publishing.)

HZSM-5	Al Concentration 10^{+4}/mol*g^{-1}		Amt. NH_3 Desorbed[b] 10^{+4}/mol*g^{-1}
	Bulk	Framework[a]	
Ordinary	1.7	1.7	1.8
Aluminated	5.7	2.0	3.3

a. Determined by 27Al MAS NMR
b. Evaluated from TPD spectra

groups at the defect sites. But a complete understanding of the contribution made by the spuriously deposited aluminum species to both the physical properties and catalytic activity has yet to be systematically examined. The insertion of other metals into the framework using metal halide vapors has not yet been established.

CONCLUSION

The modification of zeolites through secondary or post-synthesis methods as a way of controlling properties specific to a desired reaction has been successfully utilized since the first experiments in steam stabilization of zeolite type Y. The last twenty years have seen advances on two fronts: new methods of modification and application of a battery of characterization techniques to the understanding of the nature of these materials. As the utilization of advanced spectroscopic techniques becomes more routine, even greater strides in understanding the nature of the modified materials based on the modification method should come to fruition. With such understanding, a more complete integration of a specific catalyst and process should follow.

ACKNOWLEDGEMENT

The author gratefully acknowledges the helpful comments and lively discussions of Drs. Gary Skeels and Janus B.Nagy generated by this manuscript and the dedication and perseverance of Messrs. David Gwinup, Rick Kuvadia, and Scott Simmons, Ms. Alicia Long, Ms. Allison Longaker, and Mrs. Sheron Meyers for their assistance in preparing this document.

REFERENCES

1. Kunin, R., "Elements of Ion Exchange" Reinhold Publ., New York 1960, 25.

2. Keough, A.H., L.B. Sand, "A New Intracrystalline Catalyst", J.Amer.Chem.Soc. 83, 1961, 3536

3. Barrer, R.M., B. Coughlan, "Molecular Sieves Derived from Clinoptilolite by Progressive Removal of Framework Charge: Characterization of sorption of CO_2 and Krypton" Society of Chemical Industry, London, 1968, 141.

4. Kranich, W.L., Y.H. Ma, L.B. Sand, A.H. Weiss, I. Zwiebel, "Properties of Aluminum-Deficient Large-Port Mordenites", Adv.Chem.Ser., 101, 1971, 502

5. Scherzer, J., "The Preparation and Characterization of Aluminum Deficient Zeolites" in Catalytic Materials, American Chemical Society, 1984, 157

6. Kerr, G.T., "The Intracrystalline Rearrangement of Constitutive Water in Hydrogen Zeolite Y", J.Phys.Chem., 71, 1967, 4155

7. Kerr, G.T., "Chemistry of Crystalline Aluminosilicates. V. Preparation of Aluminum-Deficient Faujasites", J.Phys.Chem.72, 1968, 2594

8. Kerr, G.T., G.F. Shipman, "The Reaction of Hydrogen Zeolite Y with Ammonia at Elevated Temperatures", J.Phys.Chem., 72, 1968, 3071

9. Kerr, G.T., "Chemistry of Crystalline Aluminosilicates. VI. Preparation and Properties of Ultrastable Hydrogen Zeolite Y" J.Phys.Chem.,73, 1969, 2780

10. Kerr, G.T., "Chemistry of Crystalline Aluminosilicates: VII. Thermal Decomposition Products of Ammonium Zeolite Y", J.Catal.,15, 1969, 200

11. Kerr, G.T., J. Cattanach, E.L. Wu, "A Comment on The Nature of Tung and McIninch's "Decationized" Y Zeolite", J.Catal., 13, 1969, 114

12. Kerr, G.T., J.N. Miale, R.J. Mikovsky., U.S.Pat 3,493,519 (1970)

13. Kerr, G.T., "Hydrogen Zeolite Y, Ultrastable Zeolite Y and Aluminum-Deficient Zeolites" Adv.Chem.Ser.121, 1973, 219

14. Beyer, H.K., I. Belenkaya.,"A New Method for Dealumination of Faujasite-Type Zeolites", in Catalysis by Zeolites, B. Imelik, et.al. eds., Elsevier, Amsterdam 1980, 203

15. Skeels, G.W., D.W Breck, "Zeolite Chemistry V: Substitution for Aluminum in Zeolites Via Reaction With Aqueous Ammonium Fluorosilicate" in Proceedings of the Sixth International Zeolite Conference, Eds., D.Olson, A.Bisio, Butterworths, Guildford, UK, 1984, 87

16. Engelhardt, G., U. Lohse, V. Patzelova, M. Magi, E. Lippmaa, "High Resolution 29Si NMR of Dealuminated Y Zeolites. 1. The Dependence of The Extent of Dealumination on The Degree of Ammonium Exchange and The Temperature and Water Vapor of The Thermochemical Treatment", Zeolites, 3, 1983, 233

17. Engelhardt, G., U. Lohse, V. Patzelova, M. Magi, E. Lippmaa, High Resolution 29Si NMR of Dealuminated Y Zeolites. 2. "Silicon, Aluminum Ordering in The Tetrahedral Zeolite Lattice" Zeolites, 3, 1983, 239

18. Beaumont, R., D. Barthomeuf, "X, Y, Aluminum-Deficient and Ultrastable Faujasite-Type Zeolites: II. Acidic and Structural Properties" J.Catal., 26, 1972, 218

19. Beaumont, R., D. Barthomeuf, "X, Y, Aluminum-Deficient and Ultrastable Faujasite-Type Zeolites", J.Catal.,27, 1972, 45

20. Maher, P.K., F.D. Hunter, J. Scherzer, "Crystal Structures of Ultrastable Faujasites", Adv.Chem.Ser., 101, 1971, 266

21. Gallezot, P., R. Beaumont, D. Barthomeuf, "Crystal Structure of a Dealuminated Y-Type Zeolite", J.Phys.Chem., 78, 1974, 1550

22. Breck, D.W. "Zeolite Molecular Sieves" , Wiley, New York, 1974, 501

23. McDaniel, C.V., P.K. Maher, "Zeolite Stability and Ultrastable Zeolites" in Zeolite Chemistry and Catalysis, J.A. Rabo, ed., ACS, Washington, D.C., 1974, 285.

24. Beyer, H.K., Y.M. Belenkaya, F. Hange, M. Tielen, P.J. Grobet, P.A. Jacobs, "Preparation of High-Silica Faujasites by Treatment with Silicon Tetrachloride" J. Chem. Soc. Faraday Trans. 1, 81, 1985, 2889

25. Skeels, G.W., Breck, D.W., Proceedings of the Sixth International Zeolite Conference, D.Olson, A.Bisio, eds., Butterworths, 1984, 87.

26. Garralon, G., V. Fornes, A. Corma, "Faujasites Dealuminated with Smmonium Hexafluorosilicate: Variables Affecting The Method of Preparation" Zeolites, 8, 1988, 268

27. Rabo, J.A., R.J. Pellet, J.S. Magee, B.R. Mitchell, J.W. Moore, W.S. Letzsch, L.L. Upson, J.E. Magnusson, Presented at the NPRA Annual Meeting; Los Angeles; March 23, paper AM-86-27.

28. Kerr, G.T., "Determination of Framework Aluminum Content in Zeolites X, Y and Dealuminated Y Using Unit Cell Size", Zeolites, 9, 1989, 350

29. Breck, D.W., E.M. Flanigen, Molecular Sieves, Soc. of Chem. Ind., London, 1968, 47

30. Dempsey, E. "Acid Site Strength and Aluminum Site Reactivity of Y Zeolites", J.Catal., 33, 1974, 497

194

31. Sohn, J.R., S.J. Decanio, J.H. Lunsford, D.J. O'Donnell, "Determination of Framework Aluminum Content in Dealuminated Y Type Zeolites: A Comparison Based on Unit Cell Size and Wavenumber of I.R. Bands" Zeolites, 6, 1986, 225

32. B.Nagy, J., E.G. Derouane, "NMR Spectroscopy and Zeolite Chemistry" in Perspectives in Molecular Sieve Science, Flank and Whyte, eds., ACS Symp. Series, 1989, 2

33. Hanke, W., K. Moller, "Near-Infrared Study of The Dealumination and Water Desorption from Zeolites", Zeolites, 4, 1984, 244.

34. Zi, G., T. Yi, "Influence of Si/Al Ratio on The Properties of Faujasites Enriched in Silicon", Zeolites, 8, 1988, 232

35. Pine, L.A., P.J. Maher, W.A. Wachter, "Prediction of Cracking Catalyst Behavior by A Zeolite Unit Cell Size Model", J.Catal., 85, 1984, 466

36. Ritter, R.E., J.E. Creighton, T.G. Roberie, J.S. Chin, C.C. Wear, NPRA Annual Meeting Los Angeles, March 23, paper AM 86-45.

37. a) Corma, A., V. Fornes, A. Martinez, A.V. Orchilles, "Parameters in Addition to The Unit Cell That Determine The Cracking Activity And Selectivity of Dealuminated HY Zeolites" in Perspectives in Molecular Sieve Science, Flank and Whyte, eds, ACS Sym.Series, 1988, 542; b) Corma, A., V. Fornes, J. Perez-Pariente, E. Sastre, J.A. Martens, P.A. Jacobs, "Relation Between The Aluminum Content of The Faujasite Framework And The Isomerization And Disproportionation of m-xylene" in Perspectives in Molecular Sieve Science, Flank and Whyte, eds., ACS Sym.Series, 1988, 555

38. Bond, G.R., U.S.Pat 2,617,712 (1949)

39. Mauge, F., A.Auroux, J.C. Courcelle, Ph.Engelhard, P.Gallezot, J.Grosmangin, "Structure And Acidic Properties of High Silica Faujasites," in Catalysis by Acids and Bases, B.Imelik, et.al., eds, Elsevier, Amsterdam, 1985, 91

40. Lynch, J., F. Raatz, P. Dufresne, "Characterization of The Textural Properties of Dealuminated HY Forms", Zeolites, 7, 1987, 333

41. Patzelova, V., N.I. Jaeger, "Texture of Deep Bed Treated Y Zeolites", Zeolites, 7, 1987, 240

42. Lohse, U.,et al., "Dealuminierte Molekularsiebe Vom Type Y Bestimmung Des Mikro-und Sekundarporenvolumens Durch Adsorptions Messungen" Z. Anorg allg. Chem., 460, 1980, 179

43. a) Corma, A., V. Fornes, A .Martinez, F. Melo, O. Pallota, "Influence of The Method of Dealumination of Y Zeolites on Its Behavior for Cracking n-Heptane and Vacuum Gas-Oil" in Innovation in Zeolite Material Science, P.J. Grobet, et.al. eds., Elsevier, Amsterdam, 1988, 495; b) A. Corma, V. Fornes, F.V. Melo, J. Herrero, "Comparison of The

Information Given by Ammonia Tpd and Pyridine Adsorption-Desorption on The Acidity of Dealuminated HY and LaHY Zeolite Cracking Catalysts" Zeolites, 7,1987, 559

44. Fernandez, C., J.C. Vedrine, J. Grosmangin, G. Szabo, "Dealumination of An Offretite-Type Zeolite: Framework Modifications" Zeolites, 6, 1986, 484

45. Fernandez, C., A. Auroux, J.C. Vedrine, J. Grosmangin, G. Szabo, "The Effect of Dealumination on The Structure And Acidic Properties of Offretite" in New Developements in Zeolite Science and Technology, Murakami et al., eds., Elsevier, Amsterdam 1986, 345

46. Klinowski, J., "Solid State Si NMR and High Resolution Electron Microscopic Studies of A Silicate Analog of Faujasite" J. Chem. Soc., Chem.Commun., 1981, 570

47. Klinowski, J., J.M. Thomas, C.A. Fyfe, G.C. Gobbi, "Monitoring of Structural Changes Accompanying Ultrastabilization of Faujasitic Zeolite Catalysts", Nature, 296, 1982, 533

48. Klinowski, J., J.M. Thomas, M.W. Anderson, C.A. Fyfe, G.C. Gobbi, G.C., "Dealumination of Mordenite Using Silicon Tetrachloride Vapor", Zeolites, 3, 1983, 5

49. Zukal, A., V. Patzelova, U. Lohse, "Secondary Porous Structure of Dealuminated Y Zeolite" Zeolites, 6, 1986, 133

50. Akporiaye, D., A.P. Chapple, D.M. Clark, J. Dwyer, I.S. Elliott, D.J. Rawlence, "Faujasites Enriched in Silicon. A Comparison of Processes And Products" in New Developments in Zeolite Science and Technology, Murakami et al., eds., Elsevier, Amsterdam, 1986, 351

51. Freude, D., J. Klinowski, H. Hamdan, "Solid State NMR Studies of The Geometry of Bronsted Acid Sites in Zeolite Catalysts", Chem.Phys. Lett., 149, 1988, 355.

52. Grobet, P.J., P.A. Jacobs, H.K. Beyer, "Study of The Silicon Tetrachloride Dealumination of NaY by A Combination of NMR And IR Methods", Zeolites, 6,1986,47

53. Engelhardt, G., U. Lohse, A. Samsom, M. Magi, M. Tarmak, E. Lippmaa, "High Resolution 29Si NMR of Dealuminated And Ultrastable Y Zeolites", Zeolites, 2, 1982, 59

54. Anderson, M.W., J. Klinowski, "Zeolites Treated with Silicon Tetrachloride Vapor. IV., Acidity", Zeolites, 6, 1986, 455

55. Siegel, H., R. Schollner, B. Staudte, J.J. VanDunn, W.J. Mortier, "X-ray Structural Investigations on Hydrothermally Treated (Ca4,Na4)A Zeolites", Zeolites, 7, 1987, 372

56. Brunner, E., H. Ernst, D. Freude, M. Hunger, C.B. Krause, D. Prager, W. Reschetilowski, Schwieger, K.H. Bergk, "Solid-State NMR And Catalytic Studies of Mildly Hydrothermally Dealuminated HZSM-5" Zeolites, 9, 1989, 282

57. Engelhardt, G., H.G. Jerschkewitz, U. Lohse, P. Sarv, A. Samoson, E. Lippmaa, "500MHz 1H-MAS NMR Studies of Dealuminated HZSM-5 Zeolites" Zeolites, 7, 1987, 289

58. Ray, G.J., B.L. Meyer, C.L. Marshall, "29Si and 27Al NMR Study of Steamed Faujasites--Evidence For Nonframework Tetrahedrally Bound Aluminum" Zeolites, 7, 1987, 307

59. Garralon, G., A. Corma, V. Fornes, "Evidence for The Presence of Superacid Nonframework Hydroxyl Groups in Dealuminated Zeolites" Zeolites, 9, 1989, 84

60. Ray, G.J., A.G. Nerheim, J.A. Donohue, "Characterization of Defects in Dealuminated Faujasites" Zeolties, 8, 1988, 458

61. Feges, P., I. Hannus, I. Kiricsi, H. Pfeifer, D. Freude, W. Oehme, "Thermal Stability of Hydroxyl Groups in Dealuminated Mordenites" Zeolites, 5, 1985, 45

62. Jacobs, P.A., J.A. Martens, H.K. Beyer, "Acid-Catalyzed Conversion of n-decane Over High Silica Faujasites" Catalysis by Acids and Bases, Imelik, et al., eds., Elsevier, Amsterdam, 1985, 399

63. Hays, G.R., W.A. van Erp, N.C.M. Alma, P.A. Couperus, R. Huis, A.E. Wilson, "Solid-State Silicon NMR Studies of The Zeolite Mordenite and Its Dealumination", Zeolites 4, 1984, 377

64. Beyer, H.K., I.M. Belenykaja, I.W. Mishin, G. Borbely, "Structural Pecularities and Stabilization Phenomena of Aluminum Deficient Mordenites" in Structure and Reactivity of Modified Zeolites Jacobs, et al., eds., Elsevier, Amsterdam, 1984, 133

65. a) Lok, B.M., F.P. Gortsema, C.A. Messina, H. Rastelli, T.P.J. Izod, "Zeolite Modification -- Direct Fluorination" Intrazeolite Chemistry, Studky and Dwyer, eds., ACS Symposium Series 218, 1983, 41; b) Lok, B.M., T.P.J. Izod "Molecular Sieve Modification I -- Direct Fluorination" 1982, Zeolites, 2, 1982, 6

66. Shell, Netherlands Appl. 6,626,352 (1968)

67. Voorhies, A., Jr., U.S.Patent 3,630,965 (1971)

68. Takase, S., T. Shioiri, M. Ushio, Ger Pat. 2,219,736 (1972)

69. Szostak, R., A.H. Weiss, Y. Yang, "HF Vapor Treatment to Develop Hydrophobicity in Mordenite", 17th ACS MidAtlantic Regional Meeting April 6-8, 1983.

70. Hervert, G.L., U.S.Patent 3,467,728 (1969)

71. Fishel, N.A., U.S.Pat. 3,413,370 (1968)

72. a) Miradatos, C., D. Barthomeuf, "Superacid Sites in Zeolites", J.Chem.Soc., Chem.Commun., 1981, 39; b) Mirodatos, C., B.H. Ha, K. Otsuka, D. Barthomeuf, "Changes in Mordenite Acidity upon Various Treatments" in

Proceedings of the 5th Internationl Conference on Zeolites, Rees, ed., Heyden, London, 1980, 382

73. Lago, R.M., W.O. Haag, R.J. Mikovski, D.H. Olson, S.D. Hellring, K.D. Schmidt, G.T. Kerr, "New Developments in Zeolite Science and Technology" eds., Murakami, et al., Elsevier, Amsterdam 1986, 677

74. Haag, W.O., R.M. Lago, Eur.Pat. 34444 (1981)

75. Sendoda, Y., Y. Ono, "Effect of The Pretreatment Temperature on The Catalytic Activity of ZSM-5 Zeolites", Zeolites, 8, 1988, 101

76. Szostak, R., R. Kuvadia, T.L. Thomas, "The Role of Dendridal Aluminum in Butane Cracking Activity in ZSM-5" Report, Multi-Client Meeting, March 1988, Atlanta, Ga.

77. Barthomeuf,D. "A General Hypothesis on Zeolites Physiochemical Properties. Applications to Adsorption, Acidity, Catalysis and Electrochemistry". J.Phys.Chem., 83, 1979, 249

78. a) Ashton, A.G., S. Batmanian, D.M. Clark, J. Dwyer, F.R. Fitch, A. Hinchcliffe, F.J.Machado, "Acidity in Zeolites", in Catalysis by Acids and Bases, Imelik, et al., eds., Elsevier, Amsterdam, 1985, 101; b) Ashton, A.G., J. Dwyer, I.S. Elliott, F.R. Fitch, M. Greenwood, G. Qin, J. Speakman, "The Application of Fast Atom Bombardment Mass Spectrometry (FABMS) to The Study of Zeolites" in Proceedings of the 6 International Conference on Zeolites, Olson and Bisio, eds., Butterworths, 1984, 704

79. Ione, K.G., V.G. Stepanov, G.V. Echevskii, A.A. Shubin, E.A. Paukshtis, "Study of Nature of The Factors Determining Activity, Stability and Selectivity of Zeolite Catalysts" Zeolites, 4, 1984, 114

80. Szostak, R., V. Nair, D.K. Simmons, T.L. Thomas, R. Kuvadia, B. Dunson, D.C. Shieh, "Framework And Nonframework Ion Contribution to Molecular Sieve Catalytic Activity" in Innovations in Zeolite Materials Sciences, Grobet et al., eds., Elsevier, Amsterdam, 1988, 403

81. Nair, V., "Characterization of Ferrisilicate Molecular Sieves" PhD Thesis, School of Chemical Engineering, Georgia Institute of Technology, Atlanta, Ga, USA, 1987

82. Csencsits, R., "A Study of The Structure And Composition of An Iron Silicate Catalyst, FeZSM-5, Via Electron Microscopy" PhD Thesis, 1988, Materials and Chemical Sciences Division, Univ. of California, Berkeley, USA

83. Namba, S., A. Inaka, T. Yashima, "Effect of Selective Removal of Aluminum From External Surfaces of HZSM-5 Zeolite on Shape Selectivity" Zeolites, 6, 1986, 107

84. Hidalgo, C.V., M. Kato, T. Hattori, M. Niwa, Y. Murakami, "Modification of Mordenite by Chemical Vapor Deposition of Metal Chloride" Zeolites, 4, 1984, 175

198

85. Klinowski, J., J.M. Thomas, M.W. Anderson, C.A. Fyfe, G.C. Gobbi, "Dealumination of Mordenite Using Silicon Tetrachloride Vapor", Zeolites, 3, 1983, 5

86. Brinen, J.L., "Characterization of Modified Aluminophosphate Molecular Sieves" PhD thesis, School of Chemical Engineering, Georgia Institute of Technology, Atlanta, Ga USA, 1989

87. Theocharis, C.R., M.R. Gelsthorpe, "Modified Aluminophsopahte Microporous Solids", in Characterization of Porous Solids, Unger et al, eds., Elsevier, Amsterdam, 1988, 541

88. Brinen, J.L., M.G. White, R. Szostak, T.L. Thomas, "Development of Framework Catalytic Activity in Aluminophosphate Molecular Sieves" AICHE Meeting, Nov. 16, 1987

89. Flanigen, E.M., R.L. Patton, S.T. Wilson, "Structural, Synthetic and Physiochemical Concepts in Aluminophosphate-Based Molecular Sieves", in Innovations in Zeolite Material Science, Grobet, et al., eds., Elsevier, Amsterdam, 1988,13

90. a) Niwa, M., C.V. Hidalgo, T. Hattori, Y. Murakami, "Germanium Methoxide: New Reagent for Controlling The Pore-opening Size of Zeolite by CVD", in New Developments in Zeolite Science and Technology, Murakami, et.al.,eds., Elsevier, Amsterdam, 1986, 297; b) Niwa, M., K. Yamazaki, Y. Murakami, "Separation of Oxygen And Nitrogen Due to The Controlled Pore-opening Size of Chemical Vapor Deposition (CVD) Modified Zeolite A" Chem.Lett.,3, 1989, 441

91. Niwa, M., H. Itoh, S. Kato, T. Hattori, Y. Murakami, "Modification of H-Mordenite by Vapor-phase Deposition Method" J. Chem. Soc., Chem., Commun., 1982, 819.

92. Niwa, M., S. Kato, T. Hattori, Y. Murakami, "Fine Control of The Pore-opening Size of The Zeolite Mordenite by Chemical Vapor Deposition of Silicon Alkoxide" J.Chem.Soc., Faraday Trans., I, 80, 1984, 3135

93. a) Niwa, M., Y. Kawashima, Y. Murakami, "A Shape-selective Platinum-loaded Mordenite Catalyst for The Hydrocracking of Paraffins by The Chemical Vapor Decomposition of Silicon Alkoxide" J.Chem.Soc., Faraday Trans. I, 1985, 2757; b) Niwa, M., H. Itoh, S. Kato, T. Hattori, Y. Murakami, "Modification of H-Mordenite by A Vapor-phase Deposition Method", J.Chem.Soc., Chem.Commun., 1982, 819

94. Johns, J.R., R.F. Howe, "Preparation of Molybdenum Mordenite from MoCl5", Zeolites, 5, 1985, 251.

95. Vansant, E.F., "Pore Size Engineering in Zeolites", in Innovation in Zeolite Material Science, P.J. Grobet, et.al. eds., Elsevier, Amsterdam, 1988, 143

96. Thijs, A., G. Peeters, E.F. Vansant, I. Verhaert, P. DeBievre, "Encapsulation of Gases in H-Mordenite Modified with Silane and Diborane", J.Chem.Soc.,Far.Trans I, 79, 1983, 2835

97. Thijs, A., G. Peeters, E.F. Vansant, I. Verhaert, P. DeBievre, "Purification of Gases in H-Mordenite Modified with Silane and Diborane", J.Chem.Soc.Faraday Trans., I, 79, 1983, 2821

98. Thijs, A., G. Peeters, E.F. Vansant, I. Verhaert, P. DeBievre, "Modification of H-Mordenite with Silane And Diborane", J.Chem.Soc. Faraday Trans. I, 82, 1986, 963

99. Dessau, T.M., G.T. Kerr, "Aluminum Incorporation into High Silica Zeolites", Zeolites, 4, 1984, 315

100. Chang, C.D., C.T.W. Chu, J.N. Miale, R.F. Bridger, R.B. Calvert, "Aluminum Insertion into High Silica Zeolite Frameworks. 1. Reaction with Aluminum Halides", J.Amer.Chem.Soc., 106, 1984, 8143

101. Anderson, M.W., J. Klinowski, L. Xinsheng, "Alumination of Highly Siliceous Zeolites" J.Chem.Soc.,Chem.Commun., 1984, 1596

102. Yashima, T., K. Yamagishi, S. Nakata, S. Asaoka, "Alumination of ZSM-5 Type Zeolite with AlCl3", in Innovation in Zeolite Material Science, P.J. Grobet, et.al. eds., Elsevier, Amsterdam, 1988, 175

103. Jacobs, P.A., M. Tielen, J.B. Nagy, G. Debras, E.G. Derouane, Z. Gabelica, "Study of The Dealumination And Realumination of ZSM-5 Type Zeolites by 29Si And 27Al High Resolution Magic Angle Spinning NMR Spectroscopy" in Proceedings of the 6th International Conference on Zeolites, Olson and Bisio eds., Butterworths, 1984, 783

104. Derouane, E.G., L. Baltusis, R.M. Dessau, K.D. Schmitt, "Quantitation and Modification of Catalytic Sites in ZSM-5" in Catalysis by Acids and Bases, Imelik et al., eds., Elsevier, Amsterdam, 1985, 135

105. US Pat. 3,640,681(1972).

106. Skeels, G.W., E.M.Flanigen, "Zeolite Chemistry VII. Framework Substitution for Aluminum in Zeolites via Secondary Synthesis Treatment" in Zeolites: Facts, Figures, Future, Jacobs and van Santen, eds., Elsevier, Amsterdam, 1989, 331.

107. Freude, D., T. Frohlich, H. Pfeifer, G. Scheler, "NMR Studies of Aluminum in Zeolites," Zeolites, 1983, 3,171.

Chapter 6

CLAYS: FROM TWO TO THREE DIMENSIONS.

R. A. SCHOONHEYDT

Laboratorium voor Oppervlaktechemie, K.U.Leuven, K. Mercierlaan, 92, B-3030 Leuven, Belgium

SUMMARY
    In the first part the structures and structural chemistry of smectites are described. Charge density, cation exchange capacity, surface area, swelling and pore volumes are defined. The crystal chemistry is discussed with emphasis on the vibrations of the structural hydroxyls and the ^{29}Si MAS-NMR data. The surface chemistry is discussed in relation to the acidic and redox properties of smectitic surfaces.
    In the second part a pillared clay is defined and the different pillaring agents are summarized. The Al-pillared clays are discussed in depth as prototypes of clays with oxixdic pillars.

INTRODUCTION

The first known traces of the use of clays by mankind are 25 000 years old. Clay figures, pottery and ceramics were made by the primitive people in Europe and Asia (ref. 1). Today clay is an important industrial raw material. Hundreds of million tons are worked up in ceramics, in the paper industry, in the iron ore and metal industries, in oil drilling and in pet feeding. Clays are used as adsorbents, decoloration agents, ion exchangers, supports and as catalysts (refs. 2, 3). In agriculture the clay - humus and clay - oxide complexes determine the structure, texture, water retention and the inorganic and organic fertilizer composition of the soils.

Clay minerals are phyllosilicates: layered or two-dimensional silicates. The basic building blocks are the $Si(O,OH)_4$ tetrahedra and the $M(O,OH)_6$ octahedra with $M = Al^{3+}$, Mg^{2+} or $Fe^{2+,3+}$ (see Fig. 1). The condensation of a monolayer of tetrahedra with a layer of octahedra gives the 1:1 or TO minerals, known as kaolinites with $M = Al^{3+}$, and serpentines when $M = Mg^{2+}$. They are also respectively called dioctahedral and trioctahedral 1:1 minerals.

202

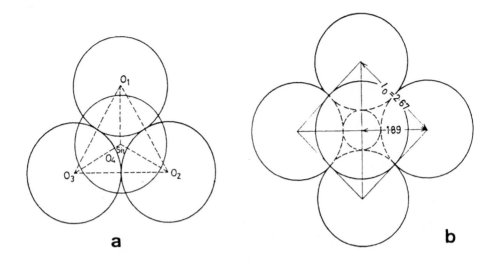

a   b

Fig. 1 (a) The SiO_4 tetrahedron. O_1, O_2 and O_3 are the basal oxygens; O_4 is the apical oxygen. (b) The AlO_6 octahedron. Distances are in Angstroms. (after Nemecz, ref. 7)

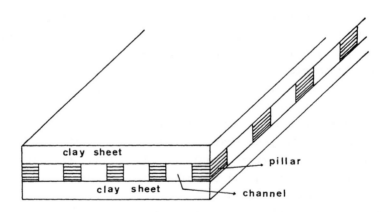

Fig. 2 Schematic drawing of a pillared clay.

When a monolayer of octahedra is sandwiched between two
monolayers of Si-tetrahedra 2:1 type clay minerals are obtained.
They are dioctahedral in the case of Al-octahedra and
trioctahedral for Mg-octahedra. In the former case two out of
three octahedral sites are occupied, in the latter case all three
are occupied. The transformation of the layered 2:1 clay minerals
into three-dimensional structures with molecular sieve properties
is our present subject. The aim is to obtain a rigid two-
dimensional channel network (see Fig. 2). The size of the
channels is preferably larger than that of the zeolitic channels:
.7 - .8 nm for a 12-membered ring of oxygens and 1.2 - 1.3 nm for
18-membered oxygen rings (refs. 4, 5). In that way materials are
obtained which extend the molecular sieving properties of
zeolites and aluminium phosphates towards larger pores, but still
with molecular size.

SMECTITES

Table 1 gives the classification of the 2:1 clay minerals and
Fig. 3 shows the ideal structures. These structures are
electroneutral. This can be easily seen by assigning ionic
charges to the atoms (Si^{4+}, Al^{3+}, O^{2-}, H^{+}, Mg^{2+}). These
electroneutral structures are found in nature as pyrophyllite
(dioctahedral) and talc (trioctahedral).

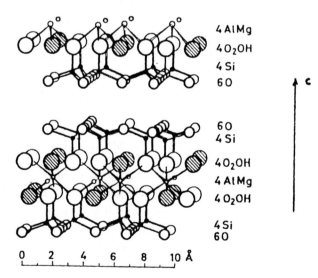

Fig. 3 Structure of a smectite

TABLE 1: Classification of 2:1 clay minerals.

Charge/ formula	group	subgroup	name
0	pyrophyllite	dioctahedral	pyrophyllite
	talc	trioctahedral	talc
0.25 –	smectite	dioctahedral	montmorillonite beiddelite
0.6		trioctahedral	saponite hectorite
0.6 –	vermiculite	dioctahedral	vermiculite
0.9		trioctahedral	vermiculite
1	mica	dioctahedral	muscovite
		trioctahedral	biotite

The other minerals of Table 1 are obtained by isomorphous substitution of Si^{4+} in the tetrahedral layer or Al^{3+} and Mg^{2+} in the octahedral layer. When the incoming cation has a lower valency than the outgoing cation, the lattice is negatively charged and electrical neutrality is maintained by the exchangeable cations. The general formula representation is

$$(Si_{8-x} M_x)^{IV} (Al_{4-y} M'_y)^{VI} (OH)_4 O_{20} M''^{n+}_{(x+y)/n} \quad mH_2O$$

for dioctahedral clays or

$$(Si_{8-x} Mx)^{IV} (Mg_{6-y} M'_y)^{VI} (OH)_4 O_{20} M''^{n+}_{(x+y)/n} \quad mH_2O$$

for trioctahedral clays. The superscripts IV and VI refer to the tetrahedral and octahedral layers respectively. M and M' have one unit of charge less than the cation they replace and M" is the exchangeable cation with valency n.

As shown in Table 1, three groups are distinguished depending on the degree of isomorphous substitution. For smectites $x + y = 0.25 - 0.6$, corresponding to a cation exchange capacity of 0.64 –

1.50 meq/g. For micas the sum x + y is at least one and vermiculites take an intermediate position.

Isomorphous substitution is mainly localized in the octahedral layer. One obtains the montmorillonites and the hectorites for dioctahedral and trioctahedral smectites respectively. If, however, isomorphous substitution is localized in the tetrahedral layer the minerals are beiddelite (dioctahedral) and saponite (trioctahedral). If the octahedral layer is composed of Fe^{3+} instead of Al^{3+} one has nontronites.

PROPERTIES OF SMECTITES.

Natural materials like smectites invariably contain impurities. The most common crystalline impurities are quartz, micas, calcite and feldspars. They can be recognized by X-ray diffraction (XRD) and removed by fractionation. When a clay material is centrifuged so as to keep the < 2 μm fraction in suspension, the precipitate mainly contains the impurities, while the clay remains in suspension. Amorphous oxidic impurities of Fe and Al are often adsorbed on the clay particles and difficult to remove. Treatment procedures with dithionite have been developed to remove them; however, not without problems (ref. 6). In any case, in the following pure smectite clay minerals are supposed and their most important properties are described in general terms. A non-exhaustive list of books with more detailed information is given in the list of references. (refs. 7-12).

1. XRD patterns and unit cell parameters.

In layer silicates the average Si-O distance is 0.162 nm and the average Al-O distance is 0.177 nm. The O-O distance in the ideal SiO_4 tetrahedron is 0.267 nm. These tetrahedra are joined together in a two-dimensional array by sharing three oxygens (Fig. 4). The basal oxygens are coplanar with the layer plane and common to two adjacent tetrahedra. The apical oxygen is free. The chemical formula of the layer is $(Si_2O_5)^{2-}{}_n$. The Si atoms at the centre of gravity of each tetrahedron form a regular hexagonal array with a = 0.530 nm and b = 0.916 nm.

The O-O distance in the ideal AlO_6 octahedron is 0.267 nm. The two-dimensional condensation of these octahedra gives a gibbsite sheet or dioctahedral layer with a = 0.462 nm and b = 0.801 nm (Fig. 4). The slight differences between the a and b dimensions of the Si- and Al-sheets indicate that (1) condensation of these sheets to form clay structures is possible; (2) the sheets are

206

distorted in the actual structures to match their dimensions. These distortions are usually represented by a rotation of the Si-tetrahedra such that the Si-O-Si angle lies between 180° (hexagonal symmetry) and 120° (ditrigonal symmetry) (Fig. 4).

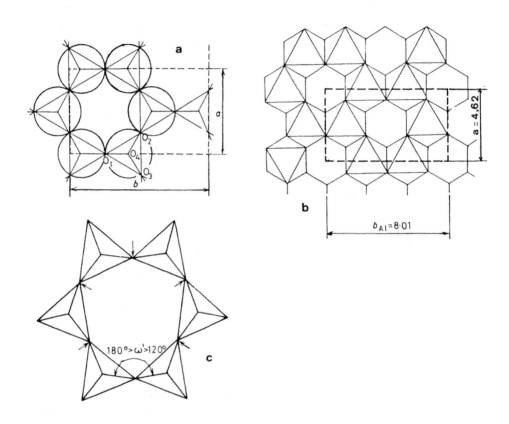

Fig. 4 (a) The $Si_2O_5)_n^{2-}$ sheet; (b) the dioctahedral Al sheet; (c) the tetrahedral array of ditrigonal symmetry. (after Nemecz, ref. 7)

The extent of the distortion depends on the isomorphic substitution (degree, type of ion in tetrahedral or in octahedral position). As a result, the values of the a and b parameters deviate from the ideal values and the dependence of the b parameter on the chemical composition as been expressed as (ref. 13):

$$b = (0.8944 + 0.0096 \, Mg + 0.0096 \, Fe^{3+} + 0.0037 \, Al^{VI}) + 0.0012 \, nm$$

$$(1)$$

Mg, Fe^{3+} and Al^{VI} are the number of Mg, Fe^{3+} and octahedral Al per half unit cell. The c parameter is variable and depends on the size of the exchangeable cation and the amount of water in the interlamellar space. The distance between two sheets is usually taken as 0.96 nm. It increases to 1.25 nm, to 1.50 - 1.55 and to 1.80 - 1.90 nm wit respectively a monolayer, a bilayer and a triple layer of water molecules in the interlamellar space (Fig. 5).

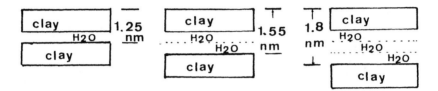

Fig. 5 One-, two- and three-layers hydrate of smectites

A typical XRD pattern of an air-dry smectite is shown in Figure 6 together with the hkl indices. The powder patterns are composed of two sets of lines: 001 lines or basal reflections, which are sensitive to the state of the minerals (hydration, exchangeable cation....); and hk0 + hkl reflections, which are diagnostic of the mineral.

Thus, for dioctahedral smectites the 060 line is at 0.149 - 0.152 nm and for trioctahedral smectites at 0.153 - 0.154 nm. More detailed identification is based on chemical composition and the Hoffmann-Klemen effect. The unit cell is monoclinic and ideal unit cell formulae are:

Montmorillonite:

$(Si_8)(Al_{3.15}Mg_{0.85})O_{20}(OH)_4X_{0.85}nH_2O$

Beiddelite:

$(Si_{7.15}Al_{0.85})(Al_4)O_{20}(OH)_4X_{0.85}nH_2O$

Saponite:

$(Si_{8-y}Al_y)(Mg_{6-z}R^{3+}_z)O_{20}(OH)_4X_{y-z}nH_2O$

Hectorite:

$(Si_8)(Mg_{6-z}Li_z)O_{20}(OH)_4X_znH_2O$

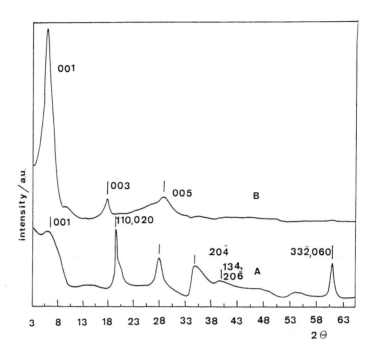

Fig. 6 X-ray diffraction pattern of a synthetic hectorite

Thus, beiddelite usually does not contain Mg, while hectorite has almost no Al. The Hoffmann-Klemen effect is the irreversible migration of Li^+ upon heating from ion exchange positions into the octahedral layers via the ditrigonal holes of the tetrahedral layers. There are two conditions to be fulfilled: the octahedral layer must have vacant sites and must carry a negative charge. Only montmorillonites fullfil these conditions. The result is a decrease in the cation exchange capacity (CEC) and the absence of swelling in water or in polyalcohols. The Hoffmann-Klemen effect is not restricted to Li^+ but to any exchangeable cation small enough to migrate through the ditrigonal holes. Examples are transition metal ions such as Cu^{2+} and Ni^{2+} (ref. 14).

<u>Charge density and cation exchange capacity</u>

The structural formulae give the amount of negative charges per unit cell generated by isomorphous substitution. They are compensated by an equal number of positive charges of the exchangeable cations. This is the so-called pH-independent part of the CEC. At the edges of the crystals the sheet structure is broken and the crystal is terminated by OH-groups. At high pH they deprotonate, the edges become negatively charged and have a cation exchange capacity. At low pH they adsorb protons and become positively charged. As a consequence there is a pH-dependent part in the experimentally determined CEC.

Maes et al. (ref. 15) proposed the following equations for the pH dependence of the CEC in the pH range 3.9 - 5.8 for clays saturated with monovalent and divalent cations respectively:

$$CEC \text{ (meq/100g)} = 79.9 + 5.04 \text{ pH} \qquad (2)$$

$$CEC \text{ (meq/100g)} = 96.1 + 3.93 \text{ pH} \qquad (3)$$

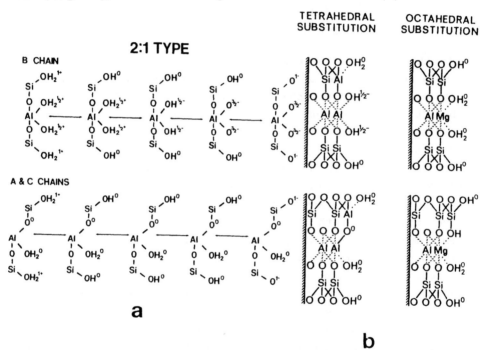

Fig. 7 (a) Effect of pH on the edge charges of smectite layers without isomorphous substitution. The pH increases from left to right. (b) Effect of octahedral and tetrahedral substitution on the edge charges at pH = 6.5.
(after White and Zelazny, ref. 16)

In Figure 7 the crystal terminating groups of dioctahedral smectites without and with isomorphous substitution are shown. For a pH of 6.5 White and Zelazny calculated a negative charge density of 5.32 mmole/m^2 (ref. 16). If we assume clay particles in the form of a square of 2 mm and 1 nm thick and a density of 2.3g/cm^3, this corresponds to 4.63 mmole/g or 0.46% of the CEC. For particles of 100 nm one obtains 92.6 mmole/g or 9.26% of the CEC.

Thus, the contribution of the edge sites to the total CEC is not negligible and will depend on the experimental conditions: crystal size and shape, type of exchangeable cation and pH.

The determination of the charge density σ, defined as CEC/S with S the surface area in m^2/g and CEC in meq/g or, better, in C/g (1 meq = 96.4 C) is subjected to the same problems as the determination of the CEC. The determination of the surface area presents particular problems (see next section) and it is better to calculate σ from the structural formula:

$$CEC \ (C/m^2) = \frac{CEC \ (meq/g) \ e \ FW}{1000 \ 2 \ a \ b} \qquad (4)$$

e is the elementary charge equal to $1.6 * 10-19$ C; FW is the formula weight of the unit cell and 2ab is the surface area of a unit cell in m^2. If the pH-dependent part of the CEC is known and subtracted, one obtains the charge density, due to isomorphous substitution.

Lagaly and Weiss (ref. 17) pioneered the alkylammonium method to determine this layer charge density. The method consists in the ion exchange of alkylammonium ions with varying chain lengths. When the average distance between the negative lattice charges is larger than the length of the alkylammonium molecules, the latter form a monolayer with $d_{001} = 1.36$ nm. When the average distance between the layer charges is smaller than the length of the molecules, a bilayer is adsorbed with $d_{001} = 1.77$ nm. The transition between both situations occurs at a critical chain length, which depends on the charge density. If all the particles of the clay have the same charge density, the transition is sharp. In reality, this is not the case (refs. 18-20). The charge density varies from one particle to another and the experimentally determined value is an average one. There are problems involved with the application of the alkylammonium method, mainly related to the estimate of the amount of chains protruding out of the clay sheets. This has made it difficult to establish relationships between charge density and particle size.

<u>Surface area, swelling and sorption</u>

The surface area S of a single sheet of a smectite can be calculated exactly if its size is known. Figure 8 illustrates that for 2 μm particles it is 800 m^2/g and for 50 nm particles 832 m^2/g. These numbers represent the sum of the planar surface areas or the interlamellar surface area and the surface area of the edges. The difference is entirely due to the difference in edge surface. These surface areas are never measured experimentally. There are several reasons for the discrepancy between theory and practice.

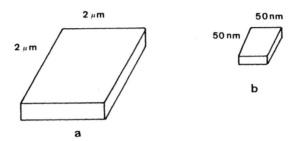

Fig. 8 Total surface areas of individual, regularly shaped clay particles: (a) 8.008 x 10^{-12} m^2/particle; 2.5 x 10^{14} particles/cm^3. (b) 5.2 x 10^{-15} m^2/particle; 4 x 10^{17} particles/cm^3.

In the first place, smectites are characterized by a distribution of particle sizes. Secondly, they donot occur as elementary sheets, but the sheets are stacked together into aggregates. Ideally, three types of stacking can be distinguished: face-to-face, edge-to-edge and face-to-edge. An air-dry powder is a collection of such aggregates of different sizes and shapes, organized into a microfabric. The result is a powder with micropores and mesopores. The surface area and the pore size distribution can only be investigated experimentally e.g. by analysis of the N$_2$ adsorption and desorption isotherms. Typical results are summarized in Table 2.

Table 2: Surface areas (m^2/g) and pore volumes (μl/g) ⠀⠀⠀⠀⠀of montmorillonites (from ref. 21)

smectite	cation	drying	S_{BET}	S_{mi}	S_{me}	V_t	V_{mi}	V_m
MO	Na$^+$	Lyo	124	69	54	116	40	76
		Sed	115	65	53	120	43	77
	Cs$^+$	Lyo	157	107	45	120	49	71
		Sed	151	95	47	118	59	59
	Ca^{2+}	Lyo	104	47	48	81	26	55
		Sed	107	76	27	86	39	47
OT	Na$^+$	Lyo	144	82	57	142	54	88
WB			44	32	10	37	18	19

MO = Moosburg montmorillonite, OT = Otay montmorillonite and WB = Wyoming bentonite; Lyo = freeze dried; Sed = dried by sedimentation. S_{BET}, S_{mi} and S_{me} are respectively the total surface area, the microporous surface area and the mesoporous surface area. V_t, V_{mi} and V_m are respectively the total pore volume, the microporous volume and the mesoporous volume.

The data of Table 2 are obtained on the size fraction < 0.5 μm. When electron micrographs are obtained on the same size fraction the average sizes of Table 3 are obtained.

Table 3: Average particle sizes (nm) of the < 0.5 μm fraction of Na$^+$-saturated smectites.

Otay	20.0 x 20.0
Moosburg	73.3 x 73.3
Camp Berteau	103.2 x 76.4
Wyoming bentonite	129.0 x 96.6
Greek white	213.9 x 160.0

The comparison of the data of both tables shows that clays with larger particles tend to have smaller surface areas. This is expected when only the "external" surface area of the aggregates is measured i.e. N$_2$ molecules do not penetrate in the

interlamellar space between the individual clay sheets of the
aggregates. It is also shown that the measured surface areas are
modulated by the drying procedure and, especially, the type of
exchangeable cation. This indicates that the number of elementary
clay sheets per aggregate and the agglomeration of the aggregates
in a powder are cation and drying procedure dependent.

The measurement of the total surface area (interlamellar surface
area included) requires a probe molecule of known size, which
penetrates the interlamellar space to form either monolayers or
bilayers. Polar molecules are preferred because of strong ion-
dipole interactions with the exchangeable cations.
Ethyleneglycol, glycerol, ethyleneglycol monoethylether (EGME),
methylene blue, cetyl pyridinium bromide (PB) and water have been
proposed (ref. 22). The main problem with these measurements is
that the packing density of these molecules is dependent on the
charge density of the clays. Consequently, the surface area
covered by one molecule varies from clay to clay.

In liquid water smectites swell. Water is adsorbed in the
interlamellar space, the distance between the elementary sheets
in the aggregate increases to such an extent that the aggregate
desintegrates into elementary sheets. Ultimately, the latter are
randomly distributed in the suspension, which is transformed into
a gel. Swelling is influenced by the charge density of the clay,
the type of exchangeable cation and the particle size. In general
the size of the aggregates in aqueous suspension increases with
the size of the exchangeable cation ($Li^+ < Na^+ < K^+ < Rb^+ < Cs^+$),
with its charge ($Ca^{2+} > Na^+$) and, for one type of exchangeable
cation, with the charge density of the clay.

Clay dispersions are not in thermodynamic equilibrium. They tend
to minimize their interfacial energy by aggregation. The
dispersion ages. This process is also called flocculation or
coagulation. This is a slow process because - mainly - two
opposing forces cooperate: electrical double layer repulsion and
London-Vander Waals attraction. They result in a secondary energy
minimum, when the mutual interaction energy of two clay sheets is
plotted versus distance (Fig. 9). The primary energy minimum is
that of the two collapsed clay sheets i.e. the distance between
the sheets is one or two layers of water molecules. This can only
be achieved by surmounting the energy barrier between the primary
and secondary energy minimum. This theory, called the Derjaguin,
Landau, Verwey, Overbeek (DLVO) theory, is now recognized as too

214

simplistic (refs. 23, 24) but is sufficient for a qualitative
discussion such as the present one.

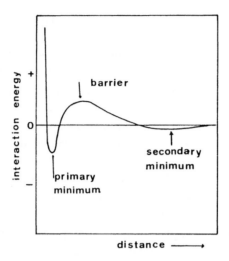

Fig. 9 The dependence of the interaction energy of two clay
sheets in aqueous suspension on the distance between the sheets
Because of its obvious industrial and economic importance the
clay-water system has attracted considerable scientific
attention. When the properties of the water phase in the double
layer next to the surface are measured spectroscopically, no
significant difference with the properties of bulk water is
observed, except for the 1 - 3 water layers, which are closest to
the surface (ref. 25). Moreover, all the measurements indicate
that something like a single clay sheet does not exist in aqueous
suspension in appreciable amount. Usually one finds aggregates.
Fripiat et al. (ref. 26) found that the average size of the
tactoids was independent of the concentration. It looks then as
if the clay particles in suspension are the same as those of the
air-dry powder. What changes is the distance between the
elementary sheets in the aggregate, due to the adsorption of
water. The clay particles swells, but do not desintegrate
completely. Somehow they "remember" their original composition.

Crystal chemistry

Our knowledge of the crystal chemistry of smectites is largely
based on results of infrared spectroscopy (IR) and magic angle
spinning nuclear magnetic resonance (MAS NMR). The stretching

vibration of the structural OH groups of the dioctahedral smectites occurs around 3630 cm^{-1}, and for trioctahedral smectites around 3680 cm^{-1} (ref. 27). These numbers vary somewhat depending on the type of exchangeable cation and the water content. This is illustrated in Table 4 for saponite. The reason is that the hydroxyls vibrate in the ditrigonal hole of the tetrahedral layer. Cations and water molecules can penetrate more or less into these holes and interact with these hydroxyls.

Table 4: IR frequencies of structural hydroxyls of saponite (from ref. 28)

cm^{-1}	assignment
3677	empty ditrigonal hole
3670	Li^+
3687	Li^+ + one monolayer of water
3682	Li^+ + two layers of water
3712	Li^+ migrated into ditrigonal hole
3719	K^+

The bending modes are much more sensitive to the chemical composition of the octahedral layer than the stretching vibrations (refs. 27 - 29). Typical bending frequencies are given in Table 5, where O represents a vacant octahedral site.

Table 5: Bending vibrational frequencies of structural hydroxyls.

cm^{-1}	assignment
920	(Al,Al,O)-OH
885	(Al,Fe,O)-OH
850	(Al,Mg,O)-OH
660	(Mg,Mg,Mg)-OH

The removal of these structural hydroxyls is an endothermic reaction, which occurs in the range 500 - 700 °C for montmorillonites and 700 - 900°C for hectorite (ref. 30). Lattice breakdown occurs after dehydroxylation and new dense crystalline phases appear above 900°C.

In any case, dehydroxylation affects the coordination of the cations in the octahedral and tetrahedral layers and these coordinations are conveniently studied by MAS-NMR. The SiO_4 tetrahedra share three oxygens of the basal plane and form the so-called Q3 structures. The fourth oxygen is shared with the octahedral layer. The ^{29}Si resonances are therefore determined in the first place by the chemical composition of the tetrahedral layers i.e. the isomorphous substitution of Si by Al, giving Q3(0Al), Q3(1Al), Q3(2Al) and Q3(3Al) with between brackets the number of nearest neighbouring Al in the tetrahedral layer. The typical resonance postions are given in Table 6 (refs. 32-37).

Table 6: ^{29}Si MAS NMR chemical shifts for smectites

clay	Q3(0Al)	Q3(1Al)	Q3(2Al)	Q3(3Al)
dioctahedral				
Wyoming	-93.5			
Cheto	-94.1			
Otay	-93.4			
Polkville	-93.3			
Texas	-93.6			
Beiddellite	-92.6			
trioctahedral				
hectorite	-94.2			
laponite	-94.4			
saponite	-95.6	-90.7	-84.8	

As for zeolites, the ^{29}Si chemical shift increases (= deshielding) with increasing amounts of neighbouring Al^{3+}. The distribution of Al^{3+} in the tetrahedral layer is based on Loewenstein's rule and on the principle of the maximum dispersion of charge i.e. Al^{3+} is homogeneously distributed over the tetrahedral layer (refs. 37, 38). The chemical shift of the trioctahedral smectites is typically 2 - 3 ppm more negative than that of the dioctahedral ones. This is due to the difference in electronegativity of the cations in the octahedral layer. Indeed, in trioctahedral smectites the apical oxygen of the Si tetrahedron is connected to 3 Mg^{2+}, while it is bonded to 2 Al^{3+} and one vacancy in the case of dioctahedral smectites. In the first case the total electronegativity of the cations is 3 x 1.31 = 3.93 and 2 x 1.61 = 3.32 in the second case. It follows also that the atom-for-atom substitution of an element with lower electronegativity in the octahedral layer results in deshielding and a less negative ^{29}Si chemical shift (ref. 36). Layer charge arises from the isomorphous substitution of one element by a less electronegative element either in the tetrahedral or in the octahedral layer. The effect is deshielding at the Si position and a less negative chemical shift. Figure 10 shows that a linear relation exists between the ^{29}Si chemical shift and the total layer charge.

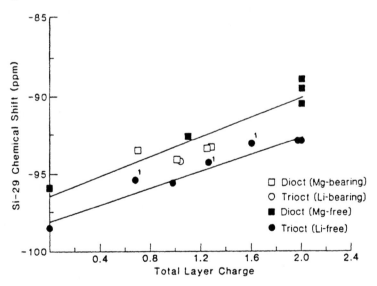

Fig. 10 Relation between the ^{29}Si chemical shift and the total layer charge of clays (after ref. 36)

Deshielding also means less electron density at the Si nucleus or weaker Si-O bonds. It is therefore no surprise to see a linear correlation between the Si(2p) binding energies and the ^{29}Si chemical shift, as shown in Figure 11 for smectites and micas (ref. 39).

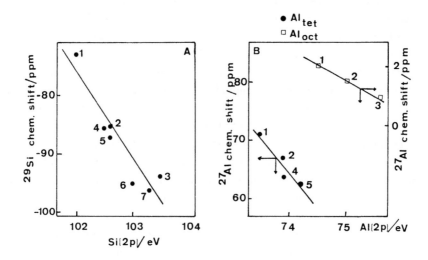

Fig. 11 Relation between the ^{29}Si and ^{27}Al chemical shifts and the binding energies of Si and Al respectively (after ref. 39)

The spectra of the quadrupolar nucleus ^{27}Al are more difficult to obtain, because of the combined effect of chemical shift anisotropy, bulk magnetic susceptibility and second order quadrupolar interaction (ref. 34). It is advisable to obtain spectra at the highest magnetic field possible. The effect of the magnetic field is shown in Figure 12. At low field the spectra are dominated by second order quadrupolar broadening and at high field by magnetic susceptibility anisotropy (due to paramagnetic ions in the lattice or present as impurities). One remarks the considerable line sharpening at high fields albeit with the production of more spinning side bands. In any case a more exact position of the lines of octahedral and tetrahedral Al^{3+} follows at respectively 4 ppm and 70 - 75 ppm.

When the smectites are dehydroxylated, octahedral Al disappears from the spectrum. Presumably Al acquires a distorted fivefold coordination. The resonance line broadens and is lost in the background (ref. 30).

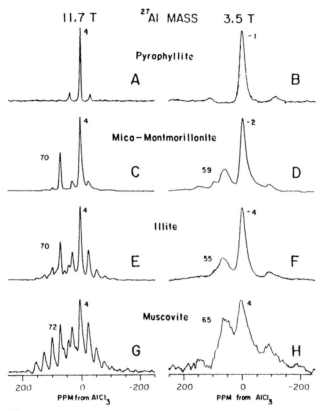

Fig. 12 ^{27}Al MAS-NMR spectra of clays (after ref. 34)

Surface chemistry

Because of their small particle size and their occurrence as aggregates smectites have an external surface area, which is not negligible in surface characterization studies. In the case of dioctahedral clays Bronsted and Lewis acid sites are present on the edges of the crystals. In Figure 7 the Bronsted acid sites are shown. A Lewis site is an exposed or three-coordinated Al^{3+}, substituting for Si^{4+} in the tetrahedral layer. The acid strength of the Bronsted sites is determined either by Hammett indicators (H_0) in an aprotic solvent or by the butylamine titration method of Benesi (ref. 40). Typical data are summarized in Table   7.

Table 7: Acid strength and acidity of smectites (from ref. 40).

smectite	H_0
Na^+-montmorillonite	+ 1.5 - - 3.0
NH_4^+-montmorillonite	+ 1.5 - - 3.0
H^+-montmorillonite	- 5.6 - - 8.2
acid activated clay	< - 8.2

clay	surface area (m^2/g)	butylamine titre (mmol/g)		
		+3.3 - +1.5	+1.5 - -3.0	<-3.0
Na^+-M	41	0.03	0.01	0.00
H^+-M	52	0.10	0.00	0.55

M = montmorillonite

The 40 mmol/g of acid sites on Na^+-M (Table 7) are not directly comparable to the theoretical number of acid sites on the edges because the particle size and the particle size distribution of Na^+-M are unknown. Also, the number of acid sites on H^+-montmorillonite, titrated with n-butylamine, is smaller than the CEC (0.98 meq/g). This means that in benzene, the solvent used for the titration, the interlamellar space is not totally available for reaction. This is a general problem for all the titration methods and not only for smectite clays. Another possibility is that the lattice of the protonated montmorillonite is partially destroyed.

The acidity of smectites is strongly influenced by the residual water content and by the type of exchangeable cation. In a pioneering study Mortland et al. (refs. 41,42) showed that the residual water in the coordination sphere of the exchangeable cations in the interlamellar space is acidic enough to protonate NH_3 with formation of NH_4^+. The higher the polarizing power of the cation the larger the amount of NH_4^+. The reaction can easily be followed by IR spectroscopy. Quantitative data are shown in Table 8.

Table 8: Amount of NH_4^+ (mmol/g) in montmorillonite

Cation	low water	high water
Li	0.23	0.17
Na	0.16	0.10
K	0.10	0.11
Ca	0.80	0.16
Mg	1.01	0.74
Al	1.01	1.00

Smectites are also electron acceptors. The electron-accepting or oxidizing sites are located at the edges and in the structure. The edge sites are supposed to be trigonal Al^{3+}, acting as Lewis acids. Fe^{3+} in the lattice sites are the structural oxidizing sites (ref.42). A nice test of the presence of these sites is the benzidine colouration test. When benzidine is adsorbed on the smectite, an electron is abstracted by the clay (Figure 13) and the blue monovalent benzidine cation is formed. The intensity of the colour is dependent on the amount of oxidizing sites. When the clay surface is acidic, the blue monovalent cation is transformed to the yellow divalent radical cation.

Redox properties may also be induced by the exchangeable cations, such as Cu^{2+}, Ag^+, Fe^{3+} and Ru^{3+} (refs. 43-45). Arene molecules, adsorbed in the coordination sphere of these cations in the interlamellar space undergo radicalar polymerization at low water contents.

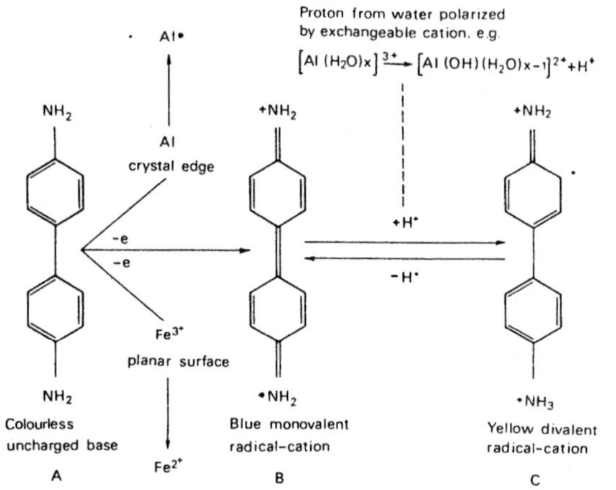

Fig. 13   The colour reaction of benzidine on clays (after B. Theng, ref. 9)

PILLARED CLAYS

Pillared clays (PILC's) are materials with permanent porosity, obtained by propping the elementary clay sheets apart with a chemical substance, called molecular prop. A cross-linked clay or CLC, in particular a cross-linked smectite or CLS, is a PILC with chemical bonds between the prop and the surface oxygens of the clay. A CLC or CLS is thus a special case of PILC.

The goal of the pillaring process is to introduce microporosity in the system, the dimensions of the micropores being complementary to those of zeolites (> .70 nm and < 2.0 nm). This is easiest to achieve by a combination of a smectite with a relatively low negative charge density and a molecular prop with a large positive charge. Then, under the condition that the props are homogeneously distributed over the surface a two-dimensional channel system is created (Fig. 2). Other requirements are that the clay sheet is rigid i.e. does not bend; that the adsorption of props on the external surface is negligible and that all the clay sheets are pillared (Fig. 14).

Pillaring agents

In principle, every molecule or particle, able to penetrate into the interlamellar space is a molecular prop. However, to create microporosity, which is complementary to that of zeolites, molecules or particles are needed with a size of at least 0.7 nm.

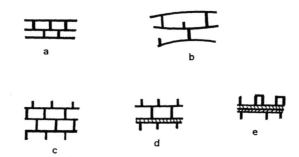

Fig. 14 Schematic drawings of possible PILC's: (a) ideal PILC; (b) PILC with bent sheets; (c) ideal PILC with pillars on the external surface; (d) incompletely pillared clay; (e) unpillared clay with microporosity created by the pillars on the external surface

1. Organic pillars. Alkylammonium ions (mono-alkyl to tetra-alkyl) with varying alkyl chain length and bicyclic amine cations were the first molecular props studied. The use of such "organoclays" as sorbents as been reviewed recently by the R.M. Barrer, the pioneer of the field (ref. 46, 47). These organoclays have an interlamellar volume, which depends on the charge density of the clay and the chain length of the molecules. They adsorb organic solvents such as alkanols to the extent that swelling occurs. This is the basis for their industrial use as gelling agents, thickeners and fillers. The adsorption of mono-alkylammonium ions of varying chain length can be used to estimate the charge density of a particular smectite (refs. 17, 48). The organoclays obviously lack, among other things, thermal stability to be used successfully in gas-phase catalysis.

2. Metal trischelates. $M(2,2'\text{-bipyridine})_3^{n+}$ and $M(o\text{-phenanthroline})_3^{n+}$ (M = transition metal ion) are extremely selectively adsorbed on smectite surfaces to the extent that the ion exchange is followed by an intersalation i.e. the adsorption of the cation and its associated anion in the interlamellar space (ref. 49). The size of the complexes is about 1.0 nm. In principle they are molecular props which can induce microporosity. On one hand these complexes lack thermal stability; on the other hand, they are rather unreactive. Nevertheless, $Ru(bipy)_3^{2+}$ and derivatives have been used

successfully on laboratory scale in photochemical catalysis, such as the dissociation of water into its elements in the presence of clays and co-catalysts (ref. 50).

3. Polyoxocations. Polyoxocations are by far the most intensively studied props, both in the open literature and in the patent literature. The cations can be prepared in solution or in situ i.e. in the interlamellar space of the clay. The first method is the preferred one, because it allows a better control of the type of polyoxocation. Al, Zr, Fe, Cr, Ti, Be oxocations have been exchanged on smectites. A special issue of CATALYSIS TODAY is devoted to the preparation and characterization of pillared clays (ref. 51). Clays with multimetallic pillars have been prepared too. There are three types: those with mixed metal oxocations; those with two (or more) different monometallic pillars and clays with monometallic pillars on which a second cation is grafted (refs. 52 - 56).
Al-PILC's are the most intensively studied. They will be used as model systems to describe the pillaring process and the properties of the pillared clays in subsequent paragraphs.

4. Organometallic pillars. This involves the adsorption of organometal complex precursors and their controlled hydrolysis in the interlamellar space. An example is $Si(acac)_3$ (acac = acetylacetonate), but also polynuclear organometallics can be envisaged (refs. 57, 58). Intercalated organometal complexes may have interesting catalytic properties as such. Thus, cationic Rh phosphine complexes, intercalated in the interlamellar space, are hydrogenation and hydroformylation catalysts. Although their activity is not always comparable to that of the homogeneous catalysts, they show some unusual selectivity, depending on the residual water content of the clay (acidity), the spacing between the sheets and the swelling in a particular solvent (ref. 40).

5. Metal cluster cations. Complexes such as $(Mo_8Cl_8)^{4+}$, $(Nb_6Cl_{12})^{n+}$ and $(Ta_6Cl_{12})^{n+}$ are ion exchanged on the clay. They contain an additional amount of water molecules which get partially hydrolyzed on the surface. Upon heating an oxidation reaction converts the Cl^--complexes to oxides, which function as pillars (ref. 59). The reactions can be written as:

$$Nb_6Cl_{12}(H_2O)_6{}^{n+} + H_2O \longrightarrow Nb_6Cl_{12}(OH)(H2O)_5 + H_3O^+$$

$$Nb_6Cl_{12}(OH)(H_2O)_5 + 9 H_2O \longrightarrow 3Nb_2O_3 + H^+ + 12HCl + 8H_2$$

6. Metal oxides. This involves the direct intercalation of oxide sols of, for instance, SiO_2, Al_2O_3 and silica-aluminas. The intercalation of the tubular imogolite, $Si_2Al_4O_6(OH)_8$ has been successfully achieved (ref. 60). The point is to create stable nanometer or subnanometer particles with a positive charge on their surface and to exchange them on the clay surface.

Al pillared clays.

The pillaring with polyoxocations is essentially an ion exchange reaction. The micropore volume will be maximized by a combination of a polyoxocation with a high charge and a smectite with a low charge density. To fix the ideas consider a clay with a negative charge density of 1 meq/g and an interlamellar surface of 700 m^2/g. Then the average surface area per charge is 1.25 nm^2. When this clay is pillared with a cation of charge +7 and a size of 1 nm the average pore size is 1 x 1.09 nm and the microporous volume is 0.264 cm^3/g. This is an ideal case and the number must be considered as an uper limit.

1. Solution and pillaring chemistry of Al. The formation of polyoxocations in solution requires a controlled hydrolysis of an Al-salt, usually $AlCl_3$. An aqueous solution of $AlCl_3$ is acidic. Water molecules are dissociated due to the highly polarizing power of the Al^{3+} cation:

$$Al(H2O)_6{}^{3+} + H_2O \longrightarrow Al(OH)(H2O)_5{}^{2+} + H_3O^+$$

The controlled neutralization of this solution with NaOH gives an whole range of polynuclear complexes and finally colloidal $Al(OH)_3$ whih settles out of the solution. The solution chemistry is usually described in terms of three species: monomeric Al (Al_{mono}) ; a low nuclearity polyoxocation, which is now believed to be mainly $[Al_{13}O_4(OH)_{24}(H_2O)_{12}]7+$, the Keggin ion, and polymeric Al (Al_{poly}).

The structure of the Keggin ion is shown in Fig. 15. It consists of a central Al^{3+}, tetrahedrally coordinated to 4 oxygens. Each of these oxygens is connected to three edge-sharing Al-octahedra, $AlO(OH)_4(H_2O)$. This complex is easily detected by ^{27}Al NMR by its characteristic tetrahedral Al peak at 62.5 ppm. It reaches a maximum concentration in solution for [OH]/[Al] in the range 1.8 to 2.4. The exact value depends on the Al concentration, the

226

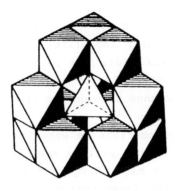

$$[Al_{13}\,O_4\,(OH)_{24}\,(H_2O)_{12}]^{+7}$$

Fig. 15 The Al Keggin ion

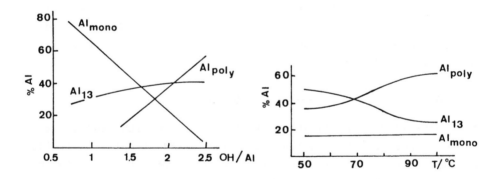

Fig. 16 Change of relative amounts of Al^{3+}, Al_{13} and polynuclear Al as a function of [OH]/[Al] at 0.1 M Al (left) and as a function of the temperature for [OH]/[Al] = 2.15 and [Al] = 1.3 M (right) (after J. Sterte, ref. 61)

temperature, the age of the solution and the source of Al. Typical data are shown in Fig. 16 (ref. 61).

This Keggin ion is believed to be the main molecular prop in the preparation of Al-PILC's (ref. 62). Indeed, it has been firmly established that the pillaring is an ion exchange reaction. The pillared clay has a typical d_{001} spacing of 1.8 - 1.9 nm, indicating that the height of the pillar is at least 0.85 - 0.95 nm, exactly what is expected for the Al_{13}^{7+} cluster. Finally the ^{27}Al NMR line at 62.5 ppm is found on the pillared products.

The average charge of Al in the Keggin ion is 0.54 e. The amount of Al, needed to neutralize the clay is then approximately twice the CEC. This is however not enough to obtain a well pillared clay. Plee et al. (ref. 63) advise a minimum of 20 meq Al/g clay for [OH]/[Al] ratios between 1.2 and 2.0. Stacey (ref. 64) noted that the amount of polymeric Al on the clay surface is simply proportional to the amount of Al in solution. This suggests that the ion exchange reaction from a partially hydrolyzed Al solution is more complex than the exchange of the Keggin ion alone. Several types of polyoxocations of Al may be exchanged and/or the Al_{13}^{7+} may be partially hydrolyzed on the clay surface.

Pillaring is also influenced by the clay slurry concentration. Depending on the clay concentration of the suspension, the average size of the clay particles, the ionic strength and the exchangeable cation the clay particles are more or less preorganized into tactoids (Fig. 17). Tactoids with preferential face-to-face aggregation are formed with clays of relatively large particle size, with Ca^{2+} as exchangeable cation and with a relatively high CEC. In these circumstances the pillared clay will be well ordered with the typical 1.8 nm spacing. Clays with extremely small particle size such as laponite form tactoids with edge-to face and edge-to-edge association. The pillaring of such clays gives an X-ray amorphous material, called delaminated PILC. It consists of clay particles composed of only a few sheets in face-to-face aggregation. These particles are associated irregularly with preferential edge-to-face and edge-to-edge associations (Fig. 18). Because of the almost complete absence of regularity in the stacking this material is X-ray amorphous (ref. 62, 65).

2. Characterization of the PILC's. Al-PILC's contain zeolitic water and Al^{3+}-coordinated water, which must be removed before use in adsorption and/or catalysis. This is not an innocent

228

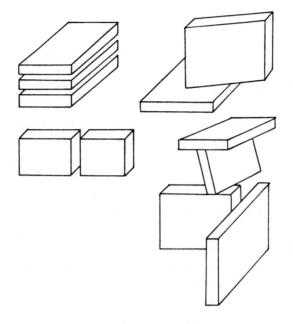

Fig. 17   Schematic drawing of tactoids

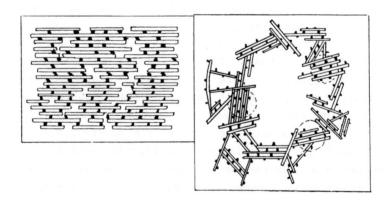

Fig. 18   Face-to-face pillared clay (left) and delaminated
pillared clay (right) (after M. Occelli, ref. 65)

process. Mononuclear Al^{3+} species create acidity through dissociation of residual water molecules in the coordination sphere, a reaction analogous to that observed in zeolites.

Also the Al_{13}^{7+} cluster, the major exchange cation, is decomposed with creation of acidity:

$$2\ [Al_{13}O_4(OH)_{24}(H_2O)_{12}]^{7+} \longrightarrow 2Al_2O_3 + 14H^+ + 41\ H_2O$$

It is clear that the fate of the protons and of Al determines the properties of the material.

Upon activation up to 773°K the d_{001} spacing decreases with about 0.1 - 0.2 nm. In the case of hectorite and laponite there is no accompanying change in the ^{29}Si MAS-NMR but the ^{27}Al signal broadens without shifting to other positions: it resonates at 59.1 - 63.5, as expected for Al_{13}; the octahedral Al resonates at 5-7 ppm. Resonances due to η-Al_2O_3 (68.1 and 8.9 ppm) or bayerite (8 ppm) are absent. Thus, the Al does not cluster into spinel-like phases and significant interaction between the cluster and the smectite is absent.

This is not the case for beiddelites and fluorhectorites (refs. 66,67). The uncalcined beiddelite has a resonance at 69.3 ppm of the tetrahedral Al in the lattice and the 62.3 ppm resonance of the Al_{13} cluster. In calcined beiddelite these two resonances are shifted to 67.5 and 56.5 ppm respectively. This shift of the Al-line of the pillar towards a chemical shift value, typical for zeolites, has been taken as evidence for the formation of a chemical bond between the pillar and the lattice oxygens.

For calcined pillared fluorhectorite the -92.3 ppm Si line splits into two components at -93.8 ppm and at -96.6 ppm (fig. 19). The last value is typical for zeolites and is taken as an indication of the beginning of the growth of a tridimensional network on the surface of the clay.

Thus, both in the cases of beiddelite and of fluorhectorite the hypothesis of the formation of a chemical bond between the pillar and the surface oxygens leads to true cross-linked clays. This is most easily visualized in terms of the Edelman-Favejee model of the clay sheet structure (Fig. 20). In this model one out of two tetrahedra of the T-sheet is inverted and terminated by an OH group. Upon calcination this OH group reacts with an OH group on the Al_{13} pillar to form an $(Si,Al)_l$-O-Al_p bridge (l = lattice; p = pillar).

From the limiting amount of data available in the literature three conditions are necessary for making CLS's: (1) the Keggin

230

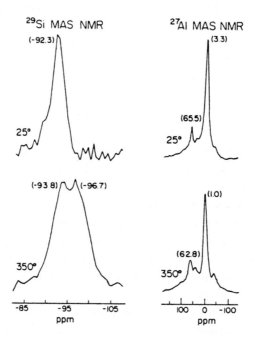

²⁹Si MAS NMR

²⁷Al MAS NMR

Fig. 19 ²⁹Si and ²⁷Al MAS-NMR of uncalcined (upper part) and calcined pillared fluorhectorite (lower part). The calcination temperature was 350°C (after T. Pinnavaia, ref. 67)

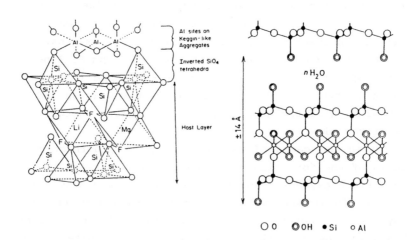

○ O   ◎ OH   ● Si   ○ Al

Fig. 20 The Edelman-Favejee model of 2:1 clay minerals (right, after E. Nemecz, ref. 7) and the cross-linked fluorhectorite (left, after T. Pinnavaia, ref. 67)

ion of Al in the interlamellar space; (2) isomorphous substitution in the tetrahedral layer or (3) F^- replacing OH groups in the octahedral layer.

This interpretation of cross-linking in terms of inversion of tetrahedra is only based on NMR evidence and must be taken with care. Tennakoon et al. (refs. 68, 69) observed shifts from -93.5 ppm to -95.5 ppm and from -95 ppm to -98 ppm in the ^{29}Si spectra Gelwhite and Hectorite respectively when calcined up to 773°K, only with Li^+, NH_4^+ and Al^{3+} as the exchangeable cations. these shifts can only be explained by the irreversible migration of the small cations Li^+, Al^{3+} and H^+ through the ditrigonal holes towards the octahedral layer; in the case of pillared clays one can envisage a similar migration of H^+, created by decomposition of the Al_{13} pillar upon calcination. The shift of the Si-line in fluorhectorite is then, at least partially, due to proton migration into the octahedral layer and not to cross-linking alone. Tennakoon et al. (refs. 68, 69) proposed a condensation of the OH groups of the octahedral layer with those of the pillaring agents to anchor the pillars. This is another kind of cross-linking, even more difficult to visualize than the inversion of tetrahedra, because it must go through the ditrigonal holes of the tetrahedral layers of the clay sheets.

In any case, calcination of pillared clays produces protons, which are the source of acidity and potentially catalytically active sites. There are then at least three mechanisms for introduction of protons in pillared clays: (1) direct exchange from the slightly acidic exchange solutions; (2) dissociation of water molecules in the coordination sphere of Al^{3+}; (3) calcination of the polynuclear Al pillars. The Al oxide pillars themselves are the source of Lewis acidity.

Occelli and Tindwa observed a band at 3700 cm^{-1} of acidic OH groups, ascribed to the pillars (ref. 70). Schutz et al. (ref. 71) found an acidic OH group vibrating at 3400 cm^{-1} on pillared beiddelites and ascribed it to Si-OH-Al bridging hydroxyls on the tetrahedral layer, similar to the bridging hydroxyls of zeolites. From pyridine adsorption and desorption experiments it follows that Lewis acidity is predominant, but the ratio Lewis:Bronsted (L/B) is strongly dependent on the type of clay, as shown in Table 9.

Table 9: Acid sites densities from pyridine desorption experiments (ref. 72)

T/°K	laponite		bentonite		alumina		HY	
	L	B	L	B	L	B	L	B
473	0.38	0.024	0.20	0.096	0.57	0.055	0.11	0.97
573	0.26	-	0.14	0.056	0.45	0.008	0.07	0.70
673	0.08	-	0.08	0.01	0.28	-	0.04	0.31
773	-	-	0.01	-	0.18	-	0.03	0.02

[1]the acid site density was obtained by dividing the integrated absorbance of the 1542 cm^{-1} band and of the 1452 cm^{-1} by the wafer density (mg cm^{-2}).

The high L/B ratio of delaminated pillared laponite is indicative for an alumina-covered laponite. This is expected in view of the delaminated structure of this material (Fig. 18). The acidity of a pillared beiddelite is closest to that of wide pore zeolites, such as HY. This is shown by the hydroconversion of n-decane on Pt-impregnated pillared beiddelites in figure 21 (ref. 73).

The activity of the pillared beiddelite is close to that of an ultrastable Y, but its isomerization yield is higher. This is taken as evidence for easy migration of branched molecules and therefore a very open pore system.

Permanent microporosity, a large surface area, molecular sieving and thermal stability are the important physical characteristics of pillared clays for catalytic applications. They are closely related to each other and can be discussed together.

The N_2 adsorption isotherms are of the Langmuir type, suggesting microporosity. Typical surface areas are in the range 200 - 250 m^2/g for montmorillonite and hectorite, 300 - 350 m^2/g for beiddelites and 400 m^2/g for pillared delaminated clays (laponite). The latter reflects the small particle size, as laponite as such has a N_2-BET surface area of 360 m^2/g (ref. 10). The surface areas strongly depend on the drying conditions of the materials. Pinnavaia et al. (ref. 62) noted systematically a larger surface area for air-dried than for freeze-dried materials, the difference being 20 - 30 m^2/g. A wet flocculated pillared clay consists of delaminated and face-to-face pillared aggregates. During freeze drying this structure is largely preserved, while air-drying promotes face-to-face aggregation. Vaughan and Lussier (ref. 74) showed that surface areas above 400 m^2/g can be obtained by spray drying a pillared clay slurry. Thus the surface area is to a large extent determined by the structure of the aggregates and not by the amount and separation of the pillars in the interlamellar space.

Nevertheless, in such a spray-dried material more than 60% of the pores have a diameter below 1.4 nm and 75 - 82% have a diameter less than 2.0 nm. 16 - 19% have a diameter in the range 4 - 6 nm, which is in the mesopore regime; the remaining 2% are macropores. These numbers suggest that the pore volume is mainly a microporous volume. Typical pore volumes are 0.15 - 0.20 cm^3/g, the beiddelites falling on the higher end, the montmorillonites on the lower end. Table 10 shows the dependence on the calcination temperature and the amount of pillaring agent.

Table 10: pore volumes of pillared smectites (cm^3/g) (ref. 64)

T/°K	Vmicro	Vtot
473	0.124	0.166
573	0.118	0.192
673	0.088	0.156
773	0.077	0.141
Al in mmole/g, activation at 473°K		
2.07	0.056	0.104
2.84	0.124	0.166
4.34	0.125	0.186
5.55	0.137	0.186

The data of Table 10 show that the total porosity is maintained up to 773°K and that at least a fivefold excess of Al with respect to the CEC is necessary to obtain maximum porosity. Again the pillared delaminated clays are the exception in that their total pore volume is as high as 0.364 cm^3/g. It consists entirely of mesopores with an average diameter of 1.8 nm and macropores. No microporosity is developed.

Microporosity is also evidenced by molecular sieving effects. They are illustrated in Table 11 (refs. 62, 64, 74).

Table 11: Molecular sieve properties of PILC's

molecule	size/nm	mmole/g	drying	mmole Al/g
cyclohexane	0.61	1.00	sp	
CCl4	0.69	0.75	sp	
1,3,5-tri-methylbenzene	0.76	0.44	sp	
		0.55	fd	
		-	ad	
1,2,3,5-tetra methylbenzene	0.80	-	sp	
$(F_9C_4)_3N$	1.04	-	sp	
		-	ad	
		0.35	fd	
neopentane	0.62	-		2.8
		1.0		5.55
		1.1		2.4(Lap)

sp = spray drying; ad = air drying; fd = freeze drying

The data of the table 11 show that there is a molecular sieving effect for molecules in the size range 0.8 - 1.0 nm. The effect is strongly dependent on the drying method, showing the importance of interparticle association in determining the microporosity of the material. This is confirmed by the absence of a molecular sieving effect on delaminated pillared laponite. One wonders then which fraction of the microporous volume is due

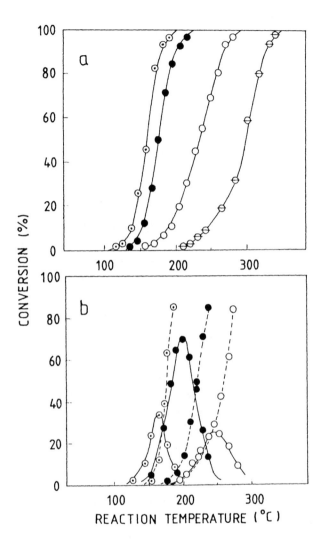

Fig. 21   (a) Evolution of the overall conversion of n-decane
versus reaction temperature for ⊙, ultrastable zeolite Y; ●,
pillared beiddelite; ⊙, pillared montmorillonite; O, silica-
alumina
(b) Evolution of the percentages of hydroisomerisation (solid
lines) and hydrocracking (dotted lines) versus reaction
temperature for ⊙, ultrastable zeolite Y; ●, pillared beiddelite;
O, pillared montmorillonite.

to the real two-dimensional pore system of the pillared clay and which fraction is due to interparticle association.

The Al-PILC's have been extensively tested in the cracking of heavy feedstocks, but they failed to meet the requirements of an efficient cracking catalyst. Because of extensive coke formation frequent regeneration is a necessity and PILC's lack the hydrothermal stability to withstand that process. First of all there is an upper limit on the thermal stability of PILC's imposed by the thermal stability of the clay itself. Smectites dehydroxylate in the range 773 - 973°K and the structure breakdown starts above 973°K. This temperature has to be considered as an upper limit for the application of PILC's. Indeed, protons, produced upon calcination of PILC's transform the clay into an acid clay which is known to be thermally unstable. The water, produced upon calcination, creates hydrothermal conditions in the pore system, similar to those in zeolites. This may lead to sintering of the Al-pillars at high temperatures. It results in a loss of the 1.8 nm spacing, of surface area and of microporosity.

CONCLUSION

Clays, especially smectites, are very versatile materials which find industrial applications in such diverse areas as oil drilling, paper coating and filling, ceramics, foundry molding, pharmaceuticals and in adsorption and ion exchange. Their use as heterogeneous catalysts is limited due to the competition of zeolites and molecular sieves in general.

Pillaring of clays is an attempt to upgrade these materials as catalysts. Although the PILC's failed to meet the industrial requirements of a fluid cracking catalyst, up to the present time at least, there are numerous other reactions, especially in the field of organic chemistry, for which they may be developed into active and selective catalysts (ref. 75). A research effort in that direction is worthwhile.

REFERENCES

1 B. Hulthén, in N.A. Shaikh and N.-G. Wik (Eds.), Clay Minerals - Modern Society, Nordic Society for Clay Research, Uppsala, 1986, pp. 99-109.

2 I.E. Odom, in L. Fowden, R.M. Barrer and P.B. Tinker (Eds.), Clay Minerals: their Structure, Behavior and Use, The Royal Society, London, 1984, pp. 171-189.

3 W.B. Jepson, in L. Fowden, R.M. Barrer and P.B. Tinker (Eds.), Clay Minerals: their Structure, Behavior and Use, The Royal Society, London, 1984, pp. 191-212.

4 D.W. Breck, Zeolite Molecular Sieves: Structure, Chemistry and Use, J. Wiley, New York, 1974.

5 M.E. Davis, C. Saldarriaga, C. Montes, J. Garces and C. Crowder, Zeolites, 8 (1988) 362-367.

6 M.S. Stul and L. Van Leemput, Clay Min., 17 (1982) 209-215.

7 E. Nemecz, Clay Minerals, Akadémiai Kiado, Budapest, 1981.

8 R.E. Grim, Clay Mineralogy, McGraw Hill, New York, 1968.

9 B.K.G. Theng, The Chemistry of Clay-Organic Reactions, Adam Hilger, London, 1974.

10 H. Van Olphen and J.J. Fripiat, Data Handbook for Clay Minerals and other Non-Metallic Materials, Pergamon, Oxford, 1979.

11 L. Fowden, R.M. Barrer and P.B. Pinker, Clay Minerals: their Structure, Behavior and Use, The Royal Society, London, 1984.

12 A.C.D. Newman (Ed.), Chemistry of Clays and Clay Minerals, Longman Scientific and Technical, Harlow, 1987.

13 see ref. 7, pp. 36-38.

14 F. Velghe, R.A. Schoonheydt and J.B. Uytterhoeven, Clays and Clay Min., 25 (1977) 375-380.

15 A. Maes, P. Peigneur and A. Cremers, in S.W. Bailey (Ed.), Proc. Int. Clay Conf., Mexico City, july 16-23, 1975, Applied Publishing, Wilmette, 1976, pp. 319-329.

16 G.N. White and L.W. Zelazny, Clays and Clay Min., 36 (1988) 141-146.

17 G. Lagaly and A. Weiss, in S.W. Bailey (Ed.), Proc. Int. Clay Conf., Mexico City, july 16-23, 1975, Applied Publishing, Wilmette, 1976, pp. 157-172.

18 M.S. Stul and W.J. Mortier, Clays and Clay Min., 22 (1974) 391-396.

19 A. Maes, M.S. Stul and A. Cremers, Clays and Clay Min., 27 (1979) 387-392.

20 D.A. Laird, A.D. Scott and T.E. Fenton, Clays and Clay Min., 37 (1989) 41-46.

21 M.S. Stul and L. Van Leemput, Surface Technology, 16 (1982) 101-112.

22 see ref. 12, pp. 245-247.

23 see ref. 12, pp. 203-224.

24 R. Kjellander, S. Marcelja and J.P. Quirk, J. Colloid and Interface Sci., 126 (1988) 194-211.

25 G. Sposito and R. Prost, Chem. Rev., 82 (1982) 553-573.

26 J.J. Fripiat, J. Cases, M. Francois and M. Letellier, J. Colloid and Interface Sci., 89 (1982) 378-400.

27 V.C. Farmer in V.C. Farmer (Ed.), The Infrared Spectra of Minerals, Mineralogical Society, London, 1974, p. 331.

28 H. Suquet, R. Prost and H. Pezerat, Clay Min., 17 (1982) 231-241.

29 G. Sposito, R. Prost and J.-P. Gaultier, Clays and Clay Min., 31 (1983) 9-16.

238

30 D. Tilak, B. Tennakoon, J.M. Thomas, W. Jones, T.A. Carpenter and S. Ramdas, J. Chem. Soc. Faraday Trans. I, 82 (1986) 545-562.

31 see ref. 8, pp. 278-352.

32 J.G. Thompson, Clay Min., 19 (1984) 229-236.

33 M. Lipsicas, R.H. Raythatha, T.J. Pinnavaia, I.D. Johnson, R.F. Giese, Jr., P.M. Costanzo and J.-L. Robert, Nature, 309 (1984) 604-607.

34 R.A. Kinsey, R.J. Kirkpatrick, J. Hower, K.A. Smith and E. Oldfield, Am. Min., 70 (1985) 537-548.

35 S. Kormarneni, C.A. Fyfe, G.J. Kennedy and H. Strobl, J. Am. Ceram. Soc., 69 (1986) C45-C47.

36 C.A. Weiss, Jr., S.P. Altaner and R.J. Kirkpatrick, Am. Miner., 72 (1987) 935-942.

37 J. Sanz and J.M. Serratosa, J. Am. Chem. Soc., 106 (1984) 4790-4793.

38 C.P. Herrero, J. Sanz and J.M. Serratosa, Solid State Comm., 53 (1985) 151-154.

39 A.R. Gonzales - Elipe, J.P. Espinos, G. Munuera, J. Sanz and J.M. Serratosa, J. Phys. Chem., 92 (1988) 3471-3476.

40 J.P. Rupert, W.T. Granquist and T.J. Pinnavaia, in A.C.D. Newman (Ed.), The Chemistry of Clays and Clay Min., Mineralogical Society, London, 1987, pp. 275-318.

41 M.M. Mortland, J.J. Fripiat, J. Chaussidon and J. Uytterhoeven, J. Phys. Chem., 67 (1963) 248-258.

42 see ref. 9, pp. 261-269.

43 M.M. Mortland and H.E. Doner, USP 3,78,457.

44 Y. Soma, M. Soma and I. Harada, J. Phys. Chem., 88 (1984) 3034-3038.

45 Y. Soma, M. Soma and I. Harada, J. Contaminant Hydrology, 1 (1986) 95-106.

46 R.M. Barrer, in L. Fowden, R.M. Barrer and P.B. Pinker (Eds.), Clay Minerals: their Structure, Behavior and Use, The Royal Society, London, 1984, pp. 113-132.

47 R.M. Barrer, Clays and Clay Min., 37 (1989) 385 - 395.

48 G. Lagaly, in L. Fowden, R.M. Barrer and P.B. Pinker (Eds.), Clay Minerals: their Structure, Behavior and Use, The Royal Society, London, 1984, pp. 95-112.

49 R.A. Schoonheydt, J. Pelgrims, Y. Heroes and J.B. Uytterhoeven, Clay Min., 13 (1978) 435-438.

50 H. Van Damme, H. Nijs and J.J. Fripiat, in T.J. Pinnavaia and R.A. Schoonheydt (Eds.), Metal Complex Catalysts in Intracrystalline Environments, Elsevier Sequoia, Lausanne, J. Molecular Catalysis, 27 (1984) 123-142.

51 R. Burch (Ed.), Pillared Clays, Elsevier, Amsterdam, Catalysis Today, 2 (1988) 185 - 367.

52 J. Shabtai and J. Fijal, USP 4 579 832 (1986).

53 J. Sterte and J. Shabtai, Clays and Clay Min. 35 (1987) 429 - 439.

54 D.E.W. Vaughan, USP 4 666 877 (1987).

55 J.R. McCauley, WO 88/06488.

56 W.Y. Lee, R.H. Raythatha and B.J. Tatarchuk, J. Catalysis, 115 (1989) 159 - 179.

57 C.G. Manos, Jr., M.M. Mortland and T.J. Pinnavaia, Clays and Clay Min.,32 (1984) 93-98.

58 R.M. Lewis, K.C. Ott and R.A. Van Santen, USP 4,510,257 (1985).

59 S.P. Christiano, J. Wang and T.J. Pinnavaia, Inorg. Chem., 24 (1985) 1222-1227.

60 T.J. Pinnavaia, Poster presentation at the 9th International Clay Conference, Strasbourg, France, august 28 - september 2, 1989.

61 J. Sterte, in R. Burch (Ed.), Pillared Clays, Elsevier, Amsterdam, Catalysis Today, 2 (1988) 219-231.

62 T.J. Pinnavaia, M.-S. Tzou, S.D. Landau and R.H. Raythatha, J. Molecular Catalysis, 27 (1984) 195-212.

63 D. Plee, L. Gatineau and J.J. Fripiat, Clays and Clay Min., 35 (1987) 81-88.

64 M.H. Stacey, Catalysis Today, 2 (1988) 621-631.

65 M.L. Occelli, J. Lynch and J.V. Senders, J. Catalysis, 107 (1987) 557-565.

66 D. Plee, F. Borg, L. Gatineau and J.J. Fripiat, J. Am. Chem. Soc., 107 (1985) 2362-2369.

67 T.J. Pinnavaia, S.D. Landau, M.-S. Tzou, I.D. Johnson and M. Lipsicas, J. Am. Chem. Soc., 107 (1985) 7222-7224.

68 D.T.B. Tennakoon, W. Jones and J.M. Thomas, J. Chem. Soc. Faraday Trans. I, 82 (1986) 3081-3095.

69 D.T.B. Tennakoon, W. Jones, J.M. Thomas, J.H. Ballantine and J.H. Purnell, Solid State Ionics, 24 (1987) 205 - 215.

70 M.L. Occelli and R.M. Tindwa, Clays and Clay Min., 31 (1983) 22-28.

71 A. Schutz, W.E.E. Stone, G. Poncelet and J.J. Fripiat, Clays and Clay Min., 35 (1987) 251-261.

72 M.L. Occelli, Catalysis Today, 2 (1988) 339-355.

73 A. Schutz, D. Plee, F. Borg, P. Jacobs, G. Poncelet and J.J. Fripiat, in L.G. Schultz, H. Van Olphen and F.A. Mumpton (Eds.), Proc. Int. Clay Conf. Denver 1985, The Clay Minerals Society, Bloomington, 1987, pp. 305-310.

74 D.E.W. Vaughan and R. Lussier, in L.V. Rees (Ed.), Proc. 5th Int. Conf. Zeolites, Naples 1980, Heyden, London, 1980, pp. 94-101.

75 P. Laszlo, Acc. Chem. Res., 19 (1986) 121-127.

Chapter 7

TECHNIQUES OF ZEOLITE CHARACTERIZATION

J.H.C. van Hooff[1]   and   J.W. Roelofsen[2]

[1] Laboratory for Inorganic Chemistry and Catalysis, Eindhoven University of Technology, P.O.Box 513, 5600 MB Eindhoven, The Netherlands.

[2] AKZO Chemicals B.V., Research Centre Amsterdam, P.O.Box 15, 1000 AA Amsterdam, The Netherlands.

CONTENTS

242

1. INTRODUCTION

Before a zeolite can be used for a certain application it is necessary to characterize this zeolite, to see if it has the desired properties for that application. If not, another synthesis method should be used or the zeolite must be modified, to meet the specifications. Zeolite synthesis, –modification, –characterization and –application thus are strongly related as schematically is indicated in Fig. 1.

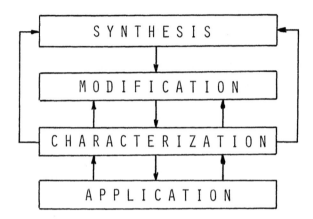

Fig. 1.   *Position of Zeolite Characterization.*

Of course not all zeolite properties are of the same importance for every application. This aspect is illustrated in Fig. 2 in which is indicated, which characteristics are of special importance for the main applications of three different zeolites.

For example for the application of zeolite A in detergents particle size and morphology are extremely important while acidity and stability play a minor role. These characteristics, however, are just of crucial importance for the application of zeolite Y in cracking catalysts.

For this reason we will discuss the different characterization methods in relation to the application for which the specific property is important.

IMPORTANT CHARACTERISTIC	TYPE OF ZEOLITE		
	ZEOLITE A	ZEOLITE Y	ZSM - 5
STRUCTURE STRUCTURAL DEFECTS		◌	●
PORE-SIZE	●	◌	●
CHEMICAL COMPOSITION	●	●	●
FRAMEWORK Si/Al RATIO		●	◌
ACIDITY		●	●
STABILITY		●	◌
MORPHOLOGY PARTICLE SIZE	●	◌	●
IMPORTANT APPLICATION	ADSORBENT ION EXCHANGER IN DETERGENTS	CRACKING CATALYST	SHAPE SELECTIVE CATALYST

Fig. 2   *Most important characteristics for three well-known zeolites and their major applications.*

244

2. ZEOLITE STRUCTURE AND STRUCTURAL DEFECTS

For the application of zeolite ZSM-5 as a shape-selective catalyst it is very important that the zeolite has the correct structure and has no structural defects. The best way to check zeolite structure is by X-Ray Diffraction.

This method is based on the fact that every crystalline material has its own characteristic X-ray diffraction pattern.

For most of the known zeolite structures the ideal diffraction patterns have been simulated and together with other crystal data these diffraction patterns are collected in the following publication of the Structure Commission of the International Zeolite Association.

Collection of Simulated XRD Powder Patterns for Zeolites
by Roland van Ballmoos
Butterworths 1984

To illustrate the kind of information that can be found in this book, we here give the diffraction pattern (Fig. 3) together with the other crystal data that are given for zeolite ZSM-5 officially named MFI (Mobil-Five).

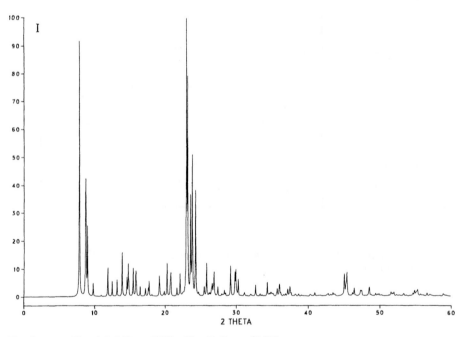

Fig. 3   *Simulated X-ray Diffraction Pattern of MFI.*

Crystal data: Space group Orthorhombic Pnma

$$a = 20.096 \text{ Å} \qquad \alpha = 90.00°$$
$$b = 19.949 \text{ Å} \qquad \beta = 90.00°$$
$$c = 13.428 \text{ Å} \qquad \Upsilon = 90.00°$$

Additional structural information can be found in a second publication of the Structure Commission of the International Zeolite Association.

Atlas of Zeolite Structure Types
by W.M. Meier and D.H. Olsen
Butterworths 1987

The book gives stereographic views for most of the known zeolite structures together with drawings of the cross-sections of the main channels.

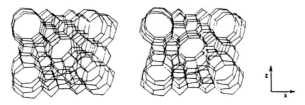

framework viewed along [010]

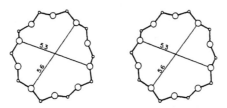

10-ring viewed along [010]
(straight channel)

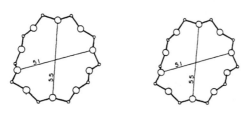

10-ring viewed along [100]
(sinusoidal channel)

Fig. 4   *Stereographic view and channel cross-sections of zeolite MFI*

Furthermore several structural data are presented; for example for ZSM-5 or MFI is given:

Framework density : 17.9 T atoms/1000 Å^3

Channels	direction	number of T-units	free aperture Å
	[010]	10 MR	5.3 x 5.6 (b)
	[100]	10 MR	5.1 x 5.5 (a)

Fault planes: (100)

Normally the first check is the comparison of the observed X-ray diffraction pattern with the simulated pattern.

If all observed peaks can be found back in the simulated pattern and have similar intensities the crystal structure of the zeolite under investigation is the same as the structure that is used for the pattern simulation. If extra peaks are observed or peaks are missing this is an indication that other crystalline phases are present, or that the structure is different.

The intensity of the diffraction peaks can be used to determine the crystallinity of the sample. For that purpose the intensity of one particular peak (or a number of peaks) is compared with the intensity of the same peak (or peaks) of a standard sample.

The X-ray crystallinity then can be calculated from: (34)

$$\text{X-ray crystallinity} = \frac{\text{Intensity of peak hkl of sample}}{\text{Intensity of peak hkl of standard}} \times 100\%$$

Furthermore the width of the diffraction peaks gives information about the average crystallite size of the investigated zeolite sample.

According to Scherrer (35) the relation between linewidth (B) and crystallite size (t) is given by the following formula.

$$t(\text{in nm}) = \frac{0,9 \; \lambda \; (\text{in nm})}{B \cos \Theta}$$

X-ray diffraction thus can give information about crystal structure, degree of crystallinity and crystallite size. It is, however, impossible to obtain information about the presence of structural defects by this method.

The presence of fault planes and/or stacking faults in the structure can be observed by high resolution electron microscopy (HREM) as shown in Fig. 5.

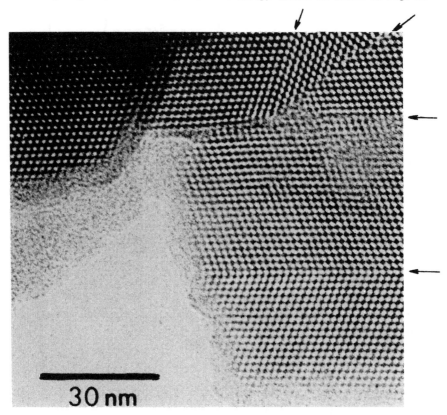

Fig. 5   Electron micrograph of zeolite Y showing the presence of structural defects, indicated by arrows.

Another type of structural defects that can be present are hydroxyl groups.

– Firstly, there are the terminal silanol groups that are always present at the outer surface of the zeolite crystals to terminate the structure. For this reason these groups are called: terminal silanol groups.

terminal
silanol groups

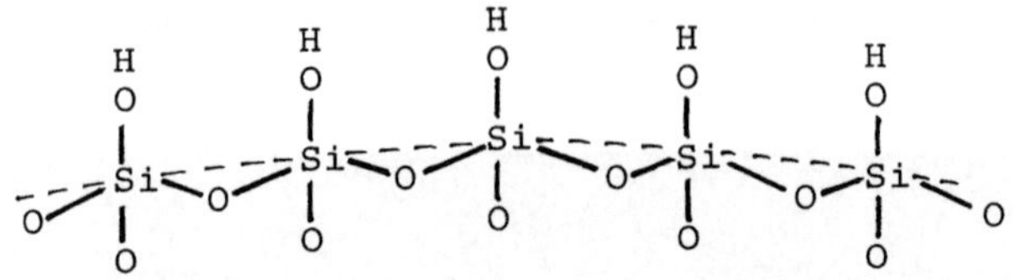

$$
\text{O}^- - \text{Si} - \text{O} - \text{Si} - \text{O} - \text{Si} - \text{O} - \text{Si} - \text{O} - \text{Si} - \text{O}
$$
(with terminal silanol –O–H groups on each Si)

- The second type of silanol groups is formed when a Si–O–Si bond is broken by a reaction with water.

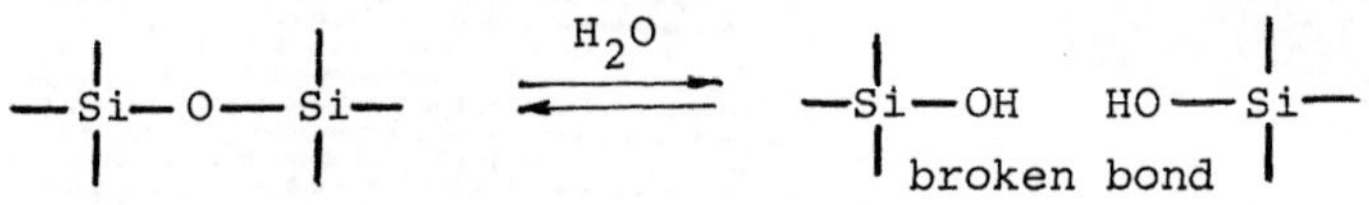

$$
-\overset{|}{\underset{|}{\text{Si}}}-\text{O}-\overset{|}{\underset{|}{\text{Si}}}- \quad \underset{\longleftarrow}{\overset{\text{H}_2\text{O}}{\longrightarrow}} \quad -\overset{|}{\underset{|}{\text{Si}}}-\text{OH} \quad \text{HO}-\overset{|}{\underset{|}{\text{Si}}}-
$$

broken bond

As indicated in the reaction equation this type of silanol groups is always present as pairs.

- Finally, silanol groups can be formed if a T-atom is missing in the structure. This type of defect leads to a cluster of four silanol groups or a so-called hydroxyl nest.

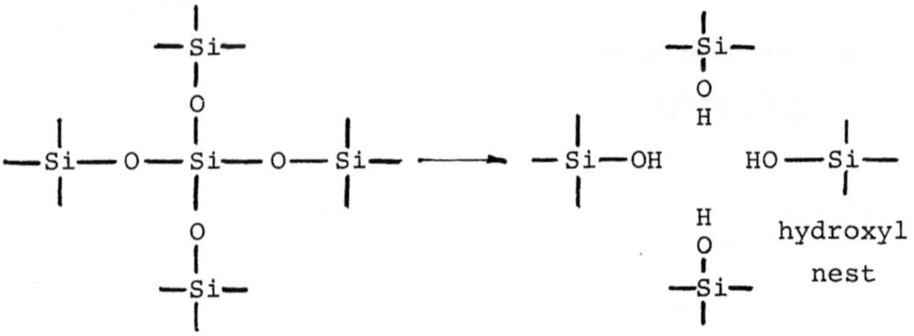

$$
-\text{Si}-\text{O}-\text{Si}-\text{O}-\text{Si}- \quad \longrightarrow \quad -\text{Si}-\text{OH} \quad \text{HO}-\text{Si}-
$$

hydroxyl nest

All these silanol groups can be observed by IR and H NMR spectroscopy. Moreover, it is possible to distinguish between the different types upon reacting these silanol groups with trimethyl-chlorosilane and subjecting the reaction products to ^{29}Si CP-MAS NMR (31).

As indicated in Fig. 6 trimethylchlorosilane will first react via the Cl ligand to form the primary product. If, however, neighbouring OH groups are present also the secondary, tertiary and quaternary products can be formed.

By ^{29}Si NMR it is possible to distinguish between these different reaction products and the presence of the signals attributed to the secondary and tertiary products then indicates the presence of pairs of silanolgroups and hydroxyl nests.

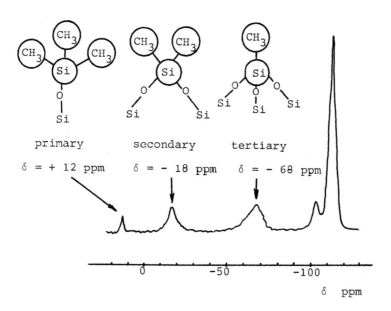

Fig. 6 ^{29}*Si CP-MAS-NMR spectrum of MFI after reaction with trimethylchlorosilane at 673K showing the presence of primary, secondary and tertiary products.*

3. PORE STRUCTURE OF ZEOLITES

For the use as molecular sieve or as shape selective catalyst it is essential to characterize the zeolite's system. Information about pore size can be obtained from:

> 3.1 – the crystal structure
> 3.2 – adsorption measurements
> 3.3 – ^{129}Xe NMR spectroscopy
> 3.4 – model compound reactions

3.1 Crystal structure

From a single crystal structure X-ray analysis of the zeolite it is possible to derive the positions of the component atomstogether with the apertures of the channels and cavities that are present in the zeolite structure, and whether the pore system is one-, two- or three-dimensional.

The pore size determined for a given zeolite form the X-ray analysis pertains to perfect crystals. Often zeolite samples do not fulfil this condition due to the presence of imperfections or minor amounts of other phases inside the pore system. Therefore it is recommendable to characterize the pore volume and pore size in a more direct way by one of the other methods.

3.2 Adsorption measurements

Adsorption measurements with molecules of different size give direct information about the dimensions of the pore system. Molecules with kinetic diameters smaller than the pore openings can be adsorbed and larger molecules can not. In boundary cases a minimum temperature of adsorption may be required.

D.W. Breck (1) presents the dimensions of a number of probe molecules, from which the following examples are compiled in Table 1.

TABLE 1. Molecular dimensions of a series of probe molecules.

	Pauling		Lennard–Jones
	length	width	kinetic diameter
	pm	pm	pm
He		300	260
H_2	310	240	289
Ne		320	275
Ar		384	340
O_2	390	280	346
N_2	410	300	364
Kr		396	360
Xe		436	396
CO	420	370	376
CO_2	510	370	330
H_2O	390	315	265
NH_3	410	380	260
SO_2	528	400	360
CH_4		420	380
C_2H_4	500	440	390
C_3H_8	650	490	430
n C_4H_{10}		490	430
i C_4H_{10}		560	500
neopentane		700	620
benzene		660	585
cyclohexane		670	600
$(C_2H_5)_3N$			780
$(C_4H_9)_3N$			810
$(C_4F_9)_3N$			1020

The Lennard–Jones kinetic diameters are perhaps most appropriate to predict accessibility of a zeolite for a given adsorbate. For comparison; the apertures of 6, 8, 10 and 12 membered oxygen rings are about 270 pm, 430 pm, 60 pm and 770 pm, respectively.

Generally, the adsorption isotherm (the amount of gas adsorbed as function of the partial pressure of the gas at constant temperature) gives information about the pore structure. This is a method often applied for characterizing materials having meso (2–50 nm) and macropores (> 50 nm).

Problems arise, however, in the characterization of micropores (< 2 nm) the dimensions of which are close to those of the adsorbing molecules. In this case the adsorption process is a filling of the micropores instead of multilayer adsorption on which the interpretation of adsorption isotherms is generally based.

This, together with the requirement to measure the adsorption isotherms at very low pressures, makes it very difficult to obtain detailed information on the pore structure of zeolites by this method.

At the moment two comparable methods are reported in literature by which it is possible to determine the micropore volume of zeolites from the N_2 adsorption isotherm. These are:

- t-plot method (2) in which the volume of the adsorbed N_2 is plotted against the statistical thickness(t) of the adsorbed layer. This results in a straight line and extrapolation to t=0 gives the micropore volume.

- α_s method (3) in which the volume of the adsorbed N_2 is plotted against the reduced standard adsorption (α_s). Extrapolation to α_s = 0 gives again the micropore volume.

3.3 ^{129}Xe NMR spectroscopy

Fraissard and co-workers (4) recently developed ^{129}Xe NMR spectroscopy as a method to obtain information on the pore system of zeolites. Due to the relatively large diameter of the xenon atom (436 pm) information is only obtained on the larger pores. For example in the faujasite type zeolites the supercages are accessible for xenon whereas the sodalite cages are not.

The method is based on the measurement of the ^{129}Xe chemical shift of the in the NMR spectrum. According to Fraissard the chemical shift of xenon adsorbed in a porous material can be expressed as:

$$\delta = \delta_o - \delta_e - \delta_m - \delta_{(Xe-Z)} \cdot \rho_z - \delta_{(Xe-Xe)} \rho_{Xe}$$

This relation shows that the observed chemical shift is the sum of the chemical shifts originating from:
- the reference (δ_o)
- the electric field (δ_e)
- the magnetic field (δ_m)

— the interaction of the xenon atoms with the zeolite $(\delta_{(Xe-Z)} \cdot \rho_z)$ $(\rho_z$ is related to the zeolite structure)

— the mutual interactions between the xenon atoms $(\delta_{(Xe-Xe)} \cdot \rho_{Xe})$ $(\rho_{Xe}$ is the concentration of the xenon atom, in the pores) (see Fig. 7).

As the exchange of Xe between different positions is fast on the NMR timescale, the NMR signal reflects the average of the different Xe positions and interactions.

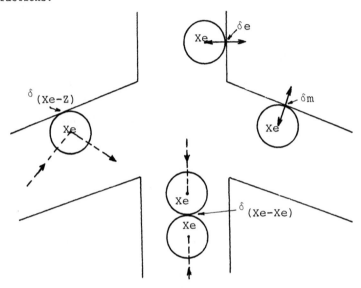

Fig. 7   Schematic representation of the different contributions to the ^{129}Xe chemical shift of xenon atoms adsorbed in the pore system of a zeolite.

Fig. 8 gives for three zeolites the ^{129}Xe chemical shift as a function of the xenon concentration in the zeolite.

254

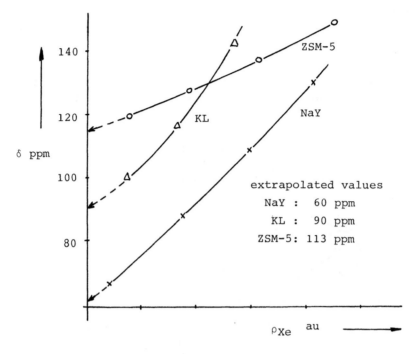

Fig. 8   The ^{129}Xe *chemical shift as function of* ρ_{Xe} *for three types of zeolite.*

The following information can be obtained:
- extrapolation of the chemical shift to zero concentration gives information about pore size and pore restrictions while
- from the slope of the curves information can be obtained about the void space available to the absorbed Xe atoms.
- The intensity of the signal gives information about the number of well-formed cages.

^{129}Xe NMR has been applied so far in various investigations such as:
- characterization of zeolite pore systems; also for zeolites with at that time unknown crystal structure (zeolite β (5))
- study of coke formation during catalytic cracking (6)
- study of the formation of Ni particles upon reduction of Ni^{2+}-exchanged Y-zeolites (7).

When, however, the electric and/or the magnetic field also contribute to the chemical shift the conclusions are less accurate.

<u>3.4 Model compound reactions</u>

The most direct method to obtain information about the dimensions of the internal space in zeolites available for catalytic reactions is of course the performance of model compound reactions.

The advantage of this method is that information is obtained under conditions that are representative for the catalytic application.

Generally, three reaction selectivity possibilities can be distinguished:
- a reactant molecule is too large to enter the pores, causing reactant selectivity
- a product molecule is too large to leave the pores, causing product selectivity
- the transition state for a certain reaction is too large to be accommodated in the pores causing transition state selectivity.

In literature several reactions have been discussed by which information on the pore system of zeolites can be obtained. Three of these reactions will be discussed here; the first is related to reactant selectivity, while the two other examples are based on transiton state selectivity.

3.4.1 The Constraint Index (CI)

Mobil Oil workers (8) introduced a test reaction in which a 1:1 mixture of n-hexane and 3-methylpentane (see Fig. 9) is cracked in a gas-flow setup.

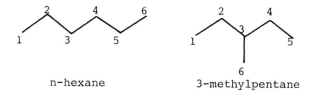

n-hexane   3-methylpentane

Fig. 9   *Molecules used for the determination of the constraint index (CI)*

256

Depending on the pore size, the relative rate of cracking of the linear versus the branched hydrocarbon molecule will change and they define:

$$\text{Constraint Index (CI)} = \frac{\text{rate constant of n-hexane cracking}}{\text{rate constant of 3-methylpentane cracking}}$$

and supposing first order kinetics for the cracking reactions this relation is equivalent with

$$\text{CI} = \frac{\log \text{(fraction n-hexane remaining)}}{\log \text{(fraction 3-methylpentane remaining)}}$$

The Constraint Index provides the following classification of the pore-system

$$\text{CI} < 1 \quad : \text{large pores}$$
$$1 < \text{CI} < 12 \quad : \text{intermediate pores}$$
$$\text{CI} > 12 \quad : \text{small pores}$$

The determation of the CI is suitable for the characterization of medium pore zeolites. The method is, however, not sensitive for large pore zeolites. Moreover, the method proved to be temperature dependent.

3.4.2 The Refined Constraint Index (RCI)

Jacobs et al. (9) introduced the bifunctional conversion (1% Pt on the zeolite in the proton form) n-decane as a test reaction. Under the conditions that 5% of the n-decane is isomerized the ratio of 2-methylnonane and 5-methylnonane which are formed in this reaction (see Fig. 10) is determined.

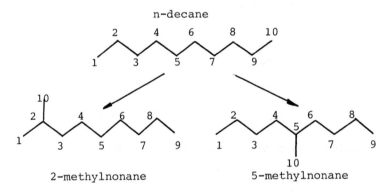

Fig. 10   Reactions used for the determination of the refined constraint index.

$$\text{Refined Constraint Index (RCI)} = \frac{[\text{2-methylnonane}]}{[\text{5-methylnonane}]}$$

It is pointed out that more space is needed for the formation of 5-methylnonane than for the formation of 2-methylnonane. So the RCI will increase with decreasing pore width.

The method proved to be especially suited for the characterization of zeolites with 10 membered ring pore systems, it is, however, not sensitive for zeolites having larger pores.

3.4.3 The Spaciousness Index (SI)

To characterize wide pore zeolites Weitkamp et al. (10) have introduced the hydrocracking of n-butylcylcohexane as a test reaction. Two of the reaction products are n-butane and isobutane and according to the authors the transition state for the formation of isobutane is larger than that for the formation of n-butane (see Fig. 11).

butylcyclohexane

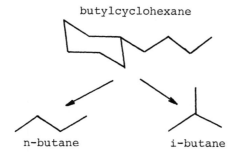

n-butane   i-butane

Fig. 11   Reaction used for the determination of the spaciousness index.

The ratio of the amounts of formed n-butane and i-butane thus can be used as an indication for the available space in the zeolite pore system leading to the following definition:

$$\text{Spaciousness Index (SI)} = \frac{[\text{formed i-butane}]}{[\text{formed n-butane}]}$$

It will be clear that SI decreases with decreasing pore width.

The method proved to be especially suited for the characterization of zeolites with 12 membered ring pore systems and so the SI is complementary to the RCI and CI discussed above.

4. CHEMICAL COMPOSITION OF ZEOLITES

There are several possibilities to express the chemical composition zeolites. For example for the sodium form of zeolite Y the following formulae can be used

$$Na_x \ [(AlO_2)_x \cdot (SiO_2)_{192-x}] \cdot z \ H_2O \qquad \text{or}$$

$$\frac{x}{2} \ Na_2O, \ \frac{x}{2} \ Al_2O_3; \ (192-x)SiO_2, \ zH_2O$$

in which the number of Al per unit cell can vary between $48 < x < 64$. Thus for dried pure zeolites the chemical composition can be characteriz by a single parameter x, the number of Al atoms per unit cell.

However, instead of the number of Al per unit cell (N_{Al}) often the Si (silicon/aluminum) atom ratio or the SiO_2/Al_2O_3 (silica/alumina) molecul ratio is used to characterize the chemical composition of zeolites. Table shows the interrelation of these parameters.

TABLE 2   *Interrelation of the different parameters used to characterize the chemical composition of NaY zeolites.*

N_{Al}	x	48	64
Si/Al ratio	$\dfrac{192-x}{x}$	3	2
SiO_2/Al_2O_3 ratio	$\dfrac{192-x}{\frac{x}{2}}$	6	4
Na_2O wt%		11.8	15.4
Al_2O_3 wt%		19.5	25.3
SiO_2 wt%		68.7	59.4

The methods for the determination of the chemical composition can be divided into two groups:

4.1   Methods based on complete dissolution of the zeolite and subsequent analysis of the solution obtained (destructive wet analysis).

4.2   Analysis of the solid material by physical methods (non-destructive dry analysis)

4.1 Complete Dissolution Method

An example of a frequently used method is given below:

Weigh accurately a sample (1 g) of dried zeolite   (g gram).

Add sulphuric acid (40 ml) and heat until SO_3 fumes.

Dilute with water (about 200 ml) and filter   (filtrate A).

Dry and ignite the residue at 1000°C   (a gram)

Add hydrofluoric acid (10 ml) evaporate carefully

and finally ignite at 1000°C   (b gram).

$$\% \ SiO_2 = \frac{a - b}{g} \times 100\%$$

Add potassium peroxodisulphate (2g) to the residue and heat until a clear melt is obtained. Dissolve the melt in water (100 ml) and combine the solution with filtrate A.

Determine Na and Al in this solution by AAS.

As can be seen this method is rather laborious and therefore several more rapid physical analysis method have been developed. For the accurate analysis of exchange of cations or substitution of T-atoms a proper chemical analysis remains a prerequisite.

4.2 Physical Analysis Methods for Zeolites

Several methods are available, all with their advantages and disadvantages (32).

260

4.2.1 X-ray Fluorescence Spectroscopy

In this method a suitable sample is radiated by high energy X-rays by which electrons can be expelled from the different atoms, leaving holes in low-lying orbitals. The main mechanism for relaxation then is, that an upper electron falls into this hole.

The energy released may result either in the generation of radiation, which is called X-ray fluorescence, or in the ejection of another electron, the secondary electron of the so-called Auger effect (see Fig. 12).

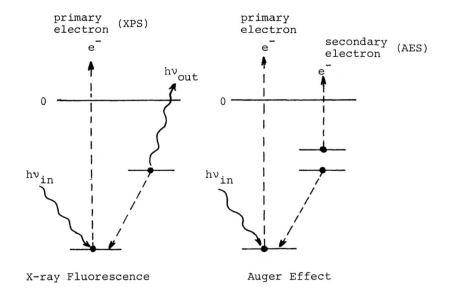

X-ray Fluorescence   Auger Effect

Fig. 12   *Schematic representation of the two possible relaxation processes after expellation of an electron from a low lying orbital.*

Both X-ray fluorescence radiation and the Auger electrons are characteristic for the emitting atoms and can be used for quantitative elemental analysis. Appropriate sample preparation and calibration is, however, essential when using this method.

A suitable method for sample preparation consists of making a solid solution of the zeolite in a $LiBO_2$ melt followed by rapid cooling to a glassy material. For calibration standard zeolite samples with known composition should be used.

4.2.2 Proton/Particle Induced X-ray Emission (PIXE)

A modification of the former method can be used if high energy particles are at one's disposal. Holes in the low-lying orbitals then can be created by a bombardment of the sample with these particles followed by a registration of the emitted X-rays. Again adequate sample preparation and calibration is essential for an accurate analysis.

4.2.3 X-ray Photoelectron Spectroscopy (XPS)

Instead of measuring the emitted X-rays or secondary electrons it is also possible to measure the energy spectrum of the primary electrons (see Fig. 12). This is done in X-ray Photoelectron Spectroscopy and the resulting spectrum can be used for qualitative analysis and after calibration also for quantitative analysis of the samples.

Because the escape depth of the electrons is limited to a few atom-layers only the surface composition of the samples is obtained. In combination with sputtering of the sample by bombarding the sample with noble gas ions it is possible the obtain a depth profile of the chemical composition of the sample.

4.2.4 Electron Probe Micro Analysis (EPMA)

Modern scanning electron microscopes can have the facility to focuss the electron beam on a small spot on the surface of the sample, which can result in the expellation of electrons from the atoms, followed by X-ray fluorescence or the emission of Auger electrons.

In this case the emitted X-rays are originating only from the small spot radiated by the electron beam and thus give information on the chemical composition of that spot (diameter about 8 nm). By scanning the electron beam over the sample it is possible to obtain information about the distribution of the elements in this way (11).

All methods discussed above can be used to determine the chemical composition. For example when the aluminum content is determined the result is the <u>total amount</u> of aluminum present in the sample (or part of the sample).

It is, however, impossible to distinguish between <u>framework</u> aluminum and <u>non-framework</u> aluminum in this way. As only the framework aluminum can give rise to the formation of Brønsted acidic sites whereas non-framework aluminum can provide Lewis acidity it is important to have methods available for the determination of both types and amounts of the aluminum. Some methods developed for this purpose will be discussed in the next paragraph.

5. DETERMINATION OF THE FRAMEWORK Si/Al RATIO.

As mentioned in the introduction the main application of zeolite Y is in catalytic cracking, and for this application the framework Si/Al ratio proved to be extremely important. So most information on the determination of the framework Si/Al ratio is based on this type of zeolite, and in this paragraph the different methods will be discussed on the basis of the results with zeolite Y.

5.1 XRD-Lattice constant

As zeolite Y possesses cubic symmetry (a=b=c) the unit cell dimensions can be characterized by the single lattice constant **a**.
Because the Al-O bond-length is larger than the Si-O bond-length this lattice constant **a** will increase with increasing (framework) Al content of the zeolite.
Based on the results of chemical and X-ray diffraction analysis of a set of hydrated NaY zeolites made by direct synthesis, having a number of framework Al atoms per unit cell (N_{Al}) in the range of 48 to 77 Breck and Flanigen (12) reported the following relation between N_{Al} and **a** (see Fig. 13).

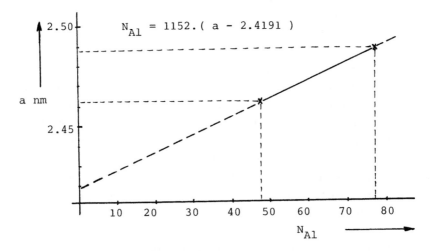

$$N_{Al} = 1152.(a - 2.4191)$$

Fig. 13   Breck-Flanigen relation between the number of framework Al atoms per unit cell N_{Al} and the lattice constant **a** for hydrated NaY zeolites.

Once this relation is known it is possible to derive N_{Al} of an unknown hydrated NaY zeolite from the experimentally determined lattice constant **a**.

A potential problem of this method is that the Breck-Flanigen relation is only valid for hydrated samples and that the lattice constant is also influenced by cation exchange and substances formed during thermal treatment of the zeolite. The zeolite framework is not a rigid construction but can accomodate itself to the requirements of cations and other phases present in the framework. This is for example illustrated by Fritz et al. (13) who studied the influence of dehydration and NH_4^+ exchange of NaY zeolites. Upon dehydration they observed an increase of the lattice constant from 2.466 nm to 2.4765 nm while NH_4^+ exchange to very low sodium content lead to an increase to 2.478 nm. Roelofsen et al. (14) found evidence that also the presence of rare earth ions causes deviations from the Breck-Flanigen relation.

Another drawback of this method is that it is restricted to zeolites with cubic structure, for other zeolites no such relations have been reported.

5.2 IR spectroscopy

In the infrared spectrum of zeolites in the range of $300 - 1300$ cm^{-1} the lattice vibrations can be observed. These vibrations can be divided in structure sensitive and structure insensitive bands, Flanigen (15) reports the following positions of these bands:

Structure insensitive vibrations
- asymmetric stretch $950 - 1250$ cm^{-1}
- symmetric stretch $650 - 720$ cm^{-1}
- T-O bending $420 - 500$ cm^{-1}

Structure sensitive vibrations
- asymmetric stretch $1050 - 1150$ cm^{-1}
- symmetric stretch $750 - 820$ cm^{-1}
- double ring vibrations $500 - 650$ cm^{-1}
- pore opening vibration $300 - 420$ cm^{-1}

For some of the structure sensitive bands a linear relation between the wave number and the number of lattice aluminum atoms is reported (16). After calibration it is possible to use this relation to derive the number of lattice aluminum atoms from the band positions. Fig. 14 shows the results obtained by Flanigen for Y-zeolites with different degree of dealumination.

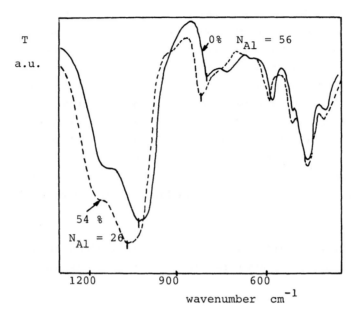

Fig. 14   *IR spectra of Y-zeolites with different degree of dealumination (% dealuminatium denoted in Figure).*

From this figure it is possible to derive the following relation between the position of the structure sensitive asymmetric stretch band (σ_1) and the number of lattice aluminum atoms (N_{Al}).

$$N_{Al} = 0.960 \ (1068 - \sigma_1)$$

Also the position of the symmetric stretch band (σ_2) can be used. Lunsford et al. (17) give the following relations to derive N_{Al} from the positions of the two bands.

$$N_{Al} = 0.766 \ (1086 - \sigma_1) \text{ and}$$
$$N_{Al} = 1.007 \ (838 \ \ - \sigma_2)$$

As no such relations are reported for other zeolite types this method is restricted to Y-zeolites.

5.3 ^{29}Si and ^{27}Al MAS–NMR

Since the introduction of the Magic Angle Spinning technique to NMR spectroscopy it is possible to obtain high resolution spectra from solid samples. Lippmaa et al. (18) were the first who applied in 1981 this method to record the ^{29}Si-spectra of zeolites. Since then the technique has been developed rapidly and at the moment is widely applied in zeolite research. The state of the art is very well described in the book of Engelhardt (19) who is also the author of one of the chapters in this book. Here only a short description of the use of this method to determine the number of framework Al atoms will be given.

Application of the ^{29}Si MAS NMR technique to zeolites results in spectra in which a number of separate Si peaks can be observed. For the faujasite structure in which all T-atoms (Si and Al) are crystallographically equivalent it is possible to distinguish up to five different peaks that can be attributed to Si connected-through oxygen- to 0,1,2,3 and 4 Al atoms, respectively (see Fig. 15).

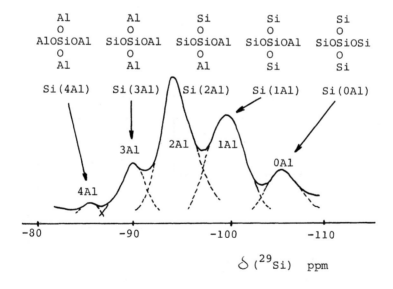

Fig. 15 ^{29}Si MAS-NMR spectrum of NaY zeolite with Si/Al = 2.5

As the intensity of a peak is proportional to the number of the Si atoms concerned, it is possible to calculate the total number of Si atoms by sommation of the peak intensities:

$$N_{Si} = C.\sum_{n=0}^{4} I(Si-nAl)$$

If the Löwenstein rule is obeyed, every framework Al atom is surrounded by 4 Si atoms. So the total number of framework Al atoms can be calculated from:

$$N_{Al} = C.\sum_{n=0}^{4} \frac{n}{4} I(Si-nAl)$$

The framework Si/Al ratio can then simply be calculated from:

$$Si/Al = \frac{N_{Si}}{N_{Al}} = \frac{\sum_{n=0}^{4} I(Si-nAl)}{\sum_{n=0}^{4} \frac{n}{4} I(Si-nAl)}$$

Table 3 shows the results of this method when applied to determine the framework Si/Al ratio of a NH_4NaY zeolite before and after steaming, clearly showing the dealumination of the framework upon steaming.

TABLE 3 ^{29}Si MAS NMR Peak Intensities of NH_4NaY before and after steaming.

Sample	Peak Intensities					Si/Al ratio
	4Al	3Al	2Al	1Al	0Al	
dried NH_4NaY	1	11.5	48.1	35.6	4.9	4.8
steamed NH_4NaY	0	4.1	5.7	30.2	60.0	14.8

Problems can be the presence of paramagnetic cations (such as some rare earth ions), that can cause appreciable broadening of the NMR peaks, as has been reported by Roelofsen (20) and Scherzer (21).

In principle it is possible to determine Al directly by ^{27}Al MAS NMR. In practice, however, several problems can arise that hamper the interpretation of the results. The main reason for these problems is the fact that the Al nucleus possesses a quadrupole moment that may give rise to strong line broadening especially in the case that the Al is present in a low symmetry environment. Because of the latter reason several authors report no problems with the determination of the tetrahedrally coordinated framework Al but the non-framework Al is not completely "visible".

Grobet et al. (22) report the addition of the Al-coordinating compound acetylacetone to the sample as a method to make visible also this part of the Al.

6. DETERMINATION OF ZEOLITE ACIDITY

The acidity of zeolites is mainly caused by the presence of Brønsted acid sites but especially after high temperature treatments also Lewis acid sites may be present (see Fig. 16).

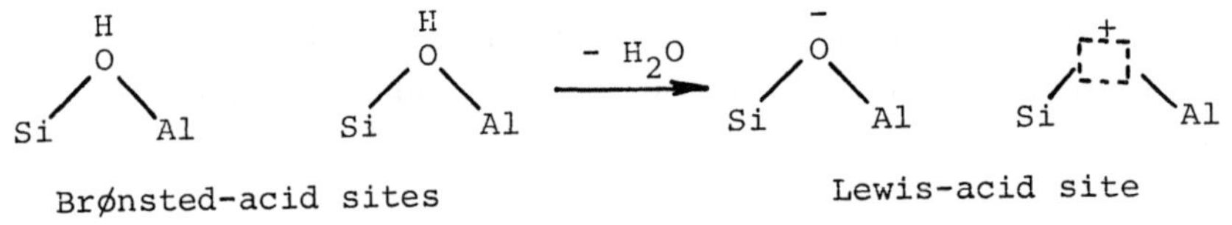

Fig. 16   *Brønsted and Lewis acid sites in zeolites.*

For a complete characterization of zeolite acidity it is, therefore, necessary to determine number and strength of both types of acid sites. Several methods have been developed for this purpose from which the most important are:

- titration methods
- adsorption and desorption of bases
- IR spectroscopy   – of OH groups
                                 – of adsorbed bases
- NMR spectroscopy   – of OH groups
                                 – of adsorbed bases

This paragraph will give a brief description of these methods.

6.1 Titration methods

The acid-strength of a Brønsted acid BH is derived from its proton donating ability and can be expressed by the equilibrium constant of the reaction

$$BH \underset{}{\overset{K_a}{\rightleftharpoons}} B^- + H^+$$

$$K_a = \frac{a_{H^+} \cdot a_{B^-}}{a_{BH}} = a_{H^+} \frac{f_{B^-} \cdot C_{B^-}}{f_{BH} \cdot C_{BH}}$$

In which f_i is the activity coefficient of species i and C_i its concentration.

Then by definition:

$$pKa = -\log Ka = -\log \left[a_{H^+} \cdot \frac{f_{B^-} \cdot C_{B^-}}{f_{BH} \cdot C_{BH}} \right]$$

Because for solid acids, as zeolites are, the activity coefficients are very difficult to quantify, Hammett introduced the acidity function H_o.

$$pK_a = -\log \left[a_{H^+} \cdot \frac{f_{B^-}}{f_{BH}} \right] - \log \frac{C_{B^-}}{C_{BH}} = H_o - \log \frac{C_{B^-}}{C_{BH}}$$

It is evident from this definition that as the limit of a dilute aqueous solution is approached the values of f_{B^-} and f_{BH} approach 1 and H_o becomes equal to the pH.

The Hammett acidity function thus behaves similar to the pH in that it becomes more negative the greater the acidity of the solution. The value reaches -10 in the most concentrated solutions of H_2SO_4 or $HClO_4$.

To determine H_o for solid acids Hammett used a series of aniline bases as indicators which required acids of different strength to give color change (see Table 4).

TABLE 4   *Series of Indicators to determine the Hammett acidity function.*

Hammett Indicator	H_o
Neutral red	+ 6.8
Phenylazonapthylamine	+ 4.0
Butter yellow	+ 3.3
Benzeneazophenylamine	+ 1.5
Dicinnamal acetone	- 3.0
Benzal acetophenone	- 5.6
Anthraquinone	- 8.2

270

To determine the acid strength (distribution) of a zeolite the solid is dispersed in a nonpolar solvent (e.g. n-hexane) and a series of indicators requiring successively stronger acids to become protonated is added until no color change is observed. When a color change is observed for a given indicator it is concluded that some of the sites on the zeolite correspond to the H_O characterized by that indicator. Then the procedure is repeated after addition of a known amount of a base (e.g. n-butylamine) that neutralizes the strongest acid sites. In this way an acid-strength distribution curve as illustrated in Fig. 17 can be obtained (23).

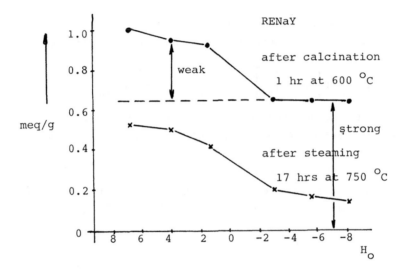

Fig. 17   *Acid-strength distribution curves of RENaY after various pretreatments.*

Problems with this method are: the impossibility to distinguish between Brønsted and Lewis acidity and the fact that the indicator molecules cannot enter the zeolite pores, and thus can react only with the acid sites at the outer surface of the zeolite.

6.2 Adsorption of bases; Calorimetric measurements

When an acid is neutralized by a reaction with a base the heat of neutralization is evolved. This heat of neutralization is larger the stronger the acid and so it can be used to characterize acid-strength. Auroux et al. (24) have demonstrated this method for several zeolites. Fig. 18 shows the results for H-ZSM-5 and Na-ZSM-5.

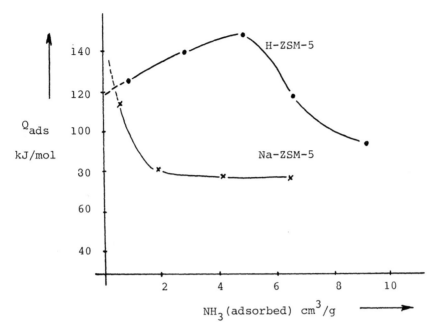

Fig. 18   *Heat evolution during the adsorption of NH₃ on H-ZSM-5 and Na-ZSM-5 at 150°C.*

As can be seen in this figure more heat is evolved when NH_3 is adsorbed on H-ZSM-5 than on Na-ZSM-5 demonstrating the presence of stronger acid-sites. A complication of this method is the accessibility of the acid sites and the long times that are needed to reach equilibrium. The fact that the heat evolved after the addition of the first dose of NH_3 to H-ZSM-5 is smaller that the heat evolved after the second and third dose in the example mentioned above is explained by the authors on that basis. The first dose of NH_3 is supposed not to react with the strongest acid sites but mainly with the more easily accessible weaker sites, and the more difficult accessible stronger acid sites are only neutralized after the addition of the second and third dose.

Furthermore, the method is rather time-consuming and so most investigators prefer the more rapid method in which the opposite reaction is used.

6.3 Desorption of bases. Temperature Programmed Desorption (TPD)

As discussed in the preceeding paragraph bases adsorb stronger on stronger than on weaker acidic sites. Thus it will cost more energy or higher temperatures to desorb bases from stronger acidic than from weaker acidic sites.

This principle can be used to characterize the acid sites present on a zeolite sample. First the zeolite is contacted with a base (NH_3 or pyridine) to neutralize the acidic sites present. Then the temperature is raised at a constant rate and the amount of desorbed base is recorded (either by mass spectrometry or by a heat-conductivity cell). In this way a desorption spectrum is obtained as presented in Fig. 19, for zeolite H-ZSM-5.

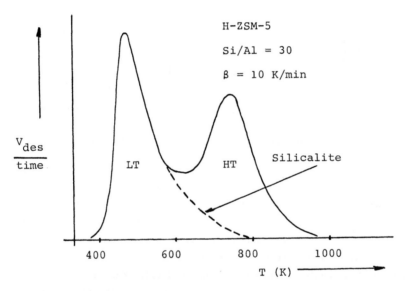

Fig. 19   *Temperature programmed desorption spectrum of NH$_3$ from H-ZSM-5 (Si/Al = 30)*

In this spectrum two peaks can be observed. A low temperature peak (LT) corresponding with NH_3 desorbing from the weaker acidic sites. This peak can also be observed when desorbing NH_3 from non-acidic silicalite. And a high temperature peak (HT) corresponding with the stronger acidic sites. The area of this peak gives information about the amount of these strong acidic sites, while the peak-maximum-temperature (T_m) gives information about its acid-strength. Cvetanovic and Amenomiya (25) have developed a theory enabling the derivation of the heat of desorption from the peak-maximum-temperature (T_m) at a certain heating rate (β).

Although temperature programmed desorption of ammonia (NH_3-TPD) is a simple and rapid method to characterize zeolite acidity, the method has its limitations. Firstly is it not possible to distinguish between ammonia desorbing from Brønsted or Lewis acidic sites, and even desorption from non-acidic sites (like in silicalite) is recorded. For Y-zeolites a second more serious problem is, however, the possibility of diffusion limitation during the desorption of ammonia from adsorption sites in the pores and cavities of the zeolite, causing desorption at higher temperatures than corresponding with the acid-strength of the desorption-sites.

6.4 IR-spectroscopy of OH groups

A more direct study of the Brønsted-acidic OH groups in zeolites is possible by IR-spectroscopy.

This method, that is extensively reviewed by Ward (26), is based on the dependence of the O-H stretching frequency on the acidity.

The weaker the O-H bond, the lower the stretching frequency and the higher the acid-strength. Fig. 20 shows a typical IR-adsorption spectrum of a thin wafer of HNaY after dehydration.

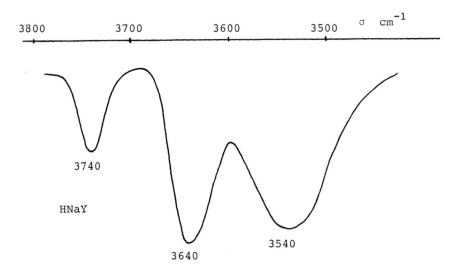

Fig. 20   *IR-Absorption spectrum of HNaY*

274

The spectrum shows three peaks in the O–H stretching region. The peak at 3740 cm^{-1} is assigned to the non-acidic silanol groups that are present at the outer surface of the zeolite crystals and at structural defects.

The peaks at 3640 cm^{-1} and 3540 cm^{-1} correspond with OH groups having lower O–H bond strength and so more acidic character. From these bands the one at 3640 cm^{-1} dissappears after adsorption of pyridine and so this band is assigned to the accessible acidic OH groups.

Other zeolites show similar spectra with small shifts in the positions of the peaks. Barthomeuf (27) has measured the exact position of the IR peaks corresponding with the acidic OH-groups for a series of zeolites with varying Si/Al ratio (see Fig. 21).

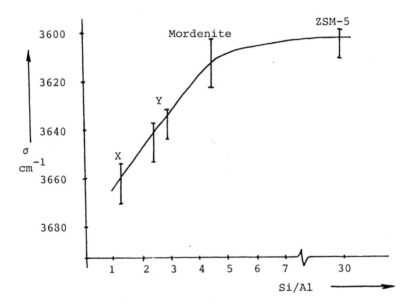

Fig. 21   *Position of the IR-band corresponding with the acidic OH-groups for a series of zeolites with varying Si/Al ratio.*

From this figure it can be seen that the peak position shifts from about 3660 cm^{-1} for zeolite X with a Si/Al ratio of 1.25 to about 3610 cm^{-1} for Mordenite with a Si/Al ratio of 4.5. Further increase of the Si/Al ratio has only a minor effect. For ZSM–5 with a Si/Al ratio of 30 the peak is observed at 3605 cm^{-1}. It must be concluded that the acid–strength of zeolites increases with increasing Si/Al ratio till a maximum value is reached at a Si/Al ratio of about 6 to 7. To explain this result it is accepted that the acid-strength of a bridging OH group in a zeolite depends on the number of neighbouring Al atoms (see Fig. 22).

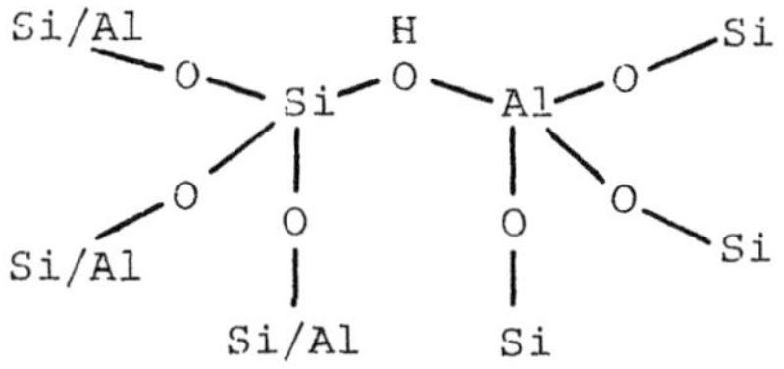

Fig. 22   Surrounding of a bridging OH group in a zeolite.

According to the Löwenstein rule Al is surrounded by 4 Si atoms, but the Si atom can be surrounded by 0 to 4 Al atom. The bridging OH thus can have 0 to 3 Al atoms as next-nearest neighbours. The probability of this will depend on the Si/Al ratio of the zeolite.

Maximum acidity is obtained with no Al atoms in the second coordination shell. This will be reached when the Si/Al ratio of the zeolite exceeds 9 or 10 (33). Further increase of Si/Al then has no more effect on the acid-strength. Thus determination of the position of the IR absorption band belonging to the acidic OH groups can give information on the acid-strength of these groups.

Because it is difficult to determine the mass of the sample from which the IR adsorption spectrum is recorded the method only gives qualitative information on the number of acid sites.

6.5 IR spectroscopy of adsorbed bases

Instead of looking to the acidic OH groups it is also possible to study the IR absorption of bases that have reacted on the acid sites of the zeolite. Frequently the reaction with pyridine is used for this purpose. As indicated in Fig. 23 pyridine can react with the Brønsted-acid sites as well as with the Lewis acid sites of the zeolites. The reaction on the Brønsted acid site results in the formation of a pyridinium ion while the pyridine molecule is coordinatively bound on the Lewis acid sites. Both the pyridinium ion and the coordinatively bound pyridine have characteristic IR absorption bands.

276

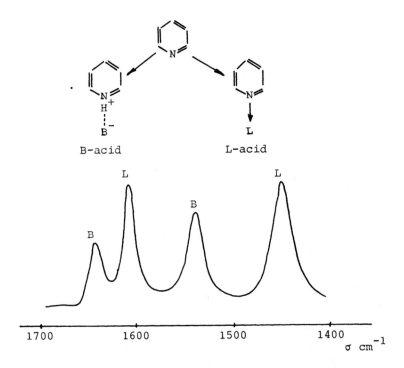

Fig. 23   *IR spectrum of pyridine adsorbed on partly dehydroxylated HY*

In this way it is possible to indicate the presence of both Brønsted and Lewis acid sites in zeolite samples whilst the intensity of the bands gives information on the number of these sites. It is, however, impossible to obtain information on the acid-strength. The presence of the absorption bands only shows that acid-sites are present with an acid-strength large enough to react with the basic probe molecule pyridine.

Jacobs et al. (28) have, therefore, modified the method by using the less basic probe molecule benzene. Reaction with Brønsted acid site then does not result in complete proton transfer to form the benzenium ion but in an interaction of the benzene with the acid proton at which the proton remains bonded to the bridging oxygen atom. As a consequence, the 0-H bond strength is weakened, which can be observed by a shift of the corresponding IR band to lower wavenumbers.

Table 5 shows the observed shifts when adsorbing benzene on different zeolites in the proton form.

TABLE 5   Observed shift of the position of the OH band in the IR spectrum when interacting with benzene for these types of zeolites

	σ_{OH} cm^{-1}	$\Delta\sigma_{OH}$ cm^{-1}	relative intensity
H–ZSM–5 Si/Al = 30	3600	300 348	50 50
H–ZSM–11 Si/Al = 30	3603	220 300 340	35 50 15
Dealuminated H–Y Si/Al = 30	3610	326	100

Two types of acidic groups are observed in H–ZSM–5. One type showing a shift of 300 cm^{-1} when interacting with benzene and a second type showing a shift of 348 cm^{-1}. The larger shift observed for the latter type is caused by a stronger interaction with the benzene molecule. The intensity ratio (50:50) indicates that equal amounts of both groups are present.

6.6 ^{1}H MAS–NMR of OH–groups

Since the invention of the magic angle spinning technique it is also possible to obtain highly resolved NMR spectra from solid samples.

Pfeifer et al. (29) have applied this technique to study the different types of protons present in zeolites. Fig. 24 shows an example of a ^{1}H MAS–NMR spectrum as obtained by them.

In the upper spectrum 4 types of protons (a,b,c,d) can be observed from which 3 types disappear after addition of deuterated pyridine. Peaks b, c and d thus must belong to acidic protons. From these, peak d can be assigned to NH$_4$$^+$ that remained in the zeolite during the calcination of the parent NH$_4$–Y. Peaks b and c, however, result form acidic OH groups and correspond with the 3640 cm^{-1} and 3540 cm^{-1} bands, respectively, in the IR spectrum (see Fig. 20). Also in this case the position of the peak in related to the O–H bond–strength (The chemical shift δ_H increases with decreasing bond–strength) and thus gives information on the acid–strength, while the peak area can be used to quantify the number of acid groups.

278

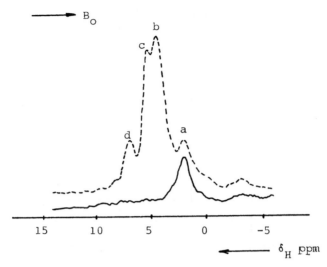

Fig. 24   *¹H-MAS-NMR spectrum of H-Y zeolite before (---) and after (—) the addition of
deuterated pyridine.*

Experimentally, the technique is rather demanding especially because the
method is extremely sensitive for the presence of H_2O.

Summarizing, it can be said that several techniques can be used to
characterize the acidity of zeolites all with their pro's and contra's (see
Table 6).

TABLE 6   *Available techniques for the characterization of zeolite acidity. Potentials and
Problems.*

METHOD	BRØNSTED	LEWIS	AMOUNT	STRENGTH	PROBLEMS
TITRATION	+	+	+	+	ACCESSIBILITY
ADSORPTION	+	+	+	+	DIFFUSION
TPD	+	+	+	±	DIFFUSION NON-ACIDIC ADSORPTION
IR OH	+	−	−	+	SAMPLE PREPARATION
IR BASES	+	+	±	±	SAMPLE PREPARATION
NMR ¹H	+	−	±	+	WATER ADSORPTION

7. ZEOLITE STABILITY

An important requirement posed on zeolites is their stability under operation conditions. This means amongst others thermal stability towards amorphization and dealumination. Various methods are available to check this stability, the most versatile of which is X-ray diffraction. By analysing the zeolite before and after a certain thermal treatment or certain reaction conditions with X-ray diffraction it can be determined whether or not changes in degree of crystallintiy, crystal structure or number of framework aluminum atoms have occurred. Of special interest is the High Temperature X-ray diffraction technqiue in which the diffraction pattern is continuously recorded while heating up the sample. An example of a pattern obtained in this way is presented in Fig. 25.

This figure clearly shows that upon heating at about 200°C the diffraction lines shift to the left indication an increase of the lattice constant from 2.477 to 2.484 nm, which is caused by dehydration of the zeolite.

At about 600°C intensity changes of lines take place and new lines appear indicating a change in crystal structure. Finally collapse of the crystal structure is observed at about 1100°C when the diffraction lines disappear.

280

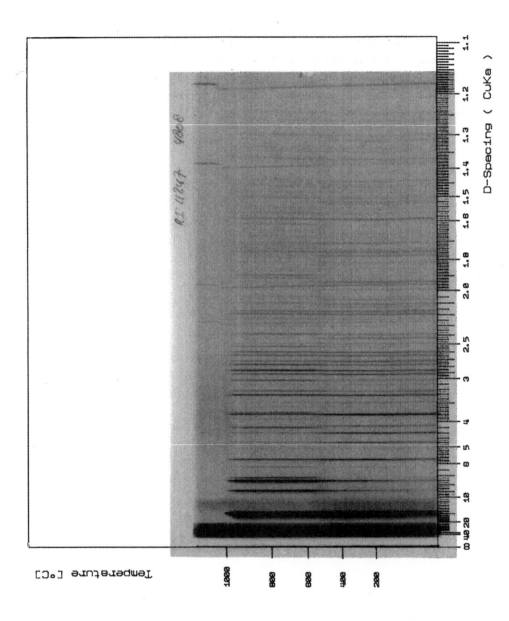

Fig. 25 *Variable temperature X-ray diffraction pattern of rare-earth exchanged zeolite Y (14% RE_2O_3).*

A second method to determine thermal stability is by differential thermal analysis (DTA) (see Fig. 26).

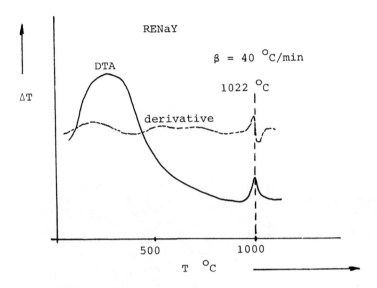

Fig. 26   *DTA pattern of RENaY (14% RE$_2$O$_3$)*

The DTA pattern shows a exothermic reaction at about 1020°C caused by the collapse of the crystal structure.

282

8. ZEOLITE MORPHOLOGY AND PARTICLE SIZE

The size and morphology of the zeolite particles will influence the properties. A typical example is the size and shape that are required for the application of NaA zeolite as a builder in detergent formulations. As reported (30) the particle size should be in the range of 0.1 to 10 μm and the edges of the cubic zeolite crystals should be rounded off. Also for other applications where the kinetics of adsorption or ion exchange play a role size and shape of the zeolite particles will be important.

Scanning electron microscopy (SEM) is the most versatile technique to study the morphology and particle size distribution of zeolites (see Fig. 27).

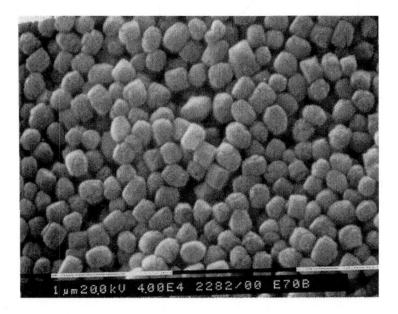

Fig. 27   Scanning Electron Micrograph of a titanium silicalite-1 sample

Other methods to determine particle size are based on sieving and sedimentation. By the dry sieving method particles larger than 5 μm can be determined, while the coulter counter sedimentation method is suitable for particles down to 0.5 μm.

REFERENCES

1. D.W. Breck; "Zeolite Molecular Sieves", John Wiley & Sons, 1974, p.636.
2. B.C. Lippens and J.H. de Boer; J. Catal. 4 (1965), 319.
3. S.J. Gregg and K.S.W. Sing; in "Adsorption, Surface Area and Porosity", 2nd Ed, Academic Press, 1982.
4. J. Fraissard and T. Ito; Zeolites 8, (1988), 351.
5. R. Benslama, J. Fraissard, A. Albizane, F. Fajula and F. Figueras; Zeolites 8, (1988), 196.
6. T. Ito, J.L. Bonardet, J. Fraissard, J.B. Nagy, C. Andre, Z. Gabelica and E.G. Derouane; Appl. Catal. 43, (1988), L5.
7. E.W. Scharpf, R.W: Crecely, B.C. Gates and C. Dybowski; J. Phys. Chem. 90, (1986), 9.
8. V.J. Frilette, W.O. Haag and R.M. Lago; J. Catal. 67, (1981), 218.
9. J.A. Martens and P.A. Jacobs; Zeolites 6, (1986), 334.
10. S. Ernst, R. Kumar, M. Neuber and J. Weitkamp; in "Characterization of Porous Solids". Ed. K.K. Unger et al, Elsevier, 1988.
11. C.E. Lyman, P.W. Betteridge and E.F. Moran; in "Intrazeolite Chemistry", Ed. G.D. Stucky and F.G. Dwyer, ACS Symp. Ser. 218, 1983.
12. D.W. Breck and E.M. Flanigen; in "Molecular Sieves", Soc. Chem. Ind. 1968, p.47.
13. P.O. Fritz, J.H. Lunsford and C.M. Fu; Zeolites 8, (1988), 205.
14. J.W. Roelofsen, H. Mathies and R.L. de Groot; in "Zeolites: Fact, Figures, Future", Elsevier 49, 1989, p.643.
15. E.M. Flanigen, H. Khatami and H.A. Szymanski; Adv. Chem. Ser., 10, (1971), 201.
16. E.M. Flanigen; in "ACS Monograph 171" Ed. J.A. Rabo, 1976.
17. J.R. Sohn, S.J. DeCanio, J.H. Lunsford and D.J.O. Donnell; Zeolites, 6, (1986), 225.
18. E. Lippmaa, M. Magi, A. Samoson, M. Tarmak and G. Engelhardt; J.Am. Chem.Soc. 103, (1981) 4992.
19. G. Engelhardt and D. Michel; in "High Resolution Solid State NMR of Silicates and Zeolites", John Wiley & Sons, 1987.
20. J.W. Roelofsen, H. Mathies, R.L. de Groot, H. Angad Gaur and P.C.W. van Woerkom; in "New developments in Zeolite Science and Technology" Ed. Y Murakami, A. Iijima and J.W. Ward; Kondansha Elsevier 1986, p.337.
21. P.S. Iyer, J. Scherzer and Z.C. Mester; in "Perspectives in Molecular Sieve Science", Ed. W.H. Flank and T.E. White jr., ACS Symp. Ser., 368 (1988), 48.
22. P.J. Grobet, H. Geerts, J.A. Martens and P.A. Jacobs; J. Chem. Soc. Chem. Comm. 22, (1987), 1688.
23. L. Moscou and R. Moné; J. Catal. 30, (1973), 417.
24. A. Auroux and J.C. Vedrine; in "Catalysis by acids and bases" Elsevier, 20, 1985, p. 311.
25. R.J. Cvetanovic and Y. Amenomiya; Adv. Catal. 17, (1967), 103.
26. J.W. Ward; in "ACS Monograph 171" Ed. J.A. Rabo, 1976.
27. D. Barthomeuf; in "Catalysis by Zeolites" Elsevier, 5, 1980, p. 55.
28. P.A. Jacobs, J.A. Martens, J. Weitkamp and H.K. Beyer; in "Selectivity in Heterogeneous Catalysis", Far. Disc. Chem. Soc., 72, (1981), 351.
29. H. Pfeifer, D. Freude and M. Hunger; Zeolites, 5, (1985), 274.
30. R.A. Llenado; in "Proceedings Sixth International Zeolite Conference", Ed. D. Olson and A. Bisio, Butterworths, 1983, p.940.
31. B. Kraushaar, L.J.M. van de Ven, J.W. de Haan and J.H.C. van Hooff; in "Innovation in Zeolite Materials Science" Elsevier, 37, 1988, p. 167.
32. D.R. Corbin, B.D. Burgess Ir., A.J. Vega and R.D. Farlee; Anal. Chem., 59, (1987), 2722.
33. W.A. Wachter; in "Proceedings Sixth International zeolite Conference", Ed. D. Olson and A. Bisio, Butterworth, 1983, p. 141.
34. ASTM Standards on Catalysts, (1988), D3906-85a.
35. P. Scherrer; Göttinger Nachrichten 2, (1918), 98.

Chapter 8

SOLID STATE NMR SPECTROSCOPY
APPLIED TO ZEOLITES

G.ENGELHARDT

Faculty of Chemistry, University of Konstanz,

D-7750 Konstanz, Federal Republic of Germany

1. INTRODUCTION

Since the phenomenon of *nuclear magnetic resonance* (NMR) was discovered in 1946 by Bloch and Purcell, NMR spectroscopy has emerged as one of the most powerful tools for investigating the structure and dynamics of molecular systems. NMR spectrosocopy relies on the fact that transitions can be induced between magnetic spin energy levels of certain atomic nuclei in a magnetic field. The main application of NMR in chemistry is based on the early observation that the transition frequency measured in the NMR spectrum of an atomic nucleus in a particular chemical or structural environment is a very sensitive probe of that environment. Provided that the distinct "chemically shifted" transitions of nuclei of the same element residing in different surroundings can be resolved in the NMR spectrum, unique information on the structure, conformation and dynamics of the system can be obtained. It is this distinction of individual resonance lines, or of specific line patterns for resonance atoms in chemically or structurally distinct locations, which defines a *"high-resolution"* NMR spectrum.

Until recently the application of high-resolution NMR spectroscopy in chemistry was more or less restricted to *liquids*, since the straightforward application of the conventional NMR techniques to *solids* yields, in general, spectra with very broad and almost featureless resonance lines, masking any differences in the transition frequencies of resonance atoms in distinct structural environments. However, the development of novel sophisticated NMR techniques in the early 1970's opened the possibility of measuring NMR spectra of solids with comparable spectral resolution to those of liquids. Subsequently, high-resolution solid-state NMR spectroscopy has developed rapidly into an effective and widely used means for elucidating subtle details of the structural properties of a wide range of different kinds of solid materials (for a comprehensive survey see ref.1). Since high-resolution solid-state NMR spectroscopy gives, in general, information on the *local order* of the structure and can be applied to crystalline, microcrystalline and amorphous products, it is a valuable complement to X-ray and other diffraction techniques which probe the *long-range order* of crystalline materials.

The first application of high-resolution solid-state NMR to *zeolites* was presented at a conference in 1979 (ref.2), and a series of pioneering papers were published in 1981 (see e.g. ref.3-6). In this early work, ^{29}Si NMR and some ^{27}Al NMR (ref.7) were used to study the aluminosilicate framework of various zeolites. Subsequently, interest in the application of multinuclear solid-state NMR to zeolites and other silicates and aluminosilicates has grown very rapidly, and more than a thousand papers on the subject have been published during the last years. Progress in the field has been summarized in several reviews (ref.8-10) and a recent book (ref.11).

In this article a short introduction will be given into the basic principles and experimental techniques of high-resolution solid-state NMR, followed by a concise survey of the main information retrievable from the NMR spectra and its application to structural studies of zeolites and related systems.

2. BASIC PRINCIPLES AND METHODS OF HIGH-RESOLUTION SOLID-STATE NMR

2.1 High-resolution NMR: Liquids versus solids

High-resolution NMR spectra of *liquid* samples exhibit, in general, sharp lines characterized by distinct *chemical shifts* (line positions) and *intensities* (integrated peak areas). Both parameters can easily be extracted from the spectra and are closely related to the chemical structure of the investigated sample. The chemical shift reflects very sensitively the immediate structural surroundings of the resonance nucleus, while the intensity is directly proportional to the number of respective nuclei in the sample.

For *solids* the situation is different in that considerable line broadenings arise in the NMR spectra, because of specific interactions of the nuclear spins tightly bound in the rigid lattice of the solid sample. These interactions are averaged in liquids by the rapid thermal motions of the molecules. If the conventional NMR technique is applied to solids, the line broadening generally prevents the resolution of separate peaks of structurally distinct resonance atoms, or even renders the NMR spectra unobservable.

The critical differences between NMR of liquids and solids arise from the directional dependence of the various spin interactions. The fixed orientations of nuclear environments in the solid lattice may affect the shape and position of the NMR lines measured for different orientations of the sample with respect to the external magnetic field B_0. In highly ordered single crystals, where only a small number of symmetry-related orientations of nuclei is present, narrow lines are observed, the positions of which change with the orientation of the crystal in the external field. Microcrystalline or amorphous powders, however, are characterized by a random distribution of different orientations and the observed NMR pattern is a broad superposition of lines from randomly oriented individual nuclei. In liquids the different orientations are averaged by the fast and essentially isotropic motions of the molecules which reduce the spin interactions to a discrete average value or to zero.

2.2 Nuclear spin interactions and line broadening in solids

There are three main interactions which may give rise to line broadening in solid-state NMR spectra:

- The *dipolar coupling* which arises from the *direct dipole-dipole interaction* between the magnetic moments of the observed nucleus and those of neighbouring nuclei. The dipolar interaction depends on the magnitude of the magnetic moments of the interacting nuclei, falls off very rapidly with the internuclear distance $(1/r^3)$ and is independent of the applied magnetic field B_0. Dipolar coupling with protons, such as ^1H-(^1H), ^{13}C-(^1H) or ^{29}Si-(^1H), is the dominant line-broadening interaction in ^1H-NMR spectra, and in NMR spectra of other spin-1/2 nuclei in proton-containing systems, and can range up to about 100 kHz. The homonuclear dipolar interaction of nuclei with low natural abundance (e.g. ^{13}C or ^{29}Si), or low chemical concentration in the sample, can generally be neglected because of the large internuclear distances.

- The *chemical shift anisotropy*, which follows from the *spatial dependence of the nuclear shielding*, is determined by the symmetry of the electronic charge distribution around the NMR nucleus. The interaction increases linearly with the strength of the external magnetic field B_0. Typical line broadenings from chemical shift anisotropy range up to about 100 ppm but may be considerably larger in special cases.

- The *quadrupolar interaction* occurs in nuclei with nuclear spin quantum number $I > 1/2$ (e.g. ^{11}B, ^{17}O, ^{23}Na, ^{27}Al). Nuclei with $I > 1/2$ possess a nuclear *electric quadrupole moment* which interacts with the non-spherically symmetrical *electric field gradient* at the nucleus. The latter is determined by the charge distribution of the surrounding electrons and other nuclei. The quadrupolar interaction can range up to several MHz and completely dominates the spectrum for most quadrupolar nuclei. In general, the quadrupolar powder patterns are mainly affected by the *second order quadrupolar interaction* which decreases with increasing magnetic field strength.

Although the line broadenings resulting from the above interactions obscure the "high-resolution" characteristics of the solid-state NMR spectra of randomly oriented powders, the interactions themselves are also a potential source of information on structure and bonding. Dipolar interactions yield internuclear distances and molecular geometries. Chemical shift anisotropies give information on coordination symmetry, and from quadrupolar interactions information on the electric field gradient, thus allowing the symmetry of charge distribution around the nucleus to be derived. It is, however, often difficult to separate the different interactions and to extract the interaction parameters from the spectra.

2.3 Experimental techniques for line narrowing and sensitivity enhancement

High-resolution NMR spectra of solids which enable structurally distinct nuclei to be resolved as individual resonance lines can be obtained only when the line broadening interactions considered above are removed, or at least substantially reduced. Line narrowing may be achieved mainly by three different experimental procedures:

- *Dipolar decoupling (DD)* removes the *heteronuclear dipolar interaction* by irradiating the resonance frequency of the nucleus giving rise to the dipolar broadening (e.g. ^1H) while observing the nucleus under study (e.g. ^{13}C, ^{29}Si). Dipolar decoupling is mainly used to remove the line broadening effects of dipolar interactions with protons in the NMR spectra of other nuclei.

- *Multiple-pulse sequences (MPS)* can be used to remove *homonuclear dipolar interactions* by irradiation of precisely defined sequences of short, intense radio-frequency pulses which average the dipolar interaction by reorienting the nuclear spins. Multiple pulse sequences are the preferred method used to remove ^1H-^1H dipolar interactions in ^1H NMR spectra.

- *Magic angle spinning (MAS)*. Line broadenings from *dipolar interactions* and *chemical shift anisotropy* can be removed by fast rotation of the sample about an axis inclined at the angle $\theta = 54^o44'$ to the direction of the external magnetic field. Both the dipolar coupling and the chemical shift anisotropy contain an angular dependence of the form $(3\cos^2\theta - 1)$. Thus, if θ is chosen to be $54^o44'$, the "magic angle", this term becomes zero, the dipolar interactions vanish, and the chemical shift anisotropy is averaged to its isotropic value. The rotation frequency in MAS experiments should be of the order of the line width (in Hz) of the static spectrum. Spinning at lower rates yields a spectrum where the central peak is flanked by a series of "spinning side bands" separated by the the rotation frequency. Although rotation frequencies of several kHz are achievable, complex side band patterns may thus appear in the MAS spectra of systems characterized by large dipolar interactions and/or chemical shift anisotropies. *Quadrupolar line broadenings* can also be reduced, but normally not fully removed, by MAS.

Another important technique in high-resolution solid-state NMR which improves the *sensitivity* (i.e. the signal-to-noise ratio) of the spectra of nuclei with low natural abundance (e.g. ^{13}C or ^{29}Si) but does *not affect the resolution* of the spectrum is:

- *Cross-polarization (CP)*. In this technique an indirect excitation of dilute spins S (which are observed) by polarization transfer from abundant spins I (mostly ^{1}H) is employed. The CP experiment consists of three steps: (*i*) excitation of the I spins by a 90^0 pulse; (*ii*) polarization transfer from I to S spins by simultaneous irradiation of the rf fields B_{1I} and B_{1S} of the I and S spins according to the Hartmann-Hahn-condition $\gamma_I B_{1I} = \gamma_S B_{1S}$ (where γ_I and γ_S are the magnetogyric ratios of the nuclei I and S); and (*iii*) acquisition of the FID of the S spins with continued irradiation of the ^{1}H field for heteronuclear dipolar decoupling (see above). The gain in sensitivity is given by the ratio γ_I/γ_S, and amounts to a factor of about four for ^{1}H-^{13}C, and about five for ^{1}H-^{29}Si CP experiments. In addition, the repetition time of the CP pulse sequence is much shorter than the pulse repetition time for direct excitation of the S spins, since the former is determined by the ^{1}H relaxation time $T_1(^1$H) rather than by the normally much longer T_1 of the S spins ^{13}C or ^{29}Si. The CP experiment may thus be repeated with much shorter intervals, giving a further increase in the signal-to-noise ratio of the S nucleus spectrum in a given period of time. The CP efficiency depends on the strength of the I-S dipolar interaction, i.e. on the distance between the I and S nuclei. The CP technique may thus also be used to detect close coordinations between protons and e.g. Si or C atoms.

To achieve optimum line narrowing, and sufficient signal-to-noise ratio in the solid-state NMR spectra, the experimental procedures considered above may be used in combination, e.g. CP-MAS, DD-MAS or CRAMPS (combined rotation and multiple pulse spectroscopy). However, in zeolites and other inorganic systems often no or only a few hydrogen atoms are present. In these cases the CP technique cannot be used and dipolar proton decoupling is not necessary. Thus, application of MAS is sufficient to remove the chemical shift anisotropy and any small dipolar interactions, and to narrow quadrupolar broadened lines, if present. Provided that the chemical shift anisotropies are not too large, it is often advantageous to carry out these experiments at the highest possible magnetic field strength since sensitivity and chemical shift dispersion is maximized at high field, while the dipolar interaction is not affected and the line broadening due to (second order) quadrupolar interactions is minimized.

3. MULTINUCLEAR HIGH-RESOLUTION SOLID-STATE NMR OF ZEOLITES

3.1 Which nuclei are candidates?

In principle, each of the three basic atomic constituents of the alumino-silicate framework of zeolites - *silicon*, *aluminium* and *oxygen* - are amenable to NMR measurements by their naturally occurring isotopes ^{29}Si, ^{27}Al and ^{17}O, respectively. However, the ^{17}O isotope has a very low natural abundance of only 0.037% and a nuclear quadrupole moment giving rise to line broadening, which together make the application of ^{17}O NMR difficult. Isotopic ^{17}O enrichment is possible but hardly practicable for routine use. Therefore, ^{17}O NMR has been little used so far, although some interesting and promising ^{17}O NMR studies of of ^{17}O enriched silicates and zeolites haven been published recently (ref.12). In contrast to ^{17}O, the ^{27}Al isotope is 100% abundant, but it also has a nuclear quadrupole moment and may thus exhibit quadrupolar broadened lines. The isotope ^{29}Si has a natural abundance of 4.7% and a nuclear spin of I = 1/2, i.e. it has no quadrupole moment and thus gives rise to narrow resonance lines. Owing to their important role as framework constituents, their favourable properties in NMR experiments, and especially due to the high sensitivity of their NMR spectra to structural effects (see below), ^{29}Si and ^{27}Al nuclei play the dominant role in NMR studies of zeolites and other silicates and aluminosilicates.

Several other elements replacing silicon or aluminium in tetrahedral sites of the aluminosilicate framework of zeolites (*phosphorous*, *boron*, *gallium*, *germanium*, *beryllium* etc.), can also be profitably studied by solid-state NMR of their respective NMR-active nuclides, e.g. ^{31}P, ^{11}B, $^{69,71}Ga$, ^{73}Ge, ^{9}Be. However, except for ^{31}P which has I = 1/2 and 100% natural abundance, most of the other nuclei possess a quadrupole moment and are often characterized by low relative receptivities in the NMR experiment. ^{31}P NMR has shown to be a useful method for structural studies of aluminophosphate molecular sieves and related systems. ^{11}B NMR, which has a reasonably high sensitivity, has been successfully applied to the study of the incorporation of boron in the framework of high-silica zeolites. Only a few in-depth studies of the other nuclei has been performed so far.

Most elements acting as *charge-balancing cations* in zeolites also have isotopes suitable for NMR experiments, but except for ^{7}Li, ^{23}Na, ^{205}Tl and ^{133}Cs, only a few others have been studied so far. *Organic species* present as templates or sorbates in the zeolite cavities can be advantageously investigated by ^{13}C CP MAS NMR.

Finally, high-resolution ^{1}H-NMR of *protons in zeolites* plays an important role in investigating bridging Si(OH)Al and terminal SiOH or AlOH groups and other proton-containing constituents.

Although the potential of multinuclear solid-state NMR in zeolite science has not yet been fully explored, it is obvious that the richness of information retrievable from ^{29}Si and ^{27}Al NMR will not be matched by the nuclei mentioned above. It is nevertheless significant that an environmental probe exists for almost every constituent which may be present in the zeolite structure. Table 1 summarizes the NMR properties of selected nuclei which have been used in solid-state NMR of zeolites.

Table 1 : NMR properties of selected nuclei applied in NMR of zeolites

Isotope	Spin	NMR frequency (MHz)[a]	Natural abundance (%)	Relative receptivity [b]
^{1}H	1/2	200.00	99.985	1.000
^{7}Li	3/2	77.73	92.58	0.272
^{11}B	3/2	64.17	80.42	0.133
^{13}C	1/2	50.29	1.108	$1.76 \cdot 10^{-4}$
^{17}O	5/2	27.11	0.037	$1.08 \cdot 10^{-5}$
^{23}Na	3/2	52.90	100.00	$9.27 \cdot 10^{-2}$
^{27}Al	5/2	52.11	100.00	0.207
^{29}Si	1/2	39.73	4.70	$3.69 \cdot 10^{-4}$
^{31}P	1/2	80.96	100.00	0.0665
^{133}Cs	7/2	26.23	100.00	$4.82 \cdot 10^{-2}$
^{205}Tl	1/2	115.42	70.50	0.140

[a] at $B_0 = 4.6975$ T

[b] product of natural abundance (%) and NMR sensitivity $[\gamma^3 I(I+1)]$, relative to ^{1}H

3.2 Silicon-29 NMR

Line broadening in solid-state ^{29}Si NMR spectra of microcrystalline zeolite powders is mainly due to the chemical shift anisotropy of the ^{29}Si nucleus. In addition, heteronuclear dipolar interactions between ^{29}Si and ^{1}H or ^{27}Al, and possibly other NMR-active nuclei present in the zeolite structure, may contribute to the linewidth. The application of the MAS technique is therefore essential, and frequently sufficient, for obtaining highly resolved ^{29}Si NMR spectra of zeolites. The simultaneous application of high-power dipolar ^{1}H decoupling is sometimes helpful in removing residual dipolar interactions of ^{29}Si spins with nearby ^{1}H nuclei. The spectra obtained under these conditions show, in general, distinct signals for structurally different Si sites.

The most direct, and rather fundamental, information which follows immediately from the number of distinct lines in the spectrum is the *number of structurally inequivalent Si sites* present in the sample. Moreover, from the normalized peak intensities the *relative proportions* of Si atoms in the various sites can be determined. Information on the *local environment* of the SiO$_4$ tetrahedra forming the zeolite framework can be derived from the chemical shift data.

3.2.1 Structure correlation of ^{29}Si chemical shifts

The ^{29}Si chemical shifts of silicates and aluminosilicates depend highly sensitively on the number and type of T-atoms (T = Si, Al or other tetrahedral framework atoms) connected with a given SiO_4 tetrahedron: Characteristic *high-field shifts* are observed with the *increasing number of SiOT bridges* formed by the given SiO_4 tetrahedron (degree of SiO_4 polymerization). Typical *low-field shifts* follow from the *replacement of Si by Al* in the second coordination sphere of the central Si atom with a given number of SiOT bridges (degree of tetrahedral Al substitution). In zeolites and other aluminosilicates with a three-dimensional framework structure, all silicon atoms are connected via oxygen bridges with four other T-atoms. There are thus five different structural units of the type $Si(OSi)_{4-n}(OAl)_n$ with n = 0 - 4 [conventionally designated "Si(nAl)"].

As a first approximation, neglecting the presence of crystallographically inequivalent Si(nAl) sites, the ^{29}Si NMR spectrum of a zeolite may thus consist of one to five peaks corresponding to the five possible Si(nAl) environments in the zeolite framework. As the number n of Al atoms increases, the peaks are systematically shifted to low field, where each Al substitution results in a shift contribution of about 5 ppm. As shown in Fig.1, such shift differences can readily be resolved in the ^{29}Si MAS NMR spectra of microcrystalline zeolites, and well separated peaks of the whole series of Si(nAl) units appear in the spectrum of the aluminium-rich zeolite P1. In addition to the chemical environment, the ^{29}Si chemical shift is affected by the bonding geometry around the Si atom under study, i.e. by SiO bond lengths and SiOT bond angles. Therefore, *chemically equivalent* but *crystallographically inequivalent* Si atoms may exhibit different chemical shifts.

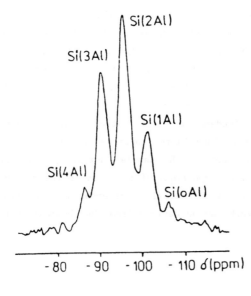

Fig.1. ^{29}Si MAS NMR spectrum of zeolite P1 (Si/Al = 1.9)

^{29}Si chemical shift data of selected zeolites are given in Table 2, and typical shift ranges for the five Si(nAl) environments established from a large body of shift data measured in various types of zeolites and other framework silicates, are shown in Fig.2. Although the different shift ranges partly overlap, the assignment of well separated peaks to the corresponding Si(nAl) environments of a particular zeolite is, in general, feasible. The largest shift range ever observed for a Si(nAl) unit is that of Si(4Al) in aluminosilicate sodalites - depending on the type of salts enclathrated in the sodalite cages, the ^{29}Si chemical shift changes by more than 20 ppm (ref.13, see also Section 3.2.2).

Table 2 : ^{29}Si chemical shifts (in ppm versus TMS) of selected zeolites[a]

Zeolite	Si/Al	Si(4Al)	Si(3Al)	Si(2Al)	Si(1Al)	Si(0Al)
LiCl-sodalite	1.0	-76.4				
NaK-X	1.0	-84.6				
Na-A	1.0	-89.6				
K-sodalite[b]	1.0	-96.7				
ZK-4	1.4	-89.1	-93.9	-99.5	-106.1	-110.7
NaK-P1	1.9	-87.6	-91.9	-97.3	-102.4	-107.0
Na-Y	2.5	-83.8	-89.2	-94.5	-100.0	-105.5
Offretite[c]	2.9		-93.5	-97.5	-101.9	-106.9
			-97.5	-101.9	-106.9	-112.5
Omega[c]	3.1		-89.1	-93.7	-98.8	-103.4
		-89.1	-93.7	-98.8	-107.0	-112.0
TMA-sodalite	4.7			-104.6	-110.5	-116.2
Na-mordenite	5.0			-100.1	-105.7	-112.1

[a] from ref.11

[b] dehydrated

[c] two crystallographically inequivalent Si sites (see text)

3.2.2 Framework structure studies

Local environment of SiO$_4$ tetrahedra. As has been shown above, distinct signals appear in the ^{29}Si NMR spectra of zeolites for the different Si(nAl) environments forming the tetrahedral zeolite framework. The relative signal intensities are directly related to the relative concentrations of the various Si(nAl) units present in the zeolite structure. Fig.3 shows the ^{29}Si NMR spectra of a series of zeolites X and Y with different Si/Al ratios (ref.4). With increasing Si/Al ratio, the decrease of the signal intensities of the aluminium-rich Si(nAl)

294

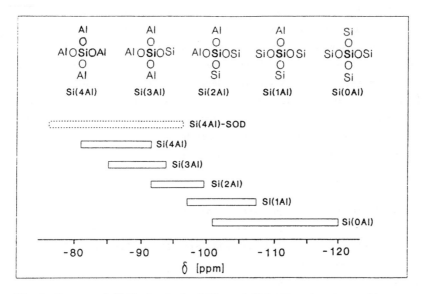

Fig.2. Ranges of ^{29}Si chemical shifts of Si(nAl) units in zeolites.
Si(4Al)-SOD denotes Si(4Al) units in sodalites (see text).

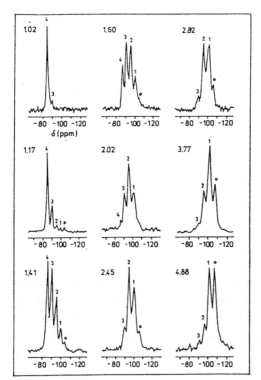

Fig.3. ^{29}Si MAS NMR spectra of zeolites X and Y. The Si/Al ratio is indicated against each spectrum, the peak assignments are given by the number n of the corresponding Si(nAl) unit. (From ref.4).

units and a corresponding increase of the aluminium-poor Si(nAl) peaks is clearly visible. Consequently, from a careful analysis of the chemical shifts and line intensities, the specific types and relative populations of the distinct Si(nAl) units present in a zeolite can, in principle, be determined. Moreover, if other tetrahedral atoms replacing aluminium and/or silicon are present in the framework, characteristic shifts of the Si(nT) lines may be observed. The gallosilicate zeolites are a pertinent example in which the Si(nGa) peaks are shifted to low field by 1.5 to 2 ppm per Ga substitution in comparison with their Si(nAl) counterparts (ref.14).

Silicon atoms bearing OH groups, e.g. $Si(OH)(OSi)_{3-n}(OAl)_n$, located in framework defects or in amorphous parts of the zeolite material exhibit peaks characteristically low field shifted from the corresponding Si(nAl) units, and may overlap with the Si[(n-1)Al] signals. The SiOH peaks can, however, be clearly identified by their intensity enhancement in the CP spectra (ref. 15).

Si/Al ratio and Si, Al ordering. Provided that the ^{29}Si NMR spectrum of a zeolite is correctly interpreted in terms of the Si(nAl) units and no AlOAl linkages are present (i.e. Loewenstein´s rule applies), the *quantitative ratio of tetrahedral Si and Al atoms* in the zeolite framework can be directly calculated from the peak intensities according to the equation

$$Si/Al = \Sigma I_n / \Sigma 0.25 \cdot n \cdot I_n \qquad (1)$$

where I_n are the intensities of the Si(nAl) peaks and summation is from $n = 0$ to $n = 4$ (ref.4). Equation (1) is independent of the specific structure of the zeolite. It includes only that aluminium which is substitutionally incorporated into the tetrahedral framework, and excludes any non-framework aluminium (which is frequently present e.g. in chemically modified zeolites, see below). Equation 1 has been tested *inter alia* in the case of the zeolites X and Y shown in Fig.3., and very good agreement was found between the Si/Al ratios obtained by chemical analysis and those calculated from the spectra.

Information on *Si, Al ordering* in the tetrahedral T-sites of the zeolite framework can be derived from comparison of the relative Si(nAl) populations obtained from NMR peak intensities with model generated populations. The latter were derived from ordering schemes constructed by distributing the respective number of Si and Al atoms on the T-sites of a certain repeating unit of the framework structure. This procedure has been used successfully to study the Si, Al ordering in faujasite-type zeolites of different Si/Al ratios (ref.4, 16) and in several other zeolite types. Fig.4 shows the ordering schemes proposed for a series of zeolites X and Y from the NMR spectra presented in Fig.3.

296

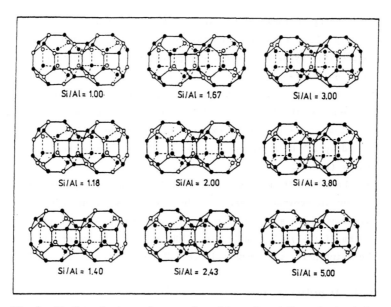

Fig.4. Si,Al ordering schemes of zeolites X and Y as derived from ^{29}Si NMR. Black dots: Si; white dots: Al. (From ref.4)

Crystallographically non-equivalent Si sites. As mentioned above, crystallographically inequivalent Si sites in chemically equivalent environments may have different chemical shifts, which lead to splits of the NMR line of a certain Si(nAl) unit. Although these splits often complicate the spectra interpretation, they may provide valuable information on the presence of crystallographically inequivalent sites and their populations in the zeolite framework. To avoid complications due to superposition of the spectral patterns for the different Si(nAl) units, it is advantegeous to investigate highly siliceous zeolites containing only Si(0Al) environments. Provided that the framework geometry is retained in the high-silica form of the zeolite, the spectrum is greatly simplified, and any multiple line pattern indicates the presence of crystallographically non-equivalent Si sites in the framework.

One of the most striking examples is represented by the essentially aluminium-free variant of the *zeolite ZSM-5* (silicalite) which shows a temperature dependent monoclinic - orthorhombic phase transition. Here, the ^{29}Si NMR spectrum at 295 K (Fig.5a) exhibits 20 well resolved lines (two with threefold intensity) representing the 24 crystallographically inequivalent Si sites of the monoclinic form, while at 393 K (Fig.5b) only 10 lines can be identified in the spectrum (two with double intensity) which correspond to the 12 inequivalent Si sites in the orthorhombic form (ref.17). Moreover, dramatic and fully reversible changes in the spectra are observed upon loading the ZSM-5 zeolite with organic sorbates, such as p-xylene, acetylacetone, pyridine (ref.17), n-decane, cyclohexane or benzene (ref.11), which, together with X-ray powder diffraction studies, provide detailed information on the sorbate-lattice interactions in these materials.

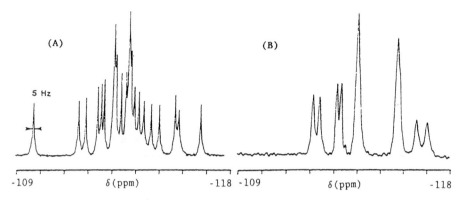

Fig.5. ^{29}Si MAS NMR spectra of highly siliceous zeolite ZSM-5 at 295 K (A) and 393 K (B) (From ref.17).

The ^{29}Si NMR spectra of the high-silica zeolites are also very helpful in the correct assignment of the spectra of the corresponding variants with low Si/Al ratio. Fig.6 shows the ^{29}Si NMR spectra of dealuminated *mordenite, offretite* and *zeolite omega* and of the corresponding low Si/Al parent materials (ref.18). Three lines with the intensity ratio 2:1:3 are observed in high-silica mordenite corresponding to the four inequivalent Si positions in the population ratio $2T_1:1T_4:(2T_2+1T_3)$, while two lines with the intensity ratio of 2:1 appear for the two inequivalent Si sites with the same population ratio in both offretite and zeolite omega. Depending on the shift differences of the crystallographically inequivalent Si sites in the high-silica form, line broadening (mordenite), deceptively simple spectra (offretite) or additional signals (omega) result from the superposition of the Si(nAl) sub-spectra in the aluminium-containing parent zeolites. Clearly, the observed peak overlap must be taken into account when the Si/Al ratios were calculated from these and similar spectra (see e.g. ref.19).

Correlations between SiOT bond angles and ^{29}Si chemical shifts. The differences in the ^{29}Si chemical shifts of Si atoms located in the same chemical environment are mainly due to changes in the bonding geometry of the Si(nAl) structural unit. It has been shown by empirical correlations (ref.13,20-22), and rationalized by theoretical considerations (ref.23,24), that the ^{29}Si chemical shift, δ, of an Si(nAl) unit is linearly correlated with the *average* value of the four SiOT bond angles, $\overline{\alpha}$, at the central silicon atom. By means of linear regression analysis, quantitative relationships between δ and $\overline{\alpha}$, $\overline{\sec \alpha}$, $\overline{\sin (\alpha/2)}$ and

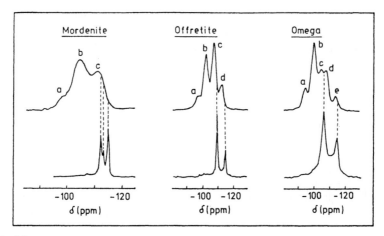

Fig.6. ^{29}Si MAS NMR spectra of mordenite [Si/Al=5], offretite [5.1] and zeolite omega [4.2] (upper spectra), and their highly siliceous (dealuminated) analogues (lower spectra). Peak assignments (upper spectra): mordenite a - Si(2Al), b - Si(1Al), c - Si(0Al); offretite a - Si$_1$(2Al), b - Si$_1$(1Al)+Si$_2$(2Al), c - Si$_1$(0Al) +Si$_2$(1Al), d - Si$_2$(0Al); omega a - Si$_1$(2Al), b - Si$_1$(1Al), c - Si$_1$(0Al)+ Si$_2$(2Al), d - Si$_2$(1Al), e - Si$_2$(0Al) (From ref.18).

cos α/(cos α - 1) have been established. The correlation works with these different functions of α since they are all approximately linear for the range of α involved. The correlation between δ and cos α/(cos α - 1) follows from theoretical considerations (ref.23). Linear regression of the data of 52 different Si(nAl) units in zeolites and silica polymorphs has led to the following relationship (ref.24) with a linear correlation coefficient of 0.90:

$$\delta = -223.9 \overline{\cos \alpha/(\cos \alpha - 1)} + 5n - 7.2 \qquad (2)$$

where n is the number of Al atoms in the Si(nAl) unit under study, and δ is given in ppm from tetramethylsilane, and α in degrees. Although eq. (2) can be applied to any Si(nAl) environment, several other relationships with correlation coefficients higher than 0.98 have been derived for Si(0Al) (ref.23) and Si(4Al) units (ref.13,22) in zeolites. The quantitative relationships between δ and α can be used to estimate mean SiOT bond angles from ^{29}Si chemical shifts or, if the bond angles are known from X-ray studies, to calculate shift data which may be helpful in the correct assignment of the spectra.

The quantitative evaluation of the various relationships between ^{29}Si chemical shifts and SiOT bond angles shows that a chemical shift change of about 0.6 ppm is to be expected for 1° change in the bond angle, with high-field shifts for larger angles. Since for highly crystalline materials a peak-to-peak resolution of 0.1 ppm can be achieved in the ^{29}Si NMR spectra (see e.g. Fig.5), differences in the bond angles of about 0.2^0 may be detected. This extremely high sensitivity of the chemical shifts to small atomic dislocations in the zeolite framework explains the dramatic changes of the ^{29}Si NMR spectra of zeolite ZSM-5 upon sorbate loading or temperature variation (considered above and shown in Fig.5. The relationship between SiOT bond angles and ^{29}Si chemical shifts also offers an explanation for the large shift differences observed for the Si(4Al) environments in sodalites with different enclathrated salts. This shift range, between -76.4 ppm and -96.7 ppm, correlates with SiOAl bond angles between 125^0 and 159^0 (ref.13).

3.2.3 Modification of the zeolite structure

The catalytic, sorptive and ion-exchange properties, and in particular the chemical and thermal stability, of zeolites depend strongly on the tetrahedral aluminium content in the framework. It is, therefore, of particular interest to modify the Si/Al ratio of the zeolite while retaining the topology and crystallinity of the framework structure. Dealumination of the zeolite framework can be achieved e.g. by hydrothermal treatment of the ammonium-exchanged form of the zeolite ("ultrastabilization"), or by chemical treatment of the zeolite with suitable reagents in solution (e.g. acids, chelating agents, salts) or in the vapour phase at high temperatures (e.g. $SiCl_4$). A review of various techniques for dealumination is given in ref.25.

The quantitative determination of the extent of framework dealumination by chemical analysis has proved to be difficult since, after removal from the tetrahedral framework, the aluminium may be retained in the zeolite pores in form of charged or neutral Al-oxo-hydroxo complexes. This non-framework aluminium can, in principle, be extracted from the zeolite by acid leaching, but its total removal cannot always be achieved and the extraction of further framework aluminium cannot be ruled out.

Application of ^{29}Si NMR has been shown to be especially useful in the characterization of the dealuminated zeolites, since it provides direct information on the composition and Si,Al distribution in the tetrahedral framework, independently from the presence of non-framework Al species. Thus, the process of framework dealumination can be monitored in detail by ^{29}Si NMR spectra. A great number of papers on this topic have appeared, which are summarized in Chapter V.3.4 of ref.11.

As an example, Fig.7 shows the ^{29}Si NMR spectra of a series of hydrothermally dealuminated Y zeolites, measured without (upper spectra) and with (lower spectra) cross-polarization (ref.15). Using the signal intensities of the spectra registered without CP, the Si/Al ratio of the zeolite framework can be calculated by eq. (1). The Si/Al ratio amounts to 2.5 for both the parent NaY (sample A) and ammonium-exchanged NaNH$_4$Y zeolites (sample B), but increases to 4.9 after

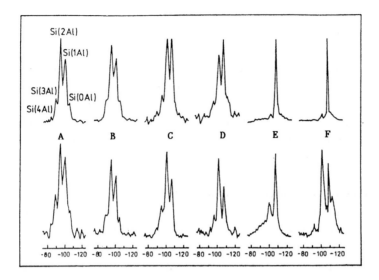

Fig.7. ^{29}Si MAS NMR spectra of dealuminated zeolites Y measured without (upper spectra) and with cross polarization (lower spectra). A - parent NaY (Si/Al=2.5); B - NaNH$_4$Y; C - hydrothermally treated for 3h at 540°C; D - acid leached with 0.1 M HCl at 100°C; E - twice NH$_4$-exchanged and subsequently hydrothermally treated for 3h at 815°C; F - acid lached with 0.1M HCl for 3.5h at 100°C (From ref.15).

hydrothermal treatment of sample B for 3h at 540°C (sample C), indicating framework dealumination with the concomitant formation of non- framework Al species. Acid treatment of sample C with 0.1 M HCl at 100°C results in a further increase of the Si/Al ratio to 6.6 which shows that not only the non- framework, but also some framework Al is extracted in sample D. After exchange of the residual Na$^+$ cations in sample D against NH$_4^+$ and another hydrothermal treatment for 3h at 815°C, most of the tetrahedral aluminium is removed from the framework, yielding a Si/Al ratio of more than 50 in sample E. Finally, sample F is obtained by acid leaching of sample E. The enhanced intensity of the signal at about -100 ppm in the *CP spectra* of samples C-F indicate clearly the formation of (SiO)$_3$SiOH groups in the zeolite framework. This results from incomplete reoccupation of the Al vacancies by Si and from an increasing number of SiOH groups at the surface of the secondary mesopores created by Si migration into the vacancies in the framework (ref.26). The coincidence of the peaks of Si(1Al) and (SiO)$_3$SiOH environments at about -100 ppm introduces some error into the determination of the Si/Al ratio from the non-CP spectra, since the intensity contribution of the latter unit must be ascribed to the Si(0Al) rather than to the Si(1Al) silicon atoms. Therefore, the Si/Al ratio determined by ^{29}Si NMR for zeolite samples containing (SiO)$_3$SiOH groups will be a minimum value. It should be noted that the

CP spectra are generally not reliable for *quantitative* evaluation, since the intensity enhancement factors by cross- polarization are not known. Nevertheless, CP technique is a very useful method for the detection of SiOH groups in zeolites. Besides the $(SiO)_3SiOH$ environments, even geminal OH groups in $(SiO)_2Si(OH)_2$ structural groups may be detected by the CP signal at -90 ppm. This is clearly visible e.g. in the CP spectrum of sample F. Moreover, the signal at -110 ppm in this spectrum can be attributed to Si(OAl) units close to surface OH groups in non-framework silica, formed by partial lattice destruction during the acid treatment.

Another important process applied in the preparation of a zeolite catalyst is *decationation*, i.e. the removal of the cations compensating the negative charge of the tetrahedral framework aluminium and their replacement by protons. Decationation is usually achieved by treatment of the zeolite with aqueous acids or by deammoniation of the NH_4-exchanged zeolite at higher temperatures. ^{29}Si NMR has shown that both reactions are usually accompanied by a certain degree of framework dealumination, even when the decationation is performed under very mild conditions [e.g. by shallow-bed calcination of $NaNH_4Y$ at 150-300°C under vacuum (ref.27)]. Fig.8 shows the ^{29}Si NMR spectra of $NaNH_4Y$ zeolite, and its protonated form obtained by shallow bed calcination at 300°C. The dealumination is evidenced

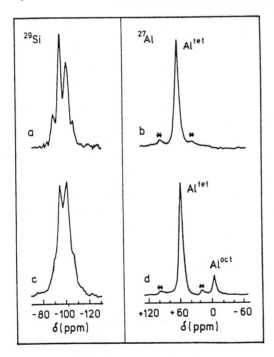

Fig.8. ^{29}Si and ^{27}Al MAS NMR spectra of (a,b) $NaNH_4Y$ zeolite and (c,d) NaHY zeolite obtained from these material by shallow bed calcination at 300°C. (From ref.27).

by the changes in the Si(nAl) peak intensities and also by the appearance of a signal for octahedral aluminium in the ^{27}Al NMR spectrum (see Section 3.3.2). Framework dealumination may also occur during the use of the zeolite in a catalytic reaction and may be a possible reason for catalyst deactivation. ^{29}Si NMR of freshly prepared, activated and used zeolite catalysts is an efficient method to monitor such dealumination processes.

Following the procedure outlined in Section 3.2.2, the *ordering of Si and Al atoms* in the tetrahedral framework of *dealuminated* Y zeolites with Si/Al ratios of 2.43 (parent zeolite), 3.0, 3.8 and 5.0 has been studied by ^{29}Si NMR (ref.28). The ordering patterns which comply with the intensity distribution of the Si(nAl) peaks in the corresponding NMR spectra indicate that, in addition to the replacement of Si for Al, a mutual site exchange of an adjacent pair of Si and Al atoms occurs at each dealumination step. This site exchange involves a rearrangement of the tetrahedral framework in the course of the dealumination reaction which is in line with the assumption of local recrystallization processes during the reinsertion of silicon atoms into the Al vacancies of the framework.

As discussed above, the framework aluminium content of a zeolite can be decreased by hydrothermal treatment of its ammonium-exchanged form, but it is also possible to reverse the dealumination process by *reinsertion* of the non-framework aluminium into tetrahedral framework positions. This Al reinsertion has been achieved by simple treatment of a dealuminated Y-zeolite (Si/Al = 4.26) with a 0.25 M aqueous KOH solution at 80°C for 24 h (ref.29). The realumination is clearly visible from the ^{29}Si NMR spectra of the zeolite before and after the treatment. In the realuminated product the intensity of the Si(0Al) signal is greatly reduced, while the intensities of the Si(2Al), Si(3Al) and Si(4Al) signals are correspondingly increased, indicating that a considerable amount of aluminium has re-entered the zeolite framework. The spectrum of the realuminated product yields a Si/Al ratio of 2.56, which is very close to that of the parent Y zeolite used in the dealumination reaction, i.e. the non-framework Al has been completely reinserted into the framework. However, the intensity distribution of the signals in the NMR spectrum of the realuminated product is different from that of the parent Y zeolite, which means that the Si,Al ordering in the framework must also be different.

3.2.4 Zeolite synthesis

The process of zeolite synthesis from reaction mixtures containing silicate, aluminate, cations and possibly organic template molecules usually involves the formation of an intermediate aluminosilicate gel from which, after an induction period, the zeolite crystallizes. The synthesis mixture thus forms a complex heterogeneous system composed of the amorphous gel, the supernatant solution and the emergent zeolite crystals. ^{29}Si NMR can be applied to both liquid and solid phases occurring in the course of zeolite synthesis and has provided valuable information on the silicate and aluminosilicate anion distribution in the starting solutions, the process of gel formation from the liquid phase, and the transformation of the gel into the crystalline zeolite.

Liquid state ^{29}Si NMR permits a detailed characterization of the type and quantitative distribution of the $Si(O^-)_{4-n}(OSi)_n$ structural units (conventionally designated Q^n) in the silicate solutions, from which a mean degree of silicate polymerization may be inferred. In addition, a variety of distinct silicate anions formed by various Q^n groups has been identified in the solutions, and dissolved aluminosilicate anions have been detected in Al-containing silicate solutions (for details see Chapter III of ref.11 and ref.30). It is, however, still only hypothesized that some of the species detected by NMR function in fact as precursor units of the zeolite structure under crystallization conditions. The distribution of the silicate species in the solutions depends on the silica concentration, the cation-to-silicon ratio, the pH-value, and the temperature. Figure 9 shows the ^{29}Si NMR spectra of two sodium silicate solutions used as starting solutions in zeolite A synthesis following two different routes,

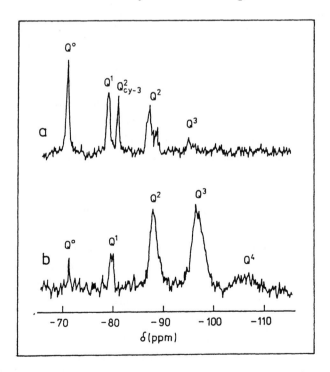

Fig.9. ^{29}Si NMR spectra of sodium silicate solutions used in zeolite NaA synthesis. (a) Solution A, c_{SiO_2} = 1.0 M, Na/Si = 1.9; (b) solution B, c_{SiO_2} = 1.65 M, Na/Si = 0.6.

A and B (ref.31). *Solution A* with c_{SiO_2} = 1 M and Na/Si = 1.9 contains mainly *low molecular weight* anion species consisting of Q^0, Q^1 and Q^2 units, while in *solution B* with c_{SiO_2} = 1.65 M and Na/Si = 0.6 *polymer* silicate anions containing branching (Q^3) and cross- linking groups (Q^4) predominate.

The intermediate aluminosilicate gels formed from the liquid reaction mixtures can be studied by solid-state ^{29}Si NMR. Owing to the amorphous structure of the gels, broad lines appear in the spectra which, however, may provide information on the distribution of the various Si(nAl) units present in the highly disordered three-dimensional framework of the gel. As an example, the above mentioned synthesis of zeolite A will be considered (ref.31,32). After mixing the sodium silicate solution A with sodium aluminate solution a gel is formed which gives a broad NMR signal at -85 ppm, indicative of Si(4Al) units in amorphous or highly disordered environments. When, however, the silicate solution B is mixed with aluminate solution (yielding the same *overall* composition of the reaction mixture as for solution A), the initial gel shows an even broader NMR peak at -93 ppm, typical of a silica-rich amorphous aluminosilicate composed mainly of Si(2Al) and Si(3Al) units. In agreement with these findings, chemical analysis yields a Si/Al ratio of 1.0 for the gel from solution A, but of 2.4 for that from solution B.

The crystallization of the zeolite from the initial gel can be followed by the ^{29}Si NMR spectra of the solid products separated from the reaction mixture after increasing periods of time. A gradual narrowing and shifting of the NMR peak to a higher field is observed for the products withdrawn from the reaction mixture A after heating to 80° for 4 and 6 hours. After 17 h at 80° a narrow line at -89.4 ppm was measured, which is characteristic of highly crystalline zeolite A. In contrast, with increasing reaction time, the NMR line of the initial gel obtained from reaction mixture B shifts to low field, indicating transformation of the Si(3Al) and Si(2Al) units into Si(4Al) by incorporation of aluminium in the gel from the solution. After about 1 h at 8o° the spectrum is very similar to that of the initial gel of mixture A, and the subsequent zeolite crystallization occurs in the same way as described above. The ^{29}Si NMR spectrum of the final sample (after 22 h at 80°) is again typical of highly crystalline zeolite A. Thus, reaction mixtures with the same overall composition, but prepared from sodium silicate solutions of different degrees of silicate condensation give rise to very different intermediate aluminosilicate gels. Finally, however, crystalline zeolite A material is formed from both types of intermediate, but after different total times of crystallization.

Similar ^{29}Si NMR investigations as discussed above for the synthesis of zeolite A have been performed for several other zeolite types, such as *zeolite Y* (ref.33,34), *mordenite* (ref.34), *hydroxysodalite* (ref.35) and *ZSM-5* (ref.34,36).

3.3 Aluminium-27 NMR

3.3.1 Quadrupole interaction, ^{27}Al chemical shifts and local environments of Al sites

Unlike ^{29}Si which has a nuclear spin I = 1/2, ^{27}Al has I = 5/2 and therefore a nuclear quadrupole moment (eQ) which interacts with an electric field gradient (eq) produced by a non-spherically symmetric charge distribution around the ^{27}Al nucleus. Those quadrupole interactions may give rise to strong line broadening and shifts of the line position (centre of gravity) in the spectrum to high field from the isotropic chemical shift. Both quadrupolar line broadening and line shifts render the interpretation of the ^{27}Al NMR spectra, with respect to chemical shifts and quantitative evaluation, more difficult. On the other hand, valuable information on the symmetry of the charge distribution at the Al nucleus may be obtained from the quadrupole interaction, described in terms of the quadrupole coupling constant QCC = e^2qQ/h and the asymmetry parameter of the electric field gradient $\eta = (q_{xx} - q_{yy})/q_{zz}$.

In powder samples, the only transition usually observed is the central +1/2, -1/2 transition, which is not affected by quadrupolar interactions to first order but by the much smaller second order interaction. The latter decreases with increasing strength of the external magnetic field B_0 and can be reduced (but not fully averaged) by magic angle spinning. Thus, application of both high B_0 fields and MAS may result in considerable line narrowing in the ^{27}Al NMR spectra of solid powders.

The ^{27}Al NMR spectra of zeolites are, in general, much simpler than their ^{29}Si NMR counterparts since according to Loewenstein's rule (which forbids AlOAl pairings) only one tetrahedral Al environment, namely Al(OSi)$_4$, exists in the zeolite framework. Only small to moderate quadrupolar line broadenings and line shifts are usually observed owing to small deviations from tetrahedral symmetry of the AlO$_4$ units. Consequently, a single, comparatively narrow line is usually observed in the ^{27}Al NMR spectra of untreated zeolites (see e.g. Fig.8b). The isotropic ^{27}Al chemical shifts of tetrahedral framework aluminium in zeolites cover the relatively small range from about 55 to 68 ppm (from aqueous Al(NO$_3$)$_3$ solution) and no definite relationships could be found between the shifts and the Si/Al ratio or Si,Al ordering of the zeolite framework. However, similarly to ^{29}Si NMR, a linear relationship between ^{27}Al chemical shifts and mean AlOSi bond angles has been established, with shift data carefully corrected for quadrupolar shift contributions (ref.37).

Non-framework aluminium in zeolites with octahedral AlO$_6$ coordination gives rise to signals at about 0 ppm, i.e. well separated from the tetrahedral Al sites of the framework (see e.g. Fig.8d). However, if the non-framework Al exists as polymeric aluminium oxides or oxide hydrates in the zeolite cavities, strong quadrupolar line broadening may be observed owing to major distortions of the octahedral symmetry of the Al sites.

3.3.2 Framework and non-framework aluminium - quantitative aspects

The high inherent sensitivity of ^{27}Al NMR (^{27}Al has a natural abundance of 100% and generally short relaxation times) permits the detection of very small quantities of tetrahedral Al in the framework of high-silica zeolites which are not seen in the ^{29}Si NMR spectra. For example, traces of tetrahedral framework Al have been detected by a peak at about 55 ppm in the ^{27}Al NMR spectrum of silicalite, prepared without any intentional addition of aluminium to the synthesis mixture (ref.38).

As mentioned above (and shown e.g. in Fig.8d), well separated signals at about 60 ppm and about 0 ppm appear in the ^{27}Al NMR spectra for four-coordinated framework Al and six-coordinated non-framework Al in chemically or thermally treated zeolites. Provided that all aluminium is visible in the spectrum (i.e. no signal intensity is lost as a result of very strong quadrupolar line broadening), the relative proportions of framework and non-framework Al in the zeolite can be directly determined from the intensities of the signals at about 60 ppm and 0 ppm, I_{60} and I_0. Moreover, if the *total* Si/Al ratio, $(Si/Al)_{tot}$, of the sample is known from chemical analysis, the Si/Al ratio of the tetrahedral framework, $(Si/Al)_{fr}$, can be determined according to

$$(Si/Al)_{fr} = (Si/Al)_{tot}(I_{60} + I_0)/I_{60} \qquad (3)$$

However, it has been demonstrated in several papers that the quantitative results derived from ^{27}Al NMR do not always agree with those obtained from ^{29}Si NMR, and that non-framework Al is often underestimated by the former method. Although several methods have been proposed which may be helpful to render all aluminium "NMR visible" (e.g. application of very strong B_0 fields and high spinning rates, rehydration of the sample or its impregnation with acetylacetone), care must always be exercised to ensure that all aluminium is detected in the ^{27}Al NMR spectrum.

Further complications may arise in the application of ^{27}Al NMR if broad signals of Al sites subject to different quadrupolar interactions overlap in the spectrum. The separation of these lines can be achieved by the *two-dimensional quadrupole nutation MAS NMR* technique (ref.39). This technique allows the resolution of ^{27}Al sites with different quadrupole coupling constants in the F_1 dimension of the two-dimensional spectrum, while in the F_2 dimension the ordinary spectrum is presented. Two-dimensional ^{27}Al nutation MAS NMR has been applied to separate the strongly overlapping lines observed in the ^{27}Al MAS NMR spectra of hydrothermally treated Y zeolites (ref.40). Figure 10 shows the original spectra

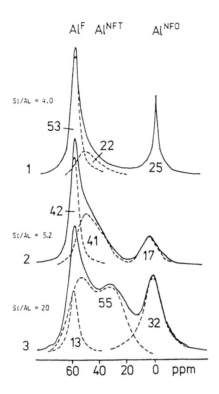

Fig.10. ^{27}Al MAS NMR spectra of framework dealuminated Y zeolites. Integral intensities (in %) of the lines are resolved using the lineshapes determined by two-dimensional ^{27}Al nutation NMR (see text). Al^F, Al^{NFT}, and Al^{NFO} denote tetrahedral framework Al, tetrahedral non-framework Al, and octahedral non-framework Al, respectively (From ref.40).

and the lines resolved by ^{27}Al nutation NMR. From the intensities of the distinct line shapes and the contents of framework and non-framework aluminium derived from the ^{29}Si NMR spectra, it has been concluded that the ^{27}Al NMR spectra are superimpositions of three lines which correspond to framework (Al^F), non-framework octahedral (Al^{NFO}) and nonframework tetrahedral (Al^{NFT}) aluminium. It was thus possible to account for all the aluminium in the samples, although the nature of the newly discovered non-framework *tetrahedral* aluminium is not fully understood.

3.4 NMR of other framework atoms

3.4.1 Oxygen-17 NMR

Since oxygen is the main constituent of the zeolite framework, the possibilities for carrying out detailed ^{17}O NMR studies in the solid state are

308

particularly attractive. However, as mentioned in Section 3.1, the ^{17}O isotope has a very low natural abundance and a quadrupole moment which renders the observation of ^{17}O NMR spectra very difficult without (expensive!) ^{17}O enrichment of the samples. Therefore, ^{17}O NMR studies of zeolites have been performed throughout on isotopically ^{17}O enriched samples. The main features of the ^{17}O NMR spectra of zeolites were demonstrated for NaA and NaY, the latter being studied with different degrees of dealumination (ref.12a). In addition, some gallosilicate zeolites and aluminophosphate molecular sieves were studied (ref.12b).

As a typical example, Figure 11 shows the static and MAS ^{17}O NMR spectra of zeolite NaY (Si/Al = 2.74). The static spectrum indicates the presence of two overlapping second-order quadrupolar broadened lines, which can be assigned to the two types of chemically inequivalent oxygen species, SiOSi and SiOAl, in the NaY framework. The line shapes of the two spectral components can be simulated using the quadrupole coupling data of zeolite NaA for SiOAl and of highly dealuminated NaY for SiOSi, and their relative intensities can be calculated from the Si/Al ratio. Addition of the corresponding component spectra results in the simulated spectrum shown in Fig.11b.

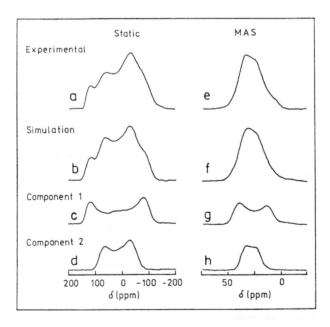

Fig.11. ^{17}O NMR spectra and spectral simulations for NaY zeolite (Si/Al = 2.74). a - static spectrum; b - simulation of (a) using component 1 (c) for Si^{17}OSi and and component 2 (d) for Si^{17}OAl; e - 5.2 kHz MAS spectrum, f - simulation of (e) using components (g) and (h) (From ref.12a).

The ^{17}O NMR studies of zeolites, gallosilicate zeolites and aluminophosphates may be summarized as follows:

- The spectra consist of characteristic resonances from chemically distinct oxygen species, i.e. SiOSi, SiOAl, SiOGa and AlOP which all exhibit a second order quadrupole broadened line shape. Crystallographically non-equivalent oxygen sites present in zeolites A and Y cannot be resolved.

- The presence of two types of SiOT fragments in a zeolite structure, e.g. SiOSi and SiOAl in Y zeolites, or SiOSi and SiOGa in gallosilicate zeolites, leads to a superposition of the resonances of either ^{17}O species, with intensities according to their relative populations in the zeolite.

- The oxygen species of the various TOT fragments are characterized by nuclear quadrupole coupling constants e^2qQ/h in the range of 3.1 to 5.7 MHz, asymmetry parameters η up to 0.3, and isotropic chemical shifts δ between 28 and 67 ppm (versus $H_2^{17}O$). The observed order of e^2qQ/h is AlOP > SiOSi > SiOGa > SiOAl, and that of δ is AlOP > SiOSi > AlOSi > GaOSi.

3.4.2 Boron-11 NMR

The NMR-related properties of the ^{11}B nucleus are in many respects similar to those of ^{27}Al. ^{11}B has a nuclear spin of 3/2 and therefore a quadrupole moment, the sensitivity of ^{11}B NMR is reasonably high, and the relaxation times are sufficiently short for spectra with good signal-to-noise ratio to be readily registered within relatively short measuring times.

In boron-containing zeolites, boron may occur as *tetrahedral* BO_4 or *trigonal* BO_3. At high magnetic field, and under MAS conditions, tetrahedral boron gives relatively narrow single lines, while trigonal boron exhibits a clear quadrupolar doublet pattern owing to its high quadrupolar interaction. The ^{11}B chemical shift range is relatively small and the signals of both coordinations may overlap. There is, however, usually sufficient resolution to differentiate the two species clearly and to determine the relative quantities present in the sample. Tetrahedral and trigonal boron has e.g. been detected during the dehydration of H-boralite, the boron containing variant of zeolite ZSM-5 (ref. 41). A sharp line of tetrahedral BO_4 appears at -3 ppm (from $BF_3 \cdot Et_2O$) in the fully hydrated sample, while in the dehydrated material a broad doublet of trigonal BO_3 predominates, which is superimposed by the line of residual BO_4 species. Rehydration leads to reconstitution of the original spectrum showing the single BO_4 line.

^{11}B NMR may also be used to prove the incorporation of tetrahedral boron into the zeolite framework and to discriminate against non-framework tetrahedral boron. $B(OSi)_4$ environments in the zeolite framework are characterized by ^{11}B chemical shifts between -1 to -4 ppm, while slightly but definitely different shifts in the range of +1 to +2 ppm are observed for other BO_4 coordinations, e.g. in borates. As with ^{27}Al NMR, the high sensitivity of ^{11}B NMR permits the detection of very small amounts of tetrahedral BO$_4$ in different environments (ref.42).

3.5 Hydrogen-1 NMR

While the proton-proton dipolar interactions in dehydrated zeolites are relatively small owing to the large proton-proton distances, ^{1}H NMR of zeolites is complicated by heteronuclear dipolar interaction of the protons with the quadrupolar ^{27}Al nuclei, and the generally narrow range of proton chemical shifts. Nevertheless, ^{1}H MAS NMR, at high magnetic fields and fast spinning rates, of carefully dehydrated samples provides sufficient spectral resolution to identify different proton sites in zeolites, and to characterize, quantitatively, their distribution in the zeolite structure. Four distinct types of protons can be identified by their chemical shifts: *terminal SiOH, bridging hydroxyl groups SiO(H)Al, AlOH at non-framework aluminium* and *ammonium ions.* Typical shift ranges of these proton sites are displayed in Figure 12. Since, as a first approximation, the ^{1}H chemical shifts can be considered as a measure of the *acid strength* of the protons, the differences in the latter may, at least qualitatively, be characterized by the chemical shift differences of the various hydroxyl protons (ref.43).

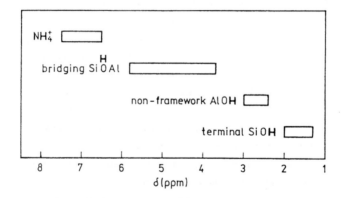

Fig.12. ^{1}H chemical shift ranges of protons in zeolites

^{1}H NMR has mainly been applied to the study of the proton distribution in hydrothermally treated zeolites Y and ZSM-5 (ref.43-45). Quantitative estimates of the concentrations of terminal, bridging and extra-framework-Al hydroxyls, and their changes by different treatment conditions, have been derived from the spectra.

As an example, Fig.13 displays the 500 MHz ^{1}H MAS NMR spectra of a series of dealuminated H-ZSM-5 zeolites with framework Si/Al ratios of 20 to 110. The spectra exhibit three lines which were attributed to terminal SiOH (at 1.8 ppm), non-framework AlOH (at 2.5 ppm) and bridging Si(OH)Al (at 4.0 ppm). The absolute intensity of the line at 4 ppm decreases with increasing Si/Al ratio of the framework and correlates linearly with the number of framework Al atoms per unit cell. This correlation shows that the intensity of the 4 ppm line is a *direct* measure of the *number of strong acid sites* (Brönstedt sites) in the zeolite. Further information on the formation of lattice defects and partial amorphization of the ZSM-5 material, and on the degree of hydroxylation of the non-framework aluminium in the course of the hydrothermal treatment, may be derived from the intensity variations of the lines of AlOH and SiOH groups.

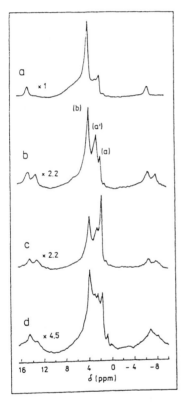

Fig.13. ^{1}H MAS NMR spectra (500 MHz) of dehydrated H-ZSM-5 zeolites prepared from NH$_4$-ZSM-5 by hydrothermal treatment. a - 500°C, 2h, shallow bed, Si/Al=22; b - 500°C, 1h, steam, Si/Al=47; c - 500°C, 3h, steam, subsequently acid leached with 1 M HNO$_3$ at 100°C for 2h, Si/Al=62; d - 500°C, 3h, steam, Si/Al=110. The weak lines at about -7 and 14 ppm are spinning side bands. (From ref.45).

3.6 NMR of cations and guest species in zeolites

3.6.1 NMR of charge compensating cations

In addition to the nature of the tetrahedral framework, the catalytic, adsorptive and ion-exchange properties of zeolites depend strongly on the kind, position and mobility of the cations balancing the negative charge brought about by the AlO$_4$$^-$ tetrahedra in the zeolite framework. Since the chemical shifts and multiplicities of the NMR lines of the cation nuclei are related to the intracrystalline environment, and the linewidths to the mobility of the cations, NMR can, in principle, provide information on site occupancy and the motion of cations in zeolites.

Of the great variety of cations which may be present in the zeolite lattice, only a small selection have been studied by NMR so far - in particular the metal cations ^{23}Na, ^{7}Li, ^{205}Tl and ^{133}Cs. Except ^{205}Tl (which has I = 1/2), all of these are quadrupolar nuclei giving rise to broad lines with or without characteristic quadrupolar line shapes, from which information on coordination and mobility of the cation sites in the zeolite cavities may be inferred.

^{23}*Na NMR* was used e.g. to study the de- and rehydration processes of ZSM-5 (ref.46) and NaA zeolites (ref.47) and sodalites (ref.48,49). Two dimensional ^{23}Na quadrupole nutation NMR has been found to be a useful technique in specifying for different sodium sites in the zeolite and investigating dynamic exchange processes (ref.47,48,50).

^{7}*Li NMR* studies of LiNaA zeolites revealed an ordered arrangement of Li and Na in the zeolite lattice and a sequential occupation of tetrahedrally situated sites I with increasing Li content (ref.51).

Siting and exchange of Cs^{+} cations in A, Y and ZSM-5 zeolites were studied by ^{133}*Cs NMR* (ref.52). From the composite line shapes of the exchanging spin systems, the exchange rates between the sites, the activation energy of the exchange process and the mean lifetime of the Cs^{+} ions at the sites were determined. ^{133}Cs NMR has also been applied to the study of cation migration between *different crystallites* of a two-component zeolite mixture, e.g CsA and CsX.

3.6.2 Carbon-13 NMR of organic templates in zeolites

Studies of the structure and position of organic species present in zeolites synthesized with organic templates are of great interest for elucidating the process of zeolite formation. High-resolution solid-state ^{13}C NMR using cross-polarization and magic angle spinning has great potential in such investigations. The isotropic ^{13}C chemical shift is highly sensitive to the environment of the carbon nucleus, and the ^{13}C NMR spectra in general display narrow and well resolved lines for each kind of distinct carbon atom of the organic guest species in the zeolite.

The high resolution of the ^{13}C NMR spectra of organic templates in zeolites is demonstrated in Figure 14 by the ^{13}C NMR spectrum of the tetrapropylammonium ion (TPA^{+}) enclathrated in the zeolite ZSM-5 structure in the course of synthesis. For comparison, the spectrum of an aqueous solution of TPAOH is also shown. The general agreement of the two spectra reveals that the occluded TPA ions are chemically intact in the ZSM-5 channels. There is, however, a clear splitting of the methyl resonances which arises from the different surroundings of the propyl chains extending in the two different channels of the ZSM-5 structure.

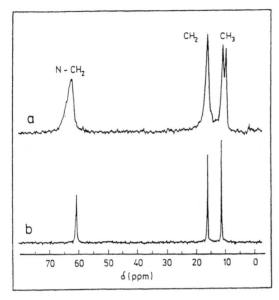

Fig.14. ^{13}C NMR spectra of (a) Na,TPA-ZSM-5 zeolite and (b) TPAOH in aqueous solution.

The ^{13}C NMR spectra of a variety of other organic templates present in as-synthesized zeolites reveal that most of the templates remain intact within the zeolite channels, although characteristic shift effects indicate some interaction of the functional group of the organic species with the zeolite framework. For tetramethylammonium ions (TMA$^+$), these shift effects are characteristic of the *size of the cavity* in which the TMA$^+$ ion is trapped (ref.53). Thus, the site of TMA ions in zeolites containing cages of different dimensions may be determined, and the size of the ion trapping cage may be estimated from the ^{13}C chemical shifts, even when the detailed structure of the zeolite is unknown.

4. NMR OF ALUMINOPHOSPHATE MOLECULAR SIEVES

Although the aluminophosphate molecular sieves, AlPO$_4$-n (n denotes a specific structure type), and their silicon or metal atoms containing variants, SAPO-n and MAPO-n, cannot be considered as belonging to the zeolite family, they have zeolite-like three-dimensional frameworks consisting of alternating AlO$_4$ and PO$_4$ tetrahedra in the AlPO$_4$-n series, and of AlO$_4$, PO$_4$ and SiO$_4$ or MeO$_4$ in the SAPO-n and MAPO-n materials.

Aluminophosphate molecular sieves are attractive materials for solid-state NMR studies since they contain two different kinds of 100% abundant nuclei, ^{31}P and

^{27}Al. The ^{31}P MAS NMR spectra generally show symmetrical lines in the shift range of -14 to -31 ppm (from 85% H_3PO_4), consistent with tetrahedral PO_4 units in the framework of these materials (ref.54,55). Three well separated lines (at -14.8, -21.4 and -26.4 ppm) have been observed in $AlPO_4$-21 for the three crystallographically inequivalent P sites in that structure (ref.55). The ^{27}Al NMR spectra are more complex and may exhibit broad and asymmetrical spectral patterns due to severe quadrupole interactions of the ^{27}Al nuclei. The latter arises from significant distortion of the AlO_4 tetrahedra caused by interaction with entrapped organic template molecules and/or H_2O in the cavities of the $AlPO_4$ structure. The ^{27}Al chemical shifts attributed to tetrahedrally coordinated Al in $Al(OP)_4$ units cover the range from about 30 to 45 ppm (from aqueous $Al(NO_3)_3$ solution). Further lines may appear in the shift range of -10 to -19 ppm, which have been assigned to octahedrally coordinated Al e.g. in $Al(OP)_4(OH_2)_2$ coordinations (ref.54).

The ^{27}Al and ^{31}P NMR spectra of SAPO's have been found to be comparable to those of the $AlPO_4$ materials. The ^{29}Si NMR spectra show single lines with chemical shifts between -89 and -92 ppm, consistent with the presence of one $Si(OAl)_4$ environment which implies that silicon substitutes for phosphorous in the aluminophosphate framework (ref.55).

REFERENCES

1   C.A. Fyfe, *Solid State NMR for Chemists*, CFC Press, Guelph, 1983.
2   G.Engelhardt, D.Kunath, M.Mägi, A.Samoson, M.Tarmak and E.Lippmaa
    Workshop on Adsorption of Hydrocarbons in Zeolites, Berlin-Adlershof, 1979
3   E.Lippmaa, M.Mägi, A.Samoson, M.Tarmak and G.Engelhardt, J.Amer.Chem.Soc.
    103 (1981) 4992
4   G.Engelhardt, U.Lohse, E.Lippmaa, M.Tarmak and M.Mägi,
    Z.anorg.allg.Chem. 482 (1981) 49
5   S.Ramdas, J.M.Thomas, J.Klinowski, C.A.Fyfe and J.S.Hartman, Nature
    292 (1981) 228
6   J.Klinowski, J.M.Thomas, C.A.Fyfe and J.S.Hartman, J.Phys.Chem. 85
    (1981) 2590
7   D.Freude and H.J.Behrens, Cryst.Res.Technol. 16 (1981) 1236
8   J.B.Nagy, G.Engelhardt and D.Michel, Adv.Colloid Interface Sci. 23 (1985) 67
9   J.M.Thomas and J.Klinowski, Adv.Catal. 33 (1985) 199
10 J.Klinowski, Ann.Rev.Mater.Sci. 18 (1988) 189
11 G.Engelhardt and D.Michel, *High-Resolution Solid-State NMR of Silicates and
    Zeolites*, Wiley, Chichester 1987
12 a - H.K.C.Timken, G.L.Turner, J.P.Gilson, L.B.Welsh and E.Oldfield,
    J.Amer.Chem.Soc. 108 (1986) 7231; b - H.K.C.Timken, N.Janes, G.L.Turner,
    S.L.Lambert, L.B.Welsh and E.Oldfield, J.Amer.Chem.Soc. 108 (1986)
    7236; c - T.H.Walter, G.L.Turner and E.Oldfield, J.Magn.Reson. 76 (1988) 106
13 G.Engelhardt, S.Luger, J.Ch.Buhl and J.Felsche, Zeolites 9 (1989) 182
14 D.E.W.Vaughan, M.T.Melchior and J.J.Jacobsen, ACS Symp.Ser. 218
    (1983) 231; J.M.Thomas, J.Klinowski, S.Ramdas, M.W.Anderson and
    C.A.Fyfe, ACS Symp.Ser. 218 (1983)159; S.Hayashi, K.Suzuki, S.Shin,
    K.Hayamizu and O.Yamamoto, Bull.Chem.Soc.Jap. 58 (1985) 52
15 G.Engelhardt, U.Lohse, A.Samoson, M.Mägi, M.Tarmak and E.Lippmaa,
    Zeolites 2 (1982) 59

16 J.Klinowski, S.Ramdas, J.M.Thomas, C.A.Fyfe and J.S.Hartman,
    J.Chem.Soc.Farad.Trans.2 78 (1982) 1025
17 C.A.Fyfe, H.Strobl, G.T.Kokotailo, G.J.Kennedy and G.E.Barlow,
    J.Amer.Chem.Soc. 110 (1988) 3373
18 C.A.Fyfe, G.C.Gobbi, G.J.Kennedy, J.D.Graham, R.S.Ozubko, W.J.Murphy,
    A.Bothner-By, J.Dadok and A.S.Chesnick, Zeolites 5 (1985) 179; C.A.Fyfe,
    G.C.Gobbi, W.J.Murphy, R.S.Ozubko, D.A.Slack, J.Amer.Chem.Soc. 106 (1984) 4435
19 R.H.Jarman, A.J.Jakobsen and M.T.Melchior, J.Phys.Chem. 88 (1984) 5784
20 J.V.Smith and C.S.Blackwell, Nature 303 (1983) 223
21 J.M.Thomas, J.Kennedy, S.Ramdas, B.K.Hunter and T.B.Tennakoon,
    Chem.Phys.Lett. 102 (1983) 158
22 J.M.Newsam, J.Phys.Chem. 91 (1987) 1259
23 G.Engelhardt and R.Radeglia, Chem.Phys.Lett. 108 (1984) 271;
24 R.Radeglia and G.Engelhardt, Chem.Phys.Lett. 114 (1985) 28
25 J.Scherzer, ACS Symp.Ser. 248 (1984) 157
26 A.Zukal, V.Patzelova and U.Lohse, Zeolites 6 (1986) 133
27 G.Engelhardt, U.Lohse, M.Mägi and E.Lippmaa, Stud.Surf.Sci.Catal. 18
    (1984) 23
28 G.Engelhardt, U.Lohse, V.Patzelova, M.Mägi and E.Lippmaa, Zeolites 3
    (1983) 329
29 H.Hamdan, B.Sulikowski and J.Klinowski, J.Phys.Chem. 93 (1989) 350
30 A.V.McCormick, A.T.Bell and C.J.Radke, Zeolites 7 (1987) 183
31 G.Engelhardt, B.Fahlke, M.Mägi and E.Lippmaa, Zeolites 5 (1985) 49
32 G.Engelhardt, B.Fahlke, M.Mägi and E.Lippmaa, Zeolites 3 (1983) 292
33 N.Dewaele, P.Bodart, Z.Gabelica and J.B. Nagy, Acta Chim.Acad.Sci.Hung.
    119 (1985) 233
34 P.Bodart, J.B.Nagy, Z.Gabelica and E.G.Derouane,
    J.Chim.Phys.Phys.Chim.Biol. 83 (1986) 777
35 S.Hayashi, K.Suzuki, S.Shin, K.Hayamizu and O.Yamamoto,
    Chem.Phys.Lett. 110 (1984) 54
36 K.F.M.G.J.Scholle, W.S.Veeman, P.Frenken and G.P.M.van der Velden,
    Appl.Catal. 17 (1985) 233
37 E.Lippmaa, A.Samoson and M.Mägi, J.Amer.Chem.Soc. 108 (1986) 1730
38 C.A.Fyfe, G.C.Gobbi, J.Klinowski, J.M.Thomas and S.Ramdas, Nature 296
    (1982) 530
39 A.Samoson and E.Lippmaa, Chem.Phys.Lett. 100 (1983) 205
40 A.Samoson, E.Lippmaa, G.Engelhardt, U.Lohse and H.G.Jerschkewitz,
    Chem.Phys.Lett. 134 (1987) 589
41 K.F.M.G.J.Scholle and W.S.Veeman, Zeolites 5 (1985) 118
42 J.-Ch.Buhl, G.Engelhardt and J.Felsche, Zeolites 9 (1989) 40
43 H.Pfeifer, D.Freude and M.Hunger, Zeolites 5 (1985) 274
44 D.Freude, M.Hunger and H.Pfeifer, Z.Phys.Chem. NF, 152 (1987) 171
45 G.Engelhardt, H.-G.Jerschkewitz, U.Lohse, P.Sarv, A.Samoson and
    E.Lippmaa, Zeolites 7 (1987) 289
46 K.F.M.G.J.Scholle, PhD Thesis, Nijmegen 1985
47 G.A.H.Tijink, R.Janssen and W.S.Veeman, J.Amer.Chem.Soc. 109 (1987) 7301
48 R.Janssen, R.E.H.Breuer, E.DeBoer and G.Geismar, Zeolites 9 (1989) 59
49 G.Engelhardt, P.Sieger, J.Felsche, in preparation
50 G.Engelhardt, J.-Ch.Buhl, J.Felsche and H.Foerster, Chem.Phys.Lett.
    153 (1988) 332
51 M.T.Melchior, D.E.W.Vaughan, A.J.Jacobsen and C.F.Pictrosky, Proc. 6th
    Intern. Zeolite Conf. (Reno 1983), p. 684, Butterworth, Guildford (1985)
52 L.E.Iton and M.-K.Ahn, Prepr.Poster Papers 7th Intern. Zeolite Conf.
    (Tokyo 1986), p.67, 69 and 115
53 R.H.Jarman and M.T.Melchior, J.Chem.Soc.Chem.Commun. 1984, 414; S.Hayashi,
    K.Suzuki, S.Shin, K.Hayamizu and O.Yamamoto, Chem.Phys.Lett. 113 (1985) 368
54 C.S.Blackwell and R.L Patton, J.Phys.Chem. 88 (1984) 6135
55 C.S.Blackwell and R.L.Patton, J.Phys.Chem. 92 (1988) 3965

Chapter 9

INTRODUCTION TO ZEOLITE THEORY AND MODELLING

R.A. VAN SANTEN

Laboratory of Inorganic Chemistry and Catalysis,
Eindhoven University of Technology,
P.O. Box 513, 5600 MB Eindhoven, The Netherlands

D.P. DE BRUYN, C.J.J. DEN OUDEN, B. SMIT

Koninklijke/Shell-Laboratorium, Amsterdam (Shell Research B.V.),
P.O.Box 3003, 1003 AA Amsterdam, The Netherlands

I. INTRODUCTION

Solid-state inorganic chemistry is undergoing a renaissance. This is partly due to important innovations in material science, such as the discovery of high-temperature superconductors, new developments in ceramics and, last but not least, the advent of new zeolites for catalysis and separation. Another significant factor is the availability of large computer power and appropriate theoretical chemistry programs that allow the computation of electronic and atomic properties with sufficiently high accuracy to be of chemical interest. Combined with graphics facilities this enables the visualization of complex chemical structures, which is very useful to the crystallographer as well as the chemist who whishes to manipulate the zeolitic materials.

In this chapter we will present an introduction to the methods currently available to study zeolites theoretically. Whereas some of the methods to be discussed have been used for more than 25 years, the applications to zeolites, with their large unit cells, are only of quite recent date. For this reason, the possibilities as well as the limitations of the use of computational chemistry in zeolite science are still being assessed.

Questions relating to the degree of ionicity and covalency of the metal-oxide bond are again the subject of major dispute. We will see how different approaches emphasize one or the other aspect and where possible we will indicate potential ways for further development.

The comparison of spectroscopic measurements on zeolites with theoretical results computed for the same system is a powerful way to progress. The many structures possible for zeolites with non-varying composition such as the silicon zeolites, make them a favoured testing material of interest to inorganic chemistry as a whole.

In the first section we will start with a discussion of theoretical methods to compute the stability of zeolite lattices. Insight into zeolite lattice stability is important to understand the dependence of zeolite topology on composition. It clearly is of relevance if one wishes to predict the stability of unknown structures. It is important to realize that because of the microporous nature of the zeolite structure, the channels and cavities will be readily filled with molecules if in contact with a liquid or gas. For instance, if a zeolite is synthesized, it will be in contact with its mother liquor. The chemical potential of the zeolite will depend on zeolite lattice stability as well as on its interaction with the molecules enclosed.

In the next sections methods to compute or estimate the interaction of the zeolite lattice and cations occluded in the zeolite channels with adsorbed molecules will be discussed.

Initially we will focus on the interaction potentials to be used; later statistical mechanics using Monte Carlo simulations will be discussed. In the section on zeolite stability, theoretical studies on the position of the channel cations neutralizing zeolite lattice charge will also be discussed. Such studies were the first to demonstrate clearly the effectiveness of solid-state chemical techniques to predict zeolite structural features.

Zeolites are of catalytic interest because they can act as solid acids. Acidity appears to be a function of zeolite composition and possibly also of zeolite structure. Several methods to predict acidity will be dealt with and their relative merits assessed. Particularly the relevance of long-range electrostatic interactions versus short-range local changes in chemical bonding environment will be discussed.

We have not included a discussion on fractals in our chapter, because zeolite micropores are crystallographically well defined and consequently regular. The concept of fractals has been defined for irregular structures. It may be of interest to study induced fractal behaviour if one blocks channels or microcavities by preadsorption, but we consider this to be outside the scope of our discussion. An analysis of fractal behaviour in zeolites can be found in ref. 1.

Computer simulation of adsorbed molecules and diffusion requires the use of molecular mechanics and molecular dynamics techniques as well as computer modelling facilities. The latter possibilities will be presented and they will be illustrated in the consecutive sections on adsorption and diffusion.

The chapter will be concluded with a brief discussion of future perspectives in zeolite theory as well as applications to zeolite catalysis.

Emphasis in this chapter will be on the introduction of a particular theoretical method in the context of a theoretical problem to be studied, rather than on an in-depth treatment of the theoretical background of such a method. For those interested we have included appropriate references.

II. THEORY ON ZEOLITE LATTICE STABILITY

Zeolite structures can be regarded as networks formed by the interconnection of silica rings consisting of four, five or six rings. Whereas hypothetical zeolite structures containing three rings have been proposed and minerals having three-ring silicate units exist, so far for zeolites they have not been found in nature, nor have they been synthesized. Using three rings, zeolite structures can be built with large channel dimensions (ref. 2), so it is in principle of interest to know whether such structures can be stable. There is a considerable body of work devoted to the question of the relative stability of silica ring systems and we will address this problem in this section.

Secondly, whereas silicalite can be crystallized with a very low Al content, this is not the case for faujasite. This structure is found in zeolites X and Y, and can only directly be made with a high aluminium content. So the question of the relative stability of silicate rings as a function of aluminium concentration is of interest. Because of charge neutrality for each Al^{3+} in substituting for Si^{4+} in the zeolite lattice, additional positive charge has to be introduced in the form of positively charged intra-channel cations. As we will see, the relative stability of these cations depends on the local environment of the cations concerned.

There are two approaches to the computational study of complex inorganic structures. One starts with the ionogenic nature of the attractive part of the chemical bonds. In addition, Born-type repulsive interactions are introduced so that equilibrium distances can be calculated. Covalent interactions are considered corrections and are accounted for by the additional introduction of Van der Waals interaction terms or by explicit incorporations of polarization. The other approach has as its starting point the quantum-chemical nature of the chemical bond. Approximate quantum-chemical approaches followed are usually limited to clusters. Only recently have practical schemes applicable to infinite lattices been implemented.

The rigid-ion and shell-model approaches to be discussed first (ref. 3) are based on the first philosophy. Bonding is assumed to be mainly ionic, so the potential used consists of a long-range electrostatic term and a short-range covalent one. The long-range term due to Coulomb interactions is calculated using an Ewald summation of the electrostatic potentials due to point charges (ref. 3c). The short-range potential that is employed is of a Buckingham type:

$$\Phi_{ij}^S = A_{ij}\exp(-r_{ij}/\tau_{ij}) - C_{ij}^{(6)}/r_{ij}^6 \tag{II.1}$$

where r_{ij} is the distance between atoms i and j, and A_{ij}, $C_{ij}^{(6)}$ and τ_{ij} are short-range parameters. The repulsive term in (II.1) is of a Born-repulsion-interaction type characterized by an exponential r_{ij} dependence. The attractive term describes the Van der Waals interaction between two polarizable atoms.

To study zeolites, it is essential, according to some authors (ref. 3b), to introduce in addition the "bond-bending" term:

$$\Phi(\theta) = k \ (\theta - \theta_o)^2 \qquad\qquad\qquad (II.2)$$

where $E(\theta)$ is the bond-bending energy, θ is the O-Si-O bond angle and θ_o is the tetrahedral angle. The term imparts a degree of "tetrahedrality" to the SiO_4 groups.

The approximations for the potentials used so far constitute the rigid-ion model. One can also include ionic polarizabilities, as is done in the shell model. According to this model the ion is thought to consist of a core and a massless shell interconnected by a "spring". The shell has a charge Y and the core has a charge equal to Z-Y, where Z is the total ionic charge. Usually, the spring is considered to be harmonic. The polarization of a free ion is related to the charges and spring constant by:

$$d = (Ye)^2 \ / \ 4\pi\epsilon_o K \qquad\qquad\qquad (II.3)$$

where d is the ionic polarizability, Y the shell charge, e the unit charge, ϵ_o the vacuum permittivity, and K the core-shell harmonic spring constant. In the shell model, the Buckingham potential and the bond-bending three-body potential are defined between shells and not between cores.

With these potentials the technique of lattice energy minimization (ref. 3a) is used to compute interatomic distances and angles and to predict some physical properties, such as the dielectric constant and the elasticity constants. The forces on the atoms are computed and an iterative procedure is used to converge atom positions to those locations where the energy is minimized and the forces on the atoms disappear.

The main difference between rigid-ion and shell-model calculations is that the shell model decreases the effective charges to be used in the Ewald summation ($\approx 80\%$) and computes a frequency-dependent dielectric constant, which also causes the Coulomb interactions to decrease (ref. 4).

The parameters to be used in eqs. (II.1) and (II.2) are fitted to give good unit cell dimensions as well as elastic constants. Full formal charges are used to compute the electrostatic interactions.

A method related to the shell-model energy-minimization method is the method of contraints investigated by No and Jhon and co-workers (ref. 5). The parameters of the potential energy function are defined by the constraint method and are obtained by minimizing the function:

$$\Sigma_{i=1}^{k} \ \Sigma_{l=1}^{N} \ \Sigma_{j=1}^{3} \ d/d\alpha_i \ (\Phi^o/dq_j^{o1}) = F \qquad\qquad\qquad (II.4)$$

α_i is the ith potential parameter and q_j^l and q_j^{ol} are the jth coordinate of the lth atom and that of the equilibrium structure, respectively. Φ^o is the potential energy function. In this method, the atom positions are assumed to be fixed and they are usually taken from experimental data. Potential parameters are then derived from eq. (II.4) by minimization of the total energy. Now in general the forces on the atoms will not disappear after minimization.

In the No-Jhon method, the Coulomb interaction is computed using charges derived from quantum-mechanical or other methods, e.g. using Hurley's electronegativity set (ref. 6a) or Sanderson's electronegativity equalization method (ref. 6b,c). The polarization energy is not computed using the shell model but from expression:

$$\Phi_{pol} = -1/2 \; \Sigma_{l=1}^N \; \alpha_l \; [\; (\Sigma_{j=1}^N E_{lj}^x)^2 + (\Sigma_{j=1}^N E_{lj}^y)^2 + (\Sigma_{j=1}^N E_{lj}^z)^2 \;] \qquad (II.5)$$

α_l is the polarizability of atom 1 and E_{lj}^u the component of the electric field along the u direction (u=x,y,z) at position 1 due to an atom on position j. Instead of a Buckingham potential, the short-range potential is calculated from a Lennard-Jones (6-12) type potential:

$$\Phi = \Sigma_l \; \Sigma_{j>l} \; \epsilon_{lj} \; [\; (\sigma_{lj}/r_{lj})^{12} - (\sigma_{lj}/r_{lj})^6 \;] \qquad (II.6)$$

The rigid-ion as well as the constraint method has been applied to the calculation of cation positions in zeolite Y (ref. 3c), zeolite X (refs. 3c,7) as well as mordenite (ref. 8).

The K^+ position in zeolite X lattices was investigated using Madelung energy calculations as well as rigid ion minimization of the cation position in an otherwise fixed structure. Comparison of different positions shows that the S_{II} site at the centre of the six-rings that project outward into the supercage is energetically favoured by most of them. The S_I and $S_{I'}$ sites, one with the hexagonal prism, the other linked to a hexagon but within the sodalite cage, appear to be equally occupied.

Other investigations of aluminium-rich zeolites concern Sr^{2+} and Na^+ (ref. 3c). Usually very good agreement with experiment is found and a clear discrimination between potential sites can be given.

Replacement of Na^+ by Ba^+ results in a destabilization (ref. 3c), as expected on the basis of a comparison of ionic radii.

The one detailed calculation that is available for low-Al-content mordenite (ref. 8) indicates that it is more important for the Ni^{2+} ion to have an environment that compensates its charge by the close approximation of two Al ions in its coordination shell than to occupy a specific extra framework position.

It should be noted that in zeolites with a high framework Al concentration the interaction between the extra framework cations themselves may significantly affect the relative stability of different cation positions. This has also been noted in a study where the Madulung energy of ZSM, mordenite and faujasite lattices was computed as a function of cation content (Na^+) (ref. 9). The compensating lattice charge was calculated by varying the average lattice cation charge, dependent on Na^+ concentration. On sodium and oxygen full formal charges were used. One finds that the electrostatic energy of the faujasite lattice is relatively insensitive to cation content, but that of mordenite or ZSM-5 strongly decreases with high Al/Si ratio. This is due to a smaller micropore volume of the channels in ZSM-5 and mordenite, which cannot accommodate as many cations as the faujasite micropores. As a result, large repulsive effects between cations appear in the high-density structures if the cation concentration is high.

Once parameters are available one can try to use similar modelling techniques to those developed for the modelling of organic molecules to predict zeolite framework or cation conformations. Free-valence geometry molecular mechanics calculations have been recently carried out on a sodalite cage cluster (ref. 10). In this method the potentials between framework atoms are essentially harmonic and an electrostatic potential, containing a dielectric constant is added. Mabilia, Pearlstein and Hopfinger (ref. 11) used charges and force constants derived from organo-silicon compounds. The structures were determined using energy minimization techniques. A study was made with varying Al/Si ratio. It appears that the conformational stability of the sodalite cage increases with aluminium content, which does not seem to agree with experiment. This indicates a need for a more extensive study of the potentials to be used, a subject which we will revert to later on.

Rigid-ion and shell-model calculations have also been used to compute the lattice stability of zeolites (ref. 12). Energy minimization of a few silica zeolites, polymorphs of quartz, shows that the stability of the most open structure (faujasite) and the most dense structure (α-quartz) differs by ≈ 40 kJ/mol SiO_2 according to rigid-ion theory and only by ≈ 20 kJ/mol SiO_2 according to the shell-model theory. Note that these results have been obtained with full formal charges on the ions.

Comparison of computed and experimental lattice infrared spectra shows reasonable agreement between shell-model and experimental spectra (refs. 4,13). The agreement is especially good for the bending modes. The vibrational models deviate on average by 10 %. An experimental probe for the long-range electrostatic field is the computed vibrational plasmon frequency. The shell-model value is lower than the rigid-ion value, but still deviates by a factor of 6 from the experimental values. Although this is probably not critical for computed lattice stabilities, it is important for theories of acidity and catalysis in zeolites as well as lattice dynamics.

Jhon and No and co-workers applied the constraint method to derive potentials for $AlPO_4-5$ (ref. 14). They applied their potentials to compute vibrational spectra for clusters derived from zeolite-A (ref. 15). However, no computations are available to test experimental and computed spectra on zeolite lattice vibrations. Therefore the accuracy of their potentials is difficult to judge.

There is an extensive body of literature on alternative potentials to be used. Here we will mention the work of Lee (ref. 16), who determined a two- as well as three-body potential for crystalline silica, and the work of Garofalini (ref. 17) on glass. Both derive empirically determined potentials. According to Lee one expands the potential as a function of two-, three-, or n-body potentials:

$$\Phi = 1/2! \ \Sigma_i^N \ \Sigma_j^N \ u(r_i,r_j) + 1/3! \ \Sigma_i^N \ \Sigma_j^N \ \Sigma_k^N \ u(r_i,r_j,r_k) + \ldots + $$
$$1/n! \ \Sigma_i^N \ \Sigma_j^N \ \ldots \ \Sigma_n^N \ u(r_i,r_j,\ldots,r_n) \qquad\qquad (II.6)$$

In all procedures discussed earlier the total potential was considered to be the sum of two body potentials $u(r_i,r_j)$ and an angular-dependent term approximating the three-body potential $u(r_i,r_j,r_k)$. These approximations imply that part of the existing dependence on three-body and higher potential terms has been incorporated in the potential constants that are empirically fitted. So a priori there is no guarantee that potentials based on such a procedure are transferable. One may even question whether it is useful to derive the transferable potentials or should limit oneself to different potentials dependent on the local environment studied. The last philosophy has been used by Clementi to determine H_2O-macromolecule interaction potentials based on ab-initio quantum chemical calculations (ref. 18).

Lee approximated $u(r_{ij})$ by a Mie-type potential:

$$\Phi(r_{ij}) = \epsilon/(m-n) \ [\ n(r_o/r_{ij})^m - m(r_o/r_{ij})^n \ ] \qquad\qquad (II.7)$$

and the three body-term as:

$$\Phi_1(r_i,r_j,r_h) = Z_1 \ (1+3\cos\theta_i\cos\theta_j\cos\theta_h) \ / \ (r_{ij}r_{ih}r_{jh})^3 \qquad\qquad (II.8)$$

This form of Φ_1 represents the dominant triple dipole interaction (ref. 19).

A number of crystalline silica properties were computed using these potentials, with parameters derived partly from Si, O_2 and O_3 and from fitting the experimental cohesive energies of low-cristobalite and low-quartz.

Using the Monte-Carlo techniques to be discussed later temperature-dependent calculations as well as surface-stability calculations were performed.

Garofalini (ref. 17) used two body potentials for molecular dynamics calculations, a method to be discussed later, on $v-SiO_2$ and glass surfaces (ref. 20).
Use was made of a modified Born–Mayer–Huggins potential function of the form:

$$\Phi_{ij} = (1+Z_i/n_i+Z_j/n_j)b_{ij}\exp((\sigma_i+\sigma_j-r_{ij})/\tau) + Z_iZ_je^2\exp(r_{ij}/(0.175L)) \qquad (II.9)$$

where Z is the electronic charge, n is the number of valence shell electrons, σ
is the atom size, and L is the box size (molecular dynamics parameter), u is an
empirical constant and b_{ij} the short–range repulsive parameter. Values of these
parameters can be found in the literature (ref. 21). Usually a correction term
is added appropriate for the Ewald summation. Some properties such as cohesive
energy are reasonably well reproduced, but internal pressures are calculated
too high, which is found to depend sensitively on the b parameter. Whereas
potential (II.9) has a form suitable for molecular dynamics calculations, it
follows from our earlier discussion that the short–range interactions are not
described very well.

A first–principle approach to derive potentials is to do quantum chemical
electronic structure calculations and compute total energies as a function of
relative coordinates.

Although techniques exist to perform electronic structure calculations on
quartz (ref. 22) and small unit cell silica polymorphs, these techniques have
not yet been applied to derive effective interatomic potentials.

Currently two studies are available that compute bulk properties of crystal–
line silica based on potentials derived from quantum–chemical cluster calculations. An early study is that of Newton, O'Keeffe and Gibbs (ref. 23). Accurate
calculations on the Hartree–Fock level have been done on the $H_6Si_2O_7$ molecule.
By computing the total energy as a function of Si–O bond length and Si–O–Si
angle, force constants for deformation of the Si–O–Si angle can be derived.
These may be expressed as an Si...Si interatomic force constant. The above
authors apply an analytical procedure to compute the bulk modulus of α–quartz
and cristobalite and find a fair agreement. This illustrates that macroscopic
properties depend mainly on proper values of short–range potential parameters.
In the calculation of, among others Newton (ref. 23) no long–range electrostatic interactions had to be included.

A more recent study concerns a first–principles interatomic potential for
silica, which also contains long–range electrostatic contributions (ref. 24).
The potential was derived from a calculation on SiO_4^{4-} surrounded by four point
charges. The potential used is similar to expression (II.1) to which a Coulomb
term is added.

The important differences from the rigid–ion calculations are firstly that
now charges of −1.2 on O and 2.4 on Si are used instead of the full formal
charges; secondly, the short–range parameters have been derived from fitting to

the potential of the SiO_4 cluster, with proper accounting of the electrostatic interaction terms which are also present. Although computed bulk properties are satisfactory, the computed lattice vibrational spectrum is inferior to that computed with the classical shell model (ref. 4).

Although promising, the quantum-chemical approach has to be extended before derived potentials can be considered acceptable.

In order to establish the presence of strain effects in the silica-tetra-hedron rings from which silica zeolites can be considered to be constructed, Van Beest et al. (ref. 25) computed energy-minimized configurations of $[SiO(OH)_2]_n$ ring systems using the GAMESS ab-initio programs. It appears that if the rings are allowed to minimize their strain, there is less than one kJ/mol energy difference between tetrahedra in four, five and six-tetrahedra rings. A three-tetrahedra ring appears to be destabilized by 10 kJ/mol tetra-hedron.

So quantum-chemical as well as shell-model calculations basically confirm each other. The energy difference between SiO_2 zeolites is very small. The absence of three-tetrahedra rings in silica zeolites appears to confirm their lower energy.

One can also study changes in energy of tetrahedra containing rings or strings, with relative coordinates derived from experimental zeolite diffrac-tion data. This has been demonstrated by, among others, Zamaraev (ref. 26), who argued that these clusters have to be considered to be embedded in the zeolite lattice, making them rigid. So far it has not been investigated how many distances can relax in zeolites, if substitutions of framework atoms occur. The driving forces are considerable. This follows from fully relaxed three-tetrahedera ring calculations, with one Si atom replaced by Al, charge compensated by a proton or Na^+ (ref. 4). The calculations show a significant weakening of the T-O bond upon substituting Al for Si.

A large deformation of the ring occurs not only if Si is replaced by Al, but also if Na^+ is replaced by a proton. The calculations also indicate that there exist large three-body potential terms, implying that effective two-body poten-tials strongly depend on the local environment.

Assuming non-relaxability of clusters and strings Derouane (ref. 27) and O'Malley (ref. 28), among others, compared total energy changes upon substitu-tion of Al for Si at specific lattice positions, simulated by the string or ring cluster. Derouane finds preferential pairing of two Al atoms in a four-tetrahedra ring cluster; O'Malley also derives preferential positions. Similar computations were done by Beran (ref. 29), who used semi-empirical CNDO-2 cal-culations.

Whereas there appears to be significant progress in the development of potentials for SiO_2 zeolites, currently there is no potential available that can reliably be used in computations not aiming to calculate lattice framework stability. This will become more apparent if we discuss the interaction of the zeolite framework with organic molecules and the acidity problem.

We mentioned earlier that methods based on a dominantly ionogenic picture of the SiO bond result in excellent predictions of alkali siting at extra framework positions in zeolites with a relatively high Al/Si ratio. Quantum-chemical calculations confirm the essential electrostatic interaction of Ni^{2+}, Mg^{2+} or K^{+}. The computed charges are close to the formal charges (refs. 30,4).

Although parameters for Al-O and P-O potentials exist no comparison with experiment (e.g. lattice vibration spectroscopy) is yet available that verifies their reliability. It is possible to derive potentials and charges also for non-purely SiO_2 systems from quantum-chemical cluster calculations. Such potentials may be expected to become available in the near future.

III. THEORY ON ADSORPTION AND ACIDITY

Bonding between the zeolite lattice and organic molecules adsorbed in the zeolite cavities is best described as Van der Waals-type weak interactions, in the case that interaction solely occurs with the silicon zeolite wall. The bonds should be considered to be hydrogen bonds if interaction occurs with the acidic protons in the zeolite. Interaction with the non-framework cations belongs to a category of bonding in between these two extremes.

Intermolecular interactions between medium-sized systems is a field of research that rapidly develops mainly in the context of interactions between molecules of biological interest. The quantum-chemical basis of these interactions with inorganic materials is relatively little explored. The most extensive work is due to Sauer (refs. 30, 31) and co-workers, which we will discuss below.

Hobza and Zahradnik (ref. 32) recently gave an extensive review on non-empirical calculations of weak interaction energies. On the basis of extensive studies they conclude that Hartree-Fock level ab-initio calculations provide reliable interaction energies, even with the use of limited basis sets, if proper corrections are made for the basis set superposition error ϵ(BSSE) and the Van der Waals interaction is explicitly included:

$$\delta E_w = \delta E(SCF) - \epsilon(BSSE) + E^D \qquad (III.1)$$

Computation of $\delta E(SCF)$ using the Huzinaga minimal basis set (ref. 33) is recommended. Using this basis set, ϵ(BSSE) arising in quantum-chemical calculations if systems interact is minimized.

The basis set superposition error of the energy can be computed using Boys and Bernardi's function counterpoise method (ref. 34).

Since the Van der Waals E^D interaction is due to the correlated motion of electrons in the two interacting fragments, it is not included in the Hartree–Fock approximation. E^D is computed from the expression:

$$E^D = -\Sigma_i^R \Sigma_j^T c_{ij}^{(6)} R_{ij}^{-6} \qquad\qquad (III.2)$$

where R and T label the interacting fragments, and i and j the atoms from which they are constructed. R_{ij} are the interatomic distances. $c_{ij}^{(6)}$ is the coefficient including e.g. the polarizabilities and ionization potentials of atoms i and j.

The atomic polarizabilities can be readily determined from the bond or total polarizabilities. It has to be determined for particular valence states of the atoms (ref. 35). In ref. 35c computed values of $c_{ij}^{(6)}$ are tabulated for CH_4, C_3H_8 and NH_3 with excellent agreement with experiment. Table 1 lists atomic polarizabilities and ionization potentials for H, C, N, O and P as a function of valence state. $c_{ij}^{(6)}$ is computed from the London expression. If one does not use modified values for the atoms bonded in the molecule, E^D is underestimated by 50 %. The London expression for $c_{ij}^{(6)}$ is:

$$c_{ij}^{(6)} = -3h/2 \, [I_i I_j/(I_i+I_j)]P_i P_j \qquad\qquad (III.3)$$

with I_i and P_i being the ionization potential and polarizability of atom i in the relevant valence state.

TABLE 1

Values of Atomic Polarizabilities (α) and Ionization Potentials (I) for the Following Atoms: H,C,N,O and P

atom	valence state[a]	$\alpha(\text{Å}^3)$	I(eV)
H	σ	0.386	13.61
C	tetetete	1.064	14.57
	trtrtrπ[b]	1.382	11.22
	trtrtrπ[c]	1.230	11.22
	trtrtrπ[d]	1.529	11.22
	didi$\pi\pi$	1.279	11.24
N	te^{2}tetete	1.094	14.31
	trtrtrπ^2	1.090	12.25
	tr^{2}trtrπ	1.030	14.51
	di^{2}di$\pi\pi$	0.852	14.47
O	te^{2}te^{2}tete	0.664	18.40
	tr^{2}tr^{2}trπ	0.460	17.25
	tr^{2}trtrπ^2	0.422	14.97
	te^{2}te^{2}te^{2}te	1.791	6.31
P	teteteπ	1.743	12.09

[a]te = tetrahedral, tri = trigonal, di = diagonal. [b]Aliphatic hydrocarbons with double bond. [c]Aromatic hydrocarbon. [d]Condensed hydrocarbon.

It is of interest to note that the $\delta E(SCF)$ value may be approximated by its electrostatic interaction term as long as the interaction energy is calculated at distances larger than the Van der Waals minimum.

A detailed account of this electrostatic approximation has been given recently by Dykstra (ref. 37). Approximating $\delta E(SCF)$ by the electrostatic part of the interaction energy requires high quality calculations with large basis sets. A useful addition to the electrostatic approximation is expression (II.1), with $c_{ij}^{(6)}$ computed according to ref. 20.

An application of expression (III.1) can be found in ref. 31a. Here the interaction of H_2O with $-SiOH$ and $-Si-O-Si$ groups is discussed. Models used are disiloxane, $H_3Si-O-SiH_3$ and silicic acid. The interaction energies and H_2O complex geometries were determined and are summarized in Table 2.

TABLE 2

Reprinted by permission from J. Phys. Chem., Vol. 85, p. 4061. Copyright (c) 1981. American Chemical Society

Energy Characteristics (in kJ/mol) for the Complex Formation

	Complex	ΔE (4-31G)	$\Delta\epsilon$	E^D	ΔE^{tot}
I	$H_2O...H_2O$	−33.47	−5.14	−6.86	−35.19
II	$(H_3Si)_2O...H_2O$	−22.51	−5.52	−14.18	−31.17
III	$Si(OH)_4...OH_2$	−38.20	−5.06	−6.70	−39.84
IV	$Si(OH)_4...H_2O$	−34.31	−10.62	−12.12	−35.81

Whereas expression (III.1) is very useful to determine local interactions between cluster models of a zeolite and interacting molecules, one requires analytical expressions for the interaction potential if one wishes to compute vibrational frequencies to compare with experiment or if one wishes to use the potentials in Monte Carlo or molecular dynamics simulation calculations.

Sauer and co-workers (ref. 31b) developed such analytical potentials for the water-silica interaction system. The method makes use of the importance of the molecular electrostatic potential (MEP) and uses the functional form of EPEN/2 (empirical potential based on interactions of electrons and nuclei (ref. 38)).

EPEN/2 potential functions V consist of a point charge interaction term V^{point} and a term $V^{n. \, spec.}$ which involves Buckingham-type potentials (e.g. eq. II.1) for describing non-specific interactions:

$$\Phi = \Phi^{point} + \Phi^{n.spec.} \tag{III.4}$$

The Coulomb interaction term Φ^{point} is computed between molecules that have non-coinciding positions of total electron charges and nuclear charges, the so-called multisite point charge model. The distribution of charges is chosen such that multipole moments or MEP curves are fitted.

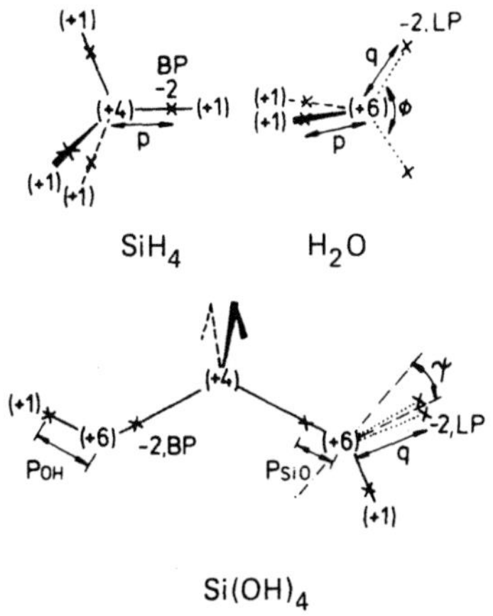

SiH₄   H₂O

Si(OH)₄

Fig. 1. Point charge models.

An example of such a charge distribution for Si(OH)$_4$ is given in Figure 1. $\Phi^{n.spec.}$ is computed between the total electron charges. Parameters in the potentials are evaluated by fitting quantum–chemical Hartree–Fock interaction energies for a sufficiently large number of relative positions of the interacting molecules. Note that in the procedure applied no basis set superposition error or a Van der Waals energy correction has been made.

In a series of papers Kiselev among others (ref. 39) developed a molecular statistical approach to the adsorption of hydrocarbons in zeolites. The potential was empirically determined, but the zeolite was not approximated by a cluster. The interaction between the hydrocarbon and zeolite lattice was written in terms of atom–atom interactions. The interaction potential of each atom with the zeolite was written as:

$$\Phi_k = -\Sigma_i \, C_i^{(k)} \, \Sigma_j \, r_{ijk}^{-6} + \Sigma_i \, B_i^{(k)} \, \Sigma_j \, r_{ijk}^{-12} - \alpha^{(k)} E^2(r)/2 \qquad (III.5)$$

In case adsorption of polar molecules is considered the corresponding contributions of the orientation electrostatic interactions of dipole and quadrupole moments of the whole molecule with the zeolite lattice ions have to be added to the total molecule potential. In eq. (III.5) r_{ijk} is the distance between the centre of the kth atom and j^{th} ion of type i, $C_i^{(k)}$ and $B_i^{(k)}$ are the constants for interaction between the kth atom of the molecule and the lattice ion of type i, $\alpha^{(k)}$ is the polarisability of C or H atoms, and E(r) is the resulting electrostatic field at position r inside the cavity (i.e. at the centre of the atom k of the molecule). The summation was carried out over all j ions of each type i inside the chosen region of the lattice.

330

The constants $C_i^{(k)}$ were not calculated using the London formula (III.3), but from the Kirkwood–Muller formula (ref. 40):

$$C_i^{(k)} = -6mc^2 \, \alpha_i\alpha_k \, / \, (\, \alpha_i/X_i + \alpha_k/X_k \,) \tag{III.6}$$

where α_i and α_k are the polarisabilities and X_i and X_k are the magnetic susceptibilities of atoms i and k, respectively, m is the mass of the electron, and c is the velocity of light. From the work of Hotza and Zahradnik it appears that expression (III.3) should be preferred.

The constants $B_i^{(k)}$ were empirically determined from the potential minimum of the pair interaction of an isolated lattice ion of type i having a charge q_i with an atom of type k of the hydrocarbon:

$$B_i^{(k)} = 1/2 \, r_o^6 \, (\, C_i^{(k)} + 1/3 \, q_i^2\alpha^{(k)}r_o^2 \,) \tag{III.7}$$

It is assumed that the equilibrium distance r_o is equal to the sum of the radius of the ith type ion of the zeolite lattice and the Van der Waals radius of the kth atom of the hydrocarbon.

A rather drastic assumption was made concerning framework atom charges as well as extra framework cation charges. The charges on Si and Al atoms were assumed to be zero; on the oxygen atoms the negative charge required to compensate for positive charge of the extra framework cations was evenly distributed (e.g. for Na–X). Some of the Na charges were +1, others close to +.5 for reasons of symmetry. Notwithstanding this approximation rather good agreement between computed and measured heats of adsorption was found. Applied to adsorption in silicalite a linear relation between isosteric heat of adsorption and mean polarizability of the adsorbed molecule was found. In this case no channel cations are present, so inspection of expressions (III.4) and (III.6) indicates the dependence on $C_i^{(k)}$, which relates to the atom polarizabilities of the molecule.

Recently Kiselev's approach has been adapted to compute the location of pyridine in K–L (ref. 41) and to compute the adsorption geometry of methane in zeolite Y (ref. 42) and also in mordenite and ZSM–5 (ref. 43) as a function of temperature and Al/Si ratio, using Monte Carlo techniques.

The interaction with H_2O was studied by the same approach (ref. 44). Charges in the molecule atoms were determined using quantum–chemical calculations. As the temperature increases one observes a change in molecular distribution from wall adsorbed positions to the middle of the cavity. The H_2O simulations show clustering of the H_2O molecules occluded in wide pores as predicted by Sauer among others (ref. 31).

Again the charges on the framework cations have been assumed to be zero and charge has been smeared out over the oxygen atoms, dependent on Na^+ concentration. It should be remembered that the parameter $B_i^{(k)}$ depends on this charge, so proper application of Kiselev's expressions requires re-evaluation of the parameters dependent on zeolite compositions. In addition the use of Lennard-Jones-type potentials may be questioned, since, as we discussed, the Buckingham potential appears to be favoured.

Notwithstanding significant understanding and progress in the development of hydrocarbon zeolite lattice potentials, it is also clear that major questions remain unsolved. The question of implementation of quantum-chemically cluster-derived potentials into the zeolite lattice properly accounting for long-range electrostatic interactions has not been studied.

The use of variations of Kiselev's potential makes practical and successful simulations possible.

Application to more general problems requires a more detailed study of the potentials to be used.

The quantum-chemical studies on acidity and interaction between organic molecules and cations in zeolites to be discussed now can be considered to be first approaches to further improvement.

As follows from the introductory part of this section, a study of the formation of hydrogen bridges, the elementary step in Bronsted acid reactions, requires careful consideration of local electrostatic interactions.

The electrostatic potential along the zeolite wall probed at the proton position should give a reasonably accurate value of the bond strength between proton and oxygen. On the basis of this philosophy Goursot, Weber and co-workers (ref. 45) provide MEP maps of clusters containing four tetrahedral units, containing a different ratio of Al and Si atoms modelling the wall of offretite. The computations are of minimal basis STO type or Extended-Huckel. No optimization of distances or angles was attempted.

The MEP determined potential for the proton positioned opposite to an O atom bridging an Si and Al atom showed a significant increase from −221 kcal/mol to −276 kcal/mol upon comparing two four-tetrahedra string, one with one Si replaced by Al and the other with two Si atoms replaced Al. This additional aluminium substitution increased the negative charge on the oxygen atoms.

This result agrees with the experimental observations (ref. 46) that the intrinsic acidity per Bronsted acid site decreases with increasing lattice Al/Si ratio.

Earlier semi-empirical calculations (ref. 46a) also indicated the increased bond strength of the bond if the Si tetrahedron becomes coordinated to tetrahedra with Al as cation. The larger degree of electron donation possible for Al compared to Si, in these calculations again give a higher oxygen charge, increasing the proton-oxygen interaction.

Using

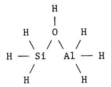

clusters, Mortier, Sauer among others (ref. 47) demonstrated the significant weakening of the OH bond in the bridging configuration compared to end-on-bonded OH. With ab-initio calculations and optimized cluster geometries they observed a decrease between bridging and end-bonded OH of 200 kJ/gat H using a reasonable quality basis set. Compared to the non-protonated cluster, the Al-O and Si-O bond lengths appeared to be considerably longer (δr_{Al-O} = 0.12 A, δr_{Si-O} = 0.09 A).

Sauer (ref. 31d) also computed the interaction energies of H_2O with such clusters. Whereas the interaction of water with siloxane is only 10.1 kJ/mol, it increases to 58.4 kJ/mol for bridging hydroxyl in the Al-Si cluster. The interaction with terminal hydroxyl in a comparable calculation gave 16.4 kJ/mol stabilization.

O'Malley and Dwyer (ref. 48) found a decrease in OH frequency when comparing the proton on the bridging oxygen of the Al-Si containing dimer with end-on hydroxyl.

These results indicate that at least qualitatively correct trends on Bronsted acidity can be derived from cluster ab-initio calculations.

Allavena (ref. 49) extended the clusters so that they were terminated with OH groups instead of H atoms and Van Santen et al. (ref. 50) studied three- and four-tetrahedra-containing rings in which one Si atom was replaced by Al. Both groups used ab-initio methods using varying basis sets. The effect of proton attachment versus compensating Na^+ ions on fully geometry-optimized clusters was studied (ref. 50). As earlier observed by Mortier et al. (ref. 47) significant changes in angle and distance are found. The high Bronsted acidity of a proton attached to a bridging O-atom was confirmed. The conformational changes found for geometry-optimized clusters indicate that the studies should be extended so that geometry constraints due to the embedding of the cluster in the zeolite lattice can be evaluated.

Allavena (ref. 49) and Catlow (ref. 50) also considered the embedding problems of clusters in an electrostatic field. The electrostatic potential on the cluster atoms contained a Madelung potential correction simulating its value in an extended zeolite lattice. Large changes in hydrogen bonding are observed in the presence and absence of such a Madelung field. It is currently unclear whether the approximate nature of the cluster models (negatively charged, or OH- or H-terminated, limited basis set) requires embedding in an effective electrostatic field or what charges should be used on the atoms modelling the zeolite lattice.

Although semi-empirical calculations (refs. 29, 46a, 51) have provided new insights into the physics of zeolite acidity, it appears that ab-initio calculations are required to enable the prediction of the detailed behaviour of molecules that become protonated. One of the main reasons is the high-quality requirement of the computed local electrostatic interactions.

Few quantum-chemical studies are available on the interaction of cations or metal atoms with a cluster modelling the zeolite framework. Sauer et al. (ref. 30) studied the interaction of Ni^{2+} and Mg^{2+} with

$$O^{(-)}$$
$$/ \ \backslash$$
$$H_3Si \quad AlH_3$$

clusters. They found that this interaction is well described with a point charge model. The computed O-M equilibrium distances are ≈ 1.85 Å. They estimate the interaction with an Ni atom to be weak and not larger than 30-50 kJ/mol.

There exists a semi-empirical calculation of ethylene interacting with a Na^+ ion attached to an alumino-silicate ring (ref. 31c). However, no ab-initio studies are available to confirm the result that the presence of the aluminosilicate ring significantly alters the binding geometry of adsorbed ethylene.

On the basis of the analysis of bonding in zeolites presented in Sections II and III we recommend the following strategy to derive zeolite-adsorbate interaction energies and geometries.

Quantum-chemical cluster calculations having a geometry representative of the complex geometry to be studied should be done to derive EPEN/2 potentials between cluster and adsorbate. In case proton transfer is studied, also the potentials between framework atoms should also be modelled.

In an independent study atom-atom potentials between lattice atoms at a distance equal to two or three atom-atom distances from the adsorption site have to be established along the lines outlined in Section II. Effective framework atom charges can be derived from quantum-chemical cluster calculations. The dielectric constant $\epsilon(\infty)$ may be estimated from the Clausius-Masotti relation using atom polarizabilities corrected as discussed for the computation of

334

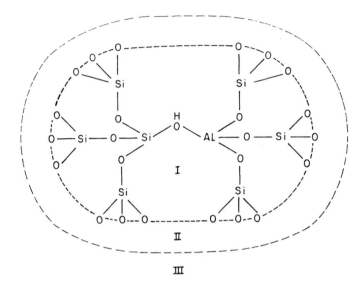

Fig. 2. Division of space around a reactive zeolite site.

the London interaction (ref. 20). These are the ingredients required for the
following approach.

As illustrated in Figure 2, we have divided the space around the adsorption
complex in three regions. In region I all interactions are described in terms
of EPEN/2-type potential functions. Covalent bonding between atoms in region I
and II can be described using potentials derived from the undisturbed lattice
potential study; the same holds for covalent bonding between atoms in regions
II and III.

Whereas the atom positions in region III are to remain fixed and are to be
the same as in the undisturbed lattice study, relaxation of atom-atom distances
is allowed to occur in regions I and II.

The electrostatic contribution to the potentials between the atoms in region
I and that in regions II and III as well as between the atoms in region II and
between the atoms of region II and III are computed using expressions such as:

$$\Phi(\text{coul})^{IJ}_{ij} = 1/\epsilon(\infty)\ q_i q_j / r_{ij}$$

$\Phi(\text{coul})^{IJ}_{ij}$ describes the Coulomb interaction between atoms in regions I and J.
Atom i is located in region I and atom j in region J. q_i and q_j are their
respective charges.

Although for different systems part of this scheme has been accomplished and
as we discussed the techniques are in principle available, no complete study
along the lines sketched is available yet.

IV. COMPUTER MODELLING TECHNIQUES

Introduction

Computer Aided Molecular Modelling or CAMM has become a well-established tool in organic chemistry within a time span of 10 years. An important application area is in the field of macromolecules, for example the design of new drug molecules by pharmaceutical industries (ref. 52). CAMM, a combination of computational chemistry and molecular graphics, has almost bridged the gap between theoretical and experimental chemistry. On the one hand, the 'workbench chemist' now has access to the insight gained by theory through his CAMM facility, aiding him to interpret his findings and moreover to select the most promising experiments in advance. On the other hand, the theoretical chemist now has a very natural graphical interface to his computational routines. Both share the benefits of an easily gained insight into the three-dimensional structure of molecules and related properties, and a better mutual communication.

In molecular modelling a wide range of computational methods is available in an integrated way (ref. 53). Many problems can adequately be tackled with empirical techniques such as molecular mechanics (ref. 54) and molecular dynamics (ref. 55), the first mainly to find the conformation of molecules and the latter to simulate the behaviour. A bottleneck often still is the availability of good atom-atom force fields, especially for metals and other 'non-organic' elements. Molecular orbital routines and even ab-initio programs (ref. 56) are also part of a standard CAMM facility. Next to this, experimental data, in the form of crystallographic databases such as the Cambridge Structural Database, and empirical search techniques applying QSAR (Quantitative Structure Activity Relationships) (ref. 57) are important ingredients.

The possibilities will be sketched to adapt and extend the techniques of CAMM, whose merits and influence have perhaps been slightly exaggerated for clarity in the first paragraph, to zeolite chemistry (ref. 58). Another source for techniques in zeolite modelling is (inorganic) crystallography, a field where graphics has already been applied for many years (ref. 59). To this end the available hardware and software, including the techniques of CAMM, will be discussed before we focus on applications in our field. Although zeolite modelling also gains its strength from combining existing computational and experimental approaches with dedicated three-dimensional graphics, we will mainly cover the latter in the present chapter.

Hardware

The developments in CAMM have hitherto clearly been hardware driven, but this situation seems to change in favour of software developments. The reason is that, from a technological point of view, for the hardware a plateau is

being approached which offers graphics of sufficient quality. The prices for
this hardware will continue to drop because there is now a sound competition
and the emergence of software standards (UNIX, PHIGS) will make it easier to
change systems. Software has not been able to keep pace with the hardware
developments, but may be able to catch up during the plateau period mentioned

The whole field of interactive three-dimensional graphics was opened up by
the Evans & Sutherland PS300 family of terminals, which was introduced in the
early eighties. The concept of dedicated and very fast hardware, addressing a
'vector display', to manipulate wireframe models was the important innovation.
Structures, once loaded from a host computer, could be rotated, translated and
zoomed in real time by turning dials. Other important facilities are hardware
clipping (making parts of the structure invisible), depth cueing (suggesting
depth by lowering the intensity of lines further away), perspective and stereo
(yielding a realistic 3-D impression with the aid of special spectacles).

Only lines and dots (Figure 3) could be displayed, the latter to create a
surface (ref. 60). The advantage of wireframe models is the possibility to look
through the structure as opposed to solid models (Figure 5) where one can view
only the outer surface of a molecule. The latter display mode will therefore
never fully replace the stick models, which are the best way to show chemical
structure. Ball-and-stick models, with dotted spheres to represent the atoms,
are also very instructive.

The last few years have shown a very rapid development of general purpose
super workstations (e.g. from Silicon Graphics or Ardent), which have now even
taken the lead in pure graphics performance. Interactive solid rendering, which
requires real-time hidden line removal and shading, has now become possible on
'affordable' (k$40-k$200) machines. This is combined with very powerful CPUs
(more than 10* VAX-780 performance) which often exhibit a parallel architecture. These are almost ideal CAMM machines because superb graphics can be done
on the same computer which performs the computational chemistry calculations.
Most of these workstations contain separate graphics processors to do part of
the display operations, but nevertheless offer new possibilities to integrate
calculations with graphics. Most of these new approaches (ref. 61), e.g. calculating and displaying interaction energies or simulating experiments (Figure 4)
at the same time the structure is altered, are still to be explored.

Software

Graphics software dedicated to zeolite modelling is still hardly available,
but the large commercial packages for CAMM are showing more and more useful
features. Since the development and maintenance of a CAMM package is a major
effort (more than 1 million lines of code is no exception), good software is
mainly obtainable through commercial suppliers. The prices are of the same

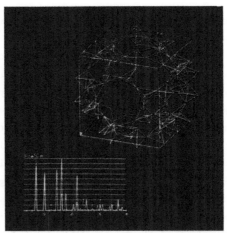

Fig. 3. The framework of ZSM-5, drawn by connecting the metal positions with (purple) lines. The inner surface of the channels is indicated by (yellow) dots, generated with the Connoly routine (ref. 60).

Fig. 4. One unit cell of zeolite A, sketched as tetrahedrons with the program PLUVA (ref. 62). Also given are the diffractograms of the known structure (in red) and the semi-real-time simulation of the displayed, modified structure (in green). With PLUVA the contents of the unit cell can be modelled interactively while the symmetry is conserved. In this demonstration, the tetrahedrons are rotated a few degrees.

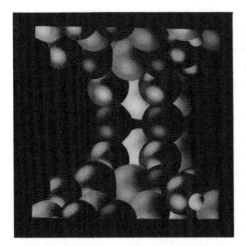

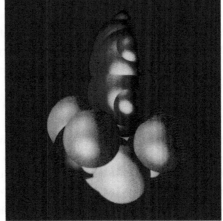

Figs. 5a and b. Solid rendering (CPK model) of faujasite, viewing from the large cage into the small sodalite cage (centre of figure a). The crystallographically different oxygen atoms have been given different colours, in order to prove that one of them is inaccessible to pyridine probe molecules as indicated in an infrared experiment. As shown in figure b, the 'yellow' oxygens (identical with the ones in the centre of figure a) are inaccessible for the 'red' pyridine molecule, which can only approach from the large cage.

order as those of the hardware, but considerable discounts are given to academic clients.

A good modelling program is capable of constructing and displaying a molecular or crystallographic structure in various ways, utilising the possibilities offered by modern hardware. Many other possibilities are available, for example quick ways to inquire atom distances, bond angles etc. Zeolite structures are mainly found in the Inorganic Crystal Structural Database (ICSD), but no commercial program offers yet a direct interface to this database as opposed to the organic databases. Molecular mechanics routines to optimise structures, even through an exhaustive search through conformational space, are normally incorporated in the display program. Possibilities to add a solvent or to apply periodic boundary conditions, to simulate a molecule as part of a crystal, are almost standard. The available force fields are optimised for macromolecules and/or small organic molecules, but can be changed by the user. As explained elsewhere in this workshop, force fields for silicalites still need improvement. A molecular dynamics 'engine' also belongs to the standard outfit of a CAMM package, which means a full integration with the graphics. For the simulation of an adsorbate molecule in a zeolite framework (see next section) this implies that the molecule can be followed on the screen while the calculation develops.

Nice interfaces to well-known programs for MO and ab-initio calculations, which are of academic origin, are often offered. This means that the input for these programs can be easily made by constructing the desired molecule and also that the output can be inspected through molecular graphics. For this purpose the program has to be capable of displaying wave functions and charge distributions. This is often done in combination with the display of the Van der Waals surface.

A basic problem with the application of CAMM programs to zeolite, or other inorganic crystals is the fact that the structure is treated as a macromolecule. Only when the crystal is constructed from crystallographic data, at the input stage, is the symmetry taken into consideration. If one wants to modify the structure at a later stage, it is only possible to change the positions of atoms individually (or to re-input a whole new structure). To circumvent this cumbersome procedure, the crystallographic modelling program PLUVA (ref. 62) has been developed. This program incorporates crystallographic symmetry in the display itself by taking advantage of the display capabilities of an Evans & Sutherland PS390 computer. Not only once, but during every display cycle is the crystal constructed by applying symmetry operations on a basic building block (the asymmetric unit). In this way the interactive displacement of one atom results in an immediate modification of the whole structure in such a way that the symmetry is preserved (Figure 5).

Zeolite modelling

As in the case of organic modelling, graphics is an indispensable tool to serve as an interface with energy and other calculations but it also has a value of its own. Gaining insight into the three-D frameworks of zeolites and the possible positions of adsorbates and cations is almost impossible without graphics. For these purposes a standard CAMM program can be very useful (ref. 63). Different display styles can be chosen and cations and organic molecules can be manipulated independent of the zeolite framework. Van der Waals or Conolly routines can be used to create the inner surface of the zeolite. Fitting a molecule in a zeolite pore bears an important resemblance to 'docking' procedures in biochemical applications, for which dedicated tools exist (e.g. distance monitors).

Areas where modelling can play an important role are:
* Modelling of zeolite-adsorbate interactions.
* The search for new zeolite structures.
* Interpretation of diffraction and spectroscopic experiments.
* The search for template molecules and synthesis intermediates.

The first item will be discussed in Section V; the others will only briefly be dealt with in the next paragraphs.

By using symmetry considerations, as attempted by for example by Smith (ref. 64), one can systematically generate geometrically possible zeolite structures. Several other approaches (ref. 65) exist, for example considering possible cage and channel structures, realising that crystals tend to form periodic minimal surfaces (ref. 66). A modelling tool such as PLUVA, capable of rearranging zeolite fragments under crystallographic symmetry limitations and having fast control over the unit cell dimensions, can be of great assistance. Molecular mechanics calculations, such as the approach chosen by Catlow and co-workers (ref. 3), can be used to estimate the stability of a hypothetical structure.

Various analytical techniques are required to investigate the structure of zeolites (X-Ray Diffraction, High Resolution Electron Spectroscopy, Nuclear Magnetic Resonance), acid strength and siting (Infrared spectroscopy) etc. Modelling offers a way to quickly check and interpret the outcomes on a micros-copic scale (refs. 67-70), for example to test steric hindering of probe molecules. A fast interface between a crystallographic modelling program and simulation programs (XRD, HREM) is a very powerful tool to elucidate zeolite structures from powder diffractograms (ref. 10). Algorithms which combine crystallographic refinement with molecular dynamics (ref. 71) have, to our knowledge, not yet been tried in this field but they may be as successful as for macromolecules.

Suitable organic molecules, added to the synthesis solution, enhance the selective crystallisation of zeolites. In a number of cases evidence is found

for a template effect, with the molecule exactly filling up the zeolite pores
(ref. 72). Molecular modelling tools, including organic databases, are very
convenient for searching such molecules. Again the resemblance with biochemical
applications, especially drug design, is very strong. Next to this, modelling
can be of assistance to unravel the synthesis process, for instance in identi-
fying synthesis intermediates.

V. COMPUTER SIMULATION OF ADSORPTION AND DIFFUSION

The interaction between organic adsorbates and a zeolitic environment plays
an important role in the extended field of zeolite science and technology.
Directly related to these interactions are adsorption and diffusion phenomena
which are thought to have a large impact on the catalytic activity and selectivity of zeolites.

In reactions catalysed by microporous solids, such as zeolites, catalytic
selectivity is not only a matter of stoichiometry, but also steric constraints
for molecular transport in the zeolite void space can be the reason for the
so-called shape selectivity (ref. 73). Shape selectivity can in principle be
achieved by adjusting the size (ref. 74) and accessibilities (ref. 75) of the
micropores. A distinction is made between three types of shape selectivity
depending on whether the pore size limits the entrance of the reacting molecule
(reactant selectivity), the departure of the product molecule (product
selectivity) or the formation of certain transition states along the reaction
coordinate (restricted transition state selectivity).

It will be obvious that the phenomenon of shape selectivity is directly
related to adsorption and diffusion characteristics. Since it is difficult to
obtain information about these characteristics on a molecular level, especially
at elevated temperatures at which the chemical conversions take place, computer
simulations might offer alternative approaches to problems encountered in
zeolite catalysis research.

In this Section, we will discuss procedures for the description of adsorption and molecular transport phenomena in zeolites. For the simulation of
adsorption, the Monte Carlo method is employed, whereas molecular dynamics
techniques are used in order to simulate diffusion. In the following sections,
we will give a brief overview of the Monte Carlo and molecular dynamics techniques. Finally, both methods will be exemplified by a review of adsorption and
diffusion simulation studies of various zeolites reported in the literature.

The Monte Carlo method

The importance of the Monte Carlo method is that it provides us with a tool
for calculating macroscopic quantities (energy, temperature, pressure, etc.) of
a system for which the intermolecular potentials are known. The reliability of

these calculated properties is therefore determined by the reliability of the intermolecular potentials.

The Monte Carlo algorithm is based on the proposals of Metropolis et al. (ref. 76) and can be characterized by the following three steps:

* give the adsorbate(s) a new configuration (random new position(s) and random new orientation(s)),

* calculate the energy difference between this new configuration and the old configuration: $\delta E = E_{new} - E_{old}$,

* accept this new configuration with a probability proportional to the Boltzmann weight factor at temperature T: $\exp(\delta E/kT)$

These steps are repeated in order to obtain a chain of configurations Γ_i (i=1,2,...,M).

Let us assume we have a system of N particles in a fixed volume V and at a constant temperature T. From statistical mechanics it can be derived that the thermodynamic average of a quantity A can be written as (ref. 77):

$$< A > = (1/Z) \; . \int d\Gamma \; A(\Gamma).\exp(-E(\Gamma)/kT) \qquad (V.1)$$

where

    Γ : the configuration of all particles

               (the integration is over all possible configurations)

    $E(\Gamma)$: energy of a configuration Γ

    $A(\Gamma)$: value of A at configuration Γ

    Z   : partition function:

$$Z = \int d\Gamma \; \exp(-E(\Gamma)/kT) \qquad (V.2)$$

The purpose of the Monte Carlo method is to calculate the integral as given by equation (V.1) numerically. One (naive) method to achieve this is to generate a set of randomly chosen configurations Γ_i (i=1,..,M). Following this procedure, the average of A can be approximated by:

$$< A > = \frac{\Sigma_{i=1}^{M} A(\Gamma_i).\exp(-E(\Gamma_i)/kT)}{\Sigma_{i=1}^{M} \exp(-E(\Gamma_i)/kT)} \qquad (V.3)$$

However, the statistics of this method will be very poor, because most values of Γ_i will be chosen in a region where $\exp(-E(\Gamma_i)/kT)$ is low (ref. 77).

To circumvent this problem, Metropolis et al. (ref. 76) proposed the method of importance sampling. In this method, the configurations will not be chosen at random but will be selected with a probability $P(\Gamma)$. The average of quantity A for this case can be written as (refs. 76,77):

$$< A > = \frac{\Sigma_{i=1}^M \; A(\Gamma_i)P^{-1}(\Gamma_i) \; \exp(-E(\Gamma_i)/kT)}{\Sigma_i^M \; P^{-1}(\Gamma_i) \; \exp(-E(\Gamma_i)/kT)} \qquad (V.4)$$

If we choose for the distribution of configurations in the chain the equilibrium distribution $P^{eq}(\Gamma)$:

$$P^{eq}(\Gamma) = \exp(-E(\Gamma)/kT) \qquad (V.5)$$

then we can write for the average of A:

$$< A > = 1/M\Sigma_{i=1}^M A(\Gamma_i) \qquad (V.6)$$

Generating a chain of configurations $(\Gamma_1,\ldots,\Gamma_M)$ with a certain distribution
can formally be described with the theory of Markov processes (ref. 78). A very
important condition to ensure a statistical reliability is that in the course
of the simulation it must be possible to reach (in principle) all accessible
configurations.

The molecular dynamics method

In this section we will discuss the basic principles of molecular dynamics.
Comparisons with Monte Carlo methods will be made in order to highlight the
differences and analogies of the two techniques.

As discussed above, Monte Carlo simulations deal with systems which are in
equilibrium and so static thermodynamic properties such as heats of adsorption
and average siting behaviour can be derived according to equation (V.6).

In a Monte Carlo simulation there is no time scale involved and successive
configurations are chosen at random. Molecular dynamics simulations, on the
other hand, simulate the time evolution explicitly. So, what you can do with a
molecular dynamics simulation is: all one can do with a Monte Carlo procedure,
i.e. the evaluation of static thermodynamic properties such as temperature and
pressure, but also – and this is of particular interest to us – the determination of diffusivities and site residence times of molecules absorbed in zeolitic pores. These dynamic properties are derived by generating a system trajectory followed by averaging over time. So, for any thermodynamic property A, the
average value of A following from a molecular dynamics simulation is given by:

$$<A> = 1/t.\Sigma_{i=1}^M A.\delta t_i \qquad (V.7)$$

where M is the number of molecular dynamics sampling points, t is the total
simulation time and δt_i is the time step length. Obviously, $\Sigma_{i=1}^M \delta t_i = t$. A is the
value of property A at a specific time along the trajectory.

Generating a molecular dynamics trajectory is purely based on classical
Newtonian mechanics. We consider an ensemble of particles moving in a force

field. The number of particles equals N, the force on particle i (i=1,N) at time t is given by $F_i(t)$, and the position of the particle at time t is represented by $r_i(t)$. A Taylor expansion of $r_i(t)$ around t (until second order) results in:

$$r_i(t+\delta t) = r_i(t) + (dr_i(t)/dt)_t.\delta t + 1/2(d^2r_i(t)/dt^2)_t.(\delta t)^2 \qquad (V.8)$$

$$r_i(t-\delta t) = r_i(t) - (dr_i(t)/dt)_t.\delta t + 1/2(d^2r_i(t)/dt^2)_t.(\delta t)^2 \qquad (V.9)$$

Combination of equations (V.8) and (V.9) yields:

$$r_i(t+\delta t) = 2r_i(t) - r_i(t-\delta t) + (d^2r_i(t)/dt^2)_t.(\delta t)^2$$

$$= 2r_i(t) - r_i(t-\delta t) + (F_i(t)/m_i).(\delta t)^2 \qquad (V.10)$$

and

$$(dr_i(t)/dt)_t = (r_i(t+\delta t) - r_i(t-\delta t)) / 2\delta t = v_i(t) \qquad (V.11)$$

where m_i is the mass of particle i and $v_i(t)$ its velocity at time t.

Equation (V.10) clearly shows that, given the forces on the particles, one is able to generate a trajectory in time without using any knowledge about the particle velocities. However, determination of the velocities might be useful for the evaluation of the particle ensemble temperature from the kinetic energy of the system:

$$E_{kin}(t) = 1/2\Sigma_{i=1}^N m_i.v_i^2(t) \qquad (V.12)$$

and hence

$$T(t) = (2/3Nk).E_{kin} = (1/3Nk)\Sigma_{i=1}^N m_i.v_i^2(t) \qquad (V.13)$$

wherein k is the Boltzmann constant.

Equations (V.10) through (V.13) define the body of a molecular dynamics algorithm. From equation (V.13) it is obvious that the temperature of the particle ensemble is not constant but fluctuates around the average temperature and hence the kinetic energy of the system is not constant either. It should be noted however, that the total energy, E_{tot}, is conserved throughout the simulation. The total energy is given by:

$$E_{tot} = E_{pot} + E_{kin} \qquad (V.14)$$

where the potential energy, E_{pot}, is related to the force via

$$F = -dE_{pot}/dr \qquad (V.15)$$

Such an algorithm is referred to as an NVE algorithm (constant number of particles, constant volume, constant total energy) (ref. 79). The simulation temperature, T_s, in an NVE simulation can be determined afterwards by integrating over the time:

$$T_s = 1/t . \int T(t) dt \qquad (V.16)$$

The NVE algorithm is employed in cases where the conservation of the total energy is required, that is, when the microcanonical (NVE) ensemble has to be sampled. However, in most practical processes, i.e. isothermic ones, a constant temperature is required. In these cases, one samples the canonical ensemble and one needs to employ an NVT algorithm (constant number of particles, constant volume, constant temperature). This algorithm is basically the same as the NVE algorithm but a temperature correction step is incorporated in order to ensure a constant temperature. Equation (V.10) is used to predict the new positions of the particles. Equations (V.11) and (V.13) are used to calculate the temperature $T(t)$ of the ensemble. Let $T_w(t) = T_w$ be the desired simulation temperature, then

$$T_w = (T_w/T(t)) . T(t) \qquad (V.17)$$

and so with the use of equation (V.13):

$$T_w = (T_w/T(t)) . (1/3Nk) \Sigma_{i=1}^{N} m_i . v_i^2(t)$$

$$= (1/3Nk) \Sigma_{i=1}^{N} m_i (v_i(t) . (T_w/T(t))^{1/2})^2 \qquad (V.18)$$

From equation (V.18) it follows that the velocities of the particles at time t need to be scaled by a factor $(T_w/T(t))^{1/2}$ in order to maintain a constant temperature. With these scaled velocities and the forces on the particles, the new (corrected) positions of the particles are calculated via:

$$v'_i(t) = (T_w/T)^{1/2} . v_i(t) \qquad (V.19)$$

$$r_i(t+\delta t) = r_i(t) + v'_i(t) . \delta t \qquad (V.20)$$

Basically, the implications of equations (V.17) and (V.18) are that the surroundings of the system under consideration are regarded as a medium with an infinite heat capacity (isothermic bath). All heat which is added to, or withdrawn from, the system comes from, or goes to, the isothermic bath. If we regard simulation of diffusion of adsorbates in zeolites, the NVT algorithm is based on the assumption of an infinite coupling between the adsorbates ensemble temperature and the zeolite energy content. On the other hand, the NVE algorithm assumes no coupling of this kind at all.

An advantage of the NVT algorithm over the NVE algorithm is that the NVT algorithm is cheaper in computer time. This is due to the fact that in the NVE algorithm the ensemble has to be equilibrated for a rather long time to establish a constant total energy which does not fluctuate too much as a result of simulation start-up effects. Furthermore, it seems logical that the zeolite lattice is influenced by the temperature of the adsorbates ensemble, so the NVT algorithm will reflect the physical reality in a better way.

Simulation of adsorption and diffusion; models

In the Monte Carlo adsorption simulations the following approximations are used:

* no adsorbate-adsorbate interactions are considered, so all calculations correspond to low coverage of adsorbate molecules.

    However, in a section to come, we will also meet studies in which the adsorbate-adsorbate interactions are not neglected and hence adsorption phenomena apart from the zero filling area are examined.

* the zeolite lattice is rigid

* the adsorbate molecule is rigid.

With these assumptions, the potentials contain only adsorbate-zeolite interactions which can be approximated by a pairwise potential Φ_{ij} (ref. 39d). Thus the average potential energy for a given configuration Γ is given by:

$$<E_{pot}>_{\Gamma} = \Sigma_{ij} \Phi_{ij} \qquad (V.21)$$

where i and j run over all the atoms in the zeolite and adsorbate, respectively. The potential Φ_{ij} is usually a combination of a (6-12) Lennard-Jones part and a coulombic part. The adsorption enthalpy following from a Monte Carlo simulation follows from averaging over all configurations that are generated during the simulation (eq. V.6):

$$<Q_s> = 1/M \ \Sigma_{i=1}^{M} A(\Gamma_i) \qquad (V.22)$$

In order to eliminate the effect of unit cell boundaries, periodic boundary conditions are used. To save computer time, the potentials are truncated at a cut off radius which is of the order of the size of one unit cell. To make an estimate of the statistical reliability, the simulation can be divided into subruns of 10,000 Monte Carlo steps each. The standard deviation of the inter-action energy (eq. V.22) can then be calculated from these subruns. One single run takes from ten up to twenty subruns.

Just as with the Monte Carlo simulations, with the molecular dynamics simu-lations both the adsorbate and the zeolite lattice have so far been assumed to be rigid. However, adsorbate-adsorbate interactions are not neglected, so the potential energy of adsorbate i can be represented by:

$$E_{pot,i} = \Sigma_\tau \; \Phi_{\tau,i}^{(1)} + \Sigma_{j \diamond i} \; \Phi_{i,j}^{(2)} \qquad (V.23)$$

where τ runs over all the atoms in the zeolite lattice and j runs over all the adsorbates except adsorbate i. $\Phi^{(1)}$ is the adsorbate–zeolite interaction (a combined Lennard–Jones/coulombic interaction) and $\Phi^{(2)}$ is the adsorbate–adsorbate interaction (purely Lennard–Jones). The total average potential energy of the ensemble at a specific time i out of the trajectory is given by:

$$<E_{pot}>_i = 1/N \; \Sigma_{i=1}^N \{ \; \Sigma_\tau \; \Phi_{\tau,i}^{(1)} + 1/2 \; \Sigma_{j \diamond i} \; \Phi_{i,j}^{(2)} \; \} \qquad (V.24)$$

with N the number of adsorbates in the zeolite. The factor 1/2 is introduced to avoid double counting.

From equation (V.23) and with the use of equation (V.15), an expression for the force on particle i can easily be derived:

$$F_i(t) = -d/dr \; \{ \; \Sigma_\tau \; \Phi_{\tau,i}^{(1)} + \Sigma_{j \diamond i} \; \Phi_{i,j}^{(2)} \; \} \qquad (V.25)$$

Note that $F_i(t)$ is a function of the time because $\Phi^{(1)}$ and $\Phi^{(2)}$ are time-dependent. Note also that $F_i(t)$ is a vector, whereas $E_{pot,i}$ is a scalar. Equation (V.25) defines a force field for molecules migrating in the pores of a zeolite. This force field can be used directly in the MD algorithms as discussed in the previous section to generate a system trajectory in time. From this trajectory, diffusivities are easily calculated via the Einstein relation (ref. 80):

$$D = 1/6 \; <r^2>/t \qquad (V.26)$$

where $<r^2>$ is the mean square distance travelled by the adsorbate molecules during a period t.

Adsorption enthalpies (Q_s) are calculated by using equation (V.7):

$$<Q_s> = 1/t \; \Sigma_{i=1}^M <E_{pot}>_i \cdot \delta t_i \qquad (V.27)$$

where $<E_{pot}>_i$ is given by equation (V.24), M is the number of time steps taken for the simulation, and δt_i is the time step length.

Discussion of computer simulation studies

Computer simulations of adsorption and diffusion in zeolites have gained much interest over the last few years. This increasing interest is probably due to the possibilities that modern computer hardware and methods developed in computational chemistry offer to zeolite catalysis research. The number of papers devoted to computer simulation of adsorption and diffusion which have appeared in the literature from the early 80's to 1988 has increased considerably compared to the foregoing years.

In this section, we intend to discuss several studies which have recently appeared in the literature. In the first instance, we will restrict ourselves to studies which are based on the theory as outlined in Section V. However, some attention will also be paid to other (computer) simulation studies.

Leherte et al. (ref. 44) extensively studied water adsorption in a ferrierite-type zeolite structure. The zeolite-water potential was a combined Lennard-Jones/coulomb potential, whereas the water-water potential was derived from Matsuoka et al. (ref. 81). They did not use the zero-filling approximation but considered a varying water occupancy of the zeolite. The amount of water adsorbed varied from 5.88 up to 11.95 molecules per unit cell (total number of adsorbates considered: N=40-80). Using Monte Carlo simulations, they were able to calculate water adsorption enthalpies which were in reasonable agreement with experimental data. Furthermore, they studied the configurations of the adsorbed water molecules. It was found that the water molecules tend to form cage structures in the zeolite cavities and remain far from the centre and the walls of the 10-membered ring channel. However, in the 8-membered ring

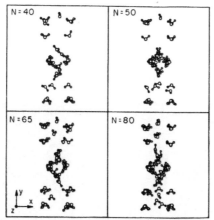

Reprinted by permission from <u>Stud. Surf. Sci. and Catal.</u>, Vol. 37, p. 293. Copyright (c) 1988. Elsevier Science Publishers B.V., Amsterdam

Fig. 6. Plots of one significant configuration for all water molecules corresponding to each density: (a) N=40, (b) N=50, (c) N=65 and (d) N=80. (Figure taken from reference 44).

channels, the molecules are closer to the walls. Figure 6 shows a plot of one significant configuration for four different densities. In these plots, the clustering of the water molecules is clearly demonstrated.

Yashonath et al. (ref. 42) simulated methane adsorption in zeolite Na-Y using a Monte Carlo technique and a Lennard-Jones/coulomb zeolite-adsorbate interaction potential. Apart from the calculation, they paid special attention to the temperature dependency of the adsorption process. Using the zero-filling

348

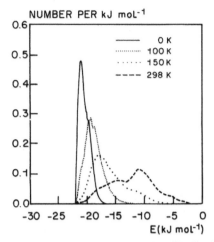

NUMBER PER kJ mol⁻¹

Fig. 7. Plots of the distribution functions of the potential energy of interaction of methane with zeolite Na-Y at different temperatures. (Figure taken from reference 42).

approximation, they generated so-called potential energy distribution functions (PEDF) which relate the potential energy of a configuration to the fraction of adsorbed molecules being in that particular configuration. Figure 7 displays a PEDF at different temperatures for methane in zeolite Na-Y.

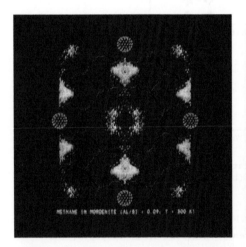
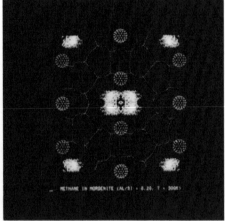

Fig. 8. Distribution of methane adsorbed in mordenite at 300 K and Al/Si=0.09 (a) and Al/Si=0.2 (b) as obtained by Monte Carlo calculations. The zeolite structure is represented by rods and the sodium ions by spheres. Each spot inside the zeolite pore represents the projection of the centre of mass of the methane molecule. (Figure taken from reference 43).

A similar Monte Carlo study was carried out by Smit et al. (ref. 43), who studied methane adsorption in the zeolites silicalite, Na-Y and Na-mordenite using the same potentials as Yashonath et al. (ref. 42). For the case of Na-mordenite, special attention was paid to the influence of the zeolitic Al/Si ratio on the adsorption process. It was found that the side-pockets, situated at the edges of the 12-membered ring channels are strong adsorption sites for methane. By increasing the Al/Si ratio, and thus increasing the sodium content, these side-pockets are gradually blocked by sodium ions. This feature results in totally different adsorption characteristics, as can be seen in Figure 8. In this figure, the dots represent the centre of mass of the methane molecule adsorbed in the mordenite pores. At low Al/Si ratios (Figure 8a) methane is preferentially adsorbed in the side-pockets (high density of dots) whereas at higher Al/Si ratios these side-pockets are blocked with sodium ions. This pore-blocking feature is more quantitatively expressed in the PEDF in Figure 9, where one adsorption site disappears with increasing Al/Si ratio.

In line with the water-adsorption Monte Carlo simulations, Leherte et al. (ref. 82) studied the diffusivity of water adsorbed in ferrierite. Using molecular dynamics techniques and the same potentials as used in their Monte Carlo study on the same system (ref. 44), they calculated the self-diffusion coefficient of water using the Einstein relation (eq. V.26). By plotting the mean square distance travelled by the water molecules against time, the diffusion coefficient was found to be 0.5×10^{-9} m^2/s. Figure 10 displays such a plot for the three different directions x (crystallographic a-axis), y (b-axis) and z (c-axis) as well as for the average displacement. In ferrierite, only a displacement in the y- and z-directions is observed due to the fact that along the x-axis the ferrierite channels are not connected. Thus, there is no way for the molecules to move from one channel to another following that direction and consequently their mobility is lowered.

Yashonath et al. (ref. 83) introduced site residence times (τ_s) and cage residence times (τ_c). They performed molecular dynamics simulations of methane self-diffusion in zeolite Na-Y using the same potentials as in their earlier Monte Carlo study on the same system (ref. 42). From these simulations, they were able to estimate the average residence times of methane molecules at adsorption sites and in the supercages of the faujasite structure. Figure 11 displays their results. At temperatures above 200 K, the mobility of the methane molecules increases considerably. It is also observed that site residence times are nearly negligible at temperatures higher than 150 K. Furthermore, the average residence time for a methane molecule in a supercage is estimated to be 2-3 ps at room temperature.

Diffusivities of methane in various all-silica zeolites, using molecular dynamics approaches, were studied by Den Ouden et al. (ref. 84). In their

350

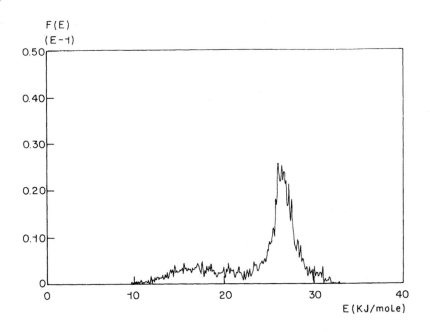

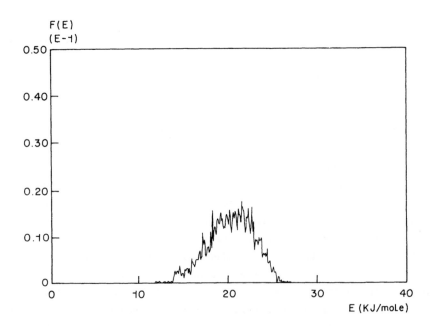

Fig. 9. Plots of the distribution functions of the potential energy of interaction of methane with zeolite Na—mordenite at different Al/Si ratios. For low Al/Si ratios (a, Al/Si=0.09), two distinct adsorption sites can be appointed. For high Al/Si ratios (b, Al/Si=0.20), only one adsorption site is left.

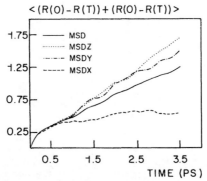

Reprinted by permission from <u>Chem. Phys. Lett.</u>, Vol. 145(3), p. 237. Copyright (c) 1988. Elsevier Science Publishers, Amsterdam

Fig. 10. Mean square deviation of the positions of the water molecules along the x-axis, y-axis and z-axis. The x, y and z directions are associated with the a, b and c cell parameters, respectively. (Figure taken from reference 82).

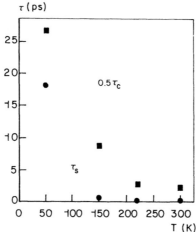

Reprinted by permission from <u>Chem. Phys. Lett.</u>, Vol. 153(6), p. 551. Copyright (c) 1988. Elsevier Science Publishers, Amsterdam

Fig. 11. Temperature dependency of the site (τ_s, dots) and cage (τ_c, squares) residence times for methane in zeolite Na-Y at a loading of six molecules per supercage. Statistical uncertainties are of order 20% for the three highest temperature points and at least twice as large for the lowest temperature. (Figure taken from reference 83).

study, special attention was paid to the influence of the zeolite topology on the diffusion process. Site residence times for methane migrating in the micropores of the zeolites silicalite and mordenite were visualised by using computer graphics. Figure 12 displays trajectories for one methane molecule migrating in the pores of mordenite and silicalite, respectively. The force field in which the molecule moves is caused by the presence of the zeolite

352

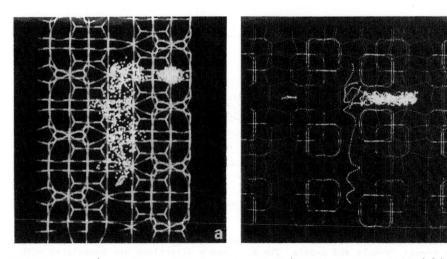

Fig. 12. Trajectory of a methane molecule migrating in mordenite (a) and silicalite (b) over a time period of 100 ps. The zeolite structure is represented by rods. The dots represent the position of the methane molecule at subsequent time intervals. For mordenite (a), the main channel is vertically displayed and the zeolite loading is two molecules/unit cell. The straight channel for silicalite (b) is vertically displayed, the sinusoidal channel horizontally. The zeolite loading is four molecules/unit cell. Both simulations were carried out using an NVT algorithm at T=300K. (Figure taken from reference 84).

lattice and the other methane molecules (not displayed in the figure). Obviously, the side pockets in mordenite (Figure 12a) are adsorption sites with a relatively long residence time (high density of dots). The same is true for methane migrating in the sinusoidal channels in silicalite (vertically displayed in Figure 12b), in which the methane molecule resides much longer than in the straight channels.

Only very recently, June et al. (ref. 85) reported a study on statistical mechanics and molecular dynamics for the modeling of sorption and transport of hydrocarbons in pentasil zeolites. Statistical mechanical relations have been employed for the determination of Henry's constants, isosteric adsorption enthalpies and adsorbate conformational properties for butane and some hexane isomers. Furthermore, molecular dynamics simulations have been employed for the determination of the self diffusivity of various hydrocarbons in silicalite.

A diffusion study based on a Monte Carlo approach was carried out by Palekar et al. (ref. 86) and Pitale et al. (ref. 87). With their approach, they simulated gravimetry experiments and tracer diffusion. They were able to establish the relation between the variation of diffusivity and sorbate concentration as obtained by experiment.

Finally, a very elegant adsorption/diffusion model which is not based on any of the methods discussed in the previous sections was developed by Derouane et al. (refs. 88,89,90). In this model, the curvature of the adsorption surface is taken into account, and it enables one to find analytical expressions for the van der Waals interaction energy and the corresponding force on the adsorbate. With this model, one is able to compare adsorption characteristics in microporous solids with adsorption on flat surfaces. Figure 13 plots the activation

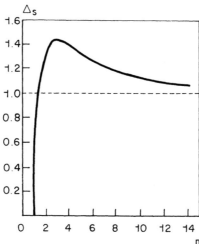

Reprinted by permission from Chem. Phys. Lett., Vol. 137(4), p. 336. Copyright (c) 1987, Elsevier Science Publishers B.V., Amsterdam

Fig. 13. Relative activation energy, δ_s, for promoting a molecule from the pore wall to the centre of the cavity as a function of the cavity size n (units of r_m). (Figure taken from reference 88).

energy δ_s needed to promote a molecule from the pore wall to the centre of the cavity as a function of the cavity size (in units of n times the adsorbate radius, r_m). δ_s is a parameter accounting for surface curvature effects that may be regarded as a relative (to the flat surface case) activation energy. Obviously, for n=1, the adsorbate radius equals the cavity radius resulting in a maximum van der Waals attraction and hence yielding a zero (sticking-) force acting on the adsorbate. This case is referred to as the floating molecule state. For n≥1.5, the activation energy always exceeds the flat surface activation energy. Hence, adsorbates in zeolites tend to be more sticked to the internal micropore surface than in the case of adsorption on a flat surface. In this respect, diffusion of small adsorbates at moderate temperatures can be seen as a hopping over the corrugation barriers of the pore wall rather than colliding with the wall and being reflected. Diffusion of chain molecules at moderate temperatures will appear as a creeping motion along the zeolite micropore walls which is referred to as creep diffusion.

From the above discussed studies, two main conclusions on Monte Carlo simulations of adsorption emerge. First of all, using empirical potentials, the
agreement of the calculated adsorption enthalpies with experimental data is
reasonable indicating a proper description of the adsorbate–zeolite and adsorbate–adsorbate interaction with respect to the particular properties studied.
It should be stressed however, that little work has been done on systems with
variations in the zeolitic Al/Si ratio. More extensive research in the area of
these potentials is certainly necessary. Bezus et al. (ref. 39a) and Kiselev
et al. (refs. 39a,b,c) already developed potentials for this case both for
organic and inorganic adsorbates.

Secondly, Monte Carlo simulations allow us to study adsorption on a molecular level. This detailed information can be considered as very valuable in
zeolite catalysis research. However, the Monte Carlo studies carried out so far
deal with rather small systems which are not of primary interest to catalysis
research. Recent studies on larger molecules indicate that an extension of
these (preliminary) studies to larger adsorbates is possible (ref. 85).

Finally, a promising future application of Molecular Dynamics techniques is
in the field of zeolite synthesis. Since the study of zeolites or zeolite
precursors in contact with the synthesis mother liquor in the presence or
absence of organic template molecules is hard to access by direct experiments,
Molecular Dynamics studies might offer a powerful approach to this problem.

Molecular dynamics simulations are indispensable techniques to understand
mass transport and chemical dynamics in porous solids on a molecular level.
Whereas applications to the computation of mass transport properties have
appeared, the development of studies on transition dynamics to determine
chemical reaction rates requires the development of potential energy surfaces
by methods as discussed in Section II.

VI. FUTURE PERSPECTIVES

Zeolite science currently is an active field for the development of mod-
ling techniques useful in solid state chemistry applications.

Because the potentials to be used have still not completely been determined
and a few basic questions remain to be solved, it is also a subject of interest
to theoretical physical chemistry.

The question of the relative importance of long range electrostatic interactions versus short range interactions cannot be considered to be definitely
solved. It may be expected that application of quantumchemical techniques and
the increasing computerpower that is becoming available will soon produce the
potentials required to study systematically the packing of molecules in the
micropores of the zeolite.

The Monte Carlo method is suitable to approach this problem important for zeolite-synthesis and the physical chemistry of cation exchange. Molecular dynamics in combination with vibrationspectroscopy will provide a sensitive test on the potentials used, by computation of dynamic properties.

In order to provide a basis to the many reaction mechanisms postulated to occur in zeolites the relative stability and potential energies of proposed transition states will have to be evaluated. This requires a solution of the problems of hydrogen bonding in zeolites. In view of the high accuracy of the computations required for the time being this problem can be best approached by studying small model clusters.

The indication that zeolite acidity is mainly determined by short range interactions provides avenues for chemical manipulation of Bronsted acidity and will make this a field of intensive experimental as well as theoretical research.

Successes of the current states of zeolite theory are the prediction and its experimental confirmation of cation positions in high Al-content zeolites as well as the predicted and experimentally confirmed small difference in cohesive energy of siliceous zeolites.

Quantum chemical cluster calculations appear to simulate trends in zeolite acidity quite well, if one changes zeolite composition.

Monte Carlo studies of the adsorption of organic molecules in zeolites appear to predict preferred adsorption positions as well as heats of adsorption quite well, not withstanding the need for improvement of the potentials employed. The agreement of computed rates of diffusion of small organic molecules using Molecular Dynamic approaches with spectroscopic data can also be considered a significant accomplishment.

REFERENCES

1 D. Avnir, D. Farin and P. Pfeifer, J. Chem. Phys. 79(7), 3566 (1983).
2 W.M. Meier, Stud. Surf. Sci Catal. 28, (1986) 13.
3a Computer Simulation of Solids, (C.R.A. Catlow and W.C. Mackrodt, Eds) Lecture Notes in Physics 166, Springer-Verlag 1982;
 b C.R.A. Catlow, M. Doherty, G.D. Price, M.J. Sanders and S.C. Parker, Materials Science Forum, 7, (1986) 163;
 c R.A. Jackson and C.R.A. Catlow, Molecular Simulation 1, (1988) 207.
4a R.A. van Santen, A. de Man and B.W. van Beest, Proc. NATO ASI "Physicochemical properties of zeolitic systems and their low dimensionality", Dourdan, France, 1989;
 b B.W. van Beest, J. van Lenthe and R.A. van Santen, in preparation.
5a K.T. No, H. Chon, T. Lee and M.S. Jhon, J. Phys. Chem. 85, (1981) 2065;
 b K.T. No, J.S. Kim, Y.Y. Huh and M.S. Jhon, J. Phys. Chem. 91, (1987) 740.
6a J.E. Hurley, J. Phys. Chem. 69, (1985) 3284;
 b R.T. Sanderson, Chemical Periodicity, Reinhold, New York, 1960;
 c W.J. Mortier, Stud. Surf. Sci. Catal. 37, (1988) 253.
7 M.K. Song, H. Chon, M.S. Jhon and K.T. No, J. Mol. Catal. 47, (1988) 73.

356

8   C.J.J. den Ouden, R.A. Jackson, C.R.A. Catlow and M.F.M. Post, to be published.

9   G. Ooms, R.A. van Santen, C.J. J. den Ouden, R.A. Jackson and C.R.A. Catlow, J. Phys. Chem. 92, (1988) 4462.

10 N.L. Allinger, J. Am. Chem. Soc. 99, (1977) 8127.

11 M. Mabilia, R.A. Pearlstein and A.J. Hopfinger, J. Am. Chem. Soc. 109, (1987) 7960.

12 R.A. van Santen, G. Ooms, C.J.J. den Ouden, B.W. van Beest and M.F.M. Post, Proc. Symposium on Advances in Zeolite Synthesis ACS Meeting, Los Angeles, 1988.

13 A.J.M. de Man, B.H.W. van Beest, M. Leslie and R.A. van Santen, J. Phys. Chem. accepted.

14 K.J. Choi, M.S. Jhon and K.T. No, Bull. Korean Chem. Soc. 8, (1987) 155.

15 K.T. No, D.H. Bal and M.S. Jhon, J. Phys. Chem. 90, (1986) 1772.

16 C. Lee, J. Phys. Chem.: Solid State Phys. 19, (1986) 5555.

17 S.H. Garofalini, J. Non–Crystalline Solids 55, (1983) 451.

18 E. Clementi, Computational Aspects of Large Chemical Systems, Springer–Verlag, Berlin, 1980.

19 B.M. Axilrod and E. Teller, J. Chem. Phys. 11, (1943) 299.

20 S.H. Garofalini, J. Am. Cer. Soc. 67, (1984) 133.

21a T.F. Soules, J. Chem. Phys. 71, (1979) 4570;
  b S.H. Garolini, J. Chem. Phys. 76, (1982) 3189.

22 C. Pisani, R. Dovesi and C. Roetti, Hartree–Fock ab–initio treatment of crystalline systems, Lecture Notes in Chemistry 48, Springer–Verlag, 1988.

23 M.D. Newton. M. O'Keeffe and G.V. Gibbs, Phys. Chem. Minerals 6, (1980) 305.

24 S. Tsuneyuki, M. Tsukada, H. Aoki and Y. Matsui, Phys. Rev. Lett. 61, (1988) 869.

25 B. van Beest and R.A. van Santen, Catal Lett. 1, (1988) 147.

26 A.G. Pelmenschikov, E.A. Paukstitis, V.S. Stepanov, K.G. Ione, G.M. Zhidoniroy and K.I. Zamaraev, Proc. 9th Int. Congr. Catal. 1, (1988) 404, The Chemical Institute of Canada

27 E.G. Derouane and J.G. Fripiat, Zeolites 5, (1985) 165.

28 P.J. O'Malley and J. Dwyer, Zeolites 8, (1988) 317.

29 S. Beran, J. Phys. Chem. 137, (1983) 89.

30 J. Sauer, H. Haberlandt, W. Schirmer and P.A. Jacobs et al. (Eds.) Proc. of the Conference on Structure and Reactivity of Modified Zeolites, Prague, 1984, Elsevier, Amsterdam, 1984.

31a P. Hotza, J. Sauer, C. Morgeneyer, J. Huych and R. Zahradnik, J. Phys. Chem. 85, (1981) 4061;
  b J. Sauer, C. Morgeneyer and R.P. Schroder, J. Phys. Chem. 88, (1984) 6375;
  c J. Sauer and R. Zahradnik, Int. J. Quantumchem. 26, (1984) 739;
  d J. Sauer, Acta Phys. Chem. 31, (1985) 19;
  e J. Sauer and K.P. Schroder, Z. Phys. Chem. Leipzig 266, (1985) 379;
  f J. Sauer, J. Phys. Chem. 91, (1987) 2315.

32 P. Hobza and R. Zahradnik, Chem. Rev. 88, (1988) 871.

33 H. Tatewaki and S. Huzinaga, J. Chem. Phys. 71, (1979) 4339.

34 S.F. Boys and F. Bernardi, Mol. Phys. 19, (1970) 553.

35a K.L. Miller, Biopolymers 18, (1979) 959;
  b J.A. Yoffe, Theor. Chim. Acta 55, (1980) 219;
  c Y.K. Kang and M.S. Jhon, Theor. Chim. Acta 61, (1982) 41.

36 G. Alagona and A. Tani, J. Chem Phys. 74, (1981) 3980.

37 C.E. Dykstra, Acc. Chem. Res. 21, (1988) 355

38 J. Snir, R.A. Nemenoff and H.A. Scheraga, J. Phys. Chem. 82, (1978) 2497.

39a A.G. Bezus, A.V. Kiselev, A.A. Lopatkin and P.Q. Du, J. Chem. Soc. Faraday Trans. II 74, (1978) 367;
  b A.V. Kiselev and P.A. Du, J. Chem. Soc. Far. Trans. II 77, (1981) 1;
  c A.V. Kiselev and P.A. Du, J. Chem. Soc. Far. Trans. II 77, (1981) 17;
  d A.V. Kiselev, A.A. Lopatkin and A.A. Schulga, Zeolites, 5, (1985) 261.

40a J.G. Kirkwood, Phys. Z. 33, (1932) 57;
  b A. Muller, Proc. Roy. Soc. A. 154, (1936) 624.

41a P.A. Wright, J.M. Thomas, A.K. Cheetham and A.K. Nowak, Nature 318, (1985) 611;

  b A.K. Cheetham, A.K. Nowak and P.W. Betteridge, Proc. Ind. Acad. Sci. (Chem. Sci.) 96, (1986) 411.

42 S. Yashonath, J.M. Thomas, A.K. Nowak and A.K. Cheetham, Nature 331, (1988) 601.

43 B. Smit and C.J.J. den Ouden, J. Phys. Chem. 92, (1988) 7169.

44 L. Leherte, D.P. Vercauteren, E.G. Derouane and J.M. Andre, Stud. Surf. Sci. Catal. 37, (1988) 293.

45a A. Guirsot, F. Fajula, C. Daul and J. Weber, J. Phys. Chem. 92, (1988) 4456;

  b J. Weber, P. Fluelinger, P.Y. Morgantini, O. Schaad, A. Guirsot and C. Daul, J. Computer Aided Molecular Design 92, (1988) 235.

46a G.M. Zhidomirov and V.B. Kazansky, Adv. Catal. 34, (1986) 131;

  b P.A. Jacobs and W.J. Mortier, Zeolites 2, (1982) 226;

  c J. Dwyer, Stud. Surf. Sci. Catal. 37, (1988) 333.

47 W.J. Mortier, J. Sauer, J.A. Lerchev and H. Noller, J. Phys. Chem. 88, (1984) 905.

48a P.J. O'Malley and J. Dwyer, J. Chem. Soc. Chem. Comm. 72 (1987);

48b P.J. O'Malley and J. Dwyer, J. Phys. Chem. 92, (1988) 3005.

49a E. Kassab, K. Seiti and M. Allavena, J. Phys. Chem. 59, to appear;

  b K. Seiti, Thesis, Univ. Paris VI (1988).

50 R. Vetrivel, C.R.A. Catlow and E.A. Colbourn, Stud. Surf. Sci. Catal. 37, (1988) 37.

51 S. Beran, J. Phys. Chem. 89, (1985) 5586.

52 K. Mueller, H.J. Ammann, D.M. Doran, P. Gerber and G. Schrepfer, Innovative Approaches in Drug Research, A.F. Harms (Ed.), Elsevier, Amsterdam, 1986.

53 T. Gund and P. Gund, Molecular Structure and Energetics Vol. 4, J.F. Liebman and A. Greenberg (Eds.), VCH Publishers, New York, 1987.

54 U. Bunkert and N.L. Allinger, Molecular Mechanics, ACS Monograph No. 177, Washington DC, 1982.

55 H.J.C. Berendsen, J. Computer Aided Molecular Design 2(3), (1988) 217.

56 C. Hansch, Molecular Structure and Energetics, Vol. 4, J.F. Liebman and A. Greenberg (Eds.), VCH Publishers, New York, 1987.

57 T. Clark, A Handbook of Computational Chemistry, Wiley, New York, 1985.

58 S. Ramdas, J.M. Thomas, P.W. Betteridge, A.K. Cheetham and E.K. Davies Angew. Chem. Int. Ed. Engl. 23, (1984) 671.

59 E. Keller, J. Appl. Cryst. 22, (1989) 19.

60 M.J. Connoly, Science 221, (1983) 709.

61 Visualization in Scientific Computing, B.C. McCormick, T.A. DeFanti and M.D. Brown (Eds.), Computer Graphics, Vol. 21, (1987).

62 R.A.J. Driessen, B.O. Loopstra, D.P. de Bruijn, H.P.C.E. Kuipers and H. Schenk, J. Computer Aided Molecular Design, 2(3), (1988) 225.

63 S. Ramdas, J. Computer Aided Molecular Design, 2(4), (1988) 137.

64 J.V. Smith and W.J. Dytrych, Nature 309, (1984) 607.

65 D.E. Akporiaye and J.M. Thomas, submitted for publication.

66a S. Andersson, S.T. Hyde and H.G. von Schnering, Z. Kristallogr. 168, (1984) 1;

  b H.G. von Schnering and R. Nesper, Angew. Chem. 99, (1987) 1097.

67 P.A. Wright, J.M. Thomas, A.K. Cheetham and A.K. Nowak, Nature 318, (1985) 611.

68 G.R. Millward, S. Ramdas, J.M. Thomas and M.T. Barlow, J. Chem. Soc. Faraday Trans. 79, (1983) 1075.

69 J.L. Schlenker, Zeolites 5 (1985), 346, 349, 352, 355.

70 M.M.J. Treacy and J.M. Newsam, Nature 332, (1988) 249.

71 M. Fujinaga, P. Gros and W.F. van Gunsteren, J. Appl. Cryst. 22, (1989) 1.

72 J.J. Keijsper, C.J.J. den Ouden and M. Post, 8th Int. Zeolite Conf., Amsterdam (1989).

73 I.E. Maxwell, J. Incl. Phenom. 4(1), (1986) 1.

74 E.F. Vansant, Proc. Int. Symp. on Innovation in Zeolite Materials Science, Elsevier, Amsterdam, 1988, p 143.

358

75 C. Mirodatos and D. Barthomeuf, J. Catalysis, 93, (1985) 246.
76 N. Metropolis, A.W. Rosenbluth, M.N. Rosenbluth, A.H. Teller and E. Teller, J. Chem. Phys. 21, (1953) 1081.
77 K. Binder, Introduction: Theory and Technical Aspects of Monte Carlo Simulations, in: Monte Carlo Methods, Springer-Verlag, Berlin, 1981.
78 N.G. van Kampen, Stochastic Processes in Physics and Chemistry North Holland, Amsterdam, 1980.
79 M.P. Allen and D.J. Tildesley, Computer Simulation of Liquids, Clarendon Press, Oxford, 1987.
80 J. Crank, The Mathematics of Diffusion, Clarendon Press, Oxford, 1975
81 O. Matsuoka, E. Clementi and M. Yoshimine, J. Chem. Phys. 77, (1982) 899.
82 L. Leherte, G.C. Lie, K.N. Swamy, E. Clementi, E.G. Derouane and J.M. Andre, Chem. Phys. Lett. 145(3), (1988) 237.
83 S. Yashonath, P. Demontis and M.L. Klein, Chem. Phys. Lett. 153(6), (1988) 551.
84 C.J.J. den Ouden, B. Smit, A.F.H. Wielers, R.A. Jackson and A.K. Nowak, Molecular Simulation, in press (1989)
85 R.L. June, A.T. Bell and D.N. Theodorou, 196th ACS National Meeting, Los Angeles, California, (September 1988).
86 M.G. Palekar and R.A. Rajadhyaksha, Chem. Eng. Sci. 40(7), (1985) 1085.
87 K.K. Pitale and R.A. Rajadhyaksha, Curr. Sci. 57(4), (1988) 172.
88 E.G. Derouane, J.M. Andre and A.A. Lucas, Chem. Phys. Lett. 137(4), (1987) 336.
89 E.G. Derouane, J.B. Nagy, C. Fernandez, Z. Gabelica, E. Laurent and P. Maljean, Appl. Catal. 40(1-2), (1988) 1.
90 E.G. Derouane, J.M. Andre and A.A. Lucas, J. Catal. 110(1), (1988) 58.

Chapter 10

ION EXCHANGE IN ZEOLITES

R.P. TOWNSEND
Unilever Research, Port Sunlight Laboratory, Bebington,
Wirral, Merseyside L63 3JW, UK.

SUMMARY

Aspects of ion exchange in zeolite are reviewed, with special reference to those properties of zeolites which give rise to characteristic, and sometimes unique, ion-exchange behaviour. As well as discussing basic principles, the thermodynamic and kinetic aspects of the theory of ion exchange are covered, with particular reference to their utility for predicting exchange behaviour in zeolites and especially for developing the use of zeolites as detergent builders. Recent studies on, and trends in, the use of zeolites in detergency are next reviewed, together with general new developments in the field of ion exchange which will help our understanding of zeolite synthesis and chemistry.

1. INTRODUCTION

Ion exchange is an intrinsic property of most zeolites. As a consequence the phenomenon has either given rise to an admittedly few but nevertheless important number of direct applications, or the phenomenon is used indirectly, as a means of "tailoring" zeolite structure and hence properties when these materials are used in other ways, such as in catalysis or gas sorption.

There are analogies which one can draw between zeolites on one hand and clay minerals on the other. Both classes of materials are mineral in origin, comprising similar elements (viz silicon, aluminium, alkali and alkaline earth metals with perhaps smaller quantities of other metals such as iron and titanium). Water is also an important component in both types of minerals. Both have an intrinsic ability to exchange cations, and this intrinsic ability arises as a consequence of isomorphous replacement. In the case of clay minerals this isomorphous replacement may be of trivalent cations by divalent, or of a tetravalent cation by a trivalent one. In the case of zeolites, the isomorphous replacement is always of the tetravalent framework cation (i.e. silicon) by a cation of lower charge (normally aluminium). As a consequence of this substitution, a net negative charge arises on the framework of the zeolite which has to be neutralised by the presence of cations within the pores. These cations may be any of the metals or complexes of the same, or alkylammonium cations, and it is the great variety in nature and extent of exchange of other cations which may occur in zeolites, which gives rise to the richness of the chemistry.

Zeolites are analogous to clay minerals in another respect, in that an ion-exchange capacity can arise not from an intrinsic property which occurs as a consequence of isomorphous substitution in the framework, but rather from unsatisfied valencies occurring at the termination of the crystal edges and faces, or from faults within the structure. The most common origin of an ion-exchange capacity arising as a consequence of such faults, dislocations or edges comes from the presence of silanol groups or hydroxyls attached to edge aluminiums. In clay minerals up to 20% of the exchange capacity may arise from these sources; in the case of zeolites the extent of exchange capacity arising from such sources is normally small relative to the intrinsic one. However in the case of high-silica zeolites, these secondary sources of ion exchange capacity plus non-homogeneous distribution of aluminium in the framework can lead to significant and unexpected effects (ref. 1).

In fact, isomorphous substitution into the framework itself can be regarded quite properly as a form of ion exchange; in recognition of this a complete statistical thermodynamic formulation for framework substitution has been drawn up (ref. 2). The phenomenon has given rise to a whole area of zeolite chemistry involving the substitution of aluminium by other trivalent metals such as iron, and recently it has been claimed that aluminium can be re-inserted into the framework subsequent to its removal by an ion-exchange process (ref. 3). However, this area more properly rests within the realm of modified zeolites, which is discussed elsewhere in this book. Here we will concentrate on the second form of exchange (i.e. of cations within the channels and cages which constitute the microporous part of the crystalline zeolite). I emphasise that this type of exchange may or may not be isomorphous in the case of zeolites. Usually, when one cation is exchanged for another within the zeolite no change in the overall structure of the zeolite occurs; however there are well documented cases where a phase transformation can occur, especially when monovalent cations are involved (refs. 4,5). In this respect zeolites have a property which is in common with some clay minerals (ref. 6), but not with ion-exchange resins.

When zeolites manifest ion-exchange properties that are distinct from clay minerals and resins, these are normally in situations where the microporous crystalline nature of the typical zeolite is manifest. This is particularly true with respect to exclusion effects: because the microporous channels in the zeolite are of comparable size to the typical cation size, cations may be excluded from all or part of the internal surface of the zeolite on the basis of their size. Alternatively, while being accepted, the size of the cation may be such that complete exchange is not possible without filling all the available space. These matters are discussed further below.

In addition to considering applications, any discussion of ion-exchange in the zeolites must consider formulations, be they thermodynamic or kinetic in nature. Formulations for ion exchange which have been developed are not unique to zeolites and are usually applicable also to other classes of ion-exchangers, such as clay minerals or resins. Because different workers who are familiar with particular types of materials have separately developed these formulations, the subject as a whole is a complicated one (refs. 7,8,9) and will only be discussed in broad outline within this chapter, with a view to emphasising the important principles. In addition, no attempt will be made to review systematically all the literature which has been published on ion exchange within zeolites. Rather, the aim will be to give the reader a broad up-to-date overview of the subject which will allow him/her to access the appropriate literature.

2. A PICTURE OF THE ION-EXCHANGE SYSTEM

Before considering zeolites in particular, it is important to understand some basic principles. One can regard ion exchange as a chemical reaction (a double decomposition) between two phases. In reality there may be any number of phases involved and any number of exchanging species, and there may be net movement of solvent(s) from one phase to another as the reaction proceeds. However, for the sake of simplicity we will restrict our considerations in this section to just two reacting cations and two phases.

There are various things which need emphasis at this point. First of all, and especially in the case of zeolites, the ion-exchange reaction is one which is best regarded as involving bulk phases. It is not essentially a surface reaction. Secondly, the reaction is stoichiometric, with electrical neutrality of each phase being maintained throughout the exchange process. For these reasons it is not helpful (as is often done) to refer to the process as adsorption, not only because adsorption is frequently non-stoichiometric, but because the term itself implies that the reaction is a surface one. In terms of the bulk phases, one usually refers to the exchanger (zeolite) and the external solution, but actually it is important to recognise that the external solution is also an exchanger phase. For the very simplest case one has therefore two exchanger phases, whether those two exchanger phases be a zeolite plus the external electrolyte solution, or a molten salt in equilibrium with a zeolite, or two zeolites together (ref. 10). In addition to the two exchanger phases, a third phase also has to be taken into account, even for the simplest system, which is the vapour, as the solvent vapour

362

(normally water) is used as the thermodynamic reference point for all three phases. The solvent can have a profound effect on various properties of the exchange reaction including the rate of attainment and the actual position of the exchange equilibrium. Thus in solid state exchange reactions, (involving two zeolites), a small amount of water can greatly speed the reaction (ref. 11). By contrast, if very dry ethanol is used in place of water as solvent the rate of reaction for (say) calcium in zeolite A is very greatly slowed (ref. 12).

Let us now "home-in" on the two exchanger phases which are present in the simplest system and compare their properties. Consider first the zeolite. At the start, and before the reaction occurs, one may regard the zeolite as (say) a salt of (e.g. $NaAlSiO_4$) in which some water is dissolved. The other phase comprises mainly water in which is dissolved a salt (e.g. KCl). After reaction has taken place we have a solution phase comprising water in which are now dissolved two salts (NaCl and KCl), while the zeolite is now a mixture of two salts $[K_xNa_{1-x}(AlSiO_4).wH_2O]$. This salt mixture may be (and usually is) a solution of one of the salts in the other, i.e. $K_x(AlSiO_4)_x.yH_2O$ dissolved in $Na_{(1-x)}(AlSiO_4)_{(1-x)}.zH_2O$, or the converse. However, as emphasised above, it is important to note that phase separation may occur, with a new phase crystallising out within the old (e.g. the Sr/Na-X system - ref. 13). (Of course, phase separation on reaction may also occur in the aqueous solution if the salt corresponding to the cation which is leaving the zeolite is insoluble in water). Thus the reaction involves cations which are initially present in each separate phase moving between the two phases until an equilibrium composition within each phase is attained.

What about the anions? Here, the more marked <u>difference</u> between the two phases may be observed. While most of the cations within the zeolite channels are normally free to move, and of course are mobile in the external solution phase, the "anions" within the zeolite are not: the "anions" in this case are the anionic framework which constitutes the zeolitic phase! However, of course, just as with the cations, the anions in solution are free to move. This has two consequences:
(a) In the absence of salt imbibition (ref. 14), the "normality" (that is the number of moles of unit charge per given volume) of a given zeolite is virtually fixed. This is <u>not</u> true for the solution phase, where (within the limits of solubility of the salts in the solvent) the normality can be varied at will. In fact, the only simple way that one can alter the "salt" concentration in the zeolite phase is to change the silicon to

aluminium ratio within the framework; in essence, this may produce a different exchange system (e.g. one involving zeolite X instead of Y).

(b) While the anions in aqueous solution can in principle move freely out and into the zeolite, the anionic framework of the zeolite cannot do the converse with respect to the solution phase. Thus the anions in solution (in order that electroneutrality within each phase be preserved) <u>cannot</u> move into the zeolite without their taking stoichiometric quantities of cations with them. If this should occur, there is then a <u>net-shift</u> in salt concentration from the solution phase to the zeolite. This phenomenon is sometimes called salt imbibition (ref. 14); in effect the ion-exchange capacity of the two phases has changed (refs. 15,16). That of the solution has declined while that of the zeolite has increased concomitantly. Donnan exclusion (ref. 17) stops the transfer of salt from one phase to another at low salt concentrations in solution, but the phenomenon can occur either at high solution concentrations or when the second exchanger phase is a molten salt.

Finally, in drawing up this picture of the ion-exchange system as a whole, let us "home-in" on the properties of the zeolite phase itself. In particular, let us concentrate on those characteristics of zeolites which are significant when it comes to applications. Zeolites are normally highly crystalline materials. This means that the anionic framework comprises an extended regular array of silicon, aluminium and oxygen ions which enclose microporous channels of specific sizes, these sizes being of molecular dimensions. This makes zeolites significantly different from either a clay mineral (where most of the exchange occurs in inter-lamellar space) or a typical (amorphous) ion exchange resin. Since the micropores within the zeolite are of molecular dimensions, various consequences may follow:

a) <u>Ion sieving</u> : because of the size of the pores, ions may be too large to enter some of the cages and/or interconnecting channels within the zeolite structure. In effect, they are therefore "sieved-out" of part of the space which is available within the crystal. This phenomenon may lead to the complete exclusion of one sort of cation from a particular zeolite (ref. 18), or partial exchange only occurring (ref. 19), with a clear maximum level of exchange for the entering cation which is less than that one would expect on the basis of the framework Si/Al ratio.

b) <u>Volume exclusion</u> : channels within the zeolite structure may be large enough for the ions to diffuse through without severe restriction, but the <u>size</u> of the cations may be such that before that particular cation can

364

fully neutralise all the negative framework charge on the zeolite, no room
is left within the channels for further cations (refs. 18,20). In effect,
the sum of the volumes of all the cations of a particular type required to
neutralise the anionic framework is greater than the available space
within the zeolite. When this occurs, partial exchange with respect to
the entering cation is again observed, not because of a sieving effect,
but because of a volume exclusion effect.

c) <u>Different exchange sites</u> : the three dimensional extended array which
constitutes the zeolite framework may have associated with it quite
clearly defined exchange sites. Thus, per unit cell, one can often speak
of particular sites within the zeolite which differ one from another in
terms of the energies of interaction associated with them. A very good
example of this is the faujasitic group of zeolites. In these materials a
large number of different sites are identified, and associated with each
type of site is a particular ion population and site energy (e.g. site I
through to site V (ref. 21)). As a consequence, it may be very difficult
to remove cations from certain of these sites even though both the leaving
and entering cations are able to move through the channels and cages in
which these sites are found. This has particular importance with respect
to the preparation of sodium-free Y catalysts for cracking (ref. 22).

d) <u>Phase changes</u> : after exchange, cations may nucleate the formation of a
new phase within the old phase (refs. 2,4,5,13). Therefore as a
consequence of carrying out the ion-exchange, a re-crystallisation may
occur resulting in a different zeolite. This has been alluded to above
and is commonly seen by the presence of a hysteresis loop within the
ion-exchange isotherm (ref. 5).

Phenomena (a), (b) and (c), listed above (all of which are characteristic of
zeolites as ion-exchangers), have a common implication, <u>viz</u> that the total
exchange capacity with respect to a particular cation may be very much less
than that which one would be led to expect on the basis of the Si/Al ratio.
This difference between the theoretical ion-exchange capacity and that which
is observed in practice is significant in many applications involving
zeolites, including catalytic cracking and detergency building, and it is
important for the synthetic chemist to be aware of these "pit-falls".

Finally, one should note the limits of stability of zeolites when ion-exchange
processes are undertaken. Many commonly used zeolites, such as zeolite A, are
hydrolytically unstable in even mildly acid pH, and removal of aluminium from
the framework will readily occur. In high aluminium zeolites this can lead

ultimately to break down of the crystalline structure. This is a phenomenon which has often been neglected in the past, has led to irreproducibility of data (ref. 22) and is of particular significance in detergency (refs. 23,24). In addition it is worth noting the limits outside which precipitation of basic salts may occur. Zeolites exhibit an alkaline reaction to their environment and in the presence of many transition metal cations basic salts may be precipitated if adequate care is not taken (refs. 25,26). Also, one should be aware that many metal ions are speciated in solution (ref. 27). If speciation occurs to a significant extent, then the exchange capacity observed can be very different to that predicted (refs. 28), and this behaviour is reported to occur in clay minerals not only with copper (ref. 28), but also with calcium and magnesium (refs. 29,30). These latter observations are however of peripheral significance for detergency since similar behaviour with zeolites appears to occur only to a very limited extent (ref. 16).

3. EQUILIBRIUM ASPECTS OF ION EXCHANGE IN ZEOLITES
Over 20 years ago, in his comprehensive book on ion exchange processes, Helfferich compared the current theoretical understanding of both the thermodynamic and kinetic aspects of ion exchange, noting how much better understood were the equilibrium aspects of ion exchange processes (ref. 31). While Helfferich considered primarily ion exchange resins, for zeolites it is also true that equilibrium processes are more comprehensively studied and better understood than are kinetics. The equilibrium properties of many experimental systems have been studied, with the usual aim of obtaining various thermodynamic parameters from the experimental data. Such studies have often been an end in themselves, and part of the purpose of this section will be to emphasise what can and cannot be done through such measurements.

3.1 BASIC CONCEPTS
In order to appreciate how the ion-exchange properties of zeolites may be important in areas such as detergency or in the preparation of catalysts, it is not necessary to discuss in depth the details of the thermodynamics. However, it is necessary to clear some common misconceptions out of the way.

3.1.1 THE ION-EXCHANGE ISOTHERM
The usual aim of those who study the equilibrium aspects of ion exchange in zeolites is to obtain a measure of the selectivity of the zeolite for one ion over either another, or a group of other, ions. For this purpose, it is normal to construct isotherms. The experimental methods by which an isotherm may be constructed have been described elsewhere (ref. 32). Essentially, one

is plotting equilibrium compositions attained between the two phases with respect to the exchanging cations. While the term "isotherm" indicates that the temperature must be kept constant, in addition both the zeolite and the solution phase are kept isonormal (see section 2 above). The isotherm is then constructed by plotting the equivalent fraction E of the incoming cation present at equilibrium within the solution phase against the equivalent fraction $\bar{E}$ of that same cation in the zeolite. The equivalent fraction may be defined as that proportion of the exchange capacity of the phase which is neutralised by the given cation A, and may be defined for the two phases as follows:

$$
\left.
\begin{aligned}
E_A &= z_A n_A / (z_A n_A + z_B n_B) \\
\bar{E}_A &= z_A \bar{n}_A / (z_A \bar{n}_A + z_B \bar{n}_B)
\end{aligned}
\right\} \tag{1}
$$

where z_A, z_B are the valencies of exchanging cations A and B, and n_A, n_B the moles per unit volume within each phase. The superscript bar indicates "zeolite phase". For reasons to do with the evaluation of the thermodynamic parameters, it is normal to plot the isotherm with the equivalent fraction of the metal ion in the solution phase E_A as the dependent variable; this is not required however, and is often not followed by workers involved with resins.

One may broadly classify isotherms into four kinds. Isotherms of the first kind are shown in Figure 1 and represent relatively simple systems in which the exchanger is either unselective for the incoming ion (case 1), non-selective (case 2), or selective (case 3). In isotherms of the second kind (Figure 2) the plot is sigmoid, indicating a change in selectivity as a function of $\bar{E}_A$. Isotherms of the third kind (Figure 3) are characterised by a plateau, and within this plateau region non-reversibility of the isotherm is seen as a "hysteresis loop" (i.e. a miscibility gap between two phases is observed). These isotherms are typical of systems where a new phase has crystallised within the old phase as a consequence of the ion-exchange reaction. Finally, isotherms of the fourth kind are often seen with zeolites (Figure 4), in which a clear limit to exchange is observed which is lower than the theoretical exchange capacity of the zeolite (see comments above in section 2). This limit may be due either to ion sieving or volume exclusion.

3.1.2 SELECTIVITY AND AFFINITY

To avoid potential confusion, it is most important to distinguish between the concepts of selectivity and affinity. When one talks of selectivity in ion

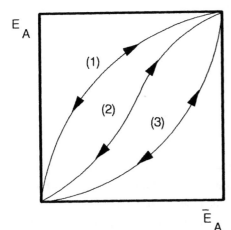

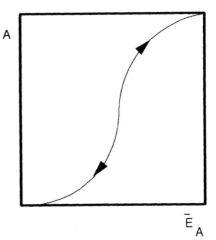

Figure 1. Ion exchange isotherms of the first kind, exhibiting unselective (case 1), non-selective (case 2) or selective (case 3) behaviour for the incoming ion. The arrows indicate reversible behaviour, i.e. that exchange of ion B for A and the converse follow the same path.

Figure 2. Ion exchange isotherms of the second kind, which are reversible but sigmoid, exhibiting a change from selective to unselective behaviour as a function of exchange level.

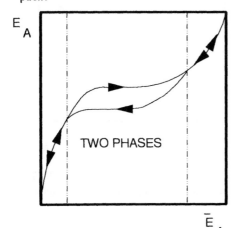

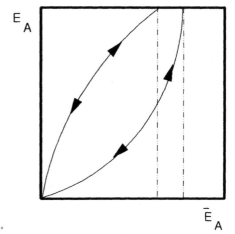

Figure 3. An ion exchange isotherm of the third kind, showing non-reversibility of exchange within the plateau region, characteristic of phase separation and the coexistence of two separate phases over the composition range corresponding to the hysteresis loop.

Figure 4. Isotherms of the fourth kind, exhibiting both unselective and selective behaviour towards the incoming ion, where clear limits to exchange are observed which are lower than that expected on the basis of the theoretical exchange capacity of the material.

368

exchange, one is referring to the preference which the exchanger shows for onecation compared with another at a given equilibrium composition. This may be expressed mathematically in terms of a quotient of two ratios, with the two ratios expressing the relative quantities of the given cations found within the two phases at equilibrium:

$$\alpha = (\bar{E}_A/E_A)/(\bar{E}_B/E_B) = (\bar{E}_A E_B/\bar{E}_B E_A) \tag{2}$$

It is evident from equation 2 that α (which is most properly called the separation factor) may be evaluated from the relative values of two areas (ref. 23) as shown on Figure 5. A few moment's inspection of Figures 1-4 will then show that α will normally vary as a function of exchanger phase composition. This point is most significant; selectivities may vary markedly and even reverse as E_A is increased (see Figure 3). Therefore, selectivity is a directly measurable composition-dependent function.

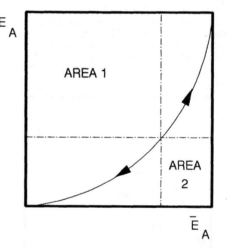

Figure 5. Graphical evaluation of the separation factor, a function which is normally highly composition dependent (see text after equation 2).

α = (Area 1)/(Area 2)

In contrast, affinity is a term which has a particular thermodynamic meaning, in terms of the chemical potentials μ for each component. The ion-exchange reaction may be written as

$$z_B A^{z_A^+} + z_A BL_{z_B} \rightleftharpoons z_A B^{z_B^+} + z_B AL_{z_A} \tag{3}$$

where the subscript 's' refers to the solution phase, and L is that portion of zeolitic framework holding unit negative charge. More concisely, eqn. (3) can be written as

$$0 = z_A z_B \sum_{\substack{B \\ (all\ phases)}} (B/z_B) \tag{3a}$$

From eqn. (3a) the definition of affinity A follows as (ref. 34)

$$A = - \sum_{\substack{B \\ (all\ phases)}} (1/z_B)(\mu_B - \mu_B^{eq}) \tag{4}$$

where μ^{eq} refers to the chemical potential of a particular component at equilibrium. The affinity defines the <u>direction</u> of ion exchange; for the reaction to proceed from left to right A must be positive, and the converse. At equilibrium, A is zero (ref. 34). It is evident that A is related to the free energy of exchange (<u>not</u> the standard function!).

3.1.3 <u>THE MASS ACTION QUOTIENT AND THE THERMODYNAMIC EQUILIBRIUM CONSTANT</u>

From eqn. 3 a mass action quotient K_m may be defined as

$$K_m = (m_B^{z_A} \bar{E}_A^{z_B})/(m_A^{z_B} \bar{E}_B^{z_A}) \tag{5}$$

where m_A, m_B are the molal concentrations (mol kg^{-1}) of ions A and B in solution.

Barrer and Klinowski have shown (ref. 33) that the mass action quotient as defined above may easily be derived from a as follows:

$$K_m = (m_A/\bar{E}_A)^{(z_A - z_B)} \cdot (z_A a/z_B)^{z_A} \tag{6}$$

Therefore, the mass action quotient is also readily obtainable from experimental data as shown on the isotherm. In contrast, the thermodynamic equilibrium constant is not readily accessible from experimental data. It is defined also using the reaction equation given above, as

$$K_a = (a_B^{z_A} \bar{a}_A^{z_B})/(a_A^{z_B} \bar{a}_B^{z_A}) \tag{7}$$

where a is the activity of the cation in either the solution or the crystal phase. One can see that the essential difference between the mass action quotient and the thermodynamic equilibrium constant is that while the former is defined in terms of concentrations the latter is defined in terms of activities; it is precisely because of this that the former is not constant with zeolite phase composition while the latter is invariant.

Why is this? Imagine the zeolite initially to be wholly in the B cation
form. Then the B cations are interacting with the framework, water in the
channels, and with each other. One can visualise the system as having a given
amount of energy and that energy will vary directly with the amount of B
zeolite which is present. One can therefore factorise the energy of the
system into a set of "packets", each packet comprising that amount of
framework (with associated water) that is required to neutralise one B
cation. If one therefore halved the amount of B zeolite then the energy of
the system would also halve. Now imagine A cations entering the zeolite
framework and replacing B cations, until the quantity of B zeolite is half
what it was originally. One cannot assume that the energy of the system
arising from B zeolite is now half what it was before, because new
interactions are present which were not present when all the zeolite was in
the B form. These new factors include interactions between A and B cations
and the effect overall on the framework of A cations interacting with it, and
the effect that A cations have on the total water content in the zeolite. The
entire converse argument can be forwarded in terms of A zeolite. Thus in
effect, because of these new interactions, the behaviour of the system will
not be predictable simply on the basis of the concentrations of A, B and water
in the zeolite. Each of these concentrations has to be multiplied by a
correction factor, which for each of these components is separately changing
with the overall composition of the zeolite, as the relative strengths of all
the different interactions with respect to the particular component change.
When we multiply each varying concentration by its appropriate correction
factor, we then obtain activities which allow for these new interactions.
Thus the thermodynamic equilibrium constant, defined in terms of activities,
is independent of exchanger composition, in contrast to the mass action
quotient, which applies to the real system and is related closely to the
composition dependent selectivity. Since (see below) the standard free
energy $\Delta G^{\ominus}$ is directly related to the thermodynamic equilibrium constant K_a,
it follows that it is misleading to attempt to use standard free energy values
to draw mechanistic conclusions about composition dependent selectivity trends
in the real system (ref. 36).

3.1.4 THE STANDARD FREE ENERGY OF EXCHANGE

The standard free energy of exchange $\Delta G^{\ominus}$ is related to the thermodynamic
equilibrium constant as follows:

$$\Delta G^{\ominus} = -(RT/z_A z_B).\ln K_a \tag{8}$$

It is evident that since the thermodynamic equilibrium constant is defined in terms of activities, then before the standard free energy of exchange can be determined it is necessary to apply the appropriate correction factors to the concentration terms. These correction factors are called <u>activity coefficients</u>. Activity coefficient values have to be determined for <u>each</u> component in each exchanger phase in order that the thermodynamic equilibrium constant can be found. The means by which this is done is complicated and has been described in detail elsewhere (refs. 8,9,32,35). What is important to recognise is that the magnitudes of the activity coefficients are determined with reference to appropriate <u>standard states</u> for components within the system. For the salts in solution, these reference states can be the so-called hypothetical ideal molal solutions (i.e. a solution containing 1 mol kg^{-1} of the salt in water with the cations and anions from which the salt is comprised still behaving as though they were in an infinitely dilute solution). For each form of the zeolite (i.e. the A form or the B form) the standard states are respectively these homoionic forms of the zeolite each respectively immersed in infinitely dilute solutions of the same cation (ref. 15). This ensures that the activity of the water within the zeolite is equal to the value found in the external solution (i.e. the activity of the water in both phases is equal to that in the vapour phase and is unity in the standard state).

Following determination of activities, the standard free energy is then determined. In terms of reaction equation (3) above, the standard free energy, is then found to be defined in terms of <u>standard</u> chemical potentials $\mu^{\ominus}$ as:

$$\Delta G^{\ominus} = \sum_{\substack{B \\ (both\ phases)}} (1/z_B)\mu_B^{\ominus} \tag{9}$$

Thus the standard free energy change for the ion-exchange reaction is that free energy change which occurs when $1/z_B$ moles of homoionic B zeolite in its standard state react with $1/z_A$ moles of A ion in its standard state solution to give entirely $1/z_B$ moles of B in its standard state solution plus $1/z_A$ moles of A zeolite also in its standard state, all at the prescribed temperature and pressure.

3.2 PREDICTION OF ION EXCHANGE EQUILIBRIA

From the discussion above, it is evident that $\Delta G^{\ominus}$ refers to the free energy change when moving <u>wholly</u> from one set of idealised standard states to another. One may ask, therefore, what use is this function? Indeed, what use is thermodynamics? It is evident that the standard free energy, which is composition independent, cannot tell us about the causes of selectivity as a function of composition, or the reasons why the selectivity may change with composition. So what use are thermodynamic functions and the equations in the context of zeolite ion exchange? McGlashan elegantly and succinctly answers this question (ref. 37): "What can...... an equation 'tell one' about one's system or process? Or, in other words, what can we learn from such an equation about the microscopic explanation of a macroscopic change? Nothing whatever. What then is the use of thermodynamic equations? They are useful because some quantities are easier to measure than others. If every equilibrium property of every kind of system had been measured with high accuracy then thermodynamics would be useless (though it would still be beautiful)."

This quotation encompasses what the purpose of ion exchange thermodynamics is for zeolite systems, or indeed for any system. Essentially its purpose is not to enable us to speculate about mechanism, but to enable us to predict the behaviour of the system over ranges of conditions (temperature, pressure, concentration) which may be difficult to measure directly because of experimental limitations. Having determined the thermodynamic equilibrium constant for one set of conditions by carrying out the appropriate activity corrections to the experimental isotherm, one can then in principle use these data to predict how the zeolite will behave under any other set of conditions. The development and testing of such predictive procedures has been a major aim of experimentalists in the last few years. Essentially, three things are required to enable one to predict ion exchange equilibria accurately. These are:

a) an adequate thermodynamic formulation;
b) very accurate experimental data for one set of conditions;
c) accurate activity coefficient data for the dissolved salts in the external solutions for the range of conditions over which one wishes to predict.

The ability to be able to predict exchange equilibria in this manner becomes increasingly important when one is considering (for example) the direct application of zeolites in ion exchange. In detergency one is frequently dealing with a range of temperatures, a range of concentrations, and a multicomponent system involving sodium, calcium, magnesium, and oxonium cations exchanging into the zeolite (ref. 22). It is clearly impossible to measure the behaviour of a zeolite under every conceivable set of conditions that one can envisage. It is in such situations that the prediction of equilibrium exchange behaviour becomes important. This is obviously a subject of great complexity; below just the basic principles are outlined below.

3.2.1 PREDICTION PROCEDURES

Numerous approaches have been developed by workers on such diverse materials as resins, clay minerals and zeolites in order to predict exchange behaviour for multicomponent systems from binary exchange data only. Many of these methods have been reviewed fairly recently (refs. 7,38). Approaches which use binary data in order to predict multicomponent equilibria can be broadly classified into two groups (ref. 38). In the first group are methods where ternary activity coefficients for the exchanger phase are derived by means of a semi-empirical mathematical procedure from experimentally determined activity coefficients for the conjugate binary systems; these derived data are then used to predict the ternary equilibrium compositions. The second type of approach is to predict ternary equilibrium compositions by graphical manipulation of the conjugate binary exchange data. Both predictive methods have the merit that they require less experimental work than does the rigorous thermodynamic approach (i.e. only the acquisition of accurate binary exchange data); however a recent detailed study on various zeolitic systems has shown that both these types of approaches are deficient compared to a rigorous thermodynamic one (ref. 38).

How then can thermodynamic procedures be used to predict exchange selectivities? To answer this question we need just to consider a little more carefully the relationship between the thermodynamic equilibrium constant and the mass action quotient (eqns. 6,7). The relationship between them is:

$$K_a = K_m \Gamma \Phi \tag{10}$$

where Γ and Φ are respectively activity corrections for the solution and crystal exchanger phases. It is common practice, and indeed convenient, to define another function K_G (commonly called the corrected selectivity quotient), which is related to K_m by

$$K_G = K_m \Gamma \tag{11}$$

Gaines and Thomas (ref. 15) demonstrated long ago that provided the variation of K_G with the crystal phase composition is known, it is a relatively easy matter to thence obtain the thermodynamic equilibrium constant:

$$ln\ K_a = (z_B - z_A) + \int_0^1 ln\ K_G d\bar{E}_A + \nabla + \Psi \tag{12}$$

Eqn. (12) is the form which one obtains when all factors are taken into account, including salt imbibition and possible changes in water content and water activity within the zeolite framework. The two functions ∇ and Ψ respectively take these two phenomena into account (for further details see eqn. 24 in ref. 16). One can readily rearrange the above equation to express it in terms of the corrected selectivity quotient at a particular composition P, to give

$$ln\ K_{G(P)} = \int_0^1 ln K_G d\bar{E}_A - ln\Phi + (z_B - z_A) + \nabla + \Psi \tag{13}$$

Salt imbibition has been shown (ref. 14) to occur to a negligible extent when the external electrolyte concentration is low. Furthermore Barrer and Klinowski (ref. 33) have also shown that the water term ∇ is usually insignificant in zeolitic ion exchange processes. As a consequence the activity correction for the crystal phase Φ should normally be virtually independent of external electrolyte concentration used (ref. 33).

It is this latter fact which is the key to enabling us to predict exchange equilibria in zeolites. If the condition holds, then it is evident from eqn. 13 above that under these circumstances, and in the absence of salt imbibition, the whole of the RHS is constant and therefore of course so is $K_{G(P)}$. It is then evident from eqn. 11 that if $K_{G(P)}$ does not vary as the total external electrolyte concentration is changed, then the only way by which K_m can change is if Γ changes (i.e. the activity coefficients for the salts in solution change). Thus, paradoxically, exchanges in exchange equilibrium compositions involving zeolites do not depend primarily on the zeolite phase at all, but rather on the way in which non-ideality in the solution phase changes as the total electrolyte concentration is altered.

I emphasised above how important it was to have accurate experimental data for one set of conditions (i.e. solution concentration, temperature). It should also now be clear how important it is to be able to accurately evaluate Γ as we change the overall external electrolyte concentration in solution at a given temperature. Procedures to do this have been worked out, and involve evaluating the ratios of activity coefficients of individual salts in the

mixed electrolyte solutions. The methods for doing this are complicated and
are described in detail elsewhere (ref. 35).

3.2.2 RESULTS AND LIMITATIONS

That the prediction procedure outlined above works has been demonstrated not
only for binary systems (ref. 36) but also more recently for some quite
complicated ternary systems (refs. 39,40). The first of these systems (ref.
39) is the Na/Ca/Mg/-A system, and because of its relevance to detergency, the
results are discussed in more detail below. Suffice to say at this point that
the procedures succeeded in predicting accurately exchange compositions at
different normalities. Success was also generally achieved for the Na/K/Cd-X
exchange equilibrium (Figure 6), although systematic errors in prediction were
detected at higher external solution concentrations (ref. 40). The fact that
these errors were systematic suggested that the cause lay not so much in the
accuracy of the original data but rather in the failure of one of the
assumptions alluded to above (i.e. that salt imbibition occurred to a
negligible extent, or that any changes in either water content or activity
were insignificant). This view was further strengthened when the exchange of
magnesium into zeolites X and Y was examined in the presence simultaneously of
both ammonium and sodium. Here, at higher external electrolyte
concentrations, serious deviations between prediction and observation were
seen for high silica Y (Figure 7) and the authors (ref. 16) suggested that the
cause of this was imbibition of ion pairs. They predicted therefore that
similar behaviour would be seen with calcium in zeolite Y at similar
electrolyte concentrations; unfortunately recent investigations seem not to
have borne out this prediction (ref. 41).

4. THE KINETICS OF ION EXCHANGE

If the formulation and consequent prediction of ion exchange equilibria have
proved to be complicated problems, understanding the dynamics of these systems
has been found to be even less tractable. As a consequence, there have been
far fewer systematic studies on exchange kinetics, despite the obvious
importance of this subject in areas such as catalyst preparation or
detergency. However, a series of such studies on mainly natural zeolites by
Barrer, Rees and co-workers (refs. 42-49) has thrown considerable light on
this subject. These studies are reviewed briefly here, but first it is
necessary to understand some of the basic principles involved in ion exchange
kinetics.

Figure 6. The prediction of exchange compositions for the Na/K/Cd-zeolite X ternary exchange system at 298 K, 1 atm pressure. A set of six equilibrium solution ($\triangle$) and zeolite ($\blacksquare$) compositions at a total solution concentration of 0.1 equiv dm^{-3} were experimentally measured. Next, and by means of activity data and an appropriate thermodynamic formulation, each corresponding pair ($\triangle$, $\blacksquare$) of datum points was used in turn to predict the equilibrium zeolite composition

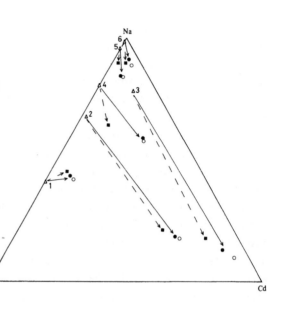

($\circ$) one would expect to correspond to the same solution composition ($\triangle$) but at a total concentration of 0.025 equiv dm^{-3}. Each predicted result ($\circ$) was checked by experiment ($\bullet$) in order to check the validity of the thermodynamic model employed. Dashed tie-lines are used to link pairs of experimental datum points at a solution normality of 0.1 equiv dm^{-3}. Solid tie-lines link the same experimental solution composition at a normality of 0.025 equiv dm^{-3} to both the predicted ($\circ$) and ($\bullet$) measured zeolite compositions. Agreement between prediction and experiment is usually good. (Taken from reference 40).

Figure 7. The prediction of exchange compositions for the Mg/Na-zeolite Y binary exchange system at 298 K, 1 atm pressure. Experimental datum points for the forward ($\bullet$) and reverse ($\circ$) exchange equilibrium at a total solution concentration of 0.1 equiv dm^{-3} are shown, together with a best-fit curve through the data. Partial exchange only was observed, to a limit just above 70% (cf. figure 4). Predicted isotherms (solid curves) at total solution concentrations of 0.025 and 0.4 equiv em^{-3} are shown also, together with experimental tests of these predictions ($\blacktriangle$ and $\blacksquare$ respectively for the two total solution concentrations). The predictions fail, especially at 0.4 equiv dm^{-3} either because of an irreversibility associated with high levels of exchange, or ion speciation, or sal timbibition (taken from reference 16).

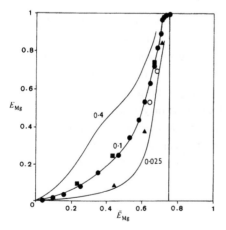

4.1 SOME BASIC CONCEPTS

It is important to recognise at the beginning that there can be more than one rate-controlling process involved in ion exchange. Furthermore, it is not necessarily true that the rate-controlling step is one mediated by diffusion processes. One can visualise a kinetic exchange process in which the rate of exchange of ions at sites within the zeolite is considerably slower than their subsequent motion through the channels into the external solution. Indeed, a formulation which takes this possibility into account has been drawn up (ref. 50).

When one is sure that the rate controlling step is diffusion mediated, there is still the possibility that the slowest diffusion process may either take place within the crystal itself (i.e. particle controlled diffusion) or through a near-static boundary layer between the external solution and the crystal surface (i.e. film diffusion). Even if the crystals are agitated very strongly a quiescent boundary layer will still remain close to the crystal surface, and therefore the possibility of film diffusion must always be considered. Indeed, as crystallite size is reduced (in order to speed up kinetic processes - and this of particular significance in detergency) so one will inevitably eventually move to a film diffusion mediated process. Film diffusion is however not determined by the properties of the zeolite itself; therefore subsequent discussion in this section will be restricted to particle controlled diffusion.

In order to appreciate the difficulties which one encounters in trying to describe the dynamics of ion exchange processes within a zeolite, it is perhaps helpful to try to picture the process which is going on within the crystal. Consider the simplest case of two types of exchanging cations within the channels and cages of the zeolite. Clearly, as a consequence of different ion sizes (and possibly also charges) the two types of cations will have different mobilities; in addition water may be present. If concentration gradients are present within the crystal, then the different types of cations will move in the directions of negative concentration gradients in order to equalise their concentrations throughout the system (this process is, of course, described by Fick's first law). However, the mobilities of the two types of cations are, as we have already noted, different. We are visualising a counter diffusion involving the two types of cations: since the mobilities are different, the faster moving cation will tend to build its concentration faster than the slower. If this process were to continue, then it is evident that charge separation will occur, and an electrical potential gradient will build as the two types of ions move towards each other within the crystal. In reality charge separation does not happen to any significant degree at all, as the electrical potential gradient which forms as charge separation takes place will operate in such a way as to slow the faster moving ion and, conversely, accelerate the slower moving ion. Thus ion exchange kinetics requires us to consider not only the elimination of the concentration gradients by diffusion (as described by Fick's first law) but also the maintenance of an absence of an electrical potential gradient throughout the crystal. Fick's first law may be written as

$$J_A = -D_A \; grad \; c_A \tag{14}$$

where J is a flux, D a diffusivity, and c a concentration. Because of the necessity to consider also the electrical potential gradient, the resulting net flux of ion exchanging species is actually described by (ref. 51)

$$J_A = -D_{AB} \; grad \; c_A - (z_A c_A F/RT) \; grad \; V \tag{15}$$

where F is the Faraday and V the electrical potential. This expression is called the Nernst-Planck equation (ref. 51). It is one which holds under ideal conditions, and a set of Nernst-Planck expressions, one for each type of exchanging ion, is required to be solved simultaneously. Helfferich (ref. 51) solved this equation for some resin systems; by combining this expression with corresponding irreversible thermodynamic formulations, Barrer and Rees derived an appropriate general expression for the inter-diffusion coefficient D_{AB}, and then applied it to ion-exchange kinetics in zeolites. Their expression is extremely complicated:

$$D_{AB} = 1/(z_1(1 + r)) \left\{ (z_1 c_1 L_{11} r - z_2 c_2 L_{12})(d \; ln \; a_1 \; / \; dc_1) + \right. \tag{16}$$
$$\left. (z_1 c_1 L_{12} r - z_2 c_2 L_{22})(d \; ln \; a_2 \; / \; dc_1) + (z_1 c_1 L_{13} r - z_2 c_2 L_{23})(d \; ln \; a_3 \; / \; dc_1) \right\}$$

where
$$r = z_2 c_2 (z_2 L_{22} + z_1 L_{12})/z_1 c_1 (z_1 L_{11} + z_2 L_{12}) \tag{17}$$

because it takes into account not only the factors which affect the mobility of each exchanging cation with respect to the other, but also the effect that changes in content of each exchanging ion have on the water present (i.e. the water flux). In eqn. (16) the L_{ii} and L_{ij} terms are phenomenological coefficients. Thus for example, L_{11} is the   coefficient for ion 1 exchanging with itself, L_{12} is a cross coefficient for ion 1 exchanging with ion 2 and L_{13} is the cross coefficient for ion 1 exchanging with water.

In their studies on natural chabazite, Barrer, Rees and co-workers (refs. 42-49) used a simplified version of the above equation. A major simplification occurs if one assumes that the cross coefficients L_{ij} are negligible. Eqn. (16) then reduces to

$$D_{AB} = \frac{D_A^* D_B^* \{c_2 z_2^{\,2}(\partial ln\, a_1/\partial ln\, c_1) + c_1 z_1^2 (\partial ln\, a_2/\partial ln\, c_2)\}}{c_1 z_1^2 D_A^* + c_2 z_2^{\,2} D_B^*} \tag{18}$$

where D^*_A, D^*_B are the so-called self-diffusion coefficients for ions A and B respectively (that is the diffusion coefficients which describe quantitatively the rate at which ions exchange with one of their own kind). Eqn. (18) follows from eqn. (16) since D^*_i equals RTL_{ii} (ref. 42).

A further simplification can be made if the exchanger behaves ideally for all exchanger compositions. In this case the terms $d\, ln\, a_i\, /\, d\, ln\, c_i = 1$, so that

$$D_{AB} = \frac{D^*_A D^*_B (c_2 z_2^{\,2} + c_1 z_1^2)}{c_1 z_1^2 D_A^* + c_2 z_2^2 D_B^*} \tag{19}$$

4.2 EXPERIMENTAL TESTS OF THEORETICAL MODELS

To test the models which have been developed, such as those above, one needs to generate curves of the fractional attainment of equilibrium versus time. Essentially, since one is dealing with transient diffusion phenomena, this requires one to solve (using the appropriate boundary conditions) Fick's so-called second law:

$$(\partial c_A/\partial t) = div\, D_{AB}\, grad\, c_A \tag{20}$$

Eqn. (20) takes into account the fact that D_{AB} may also change with concentration of diffusing species within the exchanger. Unfortunately, eqn.

(20) is not readily integrable, unless one transforms it into the simpler but more approximate form of

$$(\partial c_A / \partial t) = \tilde{D}_{AB} \; div \; grad \; c_A \tag{21}$$

and it is in this form that the equation is used with $\tilde{D}_{AB}$ being an integral diffusivity. Such calculations were carried out firstly by Barrer, Bartholomew and Rees (refs. 43,44), then by Brooke and Rees (ref. 45,46) and finally by Duffy and Rees (refs. 47-49). Various calculation methods of different sophistication were used, and comparisons were made between theory and experiment assuming either that observed self-diffusivities were not a function of composition or that they were (and this variation was measured (Figure 8) and put in to the adjusted calculations). Essentially, if cross-coefficient terms L_{ij} are not negligible (as is assumed for eqn. (18) above) then observed "self-diffusivities", found by measuring tracer diffusion coefficients for each of the exchanging ions as a function of composition, will change with composition, as observed (figure 8). Allowance was also made for non-ideality in the exchanger phase (Figure 9). Despite all these attempts, agreement between theory and experiment remained only qualitative in many cases (Figure 10).

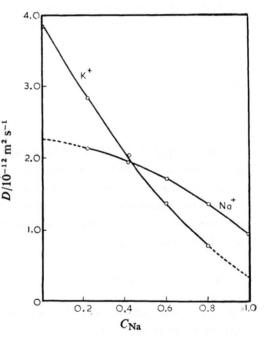

Figure 8. Measured values of the variation in magnitude of tracer diffusion coefficients for sodium and potassium ions in chabazite as a function of zeolite composition (taken from reference 47).

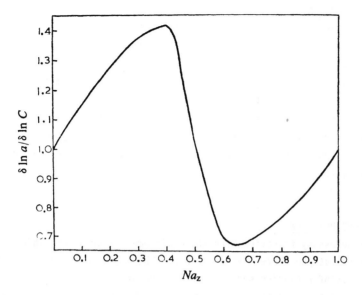

Figure 9. Plot of non-ideality term (d ln a/d ln c) for sodium in chabazite as a function of sodium concentration in chabazite (Na_Z). For ideal behaviour, the function should remain unity for all values of Na_Z. (Taken from reference 48).

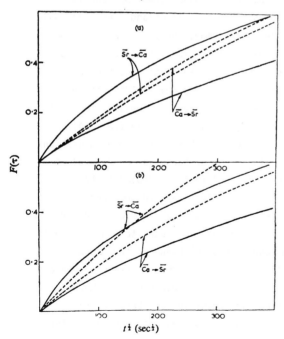

Figure 10. Effect of applying corrections for non-ideality on predicted rates of exchange (strontium for calcium and the converse) within chabazite. Broken curves represent predicted rates and solid curves the experimental data. In (a) no correction for non-ideality has been applied (i.e. d ln a/d ln c is assumed to be unity for all compositions. In (b) corrections for non-ideality have been applied. After correction, the predicted rates are reversed and now agree qualitatively with experiment, although quantitative agreement is still poor. In these studies, no corrections were applied for possible non-constancy of the observed self-diffusivities. (Taken from reference 45).

4.3 CONCLUDING COMMENTS

Bearing in mind that in studies on the kinetics of ion exchange in zeolites one is dealing with non-steady state ion transport phenomena in a concentrated solid electrolyte which contains also water, it is hardly surprising that the relatively simplistic models described above often fail when compared to experiment. It is also obvious that attempting to predict the rates of ion exchange would be rather a forlorn hope at present. With increasing availability of very fast computer systems, better progress may well be made in the future using computer simulation of the molecular and ionic dynamics within these systems.

5. ZEOLITES AND DETERGENCY

It is hardly necessary to review here the environmental pressures which led to a need to phase out phosphates in detergents. These pressures naturally led to the necessity of replacing sodium tripolyphosphate by a suitable substitute detergent "builder". Zeolite A has proved to be that suitable substitute, and indeed its current application now in detergency is the major success story in the area of the use of zeolites directly as ion exchangers. The current world wide market is well in excess of 440,000 metric tonnes (ref. 52) and is increasing all the time. Some countries, such as Switzerland (ref. 53), only allow non-phosphate detergents, and in West Germany where, since 1975, the amount of phosphate in detergents has decreased by 70%, it has recently been reported that the level of phosphate in surface waters attributable to detergents has over that same time-span dropped from 42% to 17% (ref. 53). Thus legislative and environmental pressures in favour of zeolites continue, not only in Europe and the States (ref. 53) but also in Japan (ref. 54). However, ideally a "detergent builder" should be able to remove hardness ions (calcium and magnesium) in two ways. The first of these is the removal of hardness ions from free solution; for this zeolite A has proved to be effective in the case of calcium (ref. 52). With regard to the second aspect, the situation is different. This is where calcium in particular is bound with soil on the fabric itself and is helping the soil to adhere to the surface of the textile (ref. 52). For calcium (and magnesium) bound in this form, zeolites are not particularly effective, and it is common practice to use a so-called "co-builder" in order to boost the zeolite performance. Much research has concentrated on this aspect in recent years. Despite this drawback, it has recently been emphasised that zeolites are actually better than the old phosphate detergents in overall performance because of certain other advantages, such as a decreased overall "wear and tear" of the textiles and of the washing machine itself (ref. 53).

It is logical now to see how some of the concepts and principles discussed and developed above are used to make zeolites effective in detergent building. Factors of importance are discussed below, concurrently with recent publications. In addition, recent areas in which research has concentrated will be covered, particularly where those apply to how zeolites can be used more effectively in a product, and/or in combination with other additives such as polymers or sequestrant builders.

5.1 CRITERIA FOR AN EFFECTIVE ZEOLITE BUILDER

Leaving aside obvious criteria, such as that the zeolite should be environmentally acceptable itself and of a reasonable cost, the obvious requirement for a successful zeolite builder is capacity. If this statement seems over-simplistic, consider further what is required in terms of capacity for a zeolite builder. Obviously the capacity should be the maximum possible. In terms of zeolites, this means that the silicon to aluminium ratio within the framework should be as near as possible equal to unity. However, this in itself is not sufficient, because it is also required that the zeolite be selective for calcium and (if possible) magnesium. This is because, as explained above (sections 2 and 3) the effective capacity is not just simply the overall capacity of the exchanger as determined by its silicon to aluminium ratio, but rather is also a function of the selectivity of the material for calcium and/or magnesium (i.e. the shape of the isotherm, and whether partial exchange is observed). Furthermore, when it is necessary to remove simultaneously both magnesium and calcium, the relative selectivity of the zeolite for these two ions is of importance, (i.e. we have to concern ourselves with at least a ternary exchange equilibrium).

There is another factor which also has to be taken into account. This is the kinetic factor. It is not sufficient just that the zeolite has a high intrinsic exchange capacity and that it be selective for one or both of calcium and magnesium, but also the rate at which these ions are removed from solution must be sufficiently high to ensure their removal early on in the washing cycle. Thus, in conclusion, the "capacity" of a zeolite builder is a complicated function of the framework composition of the zeolite, the selectivity manifested for the hardness ions, the levels to which those ions can exchange into the material and the rate at which they are removed.

It is hardly surprising therefore, that a considerable amount of effort has gone into examining the equilibrium and kinetic aspects of calcium and

384

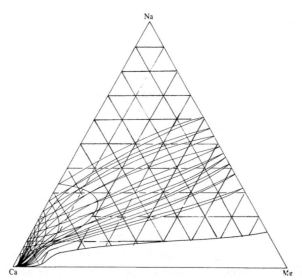

Figure 11. Ternary exchange isotherm for the Na/Ca/Mg-zeolite A system at 298 K with a total solution concentration of 0.1 equiv dm^{-3}. Whereas for a binary isotherm the equilibrium is depicted adequately by a curve (see figures 1 to 4), for a ternary system the equilibrium isotherm comprises a surface. The triangular grid lines for all compositions of the solution phase have been distorted so that all solution compositions fall on top of the corresponding equilibrium zeolite compositions (shown with an undistorted grid). Also seen is a region of zeolite compositions, low in sodium and containing more than 10% magnesium, which is unattainable. (Taken from reference 55).

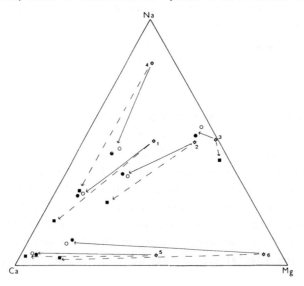

Figure 12. The prediction of exchange compositions for the Na/Ca/Mg-zeolite A ternary exchange system at 298 K, 1 atm pressure. See caption to figure 6 for detailed explanation of data. Six parts of experimental datum points were obtained for both the solution ($\diamond$) and zeolite ($\blacksquare$) phases at 0.1 equiv dm^{-3} total concentration. Predicted ($\circ$) and experimental ($\bullet$) zeolite compositions corresponding to a total solution concentration of 0.4 equiv dm^{-3} are also shown; agreement is generally satisfactory. (Taken from reference 39).

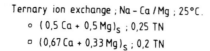

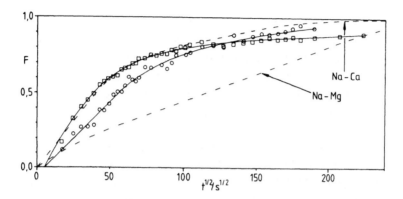

Figure 13. Examples of ternary ion exchange kinetics within the
Na/Ca/Mg-zeolite A system. The dashed lines which bound the experimental data
represent the simple binary exchange rates (viz Na/Ca and Na/Mg). When
magnesium is added to the system in progressively larger amounts (i.e. ○ > □)
then a progressive slowing of the Na/(divalent ion) exchange rate is
observed. Data were obtained using ^{22}Na as radiotracer. (Taken from
reference 56).

magnesium exchange into sodium zeolites. Because zeolite A is used as the
detergent builder, most of this effort has obviously concentrated on this
material.

5.2 RECENT ION EXCHANGE STUDIES

Franklin and Townsend (ref. 55) re-examined binary exchange equilibria
involving the ion pairs Na/Ca, Na/Mg and Ca/Mg in zeolite A. They then
constructed the first complete ternary isotherm for this system (Figure 11).
The binary exchange experiments clarified ambiguities previously seen
regarding the maximum level of magnesium exchange which can occur in zeolite A
and also demonstrated the huge preference that this material exhibits for
calcium over magnesium when only these two ions are present (ref. 55). They
also attempted to reconcile discrepancies in earlier work which appeared to be
due in part to experimental error, but were also probably due to hydrolysis
occurring, with the intervention of oxonium ions (see also refs. 23,24,56).
Subsequently, Franklin and Townsend used these experimental data (ref. 55) to
predict successfully (Figure 12) both binary and ternary exchange equilibria
for this system over a range of concentrations (ref. 39) At about the same

time as this work (refs. 39,55) another study on the prediction of ternary compositions for the Na/Ca/Mg-A system was published (ref. 57). These predictions were carried out by evaluating Wilson interaction parameters (ref. 7) for the three binary conjugate systems, then using these to predict the ternary equilibrium compositions (ref. 57).

Ion exchange studies of magnesium and calcium into precursor gels of either A (refs. 58,59) and X (ref. 59) have also been undertaken. Exchange capacities were monitored as a function of crystallinity either by XRD (refs. 58,59) or by MASNMR (ref. 58). Results obtained were similar in both cases, and showed that amorphous aluminosilicate obtained as precursor gel to A was at least comparable in performance to the crystalline product.

Detailed studies of the kinetics of ion exchange of systems relevant to detergency have been relatively few. The only major one was by Drummond, DeJonge and Rees (ref. 56). They used a earlier theoretical approach of Danes and Wolf (ref. 60), and demonstrated the inhibiting effect on the rate of calcium exchange that magnesium may exert (Figure 13) as well as the need to take into account sodium/oxonium exchange and possible hydrolysis of the zeolite (ref. 56).

In section 4, it was noted that particle controlled diffusion is normally the significant mechanism in ion exchange kinetics involving zeolites, but that as one reduces the crystallite size, the consequent increase in rate which will occur will reach a limit as film controlled diffusion takes over. In the case of self-diffusion, there is a simple relationship between the rate of exchange and the characteristic dimension of the crystal (viz the rate is inversely proportional to the square of the radius of the crystal). For exchange between two different types of ions the relationship is not so straightforward, but nevertheless decreasing crystallite size will markedly increase rate. The significance of this is seen for a country such as Japan, where typical washing conditions place stringent demands on the zeolite compared to those required in either the USA or Europe (ref. 54). Essentially, a typical Japanese wash involves short washing time cycles and cold water washings (e.g. 5° C). In order to boost performance therefore, a major aim for the Japanese market is to produce crystallites of very small size, and one way in which this problem is being overcome is to prepare zeolite A from acid-treated clay precursors (ref. 54).

5.3 SOME OTHER RECENT PRODUCT DEVELOPMENTS INVOLVING ZEOLITES

In the preceding section I alluded to the importance of crystallite size in determining the rate of exchange of calcium and magnesium into zeolite A. Recently, it has been emphasised that crystallite size can be important in another respect, namely, when one is designing a liquid detergent (ref. 61). The concern here is not primarily the rate of exchange, but rather the rate of sedimentation of the solid zeolite builder within the liquid on standing.

Unfortunately, the solution is not just simply to reduce the average crystallite size (ref. 61), because below a certain average crystallite size the rheological properties become unsuitable. Clearly, in this situation, it is important to add other ingredients to the liquid mixture which may modify the interaction between the liquid and the zeolite (ref. 61).

As mentioned above, dissolution of calcium containing precipitates which may "encrust" within the textile fibres, and removal of calcium ions which may strengthen binding between soil particles and the textile surface, are an important part of building technology. Zeolites are not effective in this respect, and studies on how to boost the performance of the zeolite with co-builders have therefore been of importance (ref. 52). These co-builders may be polymeric, such as polycarboxylate (refs. 53,62). Polycarboxylates show the additional advantage of having good ecotoxicological properties (ref. 62). Sequestrants can also be used, and there has been a recent report on the equilibrium calcium distribution which occurs between zeolite on one hand and a range of different chelating builders on the other (ref. 63). The results confirmed that calcium ion binding in binary mixtures of zeolite plus sequestrant was highly competitive, indicating continuing lability of the calcium, and, as a consequence, the order of addition of sequestrant or builder was not important (ref. 63).

6. OTHER RECENT DEVELOPMENTS AND TRENDS

Maes and Cremers (ref. 64) have published a review on highly selective ion exchange, although this concentrates fairly strongly on clay minerals rather than zeolites. Natural zeolites have continued to attract attention, either with respect to their ability to take up heavy metal ions (refs. 65,66) or because of their use in removing ammonium ion and ammonia from waste water. In this respect clinoptilolite continues to attract attention (refs. 67,68,69) and because of conflicting reports in the literature regarding the capacity and selectivity of this material for various metals, Semmens and Martin have concluded that the zeolite conditioning procedure is an important influence on overall performance of this material (ref. 70). Other interesting recent developments have either involved the introduction of exotic cationic species into zeolitic channels, or investigations into higher silica zeolites. Examples of the former are the generation of a hydrophobic form of zeolite X by exchanging the material with octadecylammonium ion (ref. 71). Ion exchange has also been carried out between zeolites A or X and non-aqueous ammonia solutions, with the consequent inclusion in the zeolite channels of ammoniated alkali metal cations (ref. 72). Concerning higher silica zeolites, ZSM-5 hasattracted some attention with a study on organic ion exchange (ref. 73) and also of monovalent and multivalent cations (ref. 74). Multivalent cations were unable to achieve 100% exchange, presumably because the negative charges on the framework arising from the isomorphous substitution of silicon for aluminium were too widely spaced to allow effective neutralisation of the multiple charges on these cations (ref. 74).

As a final comment, it is worth noting how new analytical techniques are helping us to develop our basic understanding of ion-exchange processes. Magic angle spinning nuclear magnetic resonance spectroscopy is of particular help in this regard. We have already noticed its application in elucidating isomorphous replacement of aluminium into the framework of zeolites and the nature and structure of aluminium species in the channels (ref. 3), and in monitoring solid state exchange processes (ref. 10). Two other recent studies using either ^{29}Si or ^{23}Na MASNMR (refs. 75,76) have shown the great potential of this analytical technique for elucidating the relative populations of cations in different sites within the zeolitic framework. In all these respects, MASNMR may well be able to complement appropriate x-ray diffraction studies and hence greatly increase our knowledge of ion-exchange processes in zeolites quite generally.

REFERENCES

1. G.P. Handreck and T.D. Smith, J. Chem. Soc. Faraday Trans. 1, 85(3) (1989) 645-654.
2. R.M. Barrer and J. Klinowski, Phil. Trans. R. Soc. Lond., 285 (1977), 637-680.
3. H. Hamdan, B. Sulikowski and J. Klinowski, J. Phys. Chem., 93 (1989) 350-356.
4. R.M. Barrer and B.M. Munday, J. Chem. Soc. A (1971) 2914.
5. R.M. Barrer, Bull. Soc. fr. Mineral. Cristallogr., 97 (1974) 89-100.
6. J.T. Iiyama, Bull. Soc. fr. Mineral. Cristallogr., 87 (1964) 532-541.
7. P. Fletcher, K.R. Franklin and R.P. Townsend, Phil. Trans. R. Soc. Lond., A312 (1984) 141-178.

8. R.M. Barrer and R.P. Townsend, J. Chem. Soc. Faraday Trans. 2, 80 (1984) 629-40.
9. R.P. Townsend, Pure & Appl. Chem., 58 (10) (1986) 1359-1366.
10. C.A. Fyfe, G.T. Kokotailo, J.D. Graham, C. Browning, G.C. Gobbi, M. Hyland, G.J. Kennedy and C.T. De Schutter, J. Am. Chem. Soc., 108 (1986) 522-523.
11. G.T. Kokotailo, S.L. Lawton and S. Sawruk, in J.R. Katzer (Ed.), "Molecular Sieves-II", ACS Symp. Ser., Vol 40, ACS, Washington DC, 1977 439-450.
12. A. Dyer and R.B. Gettins, J. Inorg. Nucl. Chem., 32 (1970), 2401-2410.
13. D.H. Olson and H.S. Sherry, J. Phys. Chem., 71 (12) (1968) 4095-4104.
14. R.M. Barrer and A.J. Walker, Trans. Faraday Soc., 60 (1964) 171-184.
15. G.L. Gaines and H.C. Thomas, J. Chem. Phys., 21 (4) (1953) 714-718.
16. K.R. Franklin and R.P. Townsend, J. Chem. Soc., Faraday Trans. 1, 84 (8) (1988) 2755-2770.
17. F. Helfferich, "Ion Exchange", McGraw-Hill, London, 1962 pp. 134-139.
18. R.M. Barrer, R. Papadopoulos and L.V.C. Rees, J. Inorg. Nucl. Chem., 29 (1967) 2047-2063.
19. P. Fletcher and R.P. Townsend, Zeolites, 3 (1983) 129-133.
20. B.K.G. Theng, N.Z. J. Sci., 14 (1971) 1026-1039.
21. R.M. Barrer, "Zeolites and Clay Minerals as Sorbents and Molecular Sieves", Academic Press, London, 1978 pp. 76-87.
22. K.R. Franklin, R.P. Townsend, S.J. Whelan and C.J. Adams, in Y. Murakami, A. Iijima and J.W. Ward (Eds.), Proc. 7th Int. Zeolite Conf. Tokyo, Elsevier 1986, pp. 289-296.
23. T.E. Cook, W.A. Cilley, A.C. Savitsky and B.H. Wiers, Environ. Sci. Technol., 16 (6) (1982) 344-350.
24. H.E. Allen, S.H. Cho and T.A. Neubecker, Water Res., 17 (12) (1983) 1871-1879.
25. R.M. Barrer and R.P. Townsend, J. Chem. Soc., Faraday Trans. 1, 72 (1976) 661-673.
26. P. Fletcher and R.P. Townsend, J. Chromatogr., 238 (1982) 59-68.
27. P. Fletcher and R.P. Townsend, J. Chem. Soc., Faraday Trans. 1, 91 (1985) 1731-1744.
28. G. Sposito, K.M. Holtzclaw, C.T. Johnston and C.S. Lavesque-Madore, Soil Sci. Soc. Am. J., 45 (1981) 1079.
29. G. Sposito, K.M. Holtzclaw, L. Charlet, C. Jouany and A.L. Page, Soil Sci. Soc. Am. J., 47 (1983) 51-56.
30. G. Sposito, C. Jouany, K.M. Holtzclaw and C.S. Levesque, Soil Sci. Soc. Am. J., 47 (1983) 1081-1085.
31. F. Helfferich, "Ion Exchange", McGraw-Hill, London, 1962 p. 250.
32. A. Dyer, H. Enamy and R.P. Townsend, Sep. Sci. Technol., 16 (2) (1981) 173-183.
33. R.M. Barrer and J. Klinowski, J. Chem. Soc., Faraday Trans. 1, 70 (1974) 2080-2091.
34. M.L. McGlashan, "Chemical Thermodynamics", Academic Press, London, 1979 pp. 78-79; 106-107.
35. P. Fletcher and R.P. Townsend, J. Chem. Soc., Faraday Trans. 2, 77 (1981) 2077-2089.
36. R.P. Townsend, P. Fletcher and M. Loizidou, in D. Olson and A. Bisio (Eds.), Proc. 6th Int. Zeolite Conf., Reno, 1983 pp. 110-121.
37. M.L. McGlashan, op. cit., pp. v, 111-112.
38. K.R. Franklin and R.P. Townsend, Zeolites, 8 (5) (1988) 367-375.
39. K.R. Franklin and R.P. Townsend, J. Chem. Soc., Faraday Trans. 1, 81 (1985) 3127-3141.
40. K.R. Franklin and R.P. Townsend, J. Chem. Soc., Faraday Trans. 1, 84 (1988) 687-702.
41. A. Dyer, R. Harjula and R.P. Townsend, paper in preparation.
42. R.M. Barrer and L.V.C. Rees, J. Phys. Chem. Solids, 25 (1964) 1035-1038.

43. R.M. Barrer, R.F. Bartholomew and L.V.C. Rees, J. Phys. Chem. Solids, 24 (1963) 51-62.
44. R.M. Barrer, R.F. Bartholomew and L.V.C. Rees, J. Phys. Chem. Solids, 24 (1963) 309-317.
45. N.M. Brooke and L.V.C. Rees, Trans. Faraday Soc., 64 (1968) 3383.
46. N.M. Brooke and L.V.C. Rees, Trans. Faraday Soc., 65 (1969) 2728.
47. S.C. Duffy and L.V.C. Rees, J. Chem. Soc., Faraday Trans. 1, 70 (1974) 777-786.
48. S.C. Duffy and L.V.C. Rees, J. Chem. Soc., Faraday Trans., 71 (1975) 602-609.
49. S.C. Duffy and L.V.C. Rees, J. Chromatogr., 102 (1974) 149-153.
50. L.M. Brown, H.S. Sherry and F.J. Krambeck, J. Phys. Chem., 75 (1971) 3846-3855.
51. F. Helfferich, "Ion Exchange", McGraw-Hill, London, 1962, pp. 267-273.
52. C.P. Kurzendorfer, M. Liphard, W. von Rybinski and M.J. Schwuger, Colloid and Polymer Sci., 265 (1987) 542-547.
53. H. Upadek and P. Krings, in H.G. Karge and J. Weitkamp (Eds.), "Zeolites as Catalysts, Sorbents and Detergent Builders", Elsevier, London, 1989 pp. 701-709.
54. I. Yamane and T. Nakazawa, in Y. Murakami, A. Iijima and J.W. Ward (Eds.), Proc. 7th Int. Zeolite Conf. Tokyo, Elsevier 1986, pp. 991-1000
55. K.R. Franklin and R.P. Townsend, J. Chem. Soc., Faraday Trans. 1, 81 (1985) 1071-1086.
56. D. Drummond, A. DeJonge and L.V.C. Rees, J. Phys. Chem., 87 (1983) 1967-1971.
57. E. Costa, A. de Lucas, J. Zarca and F.J. Sanz, Lat. Am. J. Chem. Eng. Appl. Chem., 17 (1987) 135-148.
58. Y. Tsuruta, T. Satoh, T. Yoshida, O. Okumura and S. Ueda, in Y. Murakami, A. Iijima and J.W. Ward (Eds.), Proc. 7th Int. Zeolite Conf. Tokyo, Elsevier 1986, pp. 1001-1007.
59. V.C. Mole and L.V.C. Rees, in P.A. Williams and M.J. Hudson (Eds.), "Recent Developments in Ion Exchange", Elsevier Appl. Science, London, 1987, pp. 264-276.
60. F. Danes and F. Wolf, Z. Phys. Chem. (Leipzig), 252 (1973) 15-32.
61. W. Leonhardt and B.M. Sax, in H.G. Karge and J. Weitkamp (Eds.), "Zeolites as Catalysts, Sorbents and Detergent Builders", Elsevier, London, 1989, pp. 691-699.
62. M.J. Schwuger and M. Liphard, in H.G. Karge and J. Weitkamp (Eds.), "Zeolites as Catalysts, Sorbents and Detergent Builders", Elsevier, London, 1989, pp. 673-690.
63. T. Mukaiyama, H. Nishio and O. Okumura, in Y. Murakami, A. Iijima and J.W. Ward (Eds.), Proc. 7th Int. Zeolite Conf. Tokyo, Elsevier 1986, pp. 1017-1023.
64. A. Maes and A. Cremers, ACS Symp. Ser., Geochem. Processes Miner. Surf., 323 (1986) 254-295.
65. M. Loizidou and R.P. Townsend, Zeolites, 7 (1987) 153-159.
66. M. Loizidou and R.P. Townsend, J. Chem. Soc., Dalton Trans., 1987 1911-1916.
67. M. Vokacova, Z. Matejka and J. Eliasek, Acta hydrochim, hydrobiol., 14 (1986) 605-611.
68. E. Czaran, A. Meszaros-Kis, E. Domokos and J. Papp, Acta Chim. Hung., 125 (1988) 201-210.
69. S-J Kang and K. Wada, Appl. Clay Sci., 3 (1988) 281-290.
70. M.J. Semmens and W.P. Martin, Water Res., 22 (1988) 537-542.
71. C.L. Renschler and C. Arnold, J. Mat. Sci. Lett., 5 (1986) 1169-1171.
72. H. Uyama, Y. Kanzaki and O. Matsumoto, Mat. Res. Bull., 22 (1987) 157-164.
73. P. Chu and F.G. Dwyer, Zeolites, 8 (1988) 423-426.
74. D.P. Matthews and L.V.C. Rees, Chem. Age India, 37 (1986) 353-357.
75. K-J Chao and J-Y Chern, J. Phys. Chem., 93 (1989) 1401-1404.
76. L.B. Welsh and S.L. Lambert, ACS Symp. Ser., Perspect. Mol. Sieve Sci., 368 (1988) 33-47.

Chapter 11

DIFFUSION IN ZEOLITE MOLECULAR SIEVES

MARTIN F.M. POST

Koninklijke/Shell-Laboratorium, Amsterdam (Shell Research B.V.), Badhuisweg 3, 1031 CM Amsterdam (The Netherlands)

ABSTRACT

In two major industrial applications of zeolites and related molecular sieves, viz. (shape-selective) catalysis and selective adsorption processes, migration or diffusion of sorbed molecules through the pores and cages within the sieve crystals plays an important role. A thorough knowledge of intra-crystalline diffusion phenomena in molecular sieves is of prime importance if current industrial applications in separation and catalysis are to be better understood - only then can they be optimized - and for the development of new applications in the above areas. During the past twenty or thirty years much emphasis has been placed on the development of reliable methods to study intracrystalline migration of a variety of sorbates in zeolites. The aim of the present chapter is to review the state-of-the-art techniques currently in use for measuring intracrystalline diffusivities, to discuss their advantages and limitations, to highlight the available information on numerical values of intracrystalline diffusivities, to rationalize the effects of sorbate type and sieve structure on diffusivity and, finally, to quantify the relationship between zeolite catalysis and intracrystalline diffusion.

CONTENTS

1. INTRODUCTION

Porous crystalline molecular sieves such as zeolites and related materials find their major application in industrial processes primarily in two areas, viz. catalysis and selective adsorption/separation. In these applications migration or diffusion of sorbed molecules through the pores and cages within the crystals plays an important and, occasionally, a dominant role. In some applications intracrystalline transport can be a rate-limiting factor. Restricted molecular migration may therefore adversely influence the efficiency of a particular process step. On the other hand, the limited mobility of one type of molecule with respect to another through the intracrystalline space can be utilized to advantage in shape-selective separation and catalysis.

Intracrystalline diffusion in molecular sieves has been put into perspective by Weisz (ref. 1) by means of a diffusivity/pore-size diagram (Fig. 1). Zeolites and related sieves with effective pore dimensions ranging from 3 to about 9 Å exhibit a region of diffusivity beyond the molecular and Knudsen regions occurring in porous media having much larger pore sizes. Weisz introduced the term "configurational diffusion" to encompass intracrystalline migration. The region of configurational diffusion spans more than 10 orders of magnitude in diffusivity: depending on the size and nature of the sorbate species, the type of molecular sieve, and temperature, intracrystalline diffusivities D_c ranging from 10^{-8} down to 10^{-20} $m^2.s^{-1}$ have been reported.

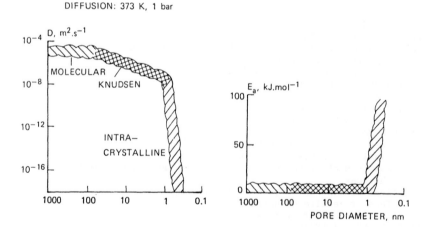

Fig. 1. Effect of pore diameter on molecular diffusivity and energy of activation of diffusion.

Basic studies on intracrystalline diffusion in molecular sieves are of relevance in commercial molecular sieve separation processes, such as the separation of normal paraffins and isoparaffins. In these processes the dynamic capacity of the adsorbent bed and the time required for regeneration are determined by intracrystalline diffusivities. The rational design and optimization of such processes requires detailed knowledge of the kinetics and equilibrium of sorption of the molecules involved (ref. 2).

For catalytic applications of molecular sieves, too, the importance of quantification of molecular migration is widely recognized. Weisz (ref. 1) has estimated that a useful reactor for catalytic transformation of hydrocarbons using zeolites must convert about one $mol.s^{-1}$ for every cubic metre invested in catalyst material. This requirement sets a lower limit to useful flow rates into and out of the zeolite crystals (ref. 3): only if the physical mass transport rate exceeds the intrinsic chemical conversion rate are limitations due to restricted mobility of reactants and/or products likely to be absent (refs. 4,5), and only then can the inner surface area of a porous catalyst be utilized to the full.

There are also many examples where limitations in transport rates have a beneficial effect on catalyst performance. We are now entering the field of "shape-selective catalysis", a term coined by Weisz and Frilette (ref. 6) almost thirty years ago. Particularly in shape-selective catalytic applications of zeolites the importance of restricted molecular migration is often emphasized (refs. 1,7-9). If a transport selectivity mechanism is operative in a particular reaction scheme, the size of the molecular sieve crystal may strongly influence the extent of selectivity attained.

It is the intention of the present contribution to provide some insight into the practice of measurement of diffusivities in zeolites and related molecular sieves. Part 2, as an introduction, briefly summarizes the general principles underlying the physical description of diffusion in porous media. In Part 3 the various experimental techniques that are commonly used to measure intracrystalline diffusivities are discussed, together with the basic theory required for the analysis and interpretation of the results. Advantages and limitations of the different methods are compared. While in Part 3 the emphasis is on methodology, Part 4 is devoted to a discussion on the numerical values of the diffusivity in various sorbate/sorbent systems. First, attention is given to the extent (or lack) of agreement between the results obtained via the different experimental techniques. In addition, the effect of sorbate (type, pressure), sorbent (type of molecular sieve, chemical composition, origin, crystal size, etc.) and temperature on intracrystalline diffusivity is discussed. Sorption from the liquid phase is also considered briefly. Finally, in Part 5, the relationship between zeolite catalysis and intracrystalline diffusion is further quantified.

2. DIFFUSION IN POROUS ZEOLITE MEDIA: GENERAL PRINCIPLES

A convenient starting point for the mathematical description of rates of adsorption and desorption in porous adsorbents such as zeolite crystals is provided by Fick's first equation, which relates the molar flux J (expressed in $mol.m^{-2}.s^{-1}$) due to diffusive transport and the concentration gradient $\partial q/\partial x$ by the following well-known expression:

$$J = -D_c(q) \frac{\partial q}{\partial x} \tag{2.1}$$

where q is the concentration of the adsorbed diffusing species in the porous zeolite medium. The intracrystalline diffusivity, or diffusion "constant", $D_c(q)$, is in general dependent on the concentration but, according to eqn. (2.1), it is independent of the intracrystalline concentration gradient $\partial q/\partial x$. The true driving force of any transport process, however, is the gradient of the chemical potential of the adsorbed species ($\partial \mu/\partial x$) rather than the intracrystalline concentration gradient. Following these lines of reasoning, the molar flux follows from

$$J = -B_c(q)q \frac{\partial \mu}{\partial x} \tag{2.2}$$

The relationship between the mobility, $B_c(q)$, and the Fickian diffusivity, $D_c(q)$, can be easily derived assuming equilibrium between the adsorbed phase (concentration q) and an ideal vapour phase (pressure p, temperature T):

$$D_c(q) = B_c(q)RT \frac{d \ln p}{d \ln q} = D_o(q) \frac{d \ln p}{d \ln q} \tag{2.3}$$

where $D_o(q) = B_c(q)RT$ is the corrected or intrinsic diffusivity. It is clear from eqn. (2.3) that the Fickian diffusivity approaches a limiting value, viz. $D_o(q)$, only when Henry's law is obeyed (d ln p/d ln q = 1). The intrinsic diffusivity, however, might still be a function of the concentration. Equation (2.3) looks very similar to, but should not be confused with, the familiar Darken equation (ref. 10) that was originally derived for the inter-diffusion of two alloys.

The intracrystalline transport diffusivities, $D_c(q)$ and $D_o(q)$, relating the flux to a concentration gradient, are in general different from the self-diffusivity $D(q)$, which defines the rate of tracer exchange of marked molecules under equilibrium conditions in a system without a net concentration gradient. It can be shown (ref. 2) that only if the cross coefficient (defined in irreversible thermodynamics: see refs. 11,12) can be neglected does $D(q) = D_0(q)$, or

$$D_c(q) = D_0(q) \frac{d \ln p}{d \ln q} = D(q) \frac{d \ln p}{d \ln q} \qquad (2.4)$$

which is the familiar Darken equation (ref. 10). In general, both D_0 and D are concentration-dependent.

The expressions thus far relate molar fluxes and gradients of concentration or chemical potential. However, a description of rates of adsorption generally requires knowledge of the change of concentration inside the porous medium as a function of time, as is represented by the differential equation:

$$\frac{\partial q}{\partial t} = D_c(q) \frac{\partial^2 q}{\partial x^2} \qquad (2.5)$$

This expression, which can be easily derived from eqn. (2.1) (see, for example, ref. 13), is known as Fick's second equation. Solutions to this equation for a diversity of physical situations are given in ref. 13, and some of these are discussed in Part 3.

3. EXPERIMENTAL MEASUREMENT OF DIFFUSIVITIES IN MOLECULAR SIEVES

Many commercial adsorbents and catalysts consist of small microporous crystals (e.g. zeolites) formed into a meso- or macroporous pellet (Fig. 2). In

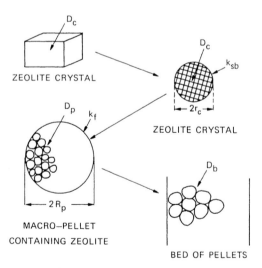

Fig. 2. Schematic diagram of a bed of composite adsorbent pellets containing zeolite, showing the principal resistances to mass transfer.

396

general, such materials may offer three distinct resistances to mass transfer: (i) the micropore resistance of the zeolite crystals (proportional to r_c^2/D_c), (ii) the resistance related to transport through the outer layer of the zeolite crystal (known as the surface barrier resistance (ref. 14), proportional to r_c/k_{sb}) and (iii) the meso- or macropore diffusional resistance of the pellet (proportional to R_p^2/D_p). In addition, a resistance due to the fluid film surrounding the pellet (mass transfer coefficient k_f) and a contribution originating from the dispersion (D_6) in a bed may be involved.

The first few sections (3.1-3.5) discuss a number of different methods of measuring transport diffusivities (D_c) inside zeolite crystals, together with the basic theory required for the analysis and interpretation of the experimental results. In the final section, techniques designed to measure self-diffusivities (D) of molecules inside zeolite crystals are discussed, and the advantages and limitations of the various methods are compared.

3.1. The gravimetric (uptake) method

3.1.1. Experimental procedure. The gravimetric method involves subjecting a zeolite sample to a step change in sorbate pressure at zero time, and recording the change in weight of the sorbent as a function of time using an accurate microbalance system (Fig. 3). During the experiment the pressure of

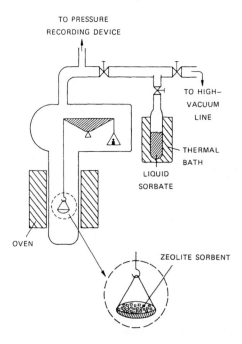

Fig. 3. Set-up for gravimetric measurements.

the sorbate is kept constant. Prior to the adsorption run, the sample is normally outgassed at elevated temperature (e.g. 673 K) in vacuo for some 5-20 h. This procedure is essential to obtain reliable experimental results.

In principle, the diffusivity can be obtained by matching the experimental uptake curves obtained in a gravimetric experiment to the appropriate transient solution of the diffusion equation (eqn. 2.5); in actual systems, however, the interpretation of the transient curves obtained by the gravimetric method might be quite complicated. In an ideal case, the intrazeolite diffusion is rate-determining (micropore diffusion is relatively slow and/or the zeolite crystals are large). In order to reduce the effects of bed diffusion and heat transfer, sample sizes should be kept as low as possible, typically in the order of 10-30 mg, or less. However, in any experimental study, it is always desirable to vary sample weight, bed height, particle size and crystallite size (if possible) to disentangle the contributions of the various diffusion resistances.

3.1.2. Kinetics of sorption

A. Isothermal single-component sorption: micropore diffusion control

The simplest case to consider is a single (hypothetical) isotropic spherical zeolite crystal of radius r_c, or an ensemble of equally-sized spherical crystals, exposed to a differential step change in sorbate concentration or pressure at the external surface of the particle at time zero. The appropriate form of the diffusion equation can be obtained from a differential mass balance over a spherical shell element:

$$\frac{\partial q}{\partial t} = \frac{1}{r^2} \frac{\partial}{\partial r} \left(r^2 D_c \frac{\partial q}{\partial r} \right) \tag{3.1}$$

This equation is Fick's second law (eqn. 2.5) expressed in radial coordinates. Provided that the intracrystalline diffusivity D_c is independent of q (and this assumption is usually an acceptable approximation if the uptake curve is measured over a small differential change in q and/or uptake studies are restricted to the Henry regime: $D_c = D_o$), the solution of eqn. (3.1) is then given by the familiar expression (ref. 13)

$$\frac{\bar{q}(t) - q_0}{q_\infty - q_0} = \gamma = 1 - \frac{6}{\pi^2} \sum_{n=1}^{\infty} \frac{1}{n^2} \exp \left(- \frac{n^2 \pi^2 D_c t}{r_c^2} \right) \tag{3.2}$$

where $\bar{q}(t)$ is the average sorbate concentration through the zeolite crystal as a function of time, and γ is the fractional approach to equilibrium. Figure 4 is a graphical representation of eqn. (3.2).

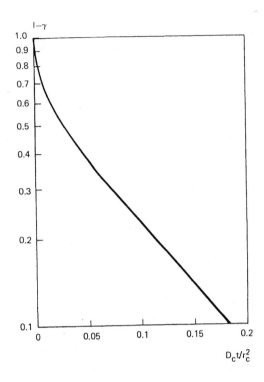

Fig. 4. Characteristic uptake curve for spherical particles (constant pressure/gravimetric method).

For small values of t or small values of γ ($\gamma < 0.3$) eqn. (3.2) reduces to the simple root-t law:

$$\gamma = \frac{6}{\sqrt{\pi}} \left(\frac{D_c}{r_c^2} \right)^{\frac{1}{2}} \sqrt{t} \qquad (3.3)$$

Thus, by plotting γ against $\sqrt{t}$, the diffusional time constant D_c/r_c^2 or (provided that r_c is known) the diffusivity D_c can be directly evaluated from the slope. The square of the ratio $\gamma/\sqrt{t}$, i.e., the expression γ^2/t, for low values of γ is often referred to as the uptake rate. For longer times, i.e., where $\gamma \geq 0.7$, eqn. (3.2) converges rapidly since the higher terms in the summation become vanishingly small. In this case we only retain the first term of the summation to obtain

$$\ln(1-\gamma) = \ln \frac{6}{\pi^2} - \frac{\pi^2 D_c t}{r_c^2} \qquad (3.4)$$

and, accordingly, plotting $\ln(1-\gamma)$ against t gives a straight line of slope $-\pi^2 D_c/r_c^2$ (Fig. 4).

The above discussion relating to spherical isotropic zeolite crystals can in principle be extended to both isotropic and non—isotropic zeolite crystals of any particular shape, but an analysis of the uptake curves is beyond the scope of the present review. In actual situations the presence of size and shape distributions of zeolite crystals will have a substantial effect on the kinetics of uptake, particularly at longer sorption times. This problem has been the subject of extensive studies by Ruthven and Loughlin (ref. 15).

B. Isothermal single—component sorption: criterion for absence of
   macropore resistance

In an actual adsorbent or catalyst pellet consisting of zeolite crystals (Fig. 2), the uptake rate may be controlled by either micropore (intracrystal—line) diffusion, or by macropore (inter— or extracrystalline) diffusion, or by both. Assuming that local equilibrium between adsorbed phase in the zeolite crystal and the fluid (gas) phase in the macropores exists and that Henry's law can be applied, the relation between sorbate concentration c in the macro—pores between the zeolite crystals and the intracrystalline sorbate concentra—tion q inside the zeolite crystals is given by $K_c = q/c$, where K_c denotes the dimensionless adsorption constant. Comparison of intra— and extracrystalline diffusion resistances should be done on the same concentration basis. Hence, in order to retain consistency, the characteristic times for micropore and macropore diffusion are defined by r_c^2/D_c and $K_c R_p^2/D_p$, respectively, and the criterion for absence of macropore diffusion control is then:

$$r_c^2/D_c \gg K_c R_p^2/D_p.$$

In many gravimetric sorption experiments at low pressure, the experimental conditions are such that the molecular motion in the macropores is determined by collisions with the pore walls rather than by collisions between diffusing molecules. Under these conditions we are in the "Knudsen" regime. The Knudsen diffusivity D_k may be estimated from

$$D_k = 97 \ \mu_p \ \left(\frac{T}{M}\right)^{\frac{1}{2}}, \ m^2 . s^{-1} \tag{3.5}$$

where μ_p is the mean pore radius (m), T is in Kelvin and M is the molecular weight of the diffusing species. The effective macropore diffusivity D_p is then of the same order of magnitude as the Knudsen diffusivity D_k (apart from a tortuosity correction, which is normally in the range 1—6). Hence, in the criterion formulated above, D_p may be replaced by the Knudsen diffusivity D_k, which can be estimated from eqn. (3.5). Accordingly, the absence or presence of macropore transport control can be easily verified by comparison of the experimentally determined total resistance, expressed as a characteristic time of diffusion (which is of the order of t/γ^2 for small values of t), with the characteristic time for Knudsen diffusion in the macropores:

$$\left.\frac{t}{\gamma^2}\right|_{\gamma < 0.3} \gg K_c \frac{R_P^2}{D_k} : \quad \begin{array}{l} \text{absence of macropore} \\ \text{limitations} \end{array} \qquad (3.6)$$

C. Non-isothermal single-component sorption

The heat effects associated with adsorption (or desorption) of molecules in molecular sieves are relatively large. Heats of sorption of the majority of sorbates studied (hydrocarbons and oxygenates with 1-10 carbon atoms) are in the range 10-100 kJ/mol. As these values are quite substantial, particularly with rapid uptake processes, the average temperature of the zeolite crystals during the sorption experiment may be significantly higher than in the initial situation. The problem of non-isothermal sorption under conditions typical of a gravimetric uptake rate experiment has been analysed by Lee and Ruthven (refs. 16-18). In this work mathematical expressions have been obtained to simulate experimental uptake curves in a variety of non-isothermal situations. Criteria have been developed to assess the absence or presence of heat effects during uptake, provided that heat transfer coefficients and the heat capacity of the sorbent are known (ref. 18).

3.1.3. Evaluation of uptake curves from gravimetric experiments: some practical comments. Particularly the early literature on uptake studies contains much conflicting information. In many cases the origin of the confusing results can be traced back to extraneous heat and macropore diffusion effects that were not taken into account, but other factors, too, have contributed to the array of erroneous results featuring in this area. It is the intention in this section to briefly summarize potential problems which may influence dramatically the reliability of determination of an intra-crystalline zeolitic diffusivity D_c from gravimetric uptake experiments.

(i) It is necessary to verify the absence of bed diffusion control by repeated experiments with different sample sizes or sample configurations, and to check the linearity of the balance.

(ii) The absence of macropore diffusion limitations in pellets composed of zeolite crystals should be tested experimentally by varying pellet size, or may be verified using eqn. (3.6).

(iii) If possible, conditions leading to non-isothermal sorption should be avoided since intrusion of thermal effects during sorption severely compli-cates the mathematical description of the uptake curve.

(iv) As the relationship between sorbate concentration in the zeolite crystal and pressure is in general non-linear, a "Darken-type" correction factor d ln p/d ln q has to be applied to the measured diffusivity. Since the applicability of this type of relation is still subject to debate, it is advisable to carry out diffusivity measurements in the Henry region, where the measured diffusivity is normally independent of sorbate concentration.

(v) Differential measurements (small pressure steps), which will result in less instability of the equipment and in a lower production of sorption heat, are preferred to integral measurements.

(vi) The occurrence of a distinct size and shape distribution of the zeolite crystals, as is normally encountered in an actual sample, has a strong impact on the kinetics of the uptake.

(vii) A definitive study of intracrystalline diffusion in zeolites should include evaluation of a series of zeolite samples with different crystal sizes; in the ideal case, diffusivity is independent of crystal size. This behaviour is not always encountered and can be attributed in many cases to differences in the origin of the particular zeolite batches or to the presence of a surface barrier at the outer rim of the zeolite crystals.

3.1.4. <u>Range of D_c that can be measured</u>. The range of intracrystalline diffusivities, or rather the range of characteristic diffusion times r_c^2/D_c that can be determined experimentally, is bounded by both upper and lower limits, estimates of which are given in Fig. 5. The lower limit of r_c^2/D_c (i.e.

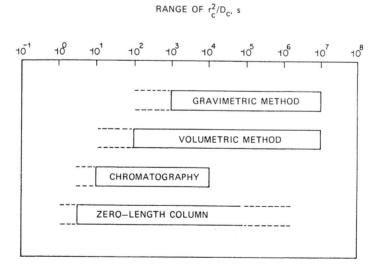

Fig. 5. Range of characteristic diffusion times that can be measured using different methods.

for rapid zeolitic diffusion) is of the order of $10^2 - 10^3$ s and this value is determined by a number of factors, the most important of which are restricted balance response, involvement of limitations due to diffusion through the extracrystalline void (macropores in pellets, bed) and problems with respect to maintaining isothermal conditions. The upper limit depends on the stability of the recording equipment and is also determined by practical considerations. An estimate of the range of intracrystalline diffusivities that can be measured for zeolites with crystal radii of 1 and 10 μm is given in Fig. 6.

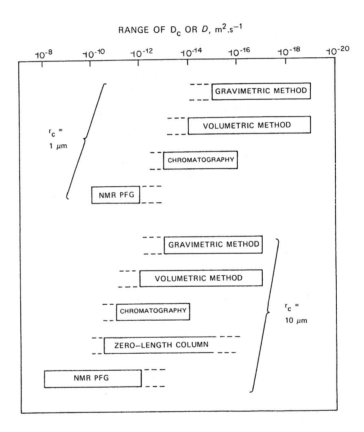

Fig. 6. Range of diffusivities that can be determined using different methods.

3.2. The volumetric (uptake) method

3.2.1. Experimental procedure. As in the gravimetric method, the volumetric method involves subjecting a zeolite sample to step change at zero time. In contrast to the gravimetric method as described in Section 3.1., however, the quantity of sorbate admitted to the sorbate buffer volume is limited. Accordingly, the pressure of the sorbate will vary during the experiment and the reliable measurement of this change in pressure as a function of time (e.g. by a sensitive differential pressure transducer) is the basis of the volumetric method. A simplified set-up for volumetric experiments is shown in Fig. 7.

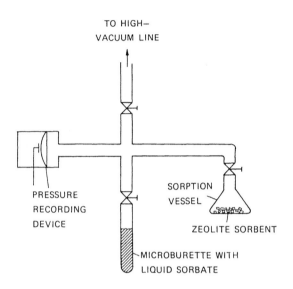

Fig. 7. Set-up for volumetric measurements.

3.2.2. <u>Kinetics of sorption</u>. The simplest case to consider is a single spherical zeolite crystal (or an ensemble of equally sized spherical crystals) of radius r_c, exposed to a differential step change in sorbate pressure at the external surface of the crystal at time zero. The starting point is Fick's second law (eqn. 3.1) expressed in spherical symmetry, assuming the validity of a linear sorption isotherm and that the diffusivity D_c is independent of the concentration q of sorbed species.

To obtain an analytical solution for the transient curve it is convenient to introduce a variable α, defined as the ratio of the amount of sorbate at equilibrium in the sorbent to the amount of sorbate in the gas phase of the sorbate buffer:

$$\alpha = (V_s q_\infty)/(V_g c_\infty) \qquad \text{or} \qquad \alpha = K_c V_s/V_g \tag{3.7}$$

where V_g is the gas volume of the sorbate buffer and V_s is the (crystal) volume of the zeolite sorbent. The solution of the uptake curve can then be

404

formulated as (refs. 13,19)

$$\gamma = 1 - \sum_{n=1}^{\infty} \frac{6(1+\alpha)}{9\alpha(1+\alpha) + p_n^2} \ \exp \ (-p_n^2 D_c t/r_c^2) \qquad (3.8)$$

with p_n as the n-th positive root of the equation $\tan p_n = p_n/(1 + 3\alpha p_n^2)$.

For small enough t, eqn. (3.8) reduces to

$$\gamma = \frac{6(1 + \alpha)}{\sqrt{\pi}} \ \left(\frac{D_c}{r_c^2}\right)^{\frac{1}{2}} \ \sqrt{t} \qquad (3.9)$$

In a variable-pressure system, too, a simple root-t law governs the initial sorption as a function of time. At the constant-pressure limit the value for α becomes zero since $V_g \longrightarrow \infty$, cf. eqn. (3.7), and eqn. (3.9) simplifies to the well-known expression for a constant-pressure system, eqn. (3.3).

3.2.3. <u>Final comments</u>. To conclude the section on the volumetric method, it is appropriate to summarize its advantages and disadvantages vis-à-vis those of the gravimetric method.

(i) The typical set-up for volumetric measurements is relatively simple, is easy to operate, and needs a much lower capital investment than required for a high-precision vacuum microbalance.

(ii) The high sensitivity and stability of the pressure transducer, together with the fast response of the system, allows constant-volume, variable-pressure measurements with extremely small sorbent sample sizes, which diminishes the risk of intercrystalline diffusion and sorption heat effects. The lower limit of characteristic time of micropore diffusion (r_c^2/D_c) that can be determined using the volumetric method is estimated to be in the range $10^1 - 10^2$ s, i.e., an order of magnitude lower (better) than with the gravimetric method (compare Section 3.1.4.; see Figs. 5 and 6).

(iii) The variation of sorbate pressure during the uptake experiment severely complicates the analytical solution of the diffusion equation. No simple, straightforward solutions are available, either in non-ideal cases where intercrystalline (macropore) diffusion and/or heat effects become involved, or in the case where the zeolite sample consists of crystals with a broad size distribution.

3.3. <u>The chromatographic method</u>

3.3.1. <u>Experimental procedure</u>. As an alternative to the conventional gravimetric and volumetric methods, a gas-solid chromatographic technique may be used to determine zeolite diffusivities. The method is based on the measurement of the response of a chromatographic column, filled with small pellets composed of zeolite material (and optionally a binder), to a change in inlet concentration caused by a pulse injection of sorbate in a carrier gas stream.

The broadening of the response peak results from the combined effects of axial dispersion and the contributions of the various mass transfer resistances originating from the zeolitic micropores, the macropores in the particles and external gas films around the pellets.

By making measurements over a range of gas velocities it is possible to separate axial dispersion and mass transfer effects. Variation of both parti-cle and zeolite crystal size is further required to determine the relative importance of various diffusional resistances and to arrive at reliable values for the micropore time constant or micropore diffusivity D_c.

3.3.2. Analytical description of the dynamics of a chromatographic column.

A solution to Fick's equation, eqn. (2.5), describing the response of a molecular sieve column subjected to pulse injection at the inlet at time zero, has been formulated by Haynes and Sarma (refs. 20,21). However, in order to maintain consistency with the nomenclature used to describe uptake studies in Sections 3.1. and 3.2., it is more useful to revert to the expressions derived by Ruthven and coworkers (refs. 2,22).

In the ideal case of a column of length L (bed porosity ϵ_b) filled with pellets (radius R_p, porosity ϵ_p) composed of uniform spherical isotropic zeolite crystals (radius r_c) subjected to a sorbate pulse in the Henry regime, the response of the column in terms of the mean retention time (μ) and variance (σ) of the response peak (Fig. 8) is given by (refs. 2,22,23):

$$\mu = \frac{L}{v_i} \left(\frac{1-\epsilon_b}{\epsilon_b} \right) K_p \tag{3.10}$$

$$\frac{\sigma^2}{2\mu^2} = \frac{D_b}{v_i L} + \left(\frac{\epsilon_b}{1-\epsilon_b} \frac{v_i}{L} \right) \left(\frac{r_c^2}{15 D_c K_p} + \frac{r_c}{3 k_{sb} K_p} + \frac{R_p^2}{15 \epsilon_p D_p} + \frac{R_p}{3 k_f} \right) \tag{3.11}$$

where v_i denotes the interstitial gas velocity, which is related to the super-ficial gas velocity (v_s) by $v_i = v_s/\epsilon_b$. K_p denotes the equilibrium constant based on sorbate concentration in a pellet of sieve and is related to the equilibrium constant K_c defined earlier by $K_p = w(1-\epsilon_p)K_c$, where w is the volume fraction of zeolite crystals present in the particle. Strictly speak-ing, eqns. (3.10) and (3.11) are only valid for strongly adsorbed species, i.e., $K_p \cong K_c \gg 1$ (ref. 22). Equation (3.11) shows clearly that the variance of the response peak is due to contributions of axial dispersion (D_b), micropore (zeolitic) diffusion (D_c), diffusion in the macropores of the pellet (D_p) and film diffusion (k_f) (refs. 2,22). In addition, a term accounting for possible transport resistance through the zeolite surface barrier (k_{sb}) has

406

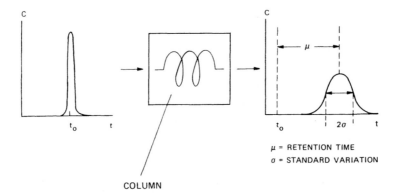

μ = RETENTION TIME
σ = STANDARD VARIATION

COLUMN

Fig. 8. The chromatographic method.

been included (ref. 23) in eqn. (3.11). Evaluation of the micropore diffusivity may be done from a series of chromatographic runs where the linear gas
velocity (v_i or v_s) is varied.

3.3.3. Final comments

(i) The mathematics required to analyse the response of a chromatographic
column is relatively simple and straightforward, as is indicated in the
previous section. However, it should be realized that, in non—ideal situations
where heat effects become involved, no simple mathematical formulation of the
response of a chromatographic column exists.

(ii) The chromatographic technique has a number of important advantages
over the conventional gravimetric technique: only a very simple, cost—effective set—up is required; a straightforward evaluation of the contribution of
intercrystalline diffusional resistances is possible; and the relatively large
sample sizes required for chromatographic evaluation (a few grams or more) are
much more representative of a commercial sorbent or catalyst than the minute
quantities normally used in gravimetric or volumetric methods (in the order of
10—30 mg or less).

(iii) As with the gravimetric and volumetric methods, the range of zeolitic
diffusivities that can be measured is limited (Figs. 5 and 6). The lower limit
of the characteristic diffusion time r_c^2/D_c is reached if the dispersion of the
chromatographic peak is controlled mainly by axial dispersion and other extracrystalline mass transfer resistances (macropores, external film). The upper
limit of the characteristic diffusion time r_c^2/D_c is reached if the intracrystalline diffusion rate is so small that the peak will not equilibrate during
its passage in the chromatographic column, leading to excessive tailing of the
chromatographic peak (refs. 21,24).

3.4. <u>The ZLC method</u>

Recently, a new experimental technique has been developed by Ruthven and coworkers (refs. 25–27), designated the ZLC (Zero–Length Column) method. The ZLC method involves following the desorption curve obtained when a very small sample (typically 1–2 mg) of zeolite crystals, previously equilibrated with sorbate at known concentration, is purged with an inert gas stream (Fig. 9).

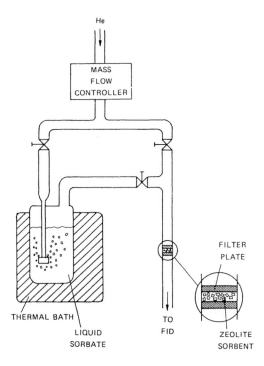

Fig. 9. Set–up for ZLC measurements. (FID = flame ionization detector).

In practice it appears possible to apply high purge gas flow rates so as to approach the ideal isothermal situation in which the external sorbate concentration during purging is maintained close to zero. Consequently, the desorption rate is controlled entirely by intracrystalline diffusion and the diffusivity D_c can be obtained from the slope of a plot of $\ln(c/c_o)$ vs t at long enough times. The absence of extracrystalline resistance can be verified by, for instance, varying the purge flow rate. The validity of the method has been confirmed by varying zeolite crystal size and by comparison with results from classical gravimetric measurements (refs. 25–27).

Because of the high purge flow rates required in order to maintain isothermal conditions, the applicability of the ZLC method is limited to zeolite samples having crystals exceeding a few micrometres in size. The high sensitivity and fast response of the FID detector allows measurements of relatively fast intracrystalline diffusion (see Figs. 5 and 6).

3.5. The membrane technique

In recent years a membrane technique has been successfully used for evaluating diffusivities (refs. 28,29). In this method a membrane is fabricated from a large single zeolite crystal (a few hundred microns in size) of thickness L mounted in a metal plate and sealed with an epoxy resin. After activation of the zeolite by heating in vacuum, the inflow side of the membrane is exposed to diffusant gas at a constant pressure (Fig. 10). Monitoring the increase in

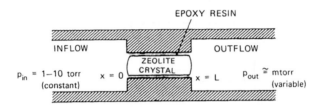

Fig. 10. Schematic diagram of membrane.

pressure with time in the constant volume cell at the outflow side of the membrane yields direct information on the quantity D_c/L^2 and hence on D_c. Since the molar fluxes through the membrane are extremely small a steady-state approach is generally valid. The membrane technique is only suitable for zeolite types allowing the synthesis of very large single crystals and it will always be a matter of debate to what extent these crystals are perfect and fully representative of commercial samples having much smaller crystallites. A distinct advantage of the membrane technique over the classical uptake techniques is the potential to discriminate between mass transfer rates in different channels in non-isotropic zeolites.

3.6. Tracer and NMR techniques

In the previous sections various methods have been discussed affording determination of transport diffusivities. Self-diffusivities (as opposed to transport diffusivities) can be measured in modified uptake rate experiments using isotopically labelled species (ref. 30). In addition to the latter methods, nuclear magnetic resonance (NMR) studies provide a more convenient and more widely used method of measuring self-diffusivities. Such studies will now be discussed in more detail.

3.6.1. The Nuclear Magnetic Resonance Pulse Field Gradient (NMRPFG) method. Nuclear Magnetic Resonance relaxation times are often closely connected with the motion of molecules, and, in principle, molecular mobility (expressed in terms of the self-diffusivity D, see Section 2) can be derived indirectly from measurement of relaxation times, provided that the average jump length is known. A direct method to determine self-diffusivities is the Pulse Field Gradient (PFG) method, which was originally developed by Stejskal and Tanner for the measurement of diffusion in liquids (ref. 31). The development and successful use of the PFG method for determining diffusivities of proton-containing molecules (such as hydrocarbons) in zeolites has been achieved largely through the work of Kärger, Pfeifer and their coworkers. A review of this method has recently been published (ref. 32).

3.6.2. Experimental procedure. In a typical NMRPFG experiment a zeolite sample loaded with (NMR-active) probe molecules is excited with a standard radio frequency (RF) echo pulse sequence in combination with two magnetic field gradient pulses which mark the spatial location of the spins in the sample (Fig. 11). In the absence of the gradient field (i.e., at a constant magnetic field B_0) the spin echo pulse sequence, consisting of a series of two RF pulses of appropriate frequency at $t = 0$ ($\pi/2$ pulse) and $t = \tau$ (π pulse), will lead to an echo signal at $t = 2\tau$ if spin-spin relaxation (characteristic time T_2) is slow compared to τ. The PFG method, however, requires application of both a strong time- (and space-) independent component, B_0, and a pulsed field, Gr, which varies linearly with a space coordinate r following the relation $B = B_0 + Gr$. The field gradient (G) is only applied during two short time intervals of identical duration (δ) separated by a time Δ (Fig. 11). After excitation of the nuclear spins with the $\pi/2$ RF pulse, the first field gradient pulse marks the various nuclear spins which precess at different angular velocity determined by the position of the molecule in the zeolite sample. If there were no diffusion, application of the π RF pulse and a second field gradient pulse would completely reverse the accumulated phase incoherence and would lead to a spin echo signal of the same intensity as would be observed in the case where the gradient pulses are absent. If molecular migration takes place in the direction of the gradient (r), however, phase coherence is partially lost and will result in an attenuation of the echo signal. Measurement of the extent of the attenuation as a function of intensity and/or duration of the gradient pulse then provides direct information on the mean square displacements of the diffusing molecules in the zeolite.

410

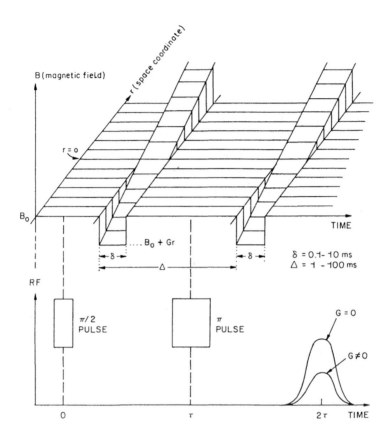

Fig. 11. Sequence of RF and field gradient pulses used in the NMRPFG method.

3.6.3. <u>Evaluation of the self-diffusivity D</u>. The self-diffusion coefficient D is related to the molecular mean square displacement $<r^2(t)>$ following the Einstein equation:

$$<r^2(t)> = 6Dt \quad \text{or} \quad <r^2(t=\Delta)> = 6D\Delta \qquad (3.12)$$

In a PFG experiment it can be shown (refs. 33,34) that the self-diffusivity is related to the ratio of the intensity of the spin echo signal in the presence, $A(G)$, and in the absence, $A(0)$, of a gradient field:

$$-\ln \frac{A(G)}{A(0)} = \gamma^2 G^2 \delta^2 \Delta D \qquad (3.13)$$

where γ denotes the gyromagnetic ratio and G is the field gradient. Usually, the gradient amplitude G is varied in a series of experiments, while δ and Δ are kept constant. Hence, the self-diffusivity D can be obtained by plotting

the left—hand side of eqn. (3.13) against G^2. In a typical PFG experiment, τ
and Δ are in the range 1—100 ms in order to satisfy the condition that
τ, $\Delta \ll T_2$. The typical length of the gradient field pulses is of the order of
0.1—10 ms while the magnitude of the gradient field G is normally of the order
of 1—10 T/m.

3.6.4. <u>Final comments</u>. Whether or not the NMRPFG method can be used
successfully depends on a number of conditions which are described below.

(i) The most decisive quantity limiting the range of applicability of the
PFG technique for slow diffusion processes is the transverse nuclear magnetic
(or spin—spin) relaxation time T_2. For molecules adsorbed on zeolites, T_2 is
typically of the order of 5—50 ms, which sets an upper limit on the experimental variables Δ and δ in order for the condition $\Delta \ll T_2$ to be satisfied. If
proton spins, having a relatively large gyromagnetic ratio and a high natural
abundance, are chosen to probe diffusion, it can be estimated that the lower
limit of intracrystalline self—diffusivity that can be measured in a PFG
experiment is of the order of $10^{-12} - 10^{-13}$ m^2.s^{-1} (Fig. 6).

(ii) The molecular mean square displacement $\langle r^2(t=\Delta)\rangle$ in eqn. (3.12) detected in a NMRPFG experiment must be smaller than the square of the average
zeolite crystal radius (r_c^2) in order to ensure that the measured self—diffusivities D are indicative of pure intracrystalline transport. Hence, the upper
limit of the intracrystalline self—diffusivity that can be measured in a
NMRPFG experiment is given by:

$$(D)_{\text{upper limit}} \cong r_c^2/6\Delta \tag{3.14}$$

Assuming the lower limit for Δ to be in the order of 1 ms (which can be easily
accomplished at low gradient fields) and an average zeolite crystal radius of
1 μm, the upper limit of D emerging from eqn. (3.14) is about 10^{-10} m^2/s. The
upper limit can be extended by using larger zeolite crystals (Fig. 6).

(iii) If intracrystalline diffusion is too slow to be measured directly
using the PFG method, a modification of this technique, known as the NMR Fast
Tracer Desorption (FTD) method, has been developed which makes it possible to
measure slower diffusion processes. Combination of the NMRPFG and NMRFTD
methods has been successfully used to provide direct evidence of a surface
barrier (ref. 35). Details of this method are reported elsewhere (refs. 32,
36—38).

4. NUMERICAL VALUES OF DIFFUSIVITIES

In the previous part, the emphasis has been upon experimental methods and
their limitations, and upon the interpretation of sorption kinetics. Consideration will now be given to the numerical value of diffusivities for sorbent/

sorbate systems. In general, the variables which influence intracrystalline diffusion include: zeolite structure type (pore geometry and size); zeolite composition (Si/Al ratio); type, charge and distribution of exchangeable cations; lattice defects in the sorbent; shape, size and polarity of the sorbate molecules; concentration of sorbate molecules; and temperature.

The past decade has seen an increased emphasis on the study of intracrystalline diffusivities and many reviews have been published (refs. 2,3,32,39, 40). Recent studies have primarily focused on intracrystalline transport of hydrocarbons in zeolites of type LTA (4A, 5A), FAU (X, Y) and MFI. One of the objectives of this part is to review the effects of sorbate size, zeolite structure and zeolite composition on diffusivity and these topics will be dealt with in Sections 4.4 and 4.5. It should be realized, however, that published information on diffusivities is fragmentary and often contradictory, particularly if different techniques are used to measure intracrystalline molecular transport. Section 4.3 is therefore devoted to a discussion on the consistency (or lack thereof) between results obtained with various methods. Many discrepancies between intracrystalline diffusivities reported are due to extracrystalline transport effects or due to intrusion of heat effects. Since variation of the size of the zeolite crystals is by far the most effective way to elucidate these discrepancies, the effect of crystal size on diffusivity will be discussed in more detail in Section 4.2. By way of introduction, Section 4.1 addresses questions related to the differences between transport, intrinsic and self-diffusivities, and the effect of sorbate concentration on diffusivity.

4.1. Effect of sorbate concentration on diffusivity

Many uptake studies have been devoted to measurement of the concentration dependence of D_0 and D_c (see, for example, refs. 2,3,25,41). The relation between D_c and D_0 is given by the Darken-type correction $D_c = D_0 (d \ln p)/(d \ln q)$ as discussed in Part 2. The correction factor follows from the sorption isotherm and equals unity if Henry's law is obeyed and exceeds unity beyond the region where Henry's law can be applied. Hence $D_c \geq D_0$. It has been generally observed that D_c increases with sorbate concentration and that in many cases the corrected diffusivity D_0 is essentially independent of sorbate concentration (Fig. 12). This observation supports the idea that D_0 is indeed an intrinsic diffusivity which is independent of concentration for many sorbate/sorbent systems.

The self-diffusivity D describes the rate of intracrystalline migration of absorbed molecules under equilibrium conditions. The relationship between self-diffusivity D and transport diffusivity D_c involves the straight and

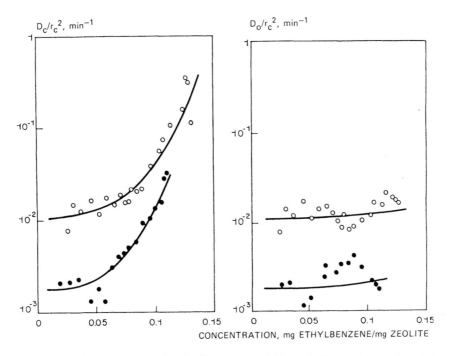

Fig. 12. Diffusion rates of ethylbenzene in 100 μm NaX crystals (o) and in 250 μm natural faujasite crystals (●) obtained via the gravimetric method. Effect of sorbate concentration. (Reproduced from ref. 41, with permission.)

cross coefficients of irreversible thermodynamics. If the cross coefficient is negligible, and this is a reasonable assumption at low sorbate concentrations, $D \cong D_o = D_c(d \ln q/d \ln p)$. Although D and D_o may refer to molecular displacement controlled by the same microdynamic processes, the difference in nature between D and D_o can be easily exemplified by their concentration dependence. While D_o is generally independent of sorbate concentration, self-diffusivities obtained from isotopically labelled tracers tend to decrease with increasing concentration (ref. 42). On the other hand, self-diffusivities measured by the NMRPFG method have been found to vary with concentration in an unpredictable way. The NMR diffusivity of benzene in Na-X crystals varies only slightly with concentration (ref. 43), while NMR self-diffusion of n-hexane decreases strongly with concentration (ref. 44; Fig. 13). NMR self-diffusivities of C_1-C_3 paraffins in Na-X and ZSM-5 decrease strongly with fractional occupation of sorbate molecules in the zeolite, while the reverse behaviour has been established in zeolite A (ref. 45; Fig. 14).

Hence, data published reveal little consistency as regard to the effect of sorbate concentration on D and D_o; and this observation raises questions as to whether simple Fickian diffusion behaviour can account for molecular transport in zeolites.

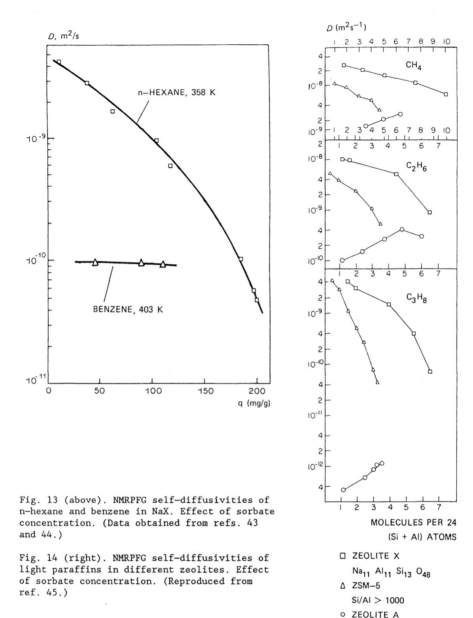

Fig. 13 (above). NMRPFG self-diffusivities of n-hexane and benzene in NaX. Effect of sorbate concentration. (Data obtained from refs. 43 and 44.)

Fig. 14 (right). NMRPFG self-diffusivities of light paraffins in different zeolites. Effect of sorbate concentration. (Reproduced from ref. 45.)

4.2. Variation of zeolite crystal size

A systematic study of the effect of zeolite crystal size on sorption kinetics and transport diffusivity is the preferred way to elucidate whether or not true intracrystalline migration is being measured. In many studies with different crystal size fractions, the uptake time constants (D_c/r_c^2) show the expected dependence on the square of the crystal radius and hence molecular transport is controlled only by intracrystalline diffusion. Examples of systems showing the above behaviour have been particularly well documented for LTA zeolite and include N_2 and light hydrocarbons in zeolite 4A (ref. 46; Fig. 15), CO_2 in zeolite 4A (ref. 47), n-butane in zeolite 5A (refs. 26,48),

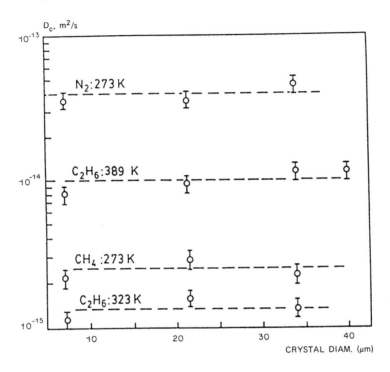

Fig. 15. Diffusivities obtained from gravimetric studies with different size fractions of zeolite 4A. (Reproduced from ref. 46, with permission.)

n-decane in zeolite 5A (ref. 49) and C_7-C_{16} linear paraffins in zeolite 5A (ref. 26). In addition, sorption studies in zeolite FAU (both X and Y type) have been reported where the emphasis was on crystal size effects. Diffusivities independent of crystal size have been reported for a variety of substances including C_8 aromatics (refs. 41,50,51), benzene (ref. 43), triethylamine (ref. 52) and C_4-C_{14} linear paraffins (ref. 26). Zeolite ZSM-5 (MFI), in particular, is very suitable for a systematic study of crystal size effects, since recipes have been developed allowing the synthesis of samples with crystal sizes from the sub-micron range to more than 100 μm. The diffusivity of 2,2-dimethylbutane in ZSM-5 was found to be independent of crystal size (ref. 53), as was confirmed by the straight-line relation between measured characteristic diffusion times and the square of the crystal radius (Fig. 16).

416

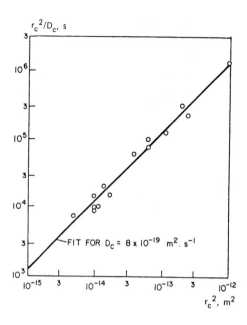

Fig. 16. Relation between the characteristic time for diffusion of
2,2-dimethylbutane in zeolite ZSM-5 and the square of the crystal radius
(gravimetric method, 373 K). (Reproduced from ref. 53, with permission.)

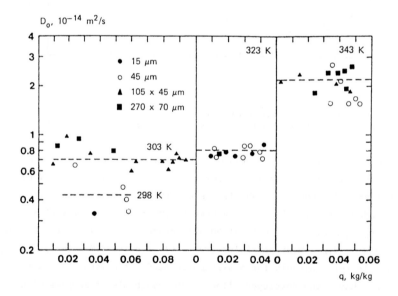

Fig. 17. Effect of crystal size, temperature and sorbate concentration on the
diffusivity of benzene in ZSM-5 zeolite. (Reproduced from ref. 54, by permission of the American Institute of Chemical Engineers.)

Similar conclusions were drawn for the diffusion of benzene (ref. 54; Fig. 17) and cyclohexane (ref. 55) in ZSM-5 samples of different crystal dimensions.

In contrast to the studies discussed above, examples exist of sorbate/sorbent systems where a decrease of zeolite crystal size led to a decrease in "apparent transport diffusivity". This behaviour is notably observed with rapid diffusion and suggests the presence of additional transport resistances (zeolite surface barrier, extracrystalline diffusion) or of effects related to non-isothermal sorption. A systematic variation of "diffusivity" with crystal size has been observed for ethane in zeolite 5A (ref. 56), methane in natural chabasite (ref. 56), n-hexane in MgA (ref. 19) and n-decane in MgA (ref. 57). These effects have been quantified in terms of a surface barrier transport mechanism prevailing in the zeolite crystals in a thin outer skin.

In many NMRPFG studies reported in literature, too, zeolite crystal size has been varied in order to check the validity of the diffusion results. Such studies include work on ethane in zeolite 5A (ref. 56; Fig. 18), methane in

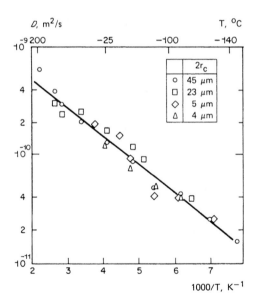

Fig. 18. Effect of crystal size and temperature on the NMRPFG self-diffusivity of ethane in NaCaA (5A) zeolite. (Reproduced from ref. 56, with permission.)

chabasite (ref. 56), benzene in zeolite NaX (ref. 43), C_6-C_{16} linear paraffins in NaX (ref. 44), C_6-C_8 aromatics in NaX (ref. 58) and water and methane in ZSM-5 (ref. 59). In the above studies the self-diffusivities were found to be independent of crystal size, and this clearly indicates the absence of any extracrystalline contribution.

4.3. <u>Comparison of diffusivities in zeolites measured by different techniques</u>

Diffusivities of sorbates in zeolite crystals can be measured using a variety of techniques. The extent of consistency of results obtained by different techniques has been a matter of debate during the past decades. Often, the discrepancies reported have trivial causes, such as: sorbate concentration effects (i.e., comparison of diffusivities at different sorbate loading); the use of "apparently identical" zeolites samples from different origins, which may show different intracrystalline diffusivities; and the presence of surface barriers, extracrystalline transport resistances and/or heat effects, which obscure determination of the true intracrystalline migration. Non—trivial origins may also exist, though. In particular, there is much dispute about the relation between corrected or intrinsic transport diffusivity obtained from transient studies, and the self—diffusivity obtained from tracer experiments or NMR—type studies.

4.3.1. <u>Transport diffusivities</u>. Transport diffusivities obtained via the gravimetric, the volumetric, the chromatographic, the ZLC, and the membrane technique relate to the same type of intracrystalline transport process and it is therefore to be expected that the diffusivities are independent of the particular technique used. Many studies have been devoted to a comparison of the gravimetric and the chromatographic technique, and the agreement is in general quite satisfactory, examples including light paraffins and cyclopropane in zeolite 5A (ref. 22), Ar and light paraffins in zeolite 4A (ref. 60), cyclopropane and cis—butene in zeolite 5A (ref. 61) and 2,2—dimethylbutane in ZSM—5 (ref. 53; Fig. 19). In addition, the agreement between results from the gravimetric and the ZLC technique has been well documented (refs. 25—27).

It is realized that the above examples have been selected in order to illustrate the agreement between the experimental results. In general terms, however, the lack of consistency between transport diffusivities obtained from various research teams using different methods is still quite obvious. As an example illustrating this problem, diffusion of benzene in ZSM—5 has been chosen, which is very well documented. Since 1981, some 13 papers have been published on this particular sorbate/sorbent system, the studies reported covering chromatographic (refs. 23,62), gravimetric (refs. 54,63—71) and volumetric (refs. 71,72) methods. Selected values of the intracrystalline transport diffusivity D_c or corrected transport diffusivity D_o as a function of temperature are plotted in Fig. 20, and it is obvious that the extent of consistency is poor. The results in the top right—hand corner of Fig. 20 may very well reflect the true intracrystalline diffusivity of benzene in ZSM—5, whereas the other data (obtained from five publications!) are presumably in error, due to extracrystalline contributions to the overall transport rate or due to heat effects. The inconsistencies observed in the above example are the rule rather than the exception in the pertinent literature on transport diffusivities. Hence, published data should be always considered with a substantial degree of scepticism.

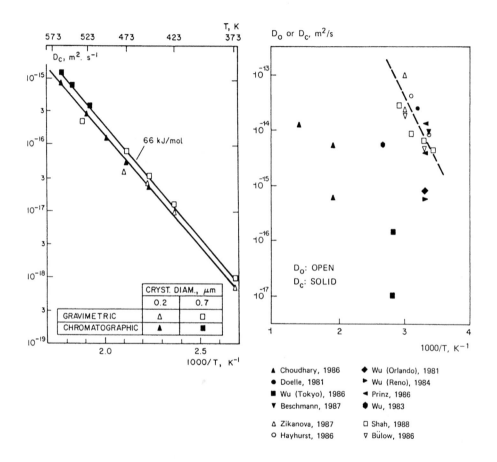

Fig. 19. Diffusivities of 2,2-dimethyl-butane in ZSM-5 (<0.01 %w Al) obtained with gravimetric and chromatographic measurements. (Reproduced from ref. 53, with permission.)

Fig. 20. Diffusivities of benzene in ZSM-5 zeolite. (Data obtained from refs. 54 and 62-72.)

4.3.2. <u>Self-diffusivities versus transport diffusivities</u>. In early studies serious discrepancies between NMR self-diffusivities and the corresponding intrinsic transport diffusivities D_0 derived from classical uptake experiments were reported (ref. 56). Data quoted in Table I of ref. 56 (Fig. 21) reveal differences of the order of 10^4–10^5. The use of very small (commercially

420

produced) zeolite crystals may have contributed to the discrepancies observed. On the one hand, the presence of extracrystalline transport resistances, heat effects and/or surface barriers in uptake experiments may have resulted in reported "intracrystalline diffusivities" that are orders of magnitude lower than the true intracrystalline diffusivity. On the other hand, the early results obtained from the NMRPFG technique may have yielded too high diffusivity values due to the presence of extracrystalline migration.

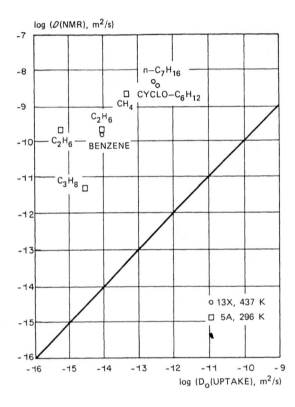

Fig. 21. Comparison of intrinsic transport diffusivities and NMRPFG self-diffusivities. (Data obtained from refs. in ref. 56)

In order to alleviate these discrepancies, much emphasis has therefore been placed on variation of crystal size in studies aimed at comparing transport diffusivities from uptake experiments and self-diffusivities from NMRPFG experiments. Diffusion of methane in natural chabasite (ref. 56) has become a classical example. The "apparent" diffusivity obtained from uptake experiments exhibits a strong dependence on zeolite crystal radius, whereas the NMR self-diffusivity was found to be independent of crystal radius (Fig. 22).

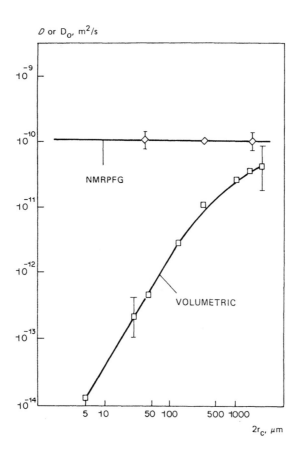

Fig. 22. Comparison of apparent transport diffusivities (volumetric method) and self-diffusivities (NMRPFG) of methane in natural chabasite at 273 K. Effect of crystal size. (Reproduced from ref. 56, with permission.)

Only for very large crystal sizes, where surface barrier effects (or extra-crystalline contributions) are of minor importance, do the results of the volumetric uptake method approximate those as predicted by the NMRPFG tech-nique. Similar behaviour to that found for methane in chabasite is observed for diffusion of ethane in zeolite 5A (ref. 56). The agreement between uptake and NMR results, however, is less apparent, even with large crystal sizes (Fig. 23). Studies on the diffusion of n-butane in zeolite 5A, on the other hand, yield consistent data (Fig. 24) if results using laboratory-grown large crystals are compared employing uptake, ZLC and NMRPFG studies (refs. 27,43).

The NMRPFG diffusivity of linear paraffins in the C_4–C_{16} range in NaX (ref. 44) is reported to be about two orders of magnitude faster than the cor-responding transport diffusivities obtained using the ZLC technique (ref. 26).

422

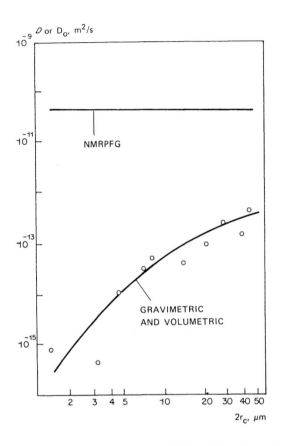

Fig. 23. Comparison of apparent transport diffusivities (uptake methods) and
self-diffusivities (NMRPFG method) of ethane in 5A zeolite at 193 K. Effect of
crystal size. (Adapted from ref. 56.)

Surprisingly, perfect agreement between volumetric and NMRPFG diffusivities
has been reported for neopentane in zeolite NaX (ref. 73) provided that large
crystals (120 μm diameter) are used in uptake experiments. For triethylamine
diffusion in zeolite NaX, the agreement between gravimetric and NMRPFG results
is also satisfactory (ref. 43).

Diffusion of aromatics in zeolite X and Y has been the subject of many
studies reported in the literature and the picture is rather confusing. NMRPFG
diffusivities of benzene in NaX (refs. 43,58) are almost two orders of magnitude larger than those measured by ZLC and gravimetric techniques (ref. 27).
Volumetric uptake studies yield values in between two extremes (ref. 74;
Fig. 25). Recent deuterium NMR relaxation measurements (ref. 75) appear to
support the results from proton NMRPFG studies. The disagreement between

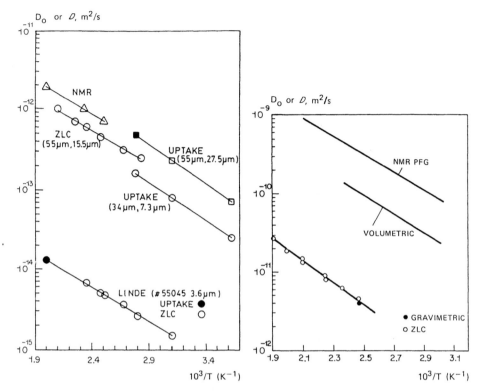

Fig. 24. Diffusivities of n–butane in
5A zeolite obtained by various techniques. (Reproduced from ref. 27,
with permission. Copyright, 1988,
American Chemical Society.)

Fig. 25. Diffusivities of benzene
in NaX zeolite obtained by various techniques. (Data obtained
from refs. 27, 43, 58 and 74.)

transport diffusivities (uptake and ZLC method) and self–diffusivities from
proton NMRPFG measurements for benzene diffusion in FAU–type zeolites has also
been established with o–xylene (ref. 27,58). On the other hand, self–
diffusivities of o–xylene in large NaX and natural faujasite single crystals,
studied by isotopically labelled tracer sorption (ref. 76), are in excellent
agreement with the corrected or limiting transport diffusivities obtained from
ZLC and uptake studies (ref. 27).

Self–diffusion studies in zeolite ZSM–5 and its high–silica analogue, silicalite, using the NMRPFG method have been reported for C_1–C_3 paraffins. NMRPFG
self–diffusivities for methane at 300 K are typically in the range 5×10^{-9} to
10^{-8} m^2/s (ref. 77). Transport diffusivities of methane at similar conditions
have been obtained both by the membrane technique (ref. 29) and by the chromatographic method (ref. 78) and yield values of ~10^{-10} m^2/s and ~10^{-14} m^2/s,

respectively. Such discrepancies have also been reported for diffusion of C_3H_8 in ZSM-5 using the chromatographic (ref. 78), membrane (ref. 29), volumetric (ref. 72), frequency response (ref. 72) and NMRPFG techniques (refs. 77, 79,80).

The discussions in this section reveal clearly that the lack of consistency between transport diffusivities and self-diffusivities (particularly those obtained via NMRPFG methods) of sorbate molecules in zeolites continues to be a matter of serious debate. In the late 1970s and early 1980s surface barriers and external mass transport limitations were shown to be mainly responsible for the erroneous results. However, during the past few years, evidence has accumulated that for a number of sorbate/zeolite systems differences in magnitude of the self-diffusivity and the intrinsic transport diffusivity still persist, the reasons for which have not yet been identified. An excellent review paper dealing with this matter has been published recently (ref. 80a).

4.4. Effect of type of sorbate on diffusivity

The previous sections (4.1-4.3) have enlarged on a number of specific problems related to the measurement and elucidation of intracrystalline diffusivities. In this section, an attempt is made to review the effects of molecular size and structure on diffusivity in a number of zeolite structures.

4.4.1. n-Paraffins: effect of chain length. The effect of chain length of a series of n-paraffins on intracrystalline diffusivity in zeolite T (an intergrowth of offretite and erionite) was well documented in early reports. The transport diffusivity of C_2-C_4 paraffins was found to decrease substantially with increasing carbon number in zeolite H-T (ref. 81). On the other hand, unexpected diffusivity behaviour of C_2-C_{14} n-alkanes was reported in a potassium form of zeolite T (ref. 82; Fig. 26). The diffusivity of C_4-C_8 n-paraffins decreases steadily with chain length but with larger molecules from C_9 up to C_{12} mobility increases and reaches values close to that of diffusivities of C_2 and C_3 paraffins. Beyond C_{12}, diffusivities decrease again with carbon number. This behaviour was rationalized in terms of a "window effect". The length of n-C_8H_{18} molecules corresponds almost exactly to the free length of the erionite cage defined by the aluminosilicate framework, and the entrapment of n-octane in the cage leads to the low mobility for this "just-fitting" molecule. Molecules that are either larger or smaller than n-C_8 do not fit in the erionite cage and will consequently have a higher mobility. Beyond n-C_{12}, however, chain length effects are apparently more important than window effects and lead to a monotonous decrease of diffusivity, which is in line with expectations.

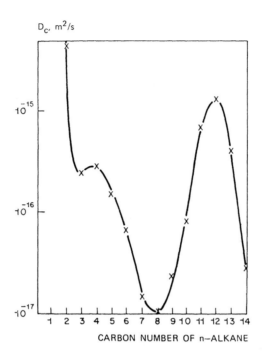

Fig. 26. Diffusivities of n—paraffins in zeolite T as a function of carbon number (gravimetric method, 573 K). (Reproduced from ref. 82, with permission.)

It should be mentioned, however, that the data referred to above may not reflect the true intracrystalline diffusivities of hydrocarbons in zeolite K—T as the samples used for ref. 82 are much too large. Moreover, the obvious checks summarized in Section 3.1.3 have not been carried out. Hence, the diffusivities reported are likely to be in error due to heat effects and extracrystalline mass transport resistances. It is therefore still very questionable whether a "window effect" really exists.

Diffusion of n—paraffins in zeolite LTA (CaA or 5A) has been studied extensively and the picture emerging is quite consistent. The data plotted in Fig. 27 are all given at 473 K and were obtained either directly from published data (refs. 14,26,37,43,48,56) or after extrapolation using quoted activation energies (refs. 80,83,84). NMR self-diffusivities of C_1-C_4 and corrected transport diffusivities of C_4-C_{16} n—paraffins in large lab—grown single crystals decrease steadily with increasing carbon number. In a commer-cial Linde 5A sample, transport diffusivities are much lower, but again a decrease with increasing chain length in the range C_1-C_{10} is observed which can be represented by a straight—line relationship in a plot of log D vs the

426

log of the carbon number. No evidence for window effects in LTA zeolite is provided by the above studies.

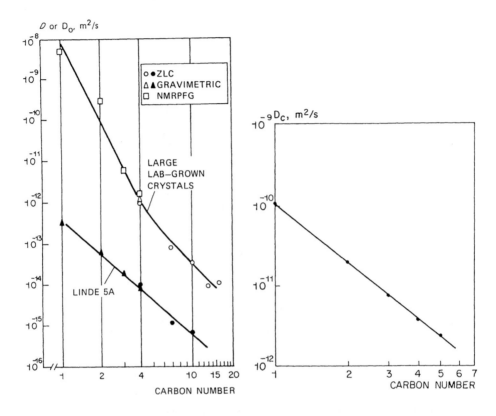

Fig. 27. Diffusivities of n-paraffins in 5A zeolite as a function of carbon number (various methods, 473 K). (Data from refs. 14,26,37,43,48,56, 80,83 and 84.)

Fig. 28. Diffusivities of n-paraffins in ZSM-5 zeolite as a function of carbon number (membrane technique, 334 K). (Reproduced from ref. 29, with permission.)

An investigation of the effect of chain length of n-paraffins in zeolite MFI has been conducted using the membrane technique (ref. 29). Transport diffusivities of C_1-C_5 n-alkanes were measured at 334 K and again diffusivity was found to decrease in a systematic manner with increasing chain length (Fig. 28).

Diffusion of a series of n-paraffins in zeolite FAU (type NaX) has been studied using both the NMRPFG (refs. 44,85) and the ZLC method (ref. 26). Although the two techniques do not yield identical results for n-alkanes of the same chain length, the data in Fig. 29 show similar diffusivity behaviour

with carbon number, irrespective of the technique used. The systematic decrease in diffusivity with chain length for a series of C_4–C_{16} n-alkanes rules
out possible window effects.

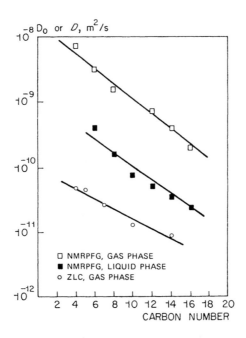

Fig. 29. Diffusivities of n-paraffins in NaX zeolite as a function of carbon
number (various methods, 423 K). (Data from refs. 26,44 and 85.)

4.4.2. Paraffins: effect of branching. Diffusion of normal butane and
isobutane in zeolite ZSM-5 has been studied using the membrane technique
(ref. 28). As expected, the diffusivity of n-butane at 334 K was found to be
larger than that of the isobutane but the differences, although significant,
are not substantial. With C_6 isomers, on the other hand, the effect of
branching is much more pronounced. The linear C_6 isomer diffuses quite rapidly
through the ZSM-5 channel system: diffusivities of the order of 10^{-13} m^2/s at
323 K are reported (ref. 86). For the mono-branched 3-methylpentane isomer,
diffusivities ranging from 10^{-15} to 10^{-16} m^2/s at 296–348 K have been found
(refs. 68,87), while intracrystalline transport of the gem-branched 2,2-
dimethylbutane through the inner space of ZSM-5 at 373 K is characterized by a
diffusivity of about 10^{-18} m^2/s (ref. 53).

4.4.3. Naphthenes. Diffusivities of cyclohexane and monosubstituted
n-alkylcyclohexanes ranging from methyl- to pentylcyclohexanes have been

428

studied in zeolite ZSM-5 (ref. 55). Sorption rates were observed to vary with chain length in an unpredictable way. The lowest diffusivities were observed with ethyl- and sec-butylcyclohexane and diffusivities increased steadily as the length of the substituent chain either increased or decreased (Fig. 30), and this effect was interpreted in terms of a window or cage effect, similarly to diffusion of n-alkanes in zeolite T (ref. 82; see Section 4.4.1.). Another unexpected result was the observation that the diffusivity of t-1,4-dimethyl-cyclohexane was two orders of magnitude larger than that of ethylcyclohexane, which has about the same molecular length and kinetic diameter as the dimethyl isomer (Fig. 30).

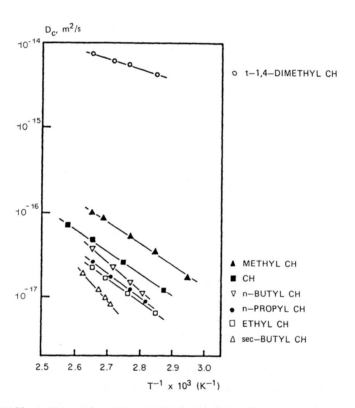

Fig. 30. Diffusivities of various (alkyl-substituted) cyclohexanes in ZSM-5 zeolite (volumetric method). (Reproduced from ref. 55, with permission.)

4.4.4. <u>Aromatics</u>. Diffusion of various alkyl-substituted benzene (and naphthalene) compounds in zeolites has been the subject of many studies. The first attempt to quantify the relationship between critical molecular diameter and diffusivity of various aromatics in zeolite NaY was published by Moore and Katzer (ref. 88), who studied liquid-phase counterdiffusion into cyclohexane-saturated NaY at 298 K (Fig. 31). The linear relationship observed between the logarithm of the effective diffusion coefficient and the critical molecular diameter of the sorbate emphasizes that one single parameter, i.e., the critical molecular diameter, correlates size effects very well. On the other hand, no effect of critical molecular diameter was observed in gas phase uptake studies with different C_8 aromatics (ref. 41). Diffusivities of the xylene isomers and ethylbenzene in large single crystals of NaX are all of the same order of magnitude. Obviously, subtle differences in molecular shape and size are more pronounced in counterdiffusion experiments than in simple uptake studies.

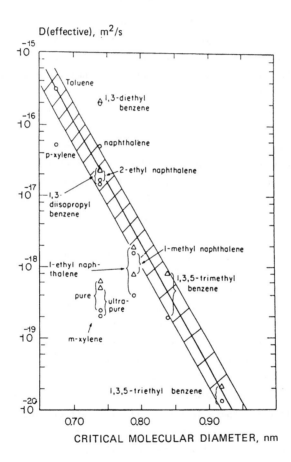

Fig. 31. Diffusivities of various aromatics in cyclohexane-saturated NaY zeolites at 298 K. (Reproduced from ref. 88, with permission.)

In zeolite ZSM-5, having a smaller pore diameter than zeolite Y or X, the diffusivities of the various xylene isomers depend strongly on their kinetic diameters. The diffusivity of o-xylene is three orders of magnitude smaller than that of p-xylene (ref. 89), while the diffusivity of the more bulky 1,3,5-trimethylbenzene is two orders of magnitude lower than that of o-xylene. The strong variation of the diffusivity of alkylaromatics with kinetic diameter is closely related to the shape-selective properties of zeolite ZSM-5 in a number of catalytic applications, and this topic is discussed in detail in Part 5.

4.5. Effect of zeolite on diffusivity

4.5.1. <u>Effect of zeolite sample source</u>. Differences in the diffusivities of a given sorbate in various zeolite samples of the same type and chemical composition have been frequently reported. Examples include CO_2 in 4A zeolite (ref. 47) and C_1-C_{10} n-paraffins in 5A zeolite (refs. 26,43,48,83; Figs. 24 and 27). In general, diffusivities in fine-crystalline commercial Linde type samples are smaller than for large laboratory-synthesized crystals. Although these differences may be partly due to surface barrier and non-isothermal sorption effects, there is also evidence that a non-ideal cation distribution in the fine-crystalline samples causes blocking of a certain fraction of windows and leads to lower apparent diffusivities than in large crystals synthesized in the laboratory under ideal conditions (ref. 2). Further support to this explanation is provided by the fact that these differences for samples from different origins are exclusively reported for Al-rich zeolites (such as zeolite LTA) and have not (yet) been found for Si-rich zeolites such as ZSM-5.

4.5.2. <u>Effect of exchangeable cations</u>. The nature and charge of the exchangeable cation may have a distinct effect on the effective pore aperture of zeolites, and, as a consequence, on the mobility of sorbed species in zeolite frameworks. The classical example is zeolite LTA where, for example, replacement of exchangeable Na^+ cations by Ca^{2+} cations leads to an effective increase in the aperture of the sieve window from about 4 to some 5 Å. This effect is illustrated in Fig. 32, which shows the variation of the corrected diffusivity of n-butane with the degree of Ca^{2+} exchange in NaCaA (ref. 90). While in the above example cation effects can be explained in terms of a change in the effective "pore diameter", chemical effects induced by the exchangeable cations have also been reported, a prime example being the diffusion of lower olefins and benzene in zeolite X, which decreases markedly upon introduction of Ag^+ in a NaX sieve (refs. 75,91). Other examples where the type of cation exerts an effect on diffusivity include benzene in NaX and BaX (ref. 45) and n-octane and n-octene in NaX, KX and BaX (ref. 92).

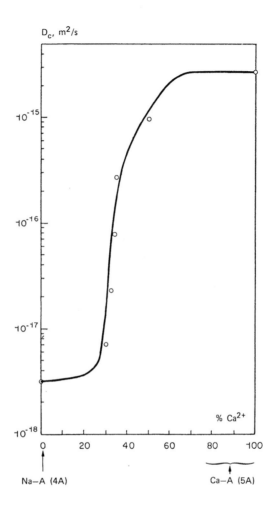

Fig. 32. Effect of the degree of cation exchange in NaCaA zeolite on the diffusivity of n–butane (gravimetric method, 364 K). (Reproduced from ref. 90, with permission.)

4.5.3. <u>Effect of silica to alumina ratio</u>. Studies on the influence of the framework composition of a particular zeolite structure on the diffusivity of a given sorbate have been primarily limited to zeolite ZSM-5. The Si/Al ratio of this zeolite can be easily varied in the range from 10 to >1000. The diffusivity of 2,2–dimethylbutane in H–ZSM-5 (ref. 53) seems to be only slightly dependent on the chemical composition of the zeolite material, as is suggested in Fig. 33, higher Al contents leading to a somewhat lower diffusivity. Similar trends have been reported for sorption of benzene in the H^+ and Na^+ forms of ZSM-5 as a function of Al content (refs. 62,71) but the influence of

432

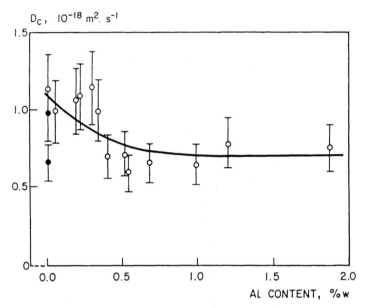

Fig. 33. Effect of the Si/Al ratio of the ZSM–5 zeolite on the diffusivity of 2,2–dimethylbutane (gravimetric method, 373 K). (Reproduced from ref. 53, with permission.)

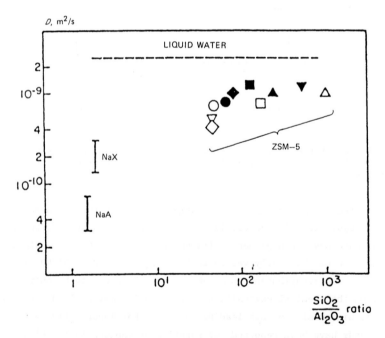

Fig. 34. Effect of the Si/Al ratio of various zeolites on the NMRPFG self–diffusivity of H_2O at 296 K. (Reproduced from ref. 59, with permission.)

zeolite composition on benzene diffusivity appears to be more pronounced than for 2,2–dimethylbutane. On the other hand, data are available which strongly suggest that benzene diffusivities in the MFI structure are virtually independent of framework composition (ref. 54).

Self–diffusivities of water in ZSM–5 measured by the NMRPFG technique tend to fall with decreasing Si/Al ratio (ref. 59) and this observation is in line with expectations since it can be easily envisaged that an increase in either Al content or hydrophilicity will have a retarding effect on the mobility of the water molecule. Interestingly, the trend observed with MFI samples of different Si/Al ratio is in line with reported NMRPFG self–diffusivities of water in the Al–rich NaX and NaA zeolites (ref. 59; Fig. 34). Apparently, zeolite framework composition (Si/Al ratio) instead of its structure determines the magnitude of the self–diffusivity of water.

4.5.4. <u>Effect of zeolite structure</u>. Although the literature on intracrystalline diffusion of sorbates in zeolites is quite extensive, there are virtually no reports on a systematic study of diffusivities of a given sorbate

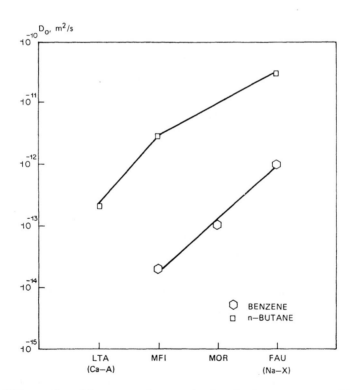

Fig. 35. Effect of zeolite pore size on the intrinsic transport diffusivity of n–butane and benzene. (Data obtained from refs. 25,26,29,54 and 95.)

in a series of zeolites of different structure types. Moore and Katzer
(ref. 88) have suggested that diffusivities in different zeolite structures
ought to be correlated in terms of the difference between effective pore
diameter and critical molecular diameter, but conclusive evidence is not
provided. Pfeifer and coworkers (ref. 45) have reported that, in line with
expectations, self-diffusivities of light alkanes at low sorbate concentrations decrease with decreasing pore size of the zeolite, i.e., in the order
NaX > ZSM-5 > NaCaA (Fig. 14). At higher sorbate concentrations, however, the
order tends to change: NaX > NaCaA > ZSM-5. Diffusivities of water in various
zeolite structures (see Section 4.5.3.) are found to correlate primarily with
zeolite composition rather than zeolite structure. It has also been suggested
that, in contrast to the qualitative "classical" reasoning, diffusion rates
should increase as the molecular and pore sizes match each other more intimately (refs. 93,94). However, as is shown in Fig. 35, comparison of literature data on transport diffusivities of benzene (refs. 25,54,95) and n-butane
(refs. 26,29) in different structural types of zeolite indicates that diffusivity decreases with pore diameter in a straightforward way.

4.6. Liquid-phase diffusivities

The discussion in the previous parts on measurement of intracrystalline
diffusivities in zeolite adsorbents and catalysts has been mainly devoted
to sorption studies carried out from the vapour phase. However, many
commercially important adsorption/separation processes, such as the UOP
Sorbex process (refs. 2,96), are operated in the liquid phase. Only a few
studies of sorption from the liquid phase using conventional uptake methods
have been reported (refs. 88,97). In recent years there has been growing
emphasis on liquid-phase sorption, with an increasing number of studies using
liquid chromatography (LC) (refs. 98,99). In contrast to gas chromatography,
LC always involves counterdiffusion of sorbate and solvent molecules within
the zeolite micropores and LC may therefore be a more suitable technique for
studying molecular transport under conditions that are more representative of
a catalytic reaction. Surprisingly, intracrystalline diffusivities of
aromatics in NaX zeolite obtained under liquid-phase conditions with saturated
solvents are of the same order as the intrinsic (corrected) diffusivities
extrapolated from vapour-phase measurements at higher temperatures (ref. 98).

5. DIFFUSION AND CATALYSIS

5.1. Introduction

In catalytic processes using zeolites, the limited intracrystalline molecular transport rate may have a significant effect on catalyst performance. If

diffusion effects of products or reactants are involved in a specific reaction scheme, variation of the zeolite crystal size will lead to changes in either observed reaction rates or product selectivities (or both). The pertinent literature provides many examples of hydrocarbon conversion reactions where intracrystalline molecular transport has a significant effect on <u>reaction rates</u>. Cracking rates of gem–branched dimethyl alkanes on zeolite ZSM–5 have been found to decrease with increasing crystal size (refs. 53,100), and simi-lar effects have been reported for cracking of 3–methylpentene and n–hexene (ref. 100) and n–hexane (refs. 101,102) in differently sized crystal fractions of zeolite ZSM–5. The effect of zeolite crystal size on <u>selectivity</u> is probab-ly best exemplified by the generation of para–xylene from either dispropor-tionation of toluene or alkylation of toluene with methanol over zeolite ZSM–5. The desired para–xylene product diffuses in the channel system a few orders of magnitude faster than the undesired ortho– and meta–xylenes (ref. 103) and, consequently, the selectivity to para–xylene increases with zeolite crystal size. Olson and Haag (ref. 104) have demonstrated that, for a series of ZSM–5 based catalysts, a unique relation exists between selectivity to para–xylene (in toluene disproportionation) and diffusion time (which is proportional to r_c^2/D_c) for ortho–xylene, the antiselective species, determined independently by sorption measurement at lower temperatures.

5.2. Quantification of diffusion limitations in zeolite catalysis

The extent to which diffusional transport limits the rate of conversion of a single reactant can be quantified using the well–known Thiele relation-ship (refs. 4,5,105), giving the effectiveness factor η of a catalyst as a function of the dimensionless Thiele modulus Φ. The effectiveness factor is defined as the ratio between observed and intrinsic rate constants, $\eta = k/k^*$. If the reaction is first order in gas phase concentration, the observed rate constant (based on zeolite crystal volume, including the micropores) generally follows from the familiar expression for an integral tube reactor:

$$k = -\ln(1-\xi) \frac{\zeta_z * WHSV * RT}{3.6 * p_o * M} \tag{5.1}$$

The Thiele modulus for a spherical isotropic zeolite crystallite of radius r_c is given by $\Phi = r_c(k^*/D_cK_c)^{\frac{1}{2}}$. The reaction rate constant is based on gas phase concentrations, whereas the zeolite diffusivity D_c is defined on the basis of intracrystalline sorbate concentrations. In order to maintain consistency, the adsorption equilibrium constant K_c defined in Part 3 is therefore included in the expression of the Thiele modulus. The relation between η and Φ depends on reaction order, type of zeolite and shape of the zeolite crystals. For a reaction following first-order kinetics in isotropic spherical zeolite crystals, the effectiveness factor function is:

$$\eta = \frac{3}{\Phi} \left[\frac{1}{\tanh \Phi} - \frac{1}{\Phi} \right] \qquad (5.2)$$

The above equations apply to diffusion limitations of reactant molecules and can therefore be used to quantify reactant selectivity in shape-selective zeolite catalysis. In the case of product selectivity, a somewhat altered approach has been developed (ref. 103).

5.3. Determination of diffusivities from catalytic measurements

As has been outlined in the previous section, the effectiveness factors η measured in catalytic experiments using a series of zeolite samples with different crystal size are uniquely related to the proper steady-state diffusivity applicable during catalysis. Determination of the effectiveness factor η yields the quantity Φ. If the intrinsic reaction rate constant k^*, zeolite crystal size and the equilibrium constant K_c are known, the diffusivity then follows from the value of Φ. Such an approach has been followed by Haag et al.

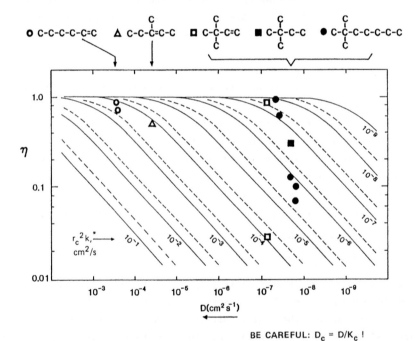

Fig. 36. Relation between the effectiveness factor and diffusivity. Graphical determination of diffusivities of C_6 and C_9 alkanes and alkenes in ZSM-5 zeolite from catalytic studies at 811 K. (Reproduced from ref. 100, with permission.)

(ref. 100), who studied cracking rates of C_6, C_8 and C_9 paraffin and olefin isomers over zeolite ZSM-5 as a function of crystal size.

To evaluate the actual diffusivities, the relation between η and Φ was recalculated to give a relation between η and diffusivity with different values of $r_c^2 k^*$ as parametric curves (Fig. 36). Sets of diffusivities of a very satisfactory consistency for each species were obtained and it was found that the diffusivities of hydrocarbons in zeolite ZSM-5 decrease by orders of magnitude as the degree of structural branching increases. Since the equilibrium constant K_c was omitted in the Thiele modulus used in ref. 100, the diffusivities reported in Fig. 36 are based on gas phase concentration and should be regarded as $K_c D_c$, i.e., the product of the equilibrium constant and the conventional intracrystalline transport diffusivity based on intracrystalline sorbate concentration.

5.4. Relation between diffusivities obtained independently from sorption and catalytic measurements

Most attempts to determine transport and self-diffusivities in zeolite crystals are based on a number of physical techniques which have been discussed extensively in Part 3. In the previous section, an alternative method has been discussed allowing elucidation of diffusivities from catalytic studies. The relation between the "physical" and "catalytic" diffusivities, however, is subject to many discussions since it is not clear to what extent intracrystalline molecular motion prevailing during uptake, chromatographic and PFGNMR studies is representative of counterdiffusion under actual catalytic conditions. Moreover, sorption measurements under catalytically relevant conditions are doomed to fail because the catalytic activity of the zeolite will convert the sorbate.

The relation between diffusion and catalytic reaction of 2,2-dimethylbutane (2,2-DMB) in ZSM-5 zeolite has been studied to bridge the gap between "physical" and "catalytic" diffusivities (ref. 53). Since the intracrystalline mobility of 2,2-DMB determined at lower temperatures (373 K) using conventional uptake studies did not vary significantly with the Si/Al ratio of the zeolite (Fig. 33), ultrapure all-silica MFI samples showing no catalytic activity were used to determine diffusivities at higher temperatures, close to normal catalytic conditions, using the chromatographic technique (Fig. 19) and the results obtained were regarded as being representative of Al-containing MFI zeolite. Simultaneously, the equilibrium constant K_c was determined for a wide range of temperatures.

In the same study, eqn. 5.1 was used to determine rates of 2,2-DMB and n-hexane cracking over a series of H-ZSM-5 catalysts of different Al content and different crystal size. While n—hexane cracking rates were proportional to Al content and independent of crystal size, 2,2—DMB cracking rates in the coarse—crystalline samples were lower than anticipated on the basis of their chemical composition due to diffusion limitations. The Thiele modulus was estimated by extrapolation of the diffusion time constants (D_c/r_c^2) measured at low temperature to actual reaction temperature, using the activation energy of the diffusivity of 2,2—DMB in all—silica MFI materials established with the chromatographic sorption studies. The adsorption constant featuring the Thiele modulus was also obtained by extrapolation of the chromatographic data available for all—silica MFI.

Figure 37 shows measured effectiveness factors η for 2,2—DMB conversion over ZSM—5 catalysts as a function of estimated Thiele moduli based on physical sorption studies. The coincidence of the experimentally determined data with the theoretical relation, eqn. (5.2), indicates that, for the system 2,2—DMB/ZSM—5, diffusivities obtained from physical sorption and catalytic studies independently show excellent agreement. It is not clear yet to what extent the results obtained with the 2,2—DMB/ZSM—5 system can be extrapolated to other sorbate/zeolite combinations.

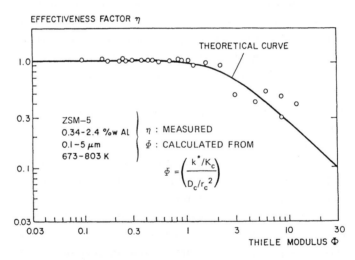

Fig. 37. Experimental verification of the Thiele concept for zeolite catalysis: conversion of 2,2—dimethylbutane over ZSM—5 zeolite. (Reproduced from ref. 53, with permission.)

6. CONCLUSIONS

The past decade has seen an increased emphasis on the study of intra—crystalline diffusivities of various sorbents in zeolite crystals. This interest is primarily fuelled by the recognition that a thorough knowledge of intracrystalline diffusion phenomena leads to a better understanding of established applications of zeolites in (shape-selective) catalysis and in selective adsorption/separation processes, and may provide the information required for (i) further optimization of existing commercial applications of zeolites and (ii) development of new applications in the above areas.

Much progress has been achieved in our understanding of single-component intracrystalline diffusion by the application of a variety of physical techniques and by careful study of the effect of zeolite crystal size. Many of the earlier discrepancies between the results obtained by different techniques have now been solved and can be traced back to the involvement of extra-crystalline transport resistances, surface barriers at the outer rim of the zeolite crystal and the involvement of heat effects. Nevertheless, for a number of sorbate/zeolite systems, substantial differences persist if transport diffusivities obtained from classical uptake studies and self-diffusivities from the NMR pulse field gradient method are compared.

While real progress has been achieved in single-component diffusion, most technologically important catalytic and adsorption processes involve multi-component systems. This area, which as yet is receiving hardly any attention, should be an important field for future research since it is counterdiffusion rather than single-component diffusion that will eventually determine zeolite performance in many commercial applications. Nevertheless, evidence has been provided that diffusivities from single-component sorption experiments can be correlated with the catalytic performance of zeolites and are useful to predict conversions and selectivities attainable.

ACKNOWLEDGEMENT
The author would like to thank Prof. Dr. Douglas M. Ruthven and Prof. Dr. Jörg Kärger for critically reading the manuscript and for providing valuable comments.

LIST OF SYMBOLS

a	$mol.kg^{-1}$ zeolite	Intracrystalline sorbate concentration
A	arb. units	Intensity of spin echo signal
B_c	$mol.Pa^{-1}.m^{-1}.s^{-1}$	Intracrystalline mobility
B, B_o	$V.s.m^{-2}$ or T(esla)	Magnetic field (1 T = 10^4 G(auss))
c	$mol.m^{-3}$	Sorbate concentration outside the zeolite crystal
c_o, c_∞	$mol.m^{-3}$	Initial and final values of c
D_b	$m^2.s^{-1}$	Axial dispersion coefficient in the bed
D_c	$m^2.s^{-1}$	Intracrystalline diffusivity
D_k	$m^2.s^{-1}$	Knudsen diffusivity
D_o	$m^2.s^{-1}$	Corrected or intrinsic diffusivity
D_p	$m^2.s^{-1}$	Effective (macro)pore diffusivity
$\mathcal{D}$	$m^2.s^{-1}$	Self-diffusivity

G	$V.s.m^{-3}$ or $T.m^{-1}$	Magnetic field gradient ($1\ T.m^{-1} = 10^2\ G.cm^{-1}$)
J	$mol.m^{-2}.s^{-1}$	Molar flux
k	s^{-1}	Observed first-order rate constant based on zeolite crystal volume (related to gas phase concentration)
k^*	s^{-1}	Intrinsic first-order rate constant based on zeolite crystal volume (related to gas phase concentration)
k_f	$m.s^{-1}$	External film mass transfer coefficient
k_{sb}	$m.s^{-1}$	Zeolite surface barrier mass transfer coefficient
K_c	–	Dimensionless adsorption constant based on zeolite crystal
K_p	–	Dimensionless adsorption constant based on pellet
L	m	Length of chromatographic column
L	m	Thickness of a zeolite crystal or zeolite membrane
M	$kg.kmol^{-1}$	Molecular weight
p	Pa	Sorbate pressure
p_o	Pa	Reactant pressure at reactor inlet
P_n	–	Defined in eqn. (3.8)
q	$mol.m^{-3}$ zeolite	Intracrystalline sorbate concentration
q_o, q_∞	$mol.m^{-3}$ zeolite	Initial and final values of q
r	m	Radial coordinate of zeolite crystal
r_c	m	Radius of zeolite crystal
r	m	Space coordinate in the direction of the magnetic field gradient
R	$m^3.Pa.K^{-1}.mol^{-1}$	Gas constant (in eqn. 2.3)
R	m	Radial coordinate of pellet
R_p	m	Radius of pellet
t	s	Time
T	K	Temperature
v_i	$m.s^{-1}$	Interstitial gas velocity
v_s	$m.s^{-1}$	Superficial gas velocity
V_g	m^3	Gas volume of sorbate buffer
V_s	m^3	Crystal volume of zeolite sorbent
w	–	Volume fraction of zeolite in pellet
$WHSV$	$kg.kg^{-1}.h^{-1}$	Weight Hourly Space Velocity of reactant
x	m	Coordinate

Greek symbols

α	–	Defined in eqn. (3.7)
γ	–	Fractional approach to equilibrium
γ	$rad.T^{-1}.s^{-1}$	Gyromagnetic ratio
δ	s	Duration of field gradient pulse
Δ	s	Time interval between the gradient pulses
ϵ_b	–	Bed porosity
ϵ_p	–	Pellet porosity
ς_z	$kg.m^{-3}$	Crystal density of zeolite
η	–	Effectiveness factor
μ	$m^3.Pa.mol^{-1}$	Chemical potential
μ	s	Mean retention time
μ_p	m	Mean pore radius
ξ	–	Conversion
σ	s	Standard variation of response peak
τ	s	Time interval between $\pi/2$ and π RF pulses
Φ	–	Thiele modulus

REFERENCES

1   P.B. Weisz, Chemtech, 3 (1973) 498–505.
2   D.M. Ruthven, Principles of Adsorption and Adsorption Processes,
    John Wiley and Sons, New York, 1984.
3   R.M. Barrer, Zeolites and Clay Minerals as Sorbents or Molecular Sieves,
    Academic Press, London, 1978, Chapter 6.
4   P.B. Weisz and C.D. Prater, Adv. Catal., 6 (1954) 143–196.
5   C.N. Satterfield and T.K. Sherwood, The Role of Diffusion in Catalysis,
    Addison/Wesley, Reading, UK, 1963.
6   P.B. Weisz and V.J. Frilette, J. Phys. Chem., 64 (1960) 382.
7   N.Y. Chen and W.E. Garwood, Catal. Rev. – Sci. Eng., 28 (1986) 185–264.
8   I.E. Maxwell, J. Inclusion Phenom., 4 (1986) 1–29.
9   E.G. Derouane, in Intercalation Chemistry, Academic Press, New York,
    1982, pp. 101–146.
10 L.S. Darken, Trans. AIME, 175 (1948) 184.
11 R. Ash and R.M. Barrer, Surf. Sci., 8 (1967) 461.
12 J. Kärger, Surf. Sci., 36 (1973) 797.
13 J. Cranck, Mathematics of Diffusion, Clarendon Press, Oxford, 1956.
14 J. Kärger, M. Bülow, G.R. Millward and J.M. Thomas, Zeolites, 6 (1986)
    146–150.
15 D.M. Ruthven and K.F. Loughlin, Chem. Eng. Sci., 26 (1971) 577.
16 L–K. Lee and D.M. Ruthven, J. Chem. Soc., Faraday Trans. 1, 75 (1979)
    2406.
17 D.M. Ruthven, L–K. Lee and H. Yucel, AIChE J., 26 (1980) 16–23.
18 D.M. Ruthven and L–K. Lee, AIChE J., 27 (1981) 654.
19 M. Bülow, P. Struve, G. Finger, C. Redszus, K. Ehrhardt, W. Schirmer and
    J. Kärger, J. Chem. Soc., Faraday Trans. 1, 76 (1980) 597–615.
20 H.W. Haynes and P.N. Sarma, AIChE J., 19 (1973) 1043–1046.
21 P.N. Sarma and H.W. Haynes, Adv. Chem. Ser., 74, vol. 133 (1974) 205–217.
22 D.B. Shah and D.M. Ruthven, AIChE J., 23 (1977) 804–809.
23 L. Forni and C.F. Viscardi, J. Catal., 97 (1986) 480–492.
24 H.W. Habgood and W.R. MacDonald, Anal. Chem., 42 (1970) 543.
25 M. Eic, M. Goddard and D.M. Ruthven, Zeolites, 8 (1988) 327–331.
26 M. Eic and D.M. Ruthven, Zeolites, 8 (1988) 472–479.
27 D.M. Ruthven and M. Eic, in W.H. Flank and T.E. Whyte, Jr. (Eds.),
    Perspectives in Molecular Sieve Science, ACS Symposium Series,
    Washington, DC, 368, 1988, pp. 362–375.
28 A.R. Paravar and D.T. Hayhurst, in D. Olson and A. Bisio (Eds.),
    Proceedings of the 6th International Zeolite Conference, Reno, Nevada,
    USA, 1983, Butterworths, Guildford, Surrey, UK, 1984, pp. 217–224.
29 D.T. Hayhurst and A.R. Paravar, Zeolites, 8 (1988) 27–29.
30 A. Quig and L.V.C. Rees, J. Chem. Soc., Faraday Trans. 1, 72 (1976) 771.
31 E.O. Stejskal and J.E. Tanner, J. Chem. Phys., 42 (1965) 288 and 49 (1968)
    1768.
32 J. Kärger and H. Pfeifer, Zeolites, 7 (1987) 90–107.
33 J. Kärger and P. Volkmer, Z. Phys. Chem. (Leipzig), 5 (1980) 900–920.
34 P. Stilbs, Prog. Nucl. Magn. Reson. Spectrosc., 19 (1987) 1–45.
35 R. Richter, R. Seidel, J. Kärger, W. Heink, H. Pfeifer, H. Fürtig, W. Höse
    and W. Roscher, Z. Phys. Chem. (Leipzig), 267 (1986) 1145–1151.
36 J. Kärger, AIChE J., 28 (1982) 417.
37 J. Kärger, H. Pfeifer, M. Rauscher, M. Bülow, M. Sumuewitc and
    S.P. Shdanow, Z. Phys. Chem. (Leipzig), 262 (1981) 567.
38 J. Kärger, H. Pfeifer and W. Heink, in D. Olson and A. Bisio (Eds.),
    Proceedings of the 6th International Zeolite Conference, Reno, Nevada,
    USA, 1983, Butterworths, Guildford, Surrey, UK, 1984, pp. 184–200.
39 M.G. Palekar and R.A. Rajadhyaksha, Catal. Rev. – Sci. Eng., 28 (1986)
    371–429.
40 D.M. Ruthven, AIChE Symp. Ser., 233, 80 (1984) 21–33.
41 M. Goddard and D.M. Ruthven, Zeolites, 6 (1986) 283–289.

442

42 K.K. Pitale, R.A. Rajadhyaksha, S.A. Pendharkar and A.C. Eapen, Curr. Sci., 57 (1988) 355–363.
43 J. Kärger and D.M. Ruthven, J. Chem. Soc., Faraday Trans. 1, 77 (1981) 1485–1496.
44 J. Kärger, H. Pfeifer, M. Rauscher and A. Walter, J. Chem. Soc., Faraday Trans. 1, 76 (1980) 717–737.
45 H. Pfeifer, J. Kärger, A. Germanus, W. Schirmer, M. Bülow and J. Caro, Adsorpt. Sci. Technol., 2 (1985) 229–239.
46 H. Yucel and D.M. Ruthven, J. Chem. Soc., Faraday Trans. 1, 76 (1980) 60–70.
47 H. Yucel and D.M. Ruthven, J. Colloid Interface Sci., 74 (1980) 186.
48 H. Yucel and D.M. Ruthven, J. Chem. Soc., Faraday Trans. 1, 76 (1980) 71–83.
49 M. Bülow, P. Struve and L.V. Rees, Zeolites, 5 (1985) 113–117.
50 M. Goddard and D.M. Ruthven, in Y. Murakami, A. Iijima and J.W. Ward (Eds.), Stud. Surf. Sci. Catal., 28, Kodansha/Elsevier, Tokyo/Amsterdam, 1986, pp. 467–473.
51 M. Eic and D.M. Ruthven, Zeolites, 8 (1988) 40–45.
52 D.M. Ruthven, A.M. Graham and A. Vavlitis, in L.V. Rees (Ed.), Proceedings of the 5th International Zeolite Conference, Naples, Italy, 1980, Heyden, London, 1980, pp. 535–544.
53 M.F.M. Post, J. van Amstel and H.W. Kouwenhoven, in D. Olson and A. Bisio (Eds.), Proceedings of the 6th International Zeolite Conference, Reno, Nevada, USA, 1983, Butterworths, Guildford, Surrey, UK, 1984, pp. 517–527.
54 D.B. Shah, D.T. Hayhurst, G. Evanina and C.J. Guo, AIChE J., 34 (1988) 1713–1717.
55 H. Chon and D.H. Park, J. Catal., 114 (1988) 1–7.
56 J. Kärger and J. Caro, J. Chem. Soc., Faraday Trans. 1, 73 (1977) 1363–1376.
57 M. Bülow, P. Struve, C. Redszus and W. Schirmer, in L.V. Rees (Ed.), Proceedings of the 5th International Zeolite Conference, Naples, Italy, 1980, Heyden, London, 1980, pp. 580–591.
58 A. Germanus, J. Kärger, H. Pfeifer, N.N. Samulevic and S.P. Zdanov, Zeolites, 5 (1985) 91–95.
59 J. Caro, S. Hocevar, J. Kärger and L. Riekert, Zeolites, 6 (1986) 213–216.
60 D.B. Shah and N.K. Oey, Zeolites, 8 (1988) 404–408.
61 N. Haq and D.M. Ruthven, J. Colloid Interface Sci., 112 (1986) 164–169.
62 V.R. Choudhary and K.R. Srinivasan, J. Catal., 102 (1986) 316–327.
63 H.-J. Doelle, J. Heering, L. Riekert and L. Marosi, J. Catal., 71 (1981) 27–40.
64 P. Wu, A. Debebe and Y.H. Ma, AIChE Winter Meeting, Orlando, Fl., USA, 1981, Paper 55a.
65 P. Wu, A. Debebe and Y.H. Ma, Zeolites, 3 (1983) 118–122.
66 P. Wu and Y.H. Ma, in D. Olson and A. Bisio (Eds.), Proceedings of the 6th International Zeolite Conference, Reno, Nevada, USA, 1983, Butterworths, Guildford, Surrey, UK, 1984, pp. 251–260.
67 C. Wu, G. Qin and Y. Xie, in Y. Murakami, A. Iijima and J.W. Ward (Eds.), Stud. Surf. Sci. Catal., 28, Kodansha/Elsevier Tokyo/Amsterdam, 1986, pp. 481–486.
68 D. Prinz and L. Riekert, Ber. Bunsen-Ges. Phys. Chem., 90 (1986) 413–417.
69 D.T. Hayhurst and G. Evanina, Chem. Express, 1 (1986) 733–736.
70 K. Beschmann, G.T. Kokotailo and L. Riekert, Chem. Eng. Process., 22 (1987) 223–229.
71 A. Zikanova, M. Bülow and H. Schlodder, Zeolites, 7 (1987) 115–118.
72 M. Bülow, H. Schlodder, L.V.C. Rees and R.E. Richards, in Y. Murakami, A. Iijima and J.W. Ward (Eds.), Stud. Surf. Sci. Catal., 28, Kodansha/Elsevier, Tokyo/Amsterdam, 1986, pp. 579–586.
73 M. Bülow, P. Lorenz, W. Mietk, P. Struve and N.N. Samulevic, J. Chem. Soc., Faraday Trans. 1, 79 (1983) 1099–1108.

74 M. Bülow, W. Mietk, P. Struve and P. Lorenz, J. Chem. Soc., Faraday Trans. 1, 79 (1983) 2457–2466.

75 B. Boddenberg and R. Burmeister, Zeolites, 8 (1988) 488–494.

76 M. Goddard and D.M. Ruthven, Zeolites, 6 (1986) 445–448.

77 J. Kärger, H. Pfeifer, D. Freude, J. Caro, M. Bülow and G. Oehlmann, in Y. Murakami, A. Iijima and J.W. Ward (Eds.), Stud. Surf. Sci. Catal., 28, Kodansha/Elsevier, Tokyo/Amsterdam, 1986, pp. 633–639.

78 A.S. Chiang, A.G. Dixon and Y.H. Ma, Chem. Eng. Sci., 39 (1984) 1461–1468.

79 J. Kärger, H. Pfeifer, J. Caro, M. Bülow, J. Richter-Mendau, B. Fahlke and L.V. Rees, Appl. Catal., 24 (1986) 187–198.

80 J. Caro, M. Bülow and J. Kärger, Chem. Eng. Sci., 40 (1985) 2169–2170.

80a J. Kärger and D.M. Ruthven, Zeolites, 9 (1989) 267–281.

81 L. Riekert, AIChE J., 17 (1971) 446–454.

82 R.L. Gorring, J. Catal., 31 (1973) 13–26.

83 D.M. Ruthven, R.I. Derrah and K.F. Loughlin, Can. J. Chem., 51 (1973) 3514–3519.

84 R. Richter, R. Seidel, J. Kärger, W. Heink, H. Pfeifer, H. Fürtig, W. Höse and W. Roscher, Z. Phys. Chem. (Leipzig), 267 (1986) 1145–1151.

85 J. Kärger, H. Pfeifer, P. Walther, A. Dyer and C.D. Williams, Zeolites, 8 (1988) 251–254.

86 M. Bülow, H. Schlodder and P. Struve, Adsorpt. Sci. Technol., 3 (1986) 229–232.

87 P. Voogd and H. van Bekkum, in H.G. Karge and J. Weitkamp (Eds.), Stud. Surf. Sci. Catal., 46, Elsevier, Amsterdam, 1989, pp. 519–531.

88 R.M. Moore and J.R. Katzer, AIChE J., 18 (1972) 816–824.

89 D.H. Olson, G.T. Kokotailo, S.L. Lawton and W.M. Meier, J. Phys. Chem., 85 (1981) 2238–2243.

90 D.M. Ruthven, Can. J. Chem., 52 (1974) 3523.

91 B. Boddenberg and R. Burmeister, Zeolites, 8 (1988) 480–487.

92 H. Herden, W.-D. Einicke, U. Messow, E. Volkmann, R. Schollner and J. Kärger, J. Colloid Interface Sci., 102 (1984) 227–231.

93 E.G. Derouane, J.-M. André and A.A. Lucas, J. Catal., 110 (1988) 58–73.

94 E.G. Derouane, J.B. Nagy, C. Fernandez, Z. Gabelica, E. Laurent and P. Maljean, Appl. Catal., 40 (1988) L1–L10.

95 W. Drachsel and K.A. Becker, Z. Phys. Chem. (Wiesbaden), 122 (1980) 91–101.

96 J.A. Johnson and A.R. Oroskar, in H.G. Karge and J. Weitkamp (Eds.), Stud. Surf. Sci. Catal., 46, Elsevier, Amsterdam, 1989, pp. 451–467.

97 C.N. Satterfield and C.S. Cheng, AIChE Symp. Ser., 67 (1971) 43.

98 F. Awum, S. Narayan and D.M. Ruthven, Ind. Eng. Chem. Res., 27 (1988) 1510–1515.

99 C.B. Ching and D.M. Ruthven, Zeolites, 8 (1988) 68–73.

100 W.O. Haag, R.M. Lago and P.B. Weisz, Faraday Discuss. Chem. Soc., 72 (1982) 317–330.

101 J. Völter, G. Lietz, U. Kürschner, E. Löffler and J. Caro, Catal. Today, 3 (1988) 407–414.

102 J. Völter, J. Caro, M. Bülow, B. Fahlke, J. Kärger and M. Hunger, Appl. Catal., 42 (1988) 15–27.

103 P.B. Weisz, in Proceedings of the 7th International Congress on Catalysis, Tokyo, 1980, Elsevier, Amsterdam, 1981, pp. 3–20.

104 D.H. Olson and W.O. Haag, ACS Symp. Ser., 248 (1984) 275–307.

105 E.W. Thiele, Ind. Eng. Chem., 31 (1939) 916.

Chapter 12

INTRODUCTION TO ACID CATALYSIS WITH ZEOLITES IN HYDROCARBON REACTIONS

Pierre A. Jacobs and Johan A. Martens

Department of Interface Science, KU Leuven, 92, Kardinaal Mercierlaan,
B-3030 Leuven, Belgium.

1. Screening of zeolite catalysts at the laboratory scale

There are many good texts on the design of catalytic reactors in general and of laboratory scale reactors in particular (1-10). Although the screening of zeolite catalysts doesnot necessarily require an in-depth knowledge of reactor design and engineering, there are certain essential principles which have been overlooked too often in the catalytic testing of zeolites. It is the aim of this first paragraph to summarize critical parameters which should be considered when a laboratory scale reactor for zeolite catalyst screening is designed.

1.a. Catalytic experiments in continuous flow reactors

Weitkamp (11) made a critical evaluation of catalytic testing of zeolites. As clearly illustrated by this author, the papers on zeolite catalysis in general are characterized by a lack of information on the details of the reactor used. The papers published in the proceedings of the 7th International Zeolite Conference were used as test sample. 49% of the contributions on catalysis do not cite the amount of catalyst used. The most popular reactor seems to be one operating in the fixed bed mode containing an amount of catalyst which varies between 0.1 and 10 g, operating near atmospheric pressure and using gases and/or vapors as reactants.

In such reactor a mixture of gases and/or vapors with constant composition in time is generated, which then flows over the catalyst bed in the piston flow mode and is converted partially or completely to products. The composition of the reactor outlet is measured by a sampling system, which preferentially should be on-line and transfers samples to (an) analytical instrument(s). Condensation of heavy products in the reactor lines or on the catalyst should be avoided but at least should be monitored by using an internal standard. A schematic representation of such reactor is shown in Fig.1.

446

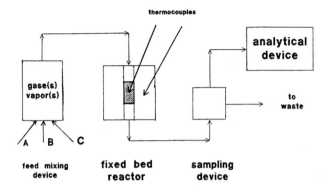

Fig. 1. Schematic representation of a fixed bed continuous flow reactor suitable for testing of zeolite catalysts.

1.a.1. Designing saturators

As the composition of the inlet mixture in a continuous flow reactor operating at pressures slightly above atmospheric should be constant in time, mixing of gases and/or vapors is critical and requires carefully designed devices. Mixing of gases can best be done with the help of mass flow controllers. In this way variable but reproducible gas compositions can be realized easily. Mixing gases with manually operated needle valves, requires that the entrance pressure in such valves is identical. This can only be done with sensitive pressure regulators. Back-mixing of gases as a result of sudden pressure changes in the reactor lines must be avoided by the use of one-way-spring-operated-valves with low cracking pressure. The use of a gas mixing volume filled with inert material such as glass beads is recommendable. The efficiency of the mixing should be checked regularly and consequently a by-pass to the analytical equipment is a requirement for accurate catalysis.

Mixing of gases and vapors is usually done in saturators. This item has been exhaustively treated (11). The design of several possible saturators is shown in Fig.2. When gases simply are bubbling through a liquid, saturation is critical when a range of partial pressures and flow rates has to be covered. Therefore, recipients with artificially lenghtened bubble path or in which condensation occurs after supersaturation are more recommendable. The preferred saturator is filled with an inert solid, thus causing enhanced mass transfer and constant composition of the mixture at the outlet for a wide range of conditions. For mixing of heavy components, one (or two) HPLC pump(s) can be very useful as they are able to deliver liquid flow rates of the order of 1 ml per hour. The critical step is the **pulsation-free vaporization of small flows of liquid**. The design of the vaporizer is critical and its behaviour should be checked via a by-pass to the analytical device.

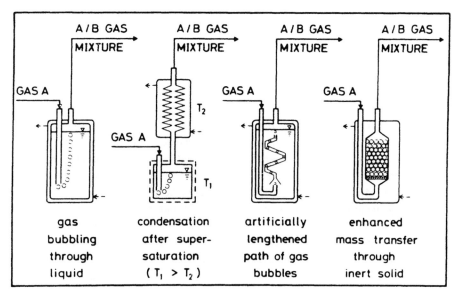

Fig. 2. Design of saturators for the generation of mixtures of vapors and gases in continuous flow reactors. From Weitkamp (ref. 11), copyright Elsevier.

1.a.2. Designing a catalyst bed

The design of the reactor for small amounts of zeolite catalyst should be such that **piston flow** goes through the bed. For gas phase piston flow over a fixed bed of catalyst Reynold numbers higher than 30 are generally required (12). However, it is not always easy to reach high Reynolds numbers in laboratory scale reactors. It is, therefore, essential, that the following two conditions for piston flow be obeyed (12):

i. $L / d_p > 50$ $\hspace{4cm}$ (1.1)

in which L, is the lenght of the catalyst bed and d_p, the diameter of the catalyst particle,

ii. $6 < D / d_p < 10$ $\hspace{3.5cm}$ (1.2)

in which D, is the reactor diameter.

Consequently, when the fixed bed reactor has an internal diameter of 10 mm, and when a catalyst particle size between 1 and 1.7 mm is used, a minimum bed lenght of 50 mm is requested for piston flow to occur through this bed. Therefore, zeolite powder has to be pelletized, crushed, and sieved and the suitable size fraction should be isolated for use in the fixed bed reactor. If this is omitted the pressure drop along the reactor will be too high, causing back-mixing and turbulent flow of the reactants. The temperature of the

448

bed should be homogeneous, thus requiring at least the presence of one thermocouple inside the bed. For highly exothermic reactions, bed dilution with inert material of the same granulometry as the catalyst particles can be of help.

Therefore, small tubular reactors for exploratory studies should consist of a tube with internal diameter between 6 and 12 mm, holding a few cm^3 of catalyst with the appropriate granulometry. The tube must be longer so as to provide volume for the preheating and the discharge section. External heating can be done with one or several independantly electronically regulated resistant wires. The present authors have been using all the time furnaces with an internal air circulation (Fig. 3), eliminating temperature gradients of over 1 ^{O}C. For very accurate kinetic work, molten salt baths or fluidized sand baths are recommended. Preheating of the feed to reaction temperature in a reactor section preceeding the actual catalyst bed and containing inert particles with the same size is necessary. A discharge section containing the same inert material should be present to minimize temperature losses at the reactor outlet.

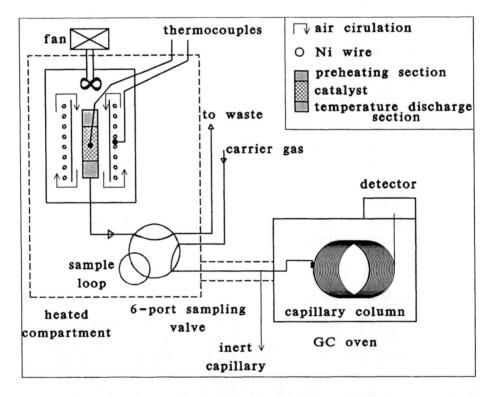

Fig. 3. Schematic representation of a fixed bed flow isothermal reactor heated with an internal recirculation of air and an on-line high temperature sampling system.

The fixed bed reactor should operate in conditions where extragranular diffusion or intragranular diffusion in non-zeolitic pores of the catalyst pellets is absent. In absence of such diffusion phenomena and for particle sizes comprized within the limitations of equation 1.2, the conversion should be independant of particle size.

The absence of external diffusion as rate limiting event can be easily detected changing the reactor volume by a factor 3 to 4, and keeping the contact time (W/F_0) constant. When in a properly dimensioned fixed bed reactor, a variation of the molar flow rate at the reactor entrance, F_0, is compensated by a proportional variation of the catalyst weight, W, the conversion should remain unchanged in absence of external diffusion limitation (14). Bad heat transfer between the fluids and the catalyst particles, may be at the origin of ignition-extinction phenomena in reactors and consequently, of irreproducible results. The thermal instability of a catalytic reactor gives rise to deviations of Arrhenius' law (4): the rate is no longer a continuous function of the reaction temperature but a definite minimum temperature is required to start the reaction.

The exclusion of intra-zeolitic pore diffusion as rate-limiting event is far less obvious. The average crystallite size of many zeolites can presently be changed during synthesis over at least half an order of magnitude. The absence of any variation in conversion when such samples are tested in the correct conditions, constitutes a strong argument for the absence of **intra-zeolitic diffusion limitation**. The observation of rather high values for the apparent activation energies (e.g. in excess of 60 kJ/mol) doesnot necessarily indicates the absence of diffusional control in zeolite pores. Carberry (17) states in this respect that for transport of molecules through zeolite cages and ports, traditional notions of diffusivity no longer apply. Transport in zeolites should be described (sic) as "chemically facilitated transport" as intramolecular collision and pore wall collision no longer can be used to describe intraphase transport. Conventional diffusion-reaction models predict intraphase effectiveness approximating zero and consequently, molecules should hardly penetrate the zeolite crystals. In Carberry's terminology, **"zeolites are catalytic membranes permeated by reactants by selective chemically facilitated transport"**.

It follows from all this that the testing of formed catalysts cannot be made in the small size fixed bed reactors described higher. Extrudates of 1/16 inch (1.6 x 3 mm) have an equivalent diameter, d_p, of 1.9 mm, as can be calculated with the following formula:

$$d_p = 3\ d_c\ l_c\ /\ (\ 2\ l_c + d_c\ ) \qquad\qquad (1.3)$$

in which d_c is the diameter of the extrudate and l_c, its lenght. It requires bed dimensions of at least 95 x 11.4 mm.

Larger catalyst pellets can be tested provided they can be broken down into particles with diameters ranging from 1 to 1.6 mm without changes in their catalytic behaviour.

1.a.3. Making mass balances

Accurate mass balances are a prerequisite to obtain good catalytic data. It is recommended, therefore, to introduce an **inert internal standard** at the reactor inlet together with the reactants or at the reactor outlet. Electronic flow control in the latter case is essential but the selection of a suitable molecule is easier. Consequently, accurate on-line analysis of the reactor outlet will be possible and mass deficiencies due to carbon deposition on the catalyst, condensation of heavy and unexpected products in the transfer lines, on the catalyst and even in the chromatographic column will be noticed immediately. In many cases methane is a sufficiently inert standard, which can be entered through the saturator.

1.a.4. Sampling at the reactor outlet

When the reactor outlet is connected via heated transfer lines to the heated sampling valve of a chromatograph, direct and programmed sampling of the reactor outlet is possible without condensation and use of a phase separator. A more sophisticated version of this technique is shown in Fig. 3. Sampling of product streams up to 350 $^\circ$C is possible. In order to avoid the heating of concentrated product streams over longer distances, the sampling valve is connected directly to the reactor exit. The reactor furnace and the sampling valve are fixed in a heated compartment and the connection between capillary column and sampling valve is made through an inert capillary. To have optimal chromatographic resolution, the sample is condensed in the cold GC oven, before analysis in the temperature programmed mode is done. This technique combines the high resolution of capillary columns with the possibility of injection of rather large amounts of sample with the help of traditional sampling valves. Sampling times are determined by the analysis times, but usually rather complex mixtures can be analyzed within the hour. A solution to this problem consists in the use of multi-port valves in which up to 16 samples can be stored and analyzed afterwards. Rapid sampling and high resolution analysis is no longer a contradiction. This technique is easy to apply and has been described in detail by Weitkamp (11).

When gases next to products boiling higher than 375 $^\circ$C are present, a more complex type of sampling has to be applied. The technique is shown in Fig. 4. Gases and volatiles are analysed on-line as in previous method, while heavier products are collected in a cooled trap as integral samples and analyzed off-line.

If previous sampling methods fail, it is still possible to do a differential sampling with glass ampoules (11). The procedure is, however, labor intensive and requires highly skilled experimentalists.

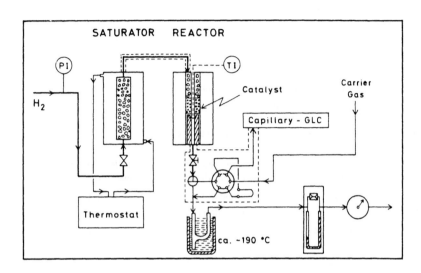

Fig. 4. Sampling via phase separation. From Weitkamp (11), copyright Elsevier.

1.a.5. Extraction of kinetic parameters from laboratory scale fixed bed reactors

In a differential reactor (with no concentration gradient) the mass balance at the reactor inlet and outlet, allows to derive the following equation (1-10):

$$F_{Ao}\, x_A = V_A\, W \qquad\qquad (1.4)$$

in which F_{Ao} is the molar flow of component A (mol/s) at the reactor inlet, x_A, the degree of conversion of A, W, the catalyst mass in the reactor, and V_A, the specific rate (mol $g^{-1}.s^{-1}$). It is generally accepted that this differential behaviour requires conversions of less than 10%. This doesnot mean that the geometry of the reactor bed should be differential, as in this case also piston flow of the reactants is required.

In order to extract rates from an integral reactor with concentration gradient, operating between 0 and 100 % conversion, a çonversion curve has to be determined at increasing values of W/F_{Ao}. Such curve can be derivatized at a given conversion, giving the molar reaction rate at that conversion.

452

1.b. The chromatographic pulse reactor

A simple reactor device is obtained when the column of a gas chromatograph is replaced by a bed of catalyst (15). Pulses of reactant(s) are then fed to the catalyst bed in chromatographic conditions, indicating that the pulse size is negligeable compared to the sorption capacity of the catalyst. As reaction occurs in the Henry part of the sorption isotherm, only **transient behaviour** of catalysts can be studied and true steady state of the catalyst will never be reached (4, 16). No kinetic data can be extracted from such reactor. It is, however, possible to establish the existence of chromatographic effects in catalytic conditions.

1.c. Other laboratory scale continuous flow reactors

The coking of a catalyst during reaction can be easily followed in a microbalance reactor (16). A schematic drawing is shown in Fig. 5. The figure is self-explanatory as far as the reactor principle is concerned. Such reactor can be considered as a differential device, which operates **free of mass and heat transfer limitations** (16). To ensure this, the usual checks on the invariance of reaction rate by changing the amount of catalyst and particle sizes should be made. The problem of intrazeolitic diffusion limitation is also not necessarily solved. The usual precautions against traditional errors, when working with thermobalances have to be taken (18).

Provided mass transfer effects are allowed to exist, such as in the case when whole catalyst pellets are tested, the **single pellet reactor** can be used to to separate the physical from the chemical phenomena in a pellet (19, 20).

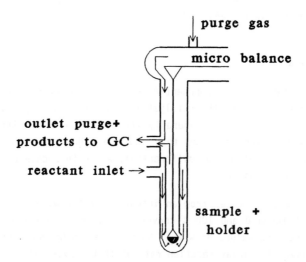

Fig. 5. Principle of operation of a microbalance reactor.

1.d. Batch laboratory reactors

Autoclaves are being used as batch reactors in the initial screening of catalysts for applications in the areas of fine chemicals, pharmaceuticals and agrochemicals. The disadvantages of such devices have been treated recently by Pratt (16). The most relevant ones in connection with zeolite catalysis seem to be:

i.   reaction and deactivation cannot be separated and thus these devices are unsteady,
ii. the device requires long heat-up and cool-down temperatures, and selectivities may be totally different from those in the continuous flow mode,
iii. temperature and pressure cannot be separated,
iv. external diffusion control is only absent at rather high stirring rates. It is the author's experience that a rate of agitation with an efficient stirrer of at least 600 rpm (rotations per minute) is required to be out of external diffusion control, and
v.   catalyst poisons may accumulate.

The activity in such devices is generally described by a first order equation and describes the rate at which the autoclave content reaches equilibrium. It gives a formal measure of catalyst activity, and does not give any indication on the phenomena occurring in the catalyst pores.

When a new reaction is started or a new type of batch reactor is used, it is essential to measure the reaction rate against the stirring rate in the vessel. A plot of a first order rate constant against the stirring rate in the range from 100 to 1000 rpm must be considered as basic information on the catalyst activity in laboratory batch reactors.

As the objective of the research on organic reactions in zeolites is to study catalyzed reactions, a minimum number of turnovers is requested. To make sure that the reaction is catalytic rather than stoechiometric, initially not more than 10% by weight of dry catalyst should be added to the reactant mixture.

The construction of each of these laboratory reactors able to work within different pressures ranges, requires the use of ancillary equipment, such as flow, and pressure meters and regulators, pumps, and heaters. This aspect has been described in detail (22).

1.e. Bench scale reactors

Bench-scale reactors require genrally a larger amount of catalyst than the 10 g quantities, which are needed at maximum in laboratory scale reactors. It is evident that all lab scale reactors can be upscaled to bench-scale models. In this paragraph only new types, which cannot be scaled down will be mentioned.

Trickle bed reactors are used in three-phase operations (solid catalyst, liquid as well as gaseous feed). They contain typically between 50 and 200 cm^3 of catalyst and have diameters between 10 and 30 mm and lenghts up to 2 m (16). Gas and liquid flow concurrently downwards over the reactor bed. Maldistribution of liquid over and incomplete wetting of the catalyst eventually is a problem (16).

The use of **riser reactors** of 11 x 1000 mm with spray-dried zeolite catalysts has been reported (23). As in such reactors a finely divided catalyst (50-90 lm) in a dilute phase is moving through a vertical tube, its use is justified for the study of rapidly decaying catalysts. Plug flow conditions and isothermal behaviour are obtained easily (16). Low volume **fluidised bed reactors** are not easy to operate, although they are ideal for the study of high thermicity reactions and rapidly desactivating catalysts. The use of a 40 x 400 mm fluid bed reactor has been reported (24).

The **Berti and Carberry continuously-fed stirred tank reactors** work in total absence of concentration and temperature gradients (16). The can be considered as recycle differential reactors and consequently, at high recyle ratios, the overall reaction rate, V, is given by equation 1.4. The catalyst volume varies between 30 and 100 cm^3 and is either fixed on the stirrer in a spinning basket (Carberry type) or is in a fixed bed of catalyst. In the latter case, the recycle is done via a propeller.

1.f. General Remarks on Catalysis with Zeolites

A general remark is necessary as far as the determination of catalytic activity on a weight basis with zeolites is concerned. Some are highly hygroscopic (high-alumina zeolites and $AlPO_4$ molecular sieves) while others are completely hydrophobic (high-silica zeolites and silica polymorphs). The hygroscopy of all zeolite catalysts imposes precautions: i., determination of the correct amount of (dry) catalyst in the reactor and ii., prevention of moisture uptake during transportation from the pretreatment to the reactor chamber. Zeolites can be saturated till constant weight with moisture when they are equilibrated in a moisture saturated atmosphere. Only in this way a predetermined amount of dry catalyst can be placed in a catalytic reactor.

Complete dehydration of zeolites without removal of chemical water in case of H-zeolites requires outgassing temperatures within a rather narrow window of temperatures (300 to 400 °C).

Generally speaking, the chemical stability of zeolites in solvents or their mixtures is restricted to the pH range between 4 and 8, depending to a certain extent on contact time and nature of the zeolite. Siliceous zeolites tend to have a better acid stability. The H-form of alumino-rich zeolites is unstable in water. The stability of zeolites in organic media saturated with acids or bases is a totally unexplored area. It should also be reminded that given the high affinity of most zeolites for protons, it behaves as an acid buffer, ultimately causing its self-destruction.

Reproducibility in the modification of zeolite catalysts is not always obvious even not when straightforward ion exchange procedures are applied. When this is done with easily hydrolyzable cations, such as Cu^{2+}, Ni^{2+}, Co^{2+} and others, severe hydrolysis occurs and precipitation of hydrolyzed species may occur in the zeolite pores and on the external crystal surface (21).

2. Reaction Mechanisms in Zeolite Catalyzed Reactions

Zeolites have been used as catalysts for the major classes of reactions, such as in Brønsted acid catalysis, metal catalysis, base catalysis and bifunctional catalysis. The generation of carboniogenic activity has been described in detail (25). All microporous crystalline molecular sieves (MCMS) which have cation exchange capacity (CEC), can in principle be converted in solid Brønsted acids by the partial or complete removal of the charge compensating cations by protons. Cation exchanged zeolites or molecular sieves, silica polymorphs and other MCMS can be used as support for metal clusters or their precursors. This subject has also been reviewed thoroughly (26, 27). When on MCMS with CEC, ion exchange of reducible cations is carried out, bifunctional catalysts with a metal function as well as Brønsted acidity are generated after reduction. In zeolites containing alkali metal clusters (28), even in trace amounts (29), base catalytic activity is generated.

Given the industrial relevance of zeolite catalysts, much work in acid and bifunctional catalysis has been carried out. Metal and certainly base catalysis didnot find any large scale industrial application till now. When hydrocarbons out of the fuel range (C_1 till say C_{18}) are contacted with large pore zeolites with 12-membered rings or larger, their behaviour as bifunctional catalysts can be rationalized with the **concepts developed in superacid chemistry** (30). The conversion of hydrocarbons involves the formation of carbocations as reaction intermediates. The carbocations that are present in the pores of zeolite catalysts are the same well-established entities as the organic cations found in superacids, for which it is known that the solvent and the anion play a nonspecific role (31). The chemistry of gaseous carbocations known from the field of mass spectrometry, and that of the carbocations in superacid solutions studied with ^{13}C and ^1H nuclear magnetic resonance is applicable to the carbocations present on the surface of working zeolite catalysts. However, the difference between the elevated working temperatures of zeolite catalysts and the moderate temperatures at which superacids are used should not be overlooked when comparing kinetic parameters of carbocation rearrangements in both systems (30).

Carbocations are generated on bifunctional zeolite catalysts through proton transfer from Brønsted sites to alkenes formed by dehydrogenation on the metal phase. The proton transfer reactions are terminated by fast hydrogenation of unsaturated species on the metal particles. Consequently, it gives rise to only primary cracking of isomerized hydrocarbons on the Brønsted acid sites.

True Brønsted acid catalysis with the same catalysts and substrates gives rise to fast hydrogen transfer reactions in the zeolite, of the type:

naphthenes + olefines <=======> aromatics + paraffines   (2.1)

This reaction is very zeolite specific and is at the basis of the major technical developments in zeolite catalyzed Fluid Catalytic Cracking (FCC) (32).

Small metal particles can be easily stabilized in zeolites and consequently, show all their traditional catalytic properties. Moreover, on acid zeolites and zeolites with oxidizing properties, they seem to have electron deficient properties (33), confining to an element the catalytic properties of the element at its left side in Mendelejefs table (34). On alkali metal ion exchanged zeolites, with basic properties, a platinum metal phase shows particular dehydrocycling properties of alkanes. Pt-on-KL zeolite is able to convert n-alkanes in the C_6-C_8 range into the corresponding aromatic with high activity and selectivity (35-37). On zeolites impregnated with alkali metal clusters, reactions seem to occur via classical **carbanion chemistry** (28).

The essential reaction mechanisms found on zeolites as acid and bifunctional catalysts will be treated in the following paragraphs. The typical role of MCMS materials is to convert substrates into products via one mechanism or the other but in a very selective manner. **Shape selective catalysis** is possible when i. one of previously mentioned catalytic functions is confined to the intracrystalline volume of the MCMS materials and ii. the size of the substrate molecules is of the same order of magnitude as the ports and cages of the catalyst. The topic of shape selective catalysis will be treated in a later paragraph.

2.a. Mechanisms of the conversions of alkanes and cycloalkanes on acid and bifunctional zeolites

Among the carbocations of concern for the conversion pathways of alkanes and cycloalkanes, a distinction has to be made between **alkylcarbenium** and **alkylcarbonium** ions (Fig.6). Alkylcarbenium ions contain a tri-coordinated positively charged carbon atom, the three substituents being alkyl groups or hydrogen atoms. Alkylcarbonium ions contain a penta-coordinated positively charged carbon atom, having the same type of substituents. In the carbonium ions that will be encountered, at least one of the five substituents is a hydrogen atom. The formal representation of alkylcarbonium ions used in Fig.6 does not violate the basic rules of organic chemistry if it is kept in mind that the five bonds of the charged carbon atom contain only eight electrons.

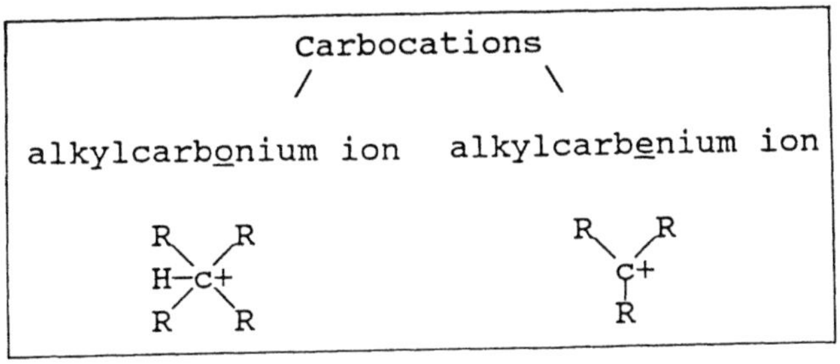

Fig.6. Representation of alkylcarbenium and alkylcarbonium ions

2.a.1. Generation of carbocations on monofunctional acid zeolites

The formation of carbocations from the feedstock molecules can occur according to different mechanisms, depending on the nature of the feedstock and the acidic properties of the catalyst (38). According to Mechanism 1, a proton from the catalyst is added to an unsaturated molecule. The protonation of an alkene (olefin) by the acid zeolite (HZ) leads to the formation of an alkylcarbenium ion:

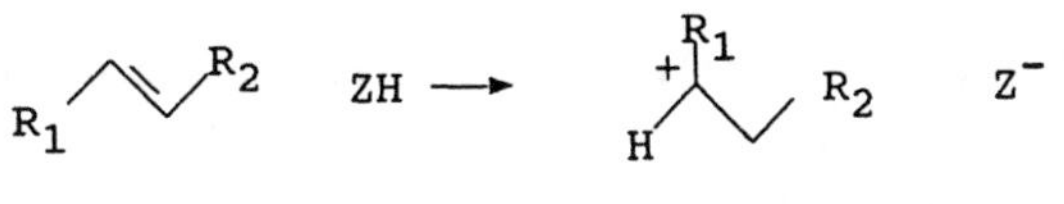

(2.2)

Mechanism 2 involves the addition of a proton to a saturated molecule. The protonation of an alkane (paraffin) leads to the formation of an alkylcarbonium ion. The alkylcarbonium ion can be transformed into a smaller alkylcarbenium ion by abstraction of an electroneutral molecule (an alkane or molecular hydrogen):

$$R_1 \diagdown\diagup\diagdown^{R_2} \quad ZH \longrightarrow \quad \begin{matrix} R_1 \diagup H \\ H \diagdown\!\!\!\!\!+ \\ H \end{matrix}\!\!\diagdown^{R_2} \quad Z^- \longrightarrow \quad H_2 \quad \begin{matrix} R_1 \\ + \end{matrix}\diagdown^{R_2} \quad Z^- \qquad (2.3)$$

Mechanisms 1 and 2 occur on zeolites posessing Brønsted acidity. Mechanism 1 is much faster than Mechanism 2 and operates at lower temperatures.

Mechanism 3 consists of the abstraction of an hydride ion from an electroneutral feed molecule. This mechanism involves acid centers of the Lewis type:

$$R_1 \diagdown\diagup\diagdown^{R_2} \quad Z \longrightarrow \quad \begin{matrix} R_1 \\ + \end{matrix}\diagdown^{R_2} \quad ZH^- \qquad (2.4)$$

A variant of Mechanism 3 can occur on a carbenium ion, R^+, adsorbed in the zeolite pores, a phenomenon denoted as bimolecular hydride transfer:

$$R_1 \diagdown\diagup\diagdown^{R_2} \quad Z^- R^+ \longrightarrow \quad \begin{matrix} R_1 \\ + \end{matrix}\diagdown^{R_2} \quad Z^- \quad RH \qquad (2.5)$$

2.a.2. Generation of carbocations on bifunctional catalysts

A bifunctional zeolite catalyst is an acid zeolite on which a metal phase is deposited. The function of the metal is to catalyse dehydrogenation and hydrogenation reactions. It is necessary to add hydrogen gas to the hydrocarbon feedstock in order to obtain a high activity and good time-on-stream stability of the metal function.

According to the classical bifunctional reaction scheme (39,40), the metal dehydrogenates the alkane feed molecules into alkenes. The alkenes are protonated over the Brønsted acid sites into alkylcarbenium ions. These cations undergo isomerisation and scission reactions, after which they desorb from the acid sites under the form of alkenes. Finally, saturated products are recovered after hydrogenation on the metal phase (Fig.7). Aluminosilicate zeolites containing platinum or palladium metal generally work according to the classical bifunctional mechanism, especially at elevated conversions, elevated hydrogen partial pressure and a sufficiently high carbon number of the feed (41). Metallic hydrocarbon conversion mechanisms always compete with the bifunctional mechanism. Their contribution becomes important under more severe reaction conditions when the zeolite has weak acidity. The presence of methane and ethane in the reaction products is evidence for the occurrence of metal catalysed cracking, termed hydrogenolysis (41).

The advantage of bifunctional catalysis over monofunctional acid catalysis is that alkylcarbenium ion formation can proceed via Mechanism 1 instead of Mechanism 2. The presence of the metal allows the acid zeolite to operate at reaction temperatures that are about 100 K lower. The rearrangements and scission reactions of the alkylcarbenium ion intermediates are the rate limiting steps in the classical bifunctional reaction scheme. On bifunctional catalysts the conversion of the feed can be stopped after the primary events by a fast desorption of the olefins (Step D in Fig.7). Thus bifunctional catalysis has allowed to get insight in the complex reaction networks encountered on acid zeolite catalysts (30).

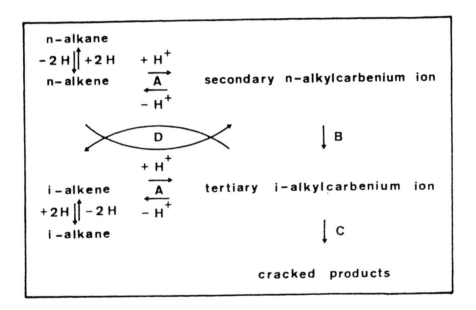

Fig. 7. Classical bifunctional conversion scheme of alkanes
A, protonation-deprotonation reaction on zeolite acid sites;
B, rearrangements of alkylcarbenium ions;
C, cracking of alkylcarbenium ions;
D, competitive adsorption-desorption of alkenes on acid sites.
(Copyright Elsevier, reproduced from ref.42).

2.a.3. The intimacy criterion

The classical bifunctional mechanism implies that the mass transport of alkenes, formed on the metal, towards the acid sites and vice-versa has to be efficient. The required intimacy of the two functions in order to obtain the maximum catalytic activity has been quantified and is known as the Weisz Intimacy Criterion (39):

$$R^2 < 1.2 \ 10^5 \ \frac{P_o \ D_o}{T \ dN/dT} \tag{2.6}$$

It determines a maximum average distance, 2R (in m), between the two functions. This distance depends on P_o, the partial pressure of the alkenes (in MPa), T, the reaction temperature (in K), D_o, the diffusivity of the alkenes in the catalyst pores (in $m^2 \ s^{-1}$) and dN/dt, the reaction rate (in $mol \ s^{-1} \ m^{-3}$). According to Weisz, the intimacy criterion is relatively insensitive to catalyst particle shape and exact kinetics (39). A comparison between zeolite Y and amorphous silica-alumina catalysts for the conversion of n-alkanes shows that the realisation of sufficient intimacy in zeolites is much more critical, due to their higher catalytic activity and lower diffusion coefficient (Table 1).

Table 1. Maximum average distance, 2R, between metal particles and acid sites in bifunctional silica-alumina (39) and zeolite Y catalysts

	silica-alumina	zeolite Y
Do ($m^2 \ s^{-1}$)	$2 \ 10^{-7}$	10^{-9}
T (K)	741	473
Po (MPa)	10^{-4}	10^{-4}
dN/dt ($mol \ s^{-1} \ m^{-3}$)	5	5
2R (m)	$50 \ 10^{-6}$	$4.5 \ 10^{-6}$

2.a.4. Rearrangements and scission reactions of acyclic alkylcarbenium ions

Only those mechanisms will be considered here that give rise to chemically distinguishable reaction products. There are two types of isomerisation reactions of alkylcarbenium ions, depending on whether the rearrangement consists of a positional shift of a side chain with respect to the main carbon chain (type A), or of the formation or vanishing of a side chain (type B).

Type A isomerisation. Type A isomerisation of long-chain methylbranched alkylcarbenium ions occurs stepwise through 1,2-methyl shifts. Only subtle rate differences are found between alternative positional shifts.

The mechanism of type A isomerisation of 2-methyl-2-pentyl cation into 3-methyl-3-pentyl cation is illustrated in Fig.8. This conversion implies a 1,2-hydride shift, three-ring closure and re-opening, and another 1,2-hydride shift to obtain a tertiary alkylcarbenium ion. The three-ring closure leads to the formation of a corner-protonated cyclopropane (CPCP) intermediate, in this case 1-protonated-2-ethyl-3-methylcyclopropane (Fig.8).

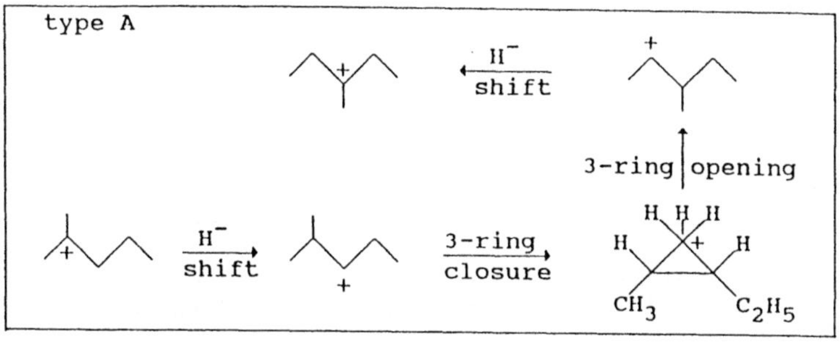

Fig. 8. Mechanism of type A isomerisation of 2-methyl-2-pentyl cation.

Type B isomerisation. Branching isomerisation is a step-by-step reaction generating or removing one side chain at the time. Methylbranching proceeds over CPCP intermediates. Fig.9 illustrates the vanishing of the methylbranching in 2-methyl-2-pentyl cation. With respect to the 1,2-methylshift mechanism, detailed in Fig.8, methylbranching involves an additional step, viz. a corner-to-corner jump of a proton on the cyclopropane ring. The corner-to-corner proton jump is the slow step of a type B rearrangement, explaining why type B isomerisation is slower than type A isomerisation (30).

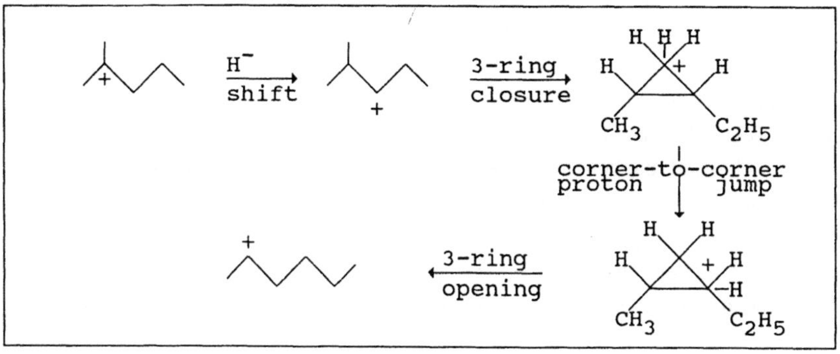

Fig. 9. Mechanism of unbranching of 2-methyl-2-pentyl cation via CPCP intermediates.

462

Type B isomerisation via substituted protonated cycloalkanes larger than cyclopropane accounts for the formation of ethyl- and larger n-alkyl side chains (30). The ethylbranching and butylbranching mechanisms of 4-tridecyl cation are shown in Figs.10 and 11, respectively. The rate of a branching rearrangement decreases the larger the size of the protonated rings involved.

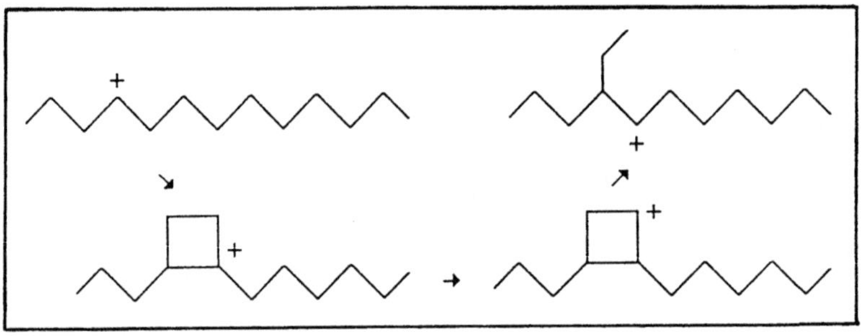

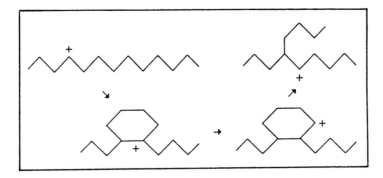

Fig. 10. Ethylbranching mechanism of 4-tridecyl cation via substituted corner-protonated cyclobutanes. (The formal representation of C-H bonds is omitted).

Fig. 11. Butylbranching mechanism of 4-tridecyl cation via substituted corner-protonated cyclohexanes. (The formal representation of C-H bonds is omitted).

β-scission. Cracking of alkylcarbenium ions occurs through β-scission. For long-chain alkylcarbenium ions, containing eight or more carbon atoms, there exist five modes of β-scission, denoted as A, B_1, B_2, C and D. The distinction between the mechanisms is based on the position of the side chains relative to the charged carbon atom (Fig.11). These specific configurations are α,γ,γ-tribranching γ,γ-dibranching α,γ-dibranching and γ-monobranching for β-scissions of types A, B_1, B_2 and C, respectively. Type D β-scissions convert unbranched ions. Further characteristics of the β-scission modes are given in Table 2. The rate of the β-scission reactions decreases in the order:

$$A \gg B_1, B_2 > C \gg D \tag{2.7}$$

The following rates were measured for the conversion of decane and its isomers over a platinum containing ultrastable Y zeolite catalyst (42):

Relative rates at 405K over Pt/USY	
A hydrocracking	1050
methyl shift	56
B_1 hydrocracking	2.8
B_2 hydrocracking	1.0
branching via CPCP	0.8
C hydrocracking	0.4

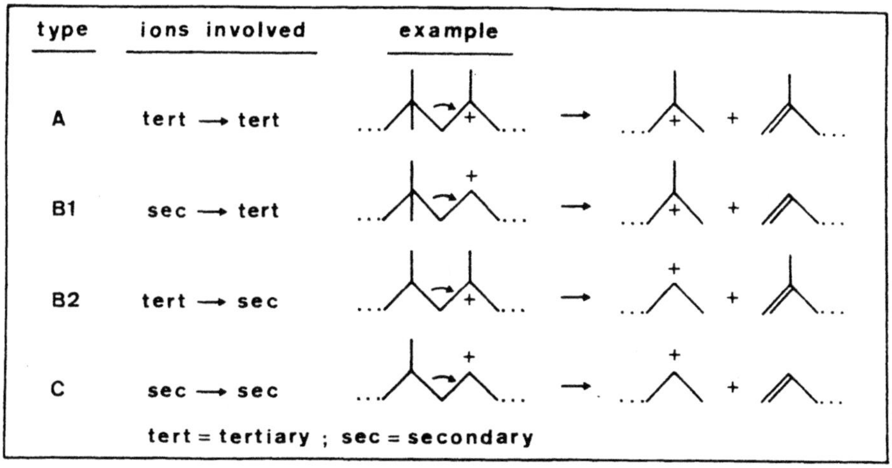

Fig. 12. Possible β-scission mechanisms on secondary and tertiary carbocations. (From ref.42, Copyright Elsevier)

Table 2. Features of β-scission mechanisms of alkylcarbenium ions

mechanism	A	B_1	B_2	C	D
Minimum CN[a] of feed	8	7	7	6	4
Minimum number of branchings in feed	3	2	2	1	0
Nature of feed cations[b]	tert.	sec.	tert.	sec.	sec.
Nature of product cations[b]	tert.	tert.	sec.	sec.	prim.
Branching of products[c]	br.	br.+lin.	br.+lin.	lin.	lin.
CN of smallest fragments generated	4	3	3	3	1[d]

a, CN = carbon number; b, tert. = tertiary; sec. = secondary; prim. = primary. c, for cations having the minimum number of branchings; br. = branched; lin. = linear. d, the generation of CH_3^+ is energetically unfavorable.

464

The susceptibility of an alkylcarbenium ion to β-scission increases with increasing degree of branching, while the generation of the branchings occurs over constant energy barriers. This explains why the yield of branched isomers from n-alkane feedstocks exhibits a maximum at a certain degree of conversion (Fig.13).

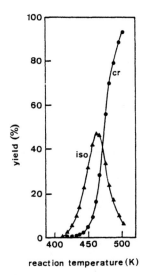

Fig. 13. Yield of branched isomers and cracked products from decane over ultrastable zeolite Y containing 1 wt-% of platinum metal.

2.a.5. Rearrangements and scission reactions of cyclic alkylcarbenium ions

Bifunctional metal-on-zeolite catalysts have been very helpful in elucidating the mechanisms of the conversion of cyclic alkylcarbenium ions. The mechanisms of the different types of rearrangements are very similar to those of acyclic cations (43-46).

External alkyl shifts. A positional shift of an alkyl substituent of a cyclic alkylcarbenium ion can proceed according to the mechanism involving CPCP intermediates. The isomerisation of 1-ethyl-4-methyl-1-cyclohexyl cation into 1-ethyl-3-methyl-1-cyclohexyl cation is shown in Fig.14.

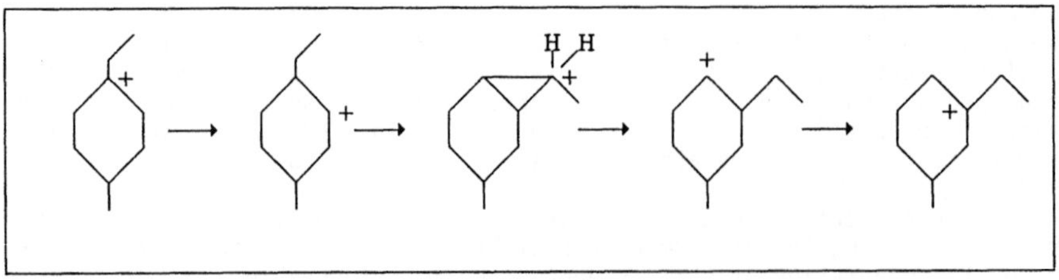

Fig. 14. Mechanism of 1,2-ethylshift in 1-ethyl-4-methyl-1-cyclohexyl cation

Internal alkyl shifts. Internal alkyl shifts result in the elongation or shortening of existing side chains with one carbon atom and concomitant contraction or expansion of the ring. Internal alkyl shifts proceed through similar reaction steps as the external alkyl shifts. The mechanism of the elongation of a methyl side chain in methylcyclohexyl cation is shown in Fig.15. In this type of rearrangement, the three carbon atoms of the CPCP intermediates involved belong to the cycloalkyl ring, whereas in external positional alkyl shifts, only two ring carbon atoms are involved (see Fig.14).

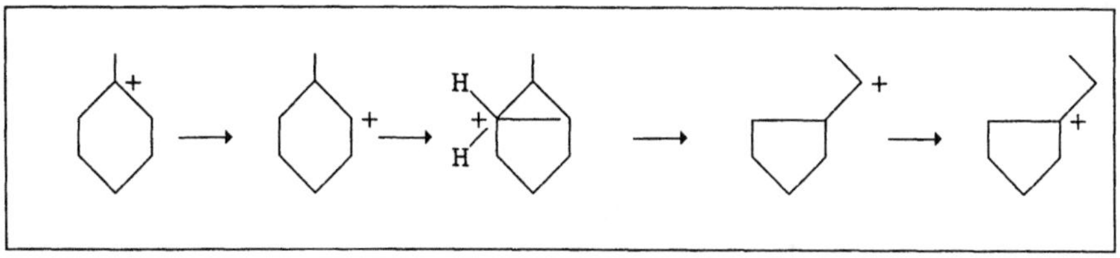

Fig. 15. Mechanism of side-chain elongation through ring contraction in methylcyclohexyl cation.

Generation and vanishing of side chains. The generation or vanishing of side chains on the ring of substituted cycloalkyl cations coincides with ring contraction and ring expansion, respectively. The formation of a methyl side chain proceeds via CPCP intermediates. An important difference between this type of reaction and side chain elongation and alkyl shifts is the necessity of a corner-to-corner proton jump on the CPCP structures. The analogy with the type A and B isomerisation mechanisms of acyclic cations is obvious. The conversion pathway of methylcyclohexyl cation into dimethylcyclopentyl cation is shown in Fig.16.

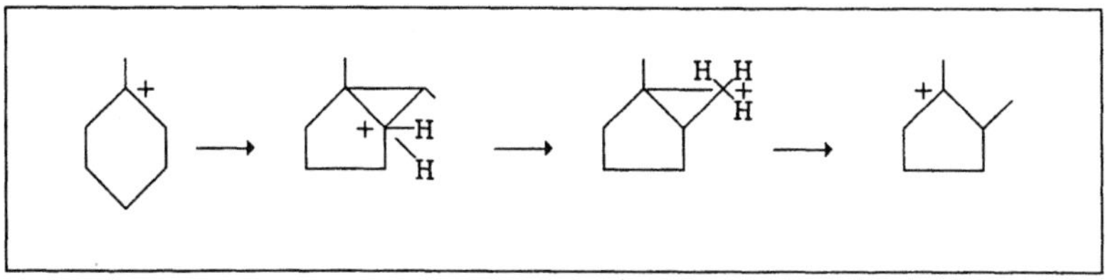

Fig. 16. Mechanism of methyl side-chain generation through ring contraction in methylcyclohexyl cation.

Internal ring alkylation. Internal ring alkylation reactions occur in unsaturated cycloalkyl cations such as cyclooctenyl and cyclodecenyl cation. As an example, the transformation of cyclooctenyl in bi [330] cyclooctyl cation is given in Fig.17.

466

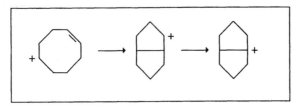

Fig. 17. Mechanism
of internal ring
alkylation in
cyclooctenyl cation.

Starting from large rings such as cyclodecane numerous parallel reaction routes are available (Fig.18)

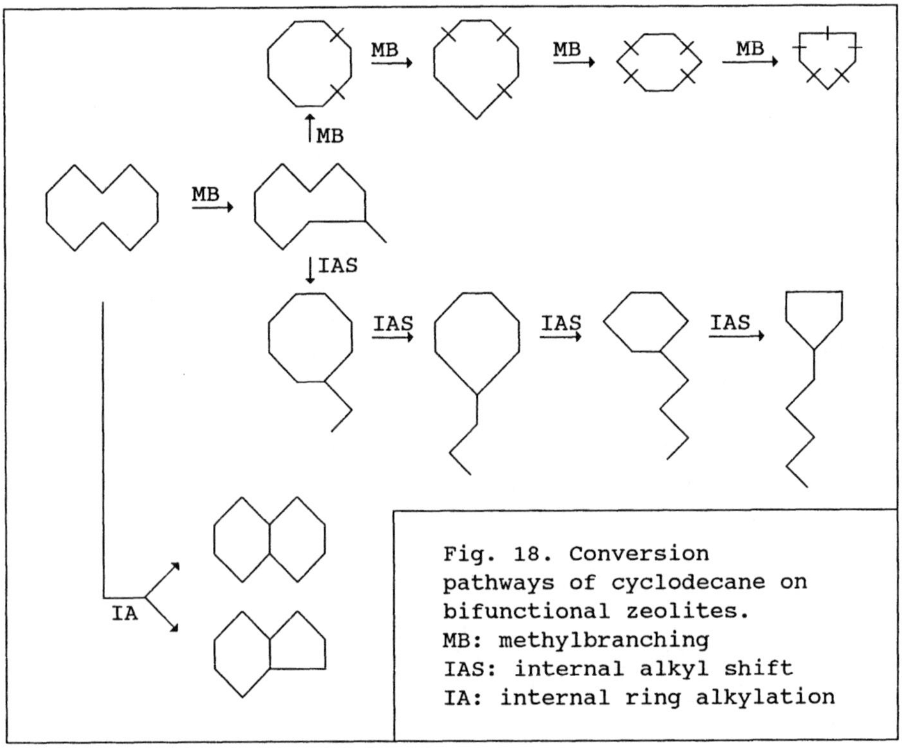

MB: methylbranching
IAS: internal alkyl shift
IA: internal ring alkylation

Fig. 18. Conversion pathways of cyclodecane on bifunctional zeolites.
MB: methylbranching
IAS: internal alkyl shift
IA: internal ring alkylation

<u>Cracking of cyclic alkylcarbenium ions.</u> The β-scission mechanisms of acyclic alkylcarbenium ions can also be operative on cyclic cations. Examples are shown in Fig.19.

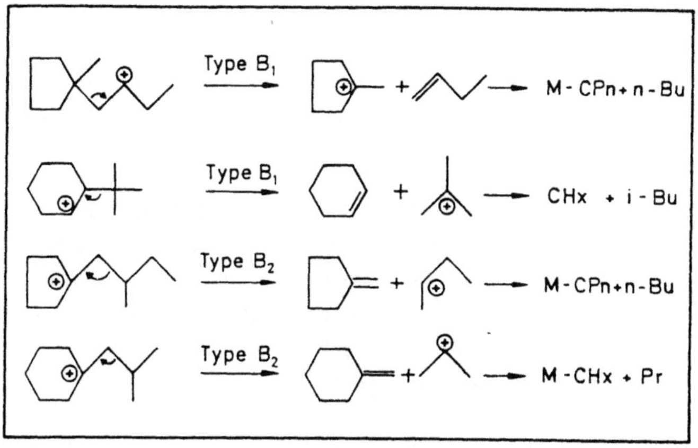

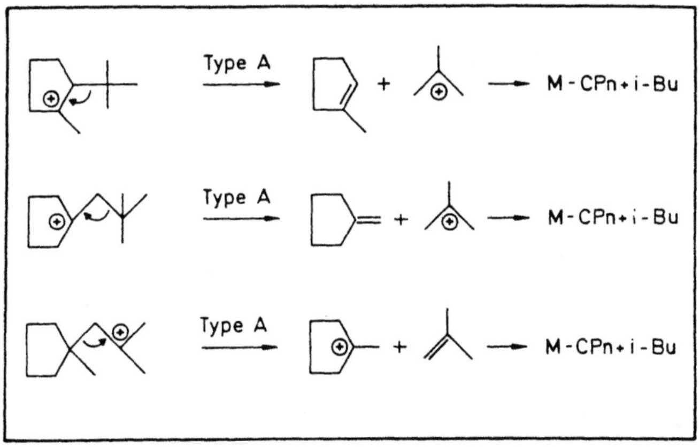

Fig. 19. Examples of β-scission of cyclic alkylcarbenium ions.
(Copyright Elsevier. Reproduced from ref.46).

2.a.6. Monofunctional acid catalysis with zeolitic solid Brønsted acids

The primary isomerization and cracking steps of alkylcarbenium ions on monofunctional acid zeolite catalysts are the same as those described in previous section for classic bifunctional catalysis. In absence of a terminating function, which in bifunctional catalysis is a metal dispersed in the zeolite pores, the propagation in mono-functional catalytic cracking occurs via bimolecular hydrogen transfer reactions. Most mechanistic evidence is based on the ability to fit and predict product distributions in detail.

There is a major difference between bifunctional and monofunctional catalytic cracking. In the former case never the formation of primary carbenium ions has to be invoked to explain experimental product distributions. Given the average higher reaction temperatures in the latter case, the occurrence of primary carbenium ions has often been invoked. Excellent reviews exist on this topic by Haag et al. (47, 48) and Rabo (49).

Although hydrogen transfer in organic reactions can occur from a neutral substrate (RH) to a cation (C$^+$), a free radical (R*) and an anion, in zeolite catalysis chain propagation only occurs via hydride transfer (48) or hydrogen abstraction (49):

$$SH + C^+ \ \text{------>} \ S^+ + CH \qquad\qquad (2.8)$$

$$SH + R^* \ \text{------>} \ S^* + RH \qquad\qquad (2.9)$$

2.a.7. Cracking mechanisms

The initial step in paraffin cracking consists in the generation of alkylcarbenium ion intermediates, through olefin protonation or of alkylcarbonium ion intermediates through alkane protonation:

The alkylcarbenium ions undergo the same transformations as those depicted for bifunctional catalysis, competing however with hydride transfer reactions. The alkylcarbonium ions transform into the corresponding alkylcarbenium ions upon release of hydrogen, or can crack into a smaller alkylcarbenium ion and an alkane fragment. This reaction path is now known as monomolecular cracking and will be discussed later.

Bimolecular cracking of alkanes occurs on monofunctional acid zeolites, but also on bifunctional ones with a very weak hydrogenation/dehydrogenation function as NiS_x (48). The alkylcarbenium ion once formed, can undergo:

* intramolecular hydride shifts, and
* skeletal isomerization, with or without change in the degree of branching as described for bifunctional catalysis;
* intermolecular hydride ion transfer from a alkylcarbenium ion to a substrate molecule, upon formation of a new alkylcarbenium ion and an olefin; this reaction is generally the slowest step and requires the strongest acid sites (50), which consequently are prone to fast deactivation; it has been estimated that a tertiary bond in an alkane is broken 16 times faster than a secondary bond (48); thus the formation of saturated products is the result of such hydrogen transfer reactions;
* alkylation with an olefin, resulting in the formation of longer alkylcarbenium ions;
* β-scission, resulting in the formation of shorter alkylcarbenium ions and olefin fragments, and ultimately
* deprotonation, which functions as chain terminating event.

All intermediately formed alkylcarbenium ions can in turn undergo all these reactions. The reactivity of the different species will mainly depend on the nature of the charged intermediates and, in fact, can be subdivided in type A, B_1, B_2, C and D type reactions. Generally, by analogy to the stability of gas phase alkylcarbenium ions, it is inferred that for ions stabilized with negative counter-ions, at least qualitatively the same reactivity differences exist between primary, secondary and tertiary ions.

The major reactions which intervene in the conversion of n-butane can be listed as follows:

bimolecular hydride transfer

branching via PCP

bimolecular hydride transfer

deprotonation

alkylation

intramolecular hydride shift

methyl shift

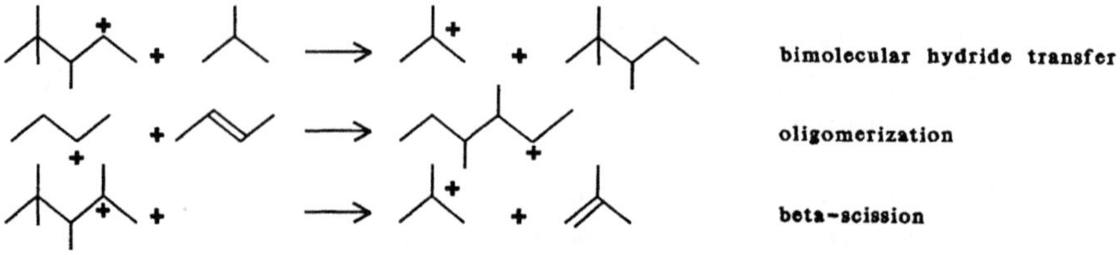

470

bimolecular hydride transfer

oligomerization

beta-scission

It is evident that with the cracking of substrates which initially donot allow the β-scission mechanism to occur (such as C_3 and C_2 substrates), alkylation or oligomerization will occur prior to β-scission:

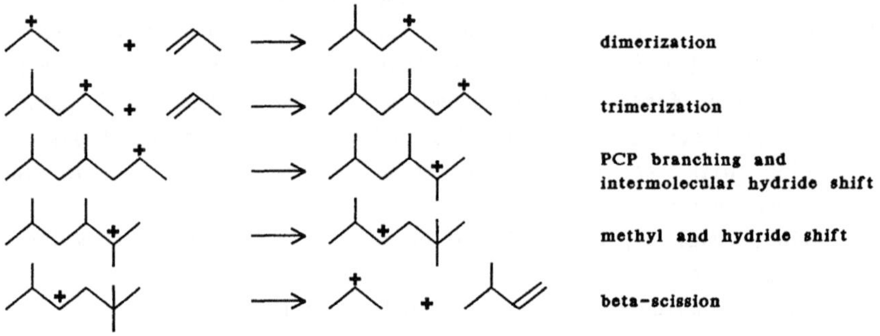

dimerization

trimerization

PCP branching and intermolecular hydride shift

methyl and hydride shift

beta-scission

From all this it follows that hydrogen transfer is a key reaction in acid zeolite catalysis as it determines the reactivity of an alkane substrate, the yield of alkanes in the cracked products and the high ratios of iso- to n-alkanes in the C_4 and C_5 cracked product fraction (49).

In constrained intracrystalline environments, as is the case in the pores of H-ZSM-5, constraints are exerted on the bimolecular cracking mechanism. Haag and Dessau (51) were the first to advance evidence for the existance of a monomolecular cracking mechanism in such zeolites. It involves direct protonation of an alkane and formation of a (penta-coordinated) alkylcarbonium ion (49). Hydrogen is a primary product as well as other smaller alkanes, as shown in the following scheme:

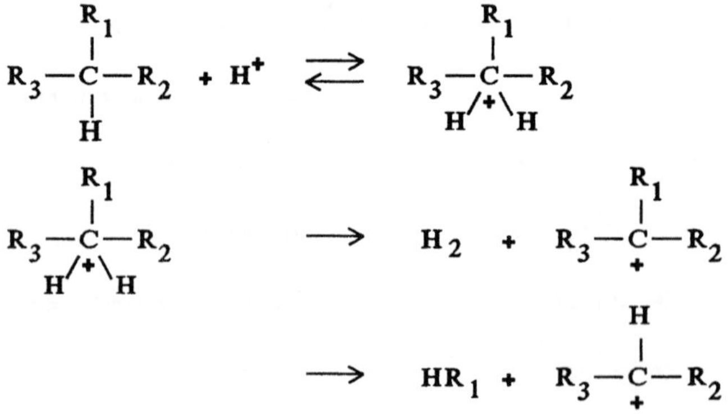

This monomolecular reaction occurs on all acidic solids at high temperatures and at low hydrocarbon partial pressure and conversion (51). In medium pore zeolites, as ZSM-5, the alkane selectivity is very much substrate dependant (51):

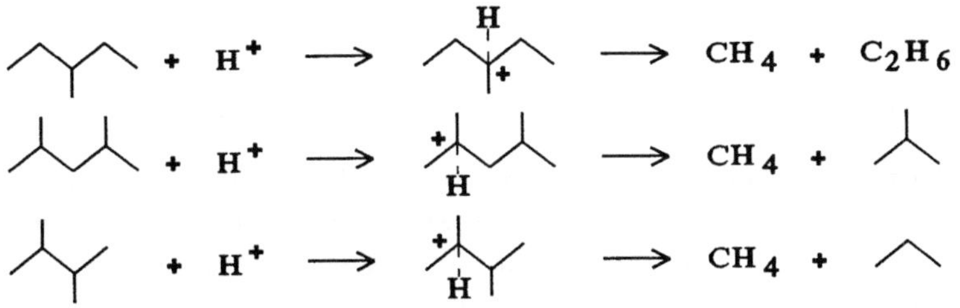

The generally accepted mechanism for the alkylation of aromatics with alkenes on H-zeolites can be depicted as follows (50):

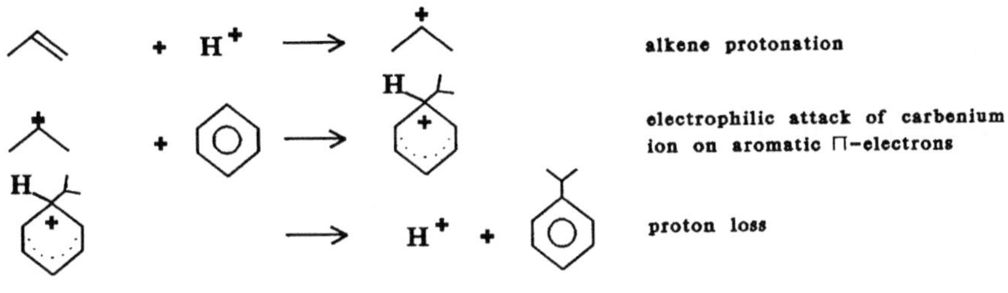

Protonation of the alkene on the Brönsted acid site gives the corresponding alkylcarbenium ion, which exerts an electrophilic attack on the π electrons of the (alkyl)aromatic, resulting in a benzenium ion which then re-aromatizes. The intermediate cation stabilizes by proton loss, while the formation of alkanes from alkylcarbenium ions requires a hydride transfer reaction from a saturated substrate (50). Proton loss in the latter case would give olefin formation. If the intermediate alkylcarbenium ion is formed from the reaction of a shorter alkylcarbenium ion with an olefin, then in case of final deprotonation the overall reaction has to be considered as an oligomerization, while in case of final hydride transfer, the overall reaction network is one of alkylation.

In the low temperature reaction of isobutane with 2-butene, Weitkamp (50) found on this basis that on CeY zeolite, initially true alkylation occurs, while on a partially deactivated catalyst pure oligomerization occurs.

It seems reasonable at this stage to advance that subsequent hydrogen tranfer reactions on acid zeolites will give rise to highly unsaturated molecules such as aromatics and multi-aromatics, which are coke precursors. Catalyst deactivation can, therefore, be manipulated by operation in a hydrogen atmosphere and adding a weak hydrogenation function to the acid catalyst. In this way coke precursors will be hydrogenated without participation in the main catalytic events. When the intracrystalline space limits the rate of bimolecular cracking and favor the monomolecular route as in H-ZSM-5, it is evident that the catalyst decay will be slower.

2.b. Zeolite effect on thermal cracking

Cracking of alkanes (in casu n-hexane) was studied in detail over non-acidic zeolites such as KY and silicalite (49). In KY, the rate of n-hexane cracking was found to be 5 times higher than in a quartz-chips filled reactor. Moreover, the product distribution was different from that in the quarts-chips reactor as well as from that from an acid Y zeolite. The zeolite effects can be summarized as follows (49):

* substantial concentration of the hydrocarbon substrate in the intracrystalline volume causes an altered product distribution, and as a result,
* the existence of an enhancement of the rates of bimolecular reactions, caused by
* the high affinity of a polarizeable substrate by zeolites with high ionic character.

Mechanistically, the altered product distribution can be explained by an enhanced ratio (R) of the rate of H-atom abstraction over β-scission or of the ratio of the rates of bimolecular over monomolecular rearrangements. In a silica polymorph as is silicalite 1, the ionic properties of the matrix are expected to be much less pronounced as in case of KY, and consequently the "zeolite effect" is less pronounced in silicalite. Indeed, the values of R^{-1} for quartz-chips, KY and silicalite 1 are 4, 0.5 and 14.8, respectively (49). In the high surface area SiO_2, the scission rate seems to be enhanced compared to quartz. This effect can be denoted as "zeolite induced thermal cracking" and is observed on all other microporous silica polymorphs.

2.c. Isomerization and disproportionation of alkylaromatics

It is clear that the isomerization/disproportionation selectivity of alkylaromatics will depend on the nature and number of alkyl-substituents on the aromatic ring, on the operation conditions in a given catalyst, and on the nature of the catalyst used (52).

The isomerization of alkylbenzenes can be rationalized using an intramolecular mechanism: during protonation a benzenium ion is formed, followed by a single (or successive) 1,2-alkylshifts in the benzenium ion (52), the alkyl-shift being the rate determining step :

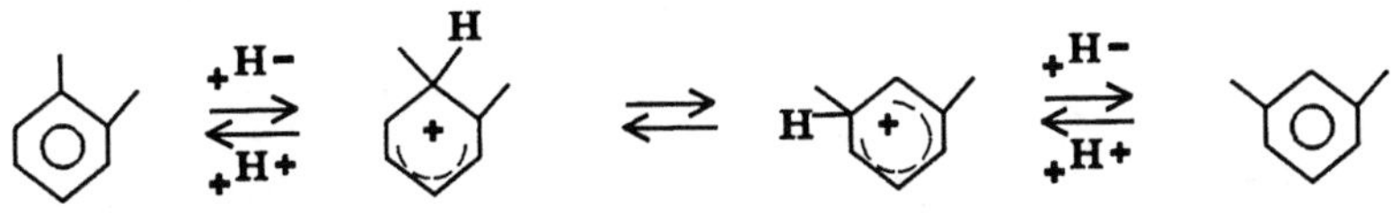

There is ample experimental proof that the disproportionation of alkylbenzenes also is a Brønsted site catalyzed reaction (52). Mechanistically, a sequence of dealkylation-realkylation would be possible at least when the alkylcarbenium ions formed are not primary. Such mechanism is, therefore, unlikely for xylenes and ethylbenzenes. At this time much evidence is available for the existence of diarylmethane intermediates (52):

*   the presence of hydrogen decreases the selectivity for toluene disproportionation on mordenite (53), which is explained by the following equilibrium:

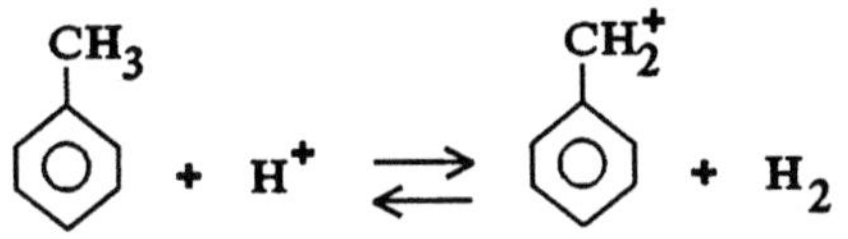

*   addition of alkanes which are able to form teriary alkylcarbenium ions, slow down the o-xylene disproportionation without affecting the isomerization. This is indicative of the different nature of intermediates for isomerization and disproportionation (53); between benzylic cations and branched alkanes the following neutralization reaction is possible (53):

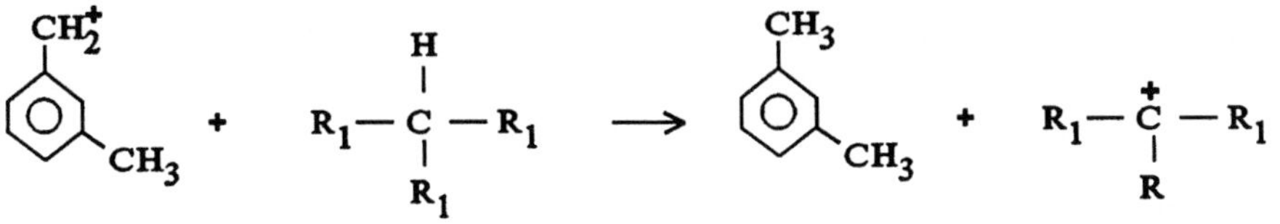

The following reaction mechanism with a diphenylmethane-type intermediate is, therefore, accepted for the disproportionation (53):

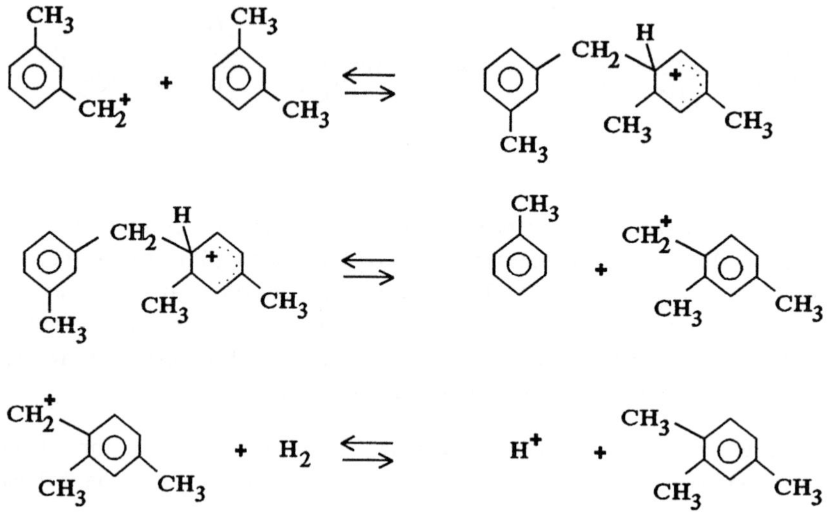

$$CH_3\text{-}(ring)\text{-}CH_2^+ \; + \; CH_3\text{-}(ring)\text{-}CH_3 \;\rightleftarrows\; (ring)\text{-}CH_2\text{-}CH(ring^+)(CH_3)(CH_3) \cdots$$

$$\rightleftarrows\; CH_3\text{-}(ring) \; + \; CH_2^+\text{-}(ring)(CH_3)(CH_3)$$

$$CH_2^+\text{-}(ring)(CH_3)(CH_3) \; + \; H_2 \;\rightleftarrows\; H^+ \; + \; CH_3\text{-}(ring)(CH_3)(CH_3)$$

3. Shape-selective catalysis in acid zeolites

3.a. Generalities on shape selective catalysis

As the major amount of protons in an acid zeolite are confined to the intracrystalline volume, the products of an acid-catalyzed reaction will be influenced by a reaction in a sterical constrained environment. The first paper on shape selective catalysis already appeared in 1960 (55). Since then the activity in this area has grown exponentially and is very well documented (56-62).

Shape selectivity consists in often a subtle matching of size and shape of reactants, transition states and products with the size and shape of pores, cages and pore volume of the intracrystalline zeolite phase. Csicsery (56) has categorized these shape selective effects as follows:

* **reactant selectivity**: it takes places when the zeolitic catalyst acts as a molecular sieve and excludes certain molecular sizes and structures from the intracrystalline voids, while other less bulky molecules are able to enter; a whole variety of zeolite types with different sizes (and shapes) of the pore orifices are available so that this critical exclusion limit can be varied over a wide range of molecular sizes;

* **product selectivity**: it is confined to the intracrystalline voids and is the result of discrete diffusivities of the reaction products in the pores

of the zeolite crystals; typical examples are the monomolecular isomerization reactions of alkylaromatics; this selectivity is not only dependant on the pore size but also on the crystal size of the catalyst particles;

*   **transition state shape selectivity**: it occurs when the spatial configuration around a transition state located in the intracrystalline volume is such that only certain configurations are possible; the transalkylation reactions of alkylaromatics are a school example of this kind of selectivity; in other cases, it is not always easy to distinguish between product and transition state shape selectivity; if the catalyst can be synthesized with crystal sizes which are significantly different, the distinction is easily made as only the former selectivity is dependant on crystal size.

Other less common expressions for molecular selectivity are in the present litterature. In general they are more difficult to identify and find less direct and straightforward illustrations:

*   **the concentration effect** (63) describes the increased concentration of hydrocarbons in zeolites, thus favoring bimolecular over monomolecular reactions and in case of acid catalysis with hydrocarbon substrates also favoring hydrogen transfer reactions; the so-called cage effect which will be treated later, is a special case of this effect; molecules with the size of heptane and octane perfectly fit into the erionite cage, thus reducing their mobility and enhancing their residence time and reactivity;

*   the existence of some kind of **"molecular traffic control"** has been invoked to describe qualitatively the transport of molecules with different shape and/or size in the intracrystalline volume of zeolites with two discrete sets of pores, as is the case in pentasil-type zeolites (64);

*   **molecular circulation** determines the way in which molecules approach pore mouths (65) and is therefore, related to reactant selectivity;

*   **energy gradient selectivity** (66) takes into account differences in tortuosity of the zeolite channels together with differences in the field gradient caused by isomorphic substitution; e.g. in a more tortuous environment, more secondary reactions occur when hydrocarbons are cracked.

3.b. Shape selective bifunctional catalysis

3.b.1. Shape selective conversion of n-alkanes

The composition of the branched isomerisation products from n-alkanes obtained on bifunctional zeolite catalysts provides a wealth of information on the zeolite intracrystalline pore architecture. The number of different branched isomers of n-alkanes and, consequently, the information contained in the product distributions increases very fast with the carbon number (CN). Since the analysis with high-resolution gas chromatography of individual isomers becomes problematic above a CN of 10, the present authors have frequently used decane to study molecular shape selectivity in bifunctional zeolites. With larger n-alkanes, the cracked product distributions provide valuable information on the nature of the parent alkylcarbenium ions which underwent β-scission.

Methylbranching selectivity. Methylbranching proceeds mainly over CPCP intermediates, as explained in paragraph 2.a.4. The contribution of mechanisms involving larger protonated rings in methylbranching is rather small (30). The CPCP mechanism predicts the following methylnonane composition from decane (30,67):

  2-methylnonane: 17%   4-methylnonane: 33%
  3-methylnonane: 33%   5-methylnonane: 17%

The initial distribution of methylnonanes obtained from decane over bifunctional faujasite-type zeolites is very close to this composition, indicating that the micropore system of faujasite is not a shape selective environment with respect to methylbranching of decane (67-69). In zeolites with smaller cavities and 10-membered ring (10-MR) ports, the formation of 2-methylnonane is favoured at the expense of especially 4- and 5-methylnonane (67,70,71). The 2-methylnonane/5-methylnonane selectivity ratio, denoted as the refined constraint index, CI^o, can be used to characterize a zeolite (67,72,73). Very small X-ray amorphous ZSM-5 crystallites, with dimensions of smaller than 10 nm, exhibit enhanced 2-methylnonane selectivities (74). Tthe phenomenon has been ascribed to transition state shape selectivy. When the size of ZSM-5 crystallites is increased from 0.5 to 15 μm, CI^o is changed from 6.8 to 9.4, indicating that the diffusional path length enhances the primary selectivity (72). CI^o values reported in literature are collected in Table 3.

The preference of the methylbranching reaction observed at one end of the hydrocarbon chain is also found at the other end (72). The dibranched isodecanes are rich in 2,7-dimethyloctane on zeolites with a high CI^o value (72).

Table 3. CI⁰ values of zeolites (73,75-77)

zeolite	CI⁰	zeolite	CI⁰	zeolite	CI⁰	zeolite	CI⁰
		12-MR zeolites				10-MR zeolites	
L	1.0	SAPO-5	1.4-1.5	ZSM-11	2.7	ZSM-22	14.4
Omega	1.2	Beta	1.4-2.9	Clinoptilolite	3.6	SAPO-11	3.6
USY	1.3	Phi	1.4	ZSM-48	5.2		
Y	1.4	Mordenite	1.8	ZSM-5	6.8-9.4		
ZSM-3	1.3	Offretite	1.8	ZSM-35	7.1-8.1		
CSZ-1	1.4	ZSM-25	2.2	Ferrierite	8.1-10.3		
SAPO-37	1.3	ZSM-12	2.2	ZSM-23	10.8		

Ethylbranching and propylbranching selectivity. Substituted protonated cyclobutanes and cyclopentanes are the reaction intermediates of ethylbranching and propylbranching, respectively. Ten-membered ring zeolites inhibit the formation of 3- and 4-ethyloctane and 4-propylheptane at low levels of conversion of decane, irrespectively of morphology and crystallite size (72). This indicates that the catalytic chemistry occurs in the intracrystalline void volume, where the formation and/or rearrangements of substituted protonated cycloalkanes larger than cyclopropane is suppressed by transition-state shape selectivity. The ethyloctane and propylheptane selectivity correlates with the dimensions of 12-MR zeolite cavities.

The mechanism of branching via substituted protonated cycloalkanes predicts the initial 3-ethyloctane/4-ethyloctane selectivity of 0.75, whereas the ratio of these isomers in the thermodynamic equilibrium is close to 0.5. The initial 3-ethyloctane/4-ethyloctane ratios on 12-MR zeolites range from 0.6 to 1.1 (67), indicating that the zeolite structures discriminate between the many reaction intermediates.

Isopentane selectivity. Isopentane is formed primarily through type A β-scission of α,γ,γ-tribranched alkylcarbenium ions, whereas pentane arises from type B_1, B_2 and C scisssions of less branched cations. The bulky nature of the α,γ,γ-tribranched cations explains why the isopentane selectivity is lower in constrained environments.

Shape selective Hydrocracking. Evidence is present that the pore geometry imposes molecular shape selectivity on the rearrangements and/or β-scissions of dibranched and tribranched alkylcarbenium ions. Deviating selectivities appear in the carbon number distributions as well as in the distributions of individual cracked products within the different carbon number fractions.

In the regular hydrocracking pattern of a n-alkane with an even CN, equal to n, the yield of the CN fractions obeys the following relationship:

$$C_3 = C_{n-3} \ll C_4 = C_{n-4} < \ldots < C_{n/2} \qquad (3.1)$$

478

For a feed with an odd CN, this relationship is:

$$C_3 = C_{n-3} \ll C_4 = C_{n-4} < \ldots < C_{(n-1)/2} = C_{(n+1)/2} \qquad (3.2)$$

The deviations encountered are the suppression of the formation of specific CN fractions, in which instances the yield sequence is reversed:

$$C_{x-1} = C_{n-x-1} > C_x = C_{n-x} < C_x \qquad (3.3)$$

For a given zeolite, deviations from the regular hydrocracking pattern set in at a specific carbon number and remain present for larger CN although new anomalies may appear. Examples are shown in Figs.20 and 21.

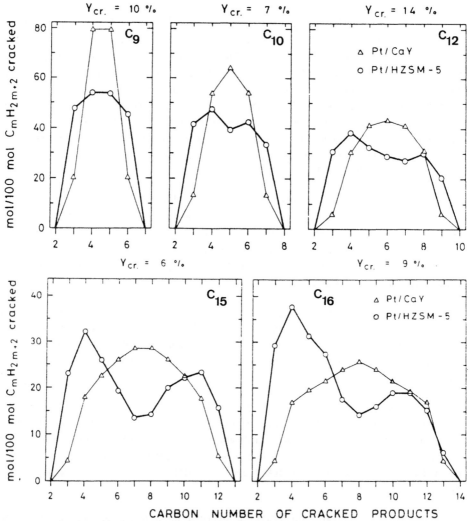

Fig.20. Carbon number distribution of the cracked products from n-alkanes over Pt/ZSM-5 and over Pt/CaY, representing the regular pattern (from ref.70, Copyright Elsevier).

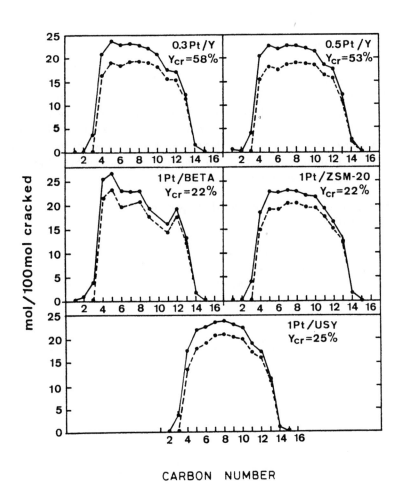

Fig.21. Carbon number distribution of the cracked products from heptadecane over 12-MR zeolites (from ref.78, Copyright Elsevier).

If for a tribranched alkylcarbenium ion the specific requirement of α,γ,γ-branching is not met, the molecule has to be cracked by the energetically less favourable mechanisms B_1, B_2 or C. The 2,3-dimethylbutane molecule is a probe for the contribution of B_1 and B_2 cracking of tribranched isomers (78).

480

3.b.2. Shape selectivity in cycloalkane conversion

Weitkamp et al. (46) have studied the bifunctional conversion of C_{10} alkylnaphthenes over bifunctional catalysts. The possible type A and B β-scissions are given in Fig. 19. Type A β-scission is highly selective and isobutane is invariably a scission product. The precursor ions for type A β-scission are bulky and will have difficulties to be formed in constrained environments, so that deviation to cracking mechanisms where less bulky transition states are involved, intervene (46). This is the case for type B β-scission. In the latter case the product distribution is much broader and in the C_4 fraction less isobutane is formed. These considerations are at the basis of the <u>spaciousness index</u> (SI) (79), which is simply the ratio of isobutane to n-butane in the products. The ratio, therefore, increases with increasing pore width and is in principle sensitive to changes in the architecture of the intracrystalline zeolitic void volume.

With some 0.3 wt % of Pd metal dispersed in the zeolite pores, SI is independant of (46):

*   the reaction temperature over a wide range of yields of cracking products,
*   the Al content of the zeolites and consequently, of the site density,
*   the crystallite size; it is therefore a true example of transition state shape selectivity.

The variation of SI with the zeolite type is given in Fig.22.

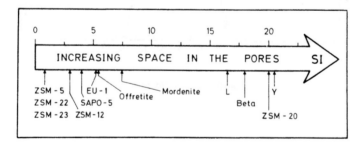

Fig.22. Spaciousness Index of some zeolites. (From ref.46; copyright Elsevier).

Fig.22 shows that SI is not particularly suitable for the characterization of 10-MR zeolites. It is, however, remarkably sensitive to 12-MR zeolites, as it discriminates easily Mordenite from ZSM-12 and EU-1 on one hand and even ZSM-20 and ordinary faujasite on the other hand.

The relation between the refined constraint index (CI^O) and the spaciousness index (SI) is given in Fig.23. The figure clearly shows that CI^O is very sensitive for 10-membered, in contrast to SI, which varies largely with the type of 12-membered ring zeolite, and vice verse.

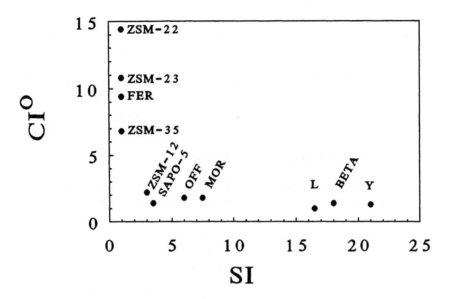

Fig.23. Relation between the refined constraint index (CI^O) and the spaciousness index (SI).

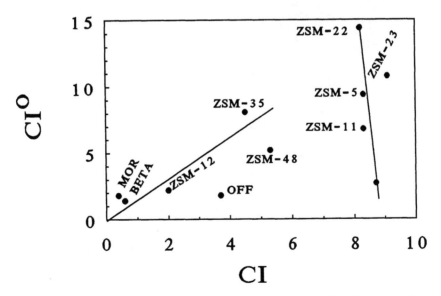

Fig.24. Relation between constraint index (CI) and refined constraint index (CI^O).

A similar correlation is made between the constraint index (CI) the refined constraint index (CI^0) in Fig.24. Generally speaking, this figure contains two discrete areas:

* one in which there exits a resonable positive correlation between both indices,
* another one, mainly involving the 10-membered ring zeolites, for which a small variation in CI is accompanied by a pronounced change in CI^0.

At this moment it is not possible to interpret this relation in more detail. As general conclusion it can be stated, however, that both indices although being adapted to characterize 10-membered ring zeolites, are not sensitive to the same variations in the pore architecture of such zeolites.

3.b.3. Bifunctional conversion of short alkanes

The bifunctional conversion of short alkanes cannot occur via the traditional mechanisms described higher as the energetically favorable conversion pathways cannot occur because the required highly branched intermediate carbenium ions cannot be formed. Van Hooff et al. (80) in a Pt/H-ZSM-5 zeolite with low Pt content (0.4 wt%) and between 200 and 350 °C, reported that from propane the C_4 alkane fraction, containing much n-butane, dominates. A multistep mechanism which involves propene oligomerization and shape-selective hydrocracking accounts for the product distribution (80):

$$C_3 \underset{-H_2}{\overset{+H_2}{\rightleftarrows}} (Pt) \quad C_3^= \underset{-H^+}{\overset{H^+}{\rightleftarrows}} C_3^+$$

$$C_3 \;+\; C_3^= \longrightarrow C_6^+$$

$$C_6^+ \;+\; C_3^= \longrightarrow C_9^+$$

$$C_9^+ \xrightarrow{\;\beta\text{-scission}\;} C_4^= \;+\; C_5^+$$

$$C_5^+ \;+\; C_3^= \longrightarrow C_8^+ \longrightarrow C_4^= \;+\; C_4^+$$
$$\xrightarrow{\;\beta\text{-scission}\;} C_3^+ \;+\; C_5^=$$

$$C_4^= \underset{+H_2}{\overset{-H_2}{\rightleftarrows}} (Pt) \quad C_4$$

The high selectivity for n-butane in the C_4 fraction in the hydrocracking of long chain alkanes, is a typical feature of shape selective bifunctional conversion of n-alkanes on ZSM-5 (see higher). When more Pt (4%) is present in the system, the metal dispersion decreases and hydrogenolysis on the metal phase is superimposed, giving rise to a very complex product mixture. It follows, that for short chain alkanes conversion via bifunctional catalysis, the establishment of a good balance between the two catalytic functions is difficult if not impossible.

Recently, much work has been published on the bifunctional conversion of propane in absence of hydrogen and with Pt and/or Ga as hydrogenation/dehydrogenation function (81-85). On such catalysts cyclisation into aromatics becomes now dominant and over 50 % of selectivity for aromatics has been reported. As to the overall reaction mechanism, there is now general agreement (81-85):

$$2\ C_3^= \xrightarrow{H^+}\ C_6^=$$

$$C_6^= \xrightarrow{Ga}\ C_6^{2=} \xrightarrow{Ga}\ C_6^{3=} \xrightarrow{Ga}\ \text{cyclo } C_6^{2=} \xrightarrow{Ga}\ C_6H_6$$

$$\text{light olefins} \xrightarrow{H^+}\ \text{aromatics}$$

In a well-balanced catalyst, the role of the acid Brønsted sites is to dimerize (oligomerize) the feed olefins. The zeolitic Ga after catalyst activation through calcination and in reaction conditions (773 K) is most probably present as Ga-oxide, whether it is impregnated or ion-exchanged in the parent H-ZSM-5 or associated with its framework during synthesis. As bulk Ga_2O_3, it has also dehydrocyclisation properties, giving rise to hexadiene, hexatriene, cyclohexadiene and benzene. The acid catalyzed conversion of light olefins to aromatics, which will be considered below, is always a competitive reaction as well as coking. The presence of the second function in the acidic ZSM-5, establishes a rapid hydrogenation-dehydrogenation equilibrium among the feed molecules and intervenes in the rapid cyclisation of C_6 and longer intermediates formed by acid oligomerization, thus enhancing considerably the aromatics yield. A good function balance is again needed, as Pt alone results in severe hydrogenolysis and light products make (81). After alloying with Cu and/or Ga, this side reaction is suppressed very much (81).

When propane is converted in H-ZSM-5, the ratio, R, of the moles of $(C_2+C_3+C_4)$ products to that of aromatics is close to 3, what is required for the following hydrogen transfer reaction (81):

$$\text{cyclohexane} + 3\ C_3^= \text{ ------> } 3\ C_3 + C_6H_6 \qquad (3.4)$$

On Ga-containing H-ZSM-5, this ratio is around 1, pointing to the strong participation of a true dehydrogenation reaction (81).

The subtle equilibration of the two functions in shape selective high-silica zeolites for the conversion of light alkanes, finds another example in the conversion of n-butane over a Pt-on-borosilicate HAMS-1B (86). In this case, the acid strength of the Si(B)-OH sites are weaker than the corresponding Si(Al)-OH sites in H-ZSM-5. It is reported that on such catalyst the C_4 fraction can be equilibrated in a selective way and depending on the temperature, H_2/butane ratio and pressure, various amounts of isobutane, isobutene, n-butenes, butadiene and n-butane are formed (86). Such system is highly selective as only hydrogenation/dehydrogenation occurs on the metal function and PCP-isomerization on the weak Brönsted acid sites.

3.c. Shape-selective acid catalysis

This kind of catalysis occurs mainly on intermediate pore zeolites of the pentasil type. In many cases, this catalytic chemistry is at the basis of new industrial processes such as selective aromatics alkylation, isomerization of alkylaromatics, methanol conversion, olefin oligomerization, olefin aromatization, catalytic dewaxing, octane boosting in fuels, etc.. However, these processes cannot be treated within the scope of the present chapter.

3.c.1. Shape-selective conversion of alkanes

The traditional cracking sequences and reactivities of alkanes, which can be derived from classical carbenium ion chemistry, are now changed by one of the abovementioned shape selective effects:

*   the rate increase with chain length is slower than in absence of steric constraints
*   the reactivity decreases with increasing degree of branching rather than increase, as in absence of constraints an energetically more favorable cracking route would be available (B instead of C, A instead of B, etc.).
*   the reactivity decreases considerably with the position of the side chain; this cannot be explained on energetic grounds alone.

A sample of reactivities of alkanes on H-ZSM-5 is given below, the relative cracking rates being in brackets (48):

(1) 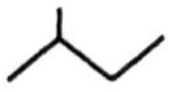   (0.04)

(3.1)    (1.65)

(4.3) 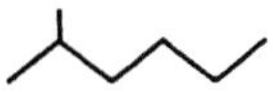   (2.26) 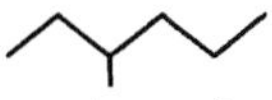   (1.65)    (0.74)    (0.35)

The relative rates are still governed by classical carbenium chemistry but disguised by steric constraints in the intracrystalline volume of the zeolite.

The definition of a **constraint index** to distinguish medium pore zeolites from large and small ones, is based on these effects. In patents assigned to Mobil Oil Corp. (87), the constraint index (CI) is defined as:

$\log_{10}$ (fraction n-C$_6$ remaining) / $\log_{10}$ (fraction 3-meC$_5$ remaining)   (3.5)

when a mixture of n-hexane (n-C$_6$) and 3-methylpentane (3-meC$_5$) are reacted over the acid zeolite in strictly defined conditions. Fig.25 representing these CI values, indicates that it allows to distinguish with certainty medium pore zeolites and rank them at least qualitatively according to their pore size. These values correspond approximately to the ratio of the cracking rate constants of n-C$_6$ to 3-meC$_5$ (88). As the relative participation of the bimolecular and monomolecular cracking mechanisms for alkanes on medium pore zeolites are temperature dependant (see higher), and the latter reaction is less spatially constrained, the value of CI will tend to reach a value of one at high temperatures (48).

An in depth discussion of CI is available from Guisnet et al. (91). The value of CI for a given zeolite structure, is dependant on (91):

*   its acidity (strength, density, presence of extracrystalline sites),
*   on the presence of dehydrogenating impurities,
*   the amount of coke present and therefore the time on stream, and for a
     given reaction time on the decay rate and absolute activity.

486

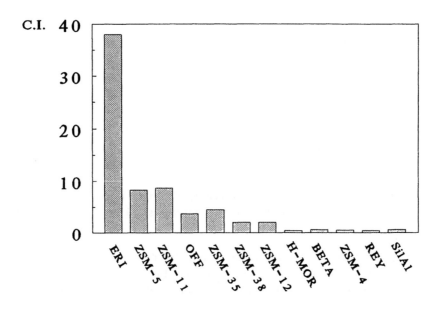

Fig.25. Mobil CI of different zeolites (data from ref. 87).

Prior to these deviations, a special type of shape selective conversion of alkanes has been detected by Gorring on erionite-type zeolite (89). This effect is known in literature as the **cage-effect**. Alkanes with chain length corresponding to that of the erionite-cage (C_7-C_9) show a strongly reduced mobility and enhanced reactivity as they are catched in an energy trap exerted by the cage size. The preferentially trapped molecules have long residence times in such cases and never desorb as such. The other alkanes with shorter or longer chain lenght show an enhanced diffusivity. Although this effect has never been reported to be reproduced by other authors, a similar effect was found by us with functional molecules (90). In this work the bimolecular ketonization of short chain aliphatic carboxylic acids has been studied. Only the reaction of butyric acid was selective, pointing to an end-to-tail adsorption and reaction in the erionite cages.

3.c.2. Shape-selective conversion of olefines

On H-ZSM-5 zeolites, true acid-catalyzed oligomerization of C_3-C_4 olefines is possible, in absence of any side reactions as hydrogen transfer (92), provided the severity is low. At higher severity and temperatures, random oligomerization or olefin interconversion starts to occur (48):

true oligomerization:

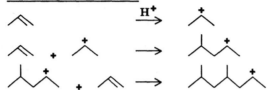

random oligomerization:

$$C_3^= + C_6^+ \;\rightleftarrows\; C_9^+ \;\rightleftarrows\; C_4^= + C_5^+$$

$$C_3^+ + C_4^= \;\rightleftarrows\; C_7^+$$

$$C_4^+ + C_4^= \;\rightleftarrows\; C_8^+ \;\rightleftarrows\; C_3^= + C_5^+$$

$$C_7^+ + C_3^= \;\rightleftarrows\; C_{10}^+ \;\rightleftarrows\; C_5^+ + C_5^=$$

$$\rightleftarrows\; C_4^= + C_6^+$$

The carbon number distribution in the products of an olefin interconversion reaction depends on (92, 93):

* the reaction temperature: higher temperatures giving more olefin interconversion, cracking and hydrogen transfer to shorter products and the appearance of successively: napthenes, cyclo-olefins, aromatics and saturates,
* the partial pressure of the olefin: an increase in partial pressure has the same effect as a temperature decrease,
* the density of the Brönsted acid sites: all other factors remaining equal, a ZSM-5 richer in Al will cause the same effects as a temperature increase.

488

3.c.3. Shape-selective aromatisation

Via essentially the same high temperature mechanism as outlined for olefines, a great number of compounds can be converted over H-ZSM-5 and related shape selective zeolites (92). The complex interplay of the different reaction types and their united view can be schematically represented as follows (92-95):

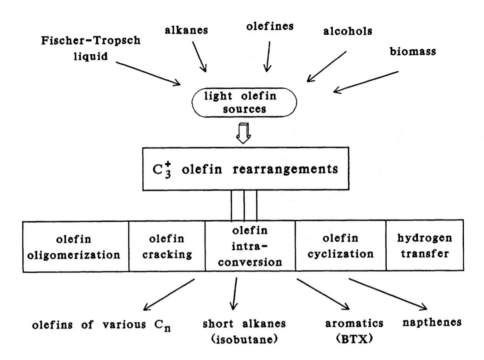

With light olefins and also from methanol, 3 mole of alkanes are formed per mole of aromatics as required by the stoechiometry of the hydrogen transfer reaction (92). The aromatics, therefore, donot stem from dehydrogenation but rather from hydrogen transfer reactions. In principle, all functionalized alkanes can be "defunctionalized" via the same reaction pathway, the severity of the conversion needed being dependant on the stability of the feed molecule.

The conversion of methanol to olefins and further via the above mechanism into gasoline compounds, is the reaction which attracted most attention on medium pore zeolites. So many authors have contributed to the development of the knowledge and catalytic chemistry of the formation of the first C-C bond an subsequent reaction steps that this alone would cover a whole chapter. A recent mechanistic overview is available from Chang (96). Every possible reactive C_1 intermediate, whether of radical, carbenic, carbocationic or ylide nature, has been advanced to explain the formation of the first C-C bond. According to Chang (96), the various intermediates are related to one another and may be interconvertable. Anyway, definitive evidence supporting any particular pathway is lacking (96).

The following initiation mechanism seems plausible, as it doesnot require the existence of strong basic sites which are required for H^+ removal from C-H, as in most other schemes occurs (96):

* generation of surface methoxyls with the Brönsted acid sites:

$$CH_3OH + ZO-H -----> CH_3-OZ + H_2O \qquad (3.6)$$

* generation of zeolite stabilized radicals by thermolysis of these surface methoxy groups:

$$CH_3O-Z + R^O -----> RH + ZO-CH_2^O \qquad (3.7)$$

* radical scission:

$$ZO-CH_2^O + CH_3-O-CH_3 -----> ZO^O - (CH_3)_2-O:-:CH_2 \qquad (3.8)$$

$$-----> ZO^O - (CH_3)_2-O^{+}-^{-}CH_2 \qquad (3.9)$$

$$ZO^O - (CH_3)_2-O:-:CH_2 -----> ZO^O + CH_3-CH_2-O-CH_3 \qquad (3.10)$$

$$ZO^O + RH -----> R^O + ZO-H \qquad (3.11)$$

The surface bound ylide, $ZO^O - (CH_3)_2-O^{+}-^{-}CH_2$, is isoelectronic with the surface carbene, $ZO^O - (CH_3)_2-O:-:CH_2$, which inserts into C-H to form the first C-C. The exact nature of such site (ZO^O) remains unspecified.

The secondary reactions in the methanol decomposition can be eliminated partially and high selectivities for C_2-C_4 olefins obtained.

*   Decoupling of olefinization (with k_1) from aromatization (with k_2 as rate constant) is possible on H-ZSM-5 at low oxygenate partial pressure (97) and allows high olefin selectivity at complete methanol conversion.
*   Apparently, the temperature dependance of k_1 (81 kJoule/mol) is higher than of k_2 (no significant change) (98). Also at a given temperature, the acid dependancy of both reactions is different as the k_1 / k_2 ratio increases with increasing Si/Al ratio or decreasing site density (98). Here again the reduction of bimolecular reactions is encountered when sites become diluted. Thus a combination of low acidity and high temperature, bring about high yields of olefins (up to 75% (98)).

*   Improvements in olefin selectivity can also be made by steric effects, so as to avoid sterically the formation of aromatics. This can be done by modifying the effective pore diameter of ZSM-5 by impregnation with P (98),

Mg (99), Zr or Ti compounds (100) or by chemical reaction with silanes (101).

* Improvements in olefin selectivity can be made in other zeolite structures as chabasites (100) including SAPO-34 (102), and erionite-type zeolites (103), which inhibit the formation of aromatics in the intracrystalline pore volume. The method of pore mouth blockage is also here effective (100) to further enhance olefin selectivity, probably by avoiding aromatization at the external surface.

* The olefin / aromatic ratio depends very much on the nature of the metal in metallo-silicates with MFI structure and Si/Me ratios of 3,200 (82). The following series for olefin selectivity at 100 % methanol conversion is obtained (82):

$$Ga > V = Cr > Sc > Mn > Ge > La > Al > Ti > Zr > Ni > Fe > Pt > Co \qquad (3.12)$$

All elements positioned to the right of Al will exhibit decreased acidity and the increased olefin formation seems logical. The elements located to the left of Al and exhibiting pronounced aromatization properties all have dehydrocyclisation properties and consequently a bifunctional mechanism is possibly operative. At this stage this explanation remains speculative, however, because of the lack of available details on the product distribution. By futher optimization of experimental conditions an 100 % yield in $C_2 + C_3$ olefines can be obtained with an ironsilicate of MFI structure (82).

The question whether ethene or propene is the initial olefin, has initiated much debate. Chu and Chang (104) clearly established that on a siliceous H-ZSM-5 at 773 K, the $C_2^= - C_3^=$ distribution is thermodynamically controlled at low conversions (and thus low partial pressures of H_2O), while with increasing conversion the distribution becomes subject to kinetic control. As the radical and autocatalytic rearrangements are controlling, ethene should be the primary reaction product. The presence of water seems to be another factor which inhibits subsequent olefin equilibration, possibly through competitive sorption (104).

3.c.4. Shape selectivity in alkylation and isomerization of (alkyl)aromatics.

The isomerization / disproportionation of m-xylene has been an often used test reaction for the determination of shape selective properties in acid zeolites. Over medium pore zeolites, these reactions can be carried out in relatively mild conditions, without severe interference of catalyst deactivation (105). The initial p/o selectivity from m-xylene has been often used as a criterion for diffusion shape selectivity in medium pore zeolites (105-112). There are a number of precautions which have to be taken into account:

* for reliable information the selectivity/conversion realtionship has to be examined (105);

* the existence of mass transport effects will lead to a distribution that also significantly deviates from thermodynamic values (109);

* the thermodynamics of allow a maximum conversion of 55 % at 432 K and 50 % at 1023 K; the corresponding p/o isomer ratios are 1.23 and 0.82, respectively (105);

* the isomerization selectivity should be as high as possible and secondary reactions of the disproportionation type should be minimized by mainly operating at lower reaction temperatures;

Thus only in absence of diffusion limitations on m-xylene mobility, initial p/o-isomer ratios significantly higher than one can be taken as evidence for shape selectivity. Minor changes such as dealumination or calcination are not neccessarily reflected, unless care is taken that activity changes as result of such treatments are compensated by change in contact times, which is impossible with a single measurement. This again points to the necessity of determing the selectivity/conversion curve at (a) given reaction temperature(s).

Only recently, it has been shown (108) that the initial p/o selectivity is strongly dependant on intracrystalline window sizes in case of 10-membered ring zeolites but not for 12-membered ring zeolites (Fig.26).

Kumar et al. (112) come to essentially the same conclusion. They were able to rank 10-membered ring zeolites according to their properties as catalysts for product shape selectivity. The following order of decreasing initial p/o isomer ratios (given in brackets) was obtained:

$$ZSM\text{-}23 \ (4.4) > ZSM\text{-}48 \ (2.8) > ZSM\text{-}50 \ (2.0) > ZSM\text{-}5 \ (1.3) \qquad (3.13)$$

Similar ranking could be obtained from methylation of toluene.

As outlined higher, in the traditional mechanism for disproportionation of alkylbenzenes, diarylalkanes are thought to be the reaction intermediates. Olson and Haag (106) were the first to correlate the isomerization / disproportionation (I/D) selectivity with the intracrystalline cavity diameter. This ratio seemed to increase steadily with decreasing cavity diameter for zeolites, irrespective of their structure. Recently, using more zeolite structure types, it was shown (108) that this relation is only valid for 10-membered ring zeolites (Fig.26) and that the bimolecular disproportionation reaction is suppressed. For the 12-membered ring structures, no relation is obtained with the window size of the structure.

492

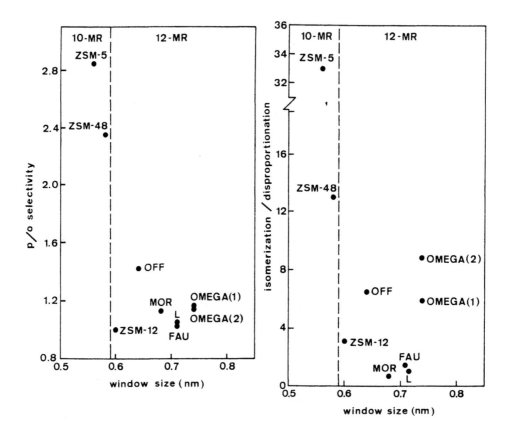

Fig.26. Initial o/p selectivity (left) and isomerization / disproportionation selectivity (right) from m-xylene over different H-zeolites. (From ref.108, copyright Elsevier).

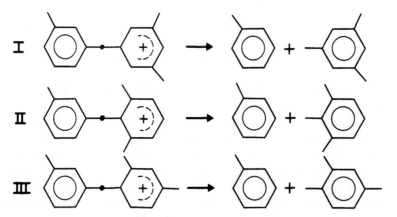

Fig.27. Possible transition state complexes for the bimolecular disproportionation of m-xylene. (From ref. 108, copyright Elsevier).

The suppression of disproportionation in 10-membered ring zeolites can be easily understood in terms of spatial restrictions on the formation of the transition states (108, 112). Fig.27 shows the 3 possible transition state complexes for the disproportionation of m-xylene. It follows that spatial restrictions on the formation of one of these intermediates, will not only influence the I/D selectivity but also the isomer distribution in the trimethylbenzenes. Changes in this selectivity seem to reflect much closer the size and shape of the intracrystalline cages (108). Moreover, the product ratio of 1,2,3-/1,3,5-trimethylbenzenes is a criterion applicable to distinguish zeolites with large 12-membered ring channels (108).

Acid medium pore aluminophosphate-based molecular sieves, such as SAPO-5 and SAPO-11, although being less active than the corresponding alumino-silicate anologues, show similar shape selective effects (113). However, as a result of presently not understood constraints, the bimolecular and thus hydrogen transfer reactions are very much reduced in SAPO's (113).

Acknowledgments

J.A.M. acknowledges the Flemish National Fund for Scientific Research for a position as Research Associate. The authors acknowledge continuous financial support from the National Fund for Scientific Research Belgium and the National Ministry of Science Policy from Belgium (in the frame of a Concerted Action on Catalysis).

References

1. J.M. Thomas and W.J. Thomas, "Heterogeneous Catalysis", Academic Press, New York and London, 1967 p. 451.
2. P. Trambouze, H. Van Landeghem and J.P. Waucquier, "Les Réacteurs Chimiques", Ed. Technip, Paris, 1984.
3. G.F. Froment and K.B. Bishop, "Chemical Reactor Analysis and Design", J. Wiley & Sons, New York, Chicester, Brisbane, Toronto, 1979.
4. J.M. Berty, "Laboratory Reactors for Catalytic Studies", in "Applied Industrial Catalysis", 1, B.E. Leach, ed., Academic Press, New York and London, 1983, p. 41.
5. A. Rodrigues, C. Costa, R. Ferreira, J. Loureiro and S. Azvedo, in "Zeolites: Science and Technology", ed. F.R. Ribeiro, A.E. Rodriques, L.D. Rollmann and C. Naccache, Martinus Nijhoff, The Hague, Boston, Lancaster, 1984, p. 423.
6. R. Prins and G. Schuit, eds., "Chemistry and Chemical Engineering of Catalytic Processes", Sijthoff & Noordhoff, 1980, Alphen a/d Rijn, Germantown, 1980.
7. J.J. Carberry, "Chemical and Catalytic Reaction Engineering", McGraw-Hill, New York, 1976.
8. C.N. Satterfield, "Heterogeneous Catalysis in Practice", McGraw-Hill, New York, 1980.
9. Y.T. Shah, "Gas-Liquid-Solid Reactor Design", McGraw-Hill, New York, 1979.
10. O. Levenspiel, "Chemical Reaction Engineering", J. Wiley & Sons, New York, Chicester, Brisbane, Toronto, 1972.
11. J. Weitkamp, Stud. Surf. Sci. Catal. 37 (1988) 515.
12. J.F. Le Page, "Applied Heterogeneous Catalysis", Technip (1987) p. 137.
13. J.F. Le Page, "Applied Heterogeneous Catalysis", Technip (1987) p. 275.
14. J.F. Le Page, "Applied Heterogeneous Catalysis", Technip (1987) p. 42.

494

15. V.R. Choudhary and L.K. Doraiswamy, Ind. Eng. Chem. Prod. Res. Dev. 10 (1971) 219.
16. K.C. Pratt, "Small Laboratory Scale Reactors", Chapt. 4 in "Catalysis Science and Technology", Vol. 8, Eds. J.R. Anderson and M. Boudart, Springer-Verlag, Berlin, 1987.
17. J.J. Carberry, idem, p.163.
18. C.H. Massen and J.A. Poulis, in "Microweighing in Vacuo and Controlled Environments", eds. A.W. Czandera, S.P. Wolsky, Vol. 4, "Methods and Phenomena-Their Application in Science and Technology", Elsevier, Amsterdam, 1980, p. 95.
19. L.L. Hegedus and E.E. Petersen, Catal. Rev. 9 (1974) 245.
20. E.E. Petersen, "Experimental Methods in Catalytic Research", II, Capt. 7, Eds. R.B. Anderson and P.T. Dawson, Academic Press, New York, 1976.
21. P.A. Jacobs, Stud. Surf. Sci. Catal. 29 (1986) 357.
22. J.R. Anderson, and K.C. Pratt, Introduction to Characterization and Testing of Catalysts, Academic Press, Sidney, 1985.
23. J.A. Paraskos, Y.T. Shah, J.D. McKinney, and N.L. Carr, Ind. Eng. Chem. Proc. Des. Dev. 15 (1976) 164.
24. E. Kikuchi, H. Adachi, T. Momoki, M. Hirose and Y. Morita, Fuel 62 (1983) 226.
25. P.A. Jacobs, "Carbonigenic Activity of Zeolites", Elsevier, Amsterdam, 1977.
26. P.A. Jacobs, Stud. Surf. Sci. Catal. 29 (1986) 357.
27. S.TS. Homeyer and W.M.H. Sachtler, Stud. Surf. Sci. Catal. 49B (975) 1989.
28. L.R.M. Martens, P.J. Grobet and P.A. Jacobs, Nature, 315 (1985) 568.
29. J.M. Garces, G.E. Vrieland, S.I. Bates and F.M. Scheidt, Stud. Surf. Sci. Catal. 20 (1985) 67.
30. J.A. Martens and P.A. Jacobs, in "Theoretical Aspects of Heterogeneous Catalysis", Ed. J.B. Moffat, Van Nostrand Reinhold, New York, 1990, Chapt. 2.
31. P. Vogel, 'Carbocation Chemistry', Stud. Org. Chem. Vol.21, Elsevier, Amsterdam, 1985.
32. H.J. Lovink and L.A. Pine, "The Hydrocarbon Chemistry of FCC Naphtha Formation", Ed. Technip, 1990.
33. F. Figueras, R. Gomez and M. Primet, Adv. Chem. Ser. 121 (1973) 490.
34. R.A. Dalla Betta and M. Boudart, Proceed. Vth Int. Congr. Catal., Palm Beach, Ed. J. Hightower, 1972, Paper 100.
35. J.R. Bernard, Proceed. Vth Int. Conf. Zeolites, Ed. L.V. Rees, Heyden, London, 1980, p.686.
36. T.R. Hughes, W.C. Buss, P.W. Tamm and R.L. Jacobson, Stud. Surf. Sci. Catal. 28 (1986) 725.
37. G. Larsen and G.L. Haller, Catal. Lett. 3 (1989) 103.
38. B.W. Wojciechowski and A. Corma, Catalytic Cracking, Catalysts, Chemistry and Kinetics, Marcel Dekker, New York, Basel (1986) p.9.
39. P.B. Weisz, Adv. Catal. 13 (1962) 137.
40. M.L. Coonradt and W.E. Garwood, Ind. Eng. Chem. Prod. Res. Dev. 3 (1964) 38.
41. J.Weitkamp, W. Gerhardt and P.A. Jacobs, Proceed. Int. Symp. Catal. Zeolites, Siofok (Hungary), May 1985, 261.
42. J.A. Martens, P.A. Jacobs and J. Weitkamp, Applied Catal. 20 (1986) 239.
43. J. Weitkamp, P.A. Jacobs and S. Ernst, in 'Structure and Reactivity of Modified Zeolites', P.A. Jacobs et al. , eds., Elsevier, Amsterdam, Stud. Surf. Sci. Catal. 18 (1984) 279.
44. P.A. Jacobs, M. Tielen, J.A. Martens and H.K. Beyer, J. Mol. Catal. 27 (1984) 11.
45. P.A. Jacobs, M. Tielen and R.C. Sosa, in 'Structure and Reactivity of Modified Zeolites', P.A. Jacobs et al. , eds., Elsevier, Amsterdam, Stud. Surf. Sci. Catal. 18 (1984) 175.
46. J. Weitkamp, S. Ernst and C.Y. Chen, Stud. Surf. Sci. Catal 49B (1989) 1115.
47. W.O. Haag, in "Heterogeneous Catalysis", Ed. B.L. Shapiro, Texas A&M University Press, 1984, p.95.

48. N.Y. Chen and W.O. Haag, in "Hydrogen Effects in Catalysis", Eds. Z. Paal and G. Menon, Marcel Dekker, New York and Basel, 1988, p. 695.
49. J.A. Rabo, in "Zeolites: Science and Technology", F.R. Ribeiro, A.E. Rodrigues, L.D. Rollmann and C. Naccache, Martinus Nijhof , The Hague , Boston, Lancaster, p. 291.
50. J. Weitkamp, Proceed. Int. Symp. Zeolite Catalysis, Siofok (Hungary), 1985, p. 271.
51. W.O. Haag and R.M. Dessau, Proceed. 8th Int. Cong. Catalysis, Berlin, 2 (1984) 305.
52. M. Guisnet, Stud. Surf. Sci. Catal. 20 (1985) 273.
53. M. Guisnet and N.S. Gnep, in "Zeolites: Science and Technology", F.R. Ribeiro, A.E. Rodrigues, L.D. Rollmann and C. Naccache, Martinus Nijhof , The Hague , Boston, Lancaster, p. 571.
54. C.W.R. Engelen, J.P. Wolthuizen and J.H.C. van Hooff, Appl. Catal. 19 (1985) 153.
55. P.B. Weisz and V.J. Frilette, J. Phys. Chem. 64 (1960) 382.
56. S.M. Csicsery, ACS Monogaph 171 (1976) 680.
57. S.M. Csicsery, Zeolites 4 (1984) 202.
58. P.B. Weisz, Pure Appl. Chem. 52 (1980) 2091.
59. E.G. Derouane, in "Intercalation Chemistry", Eds. M.S. Wittinghton and A.J. Jacobson, Academic Press, 1982, p. 101.
60. E.G. Derouane, Stud. Surf. Sci. Catal. 4 (1980) 5.
61. E.G. Derouane, Stud. Surf. Sci. Catal. 19 (1984) 1.
62. J. Dwyer, Chem. Ind. 7 (1984) 229.
63. J.A. Rabo, R. Bezman and M.L. Poutsma, Acta Phys. Chem. 24 (1987) 39.
64. E.G. Derouane and Z. Gabelica, J. Catal. 65 (1980) 486.
65. C. Mirodatos and D. Barthomeuf, J. Catal. 57 (1979) 136.
66. C. Mirodatos and D. Barthomeuf, J. Catal. 93 (1985) 246.
67. J.A. Martens, M. Tielen, P.A. Jacobs and J. Weitkamp, Zeolites 4 (1984) 98.
68. J. Weitkamp, Ind. Eng. Chem. Prod. Res. Dev. 21 (1982) 550.
69. M. Steijns, G. Froment, P. Jacobs, J. Uytterhoeven and J. Weitkamp, Ind. Eng. Chem. Prod. Res. Dev. 20 (1981) 654.
70. J. Weitkamp, P.A. Jacobs and J.A. Martens, Applied Catal. 8 (1983) 123.
71. P.A. Jacobs, J.A. Martens, J. Weitkamp and H.K. Beyer, Faraday Disc. Chem. Soc. 72 (1982) 353.
72. J.A. Martens and P.A. Jacobs, Zeolites 6 (1986) 334.
73. P.A. Jacobs and J.A. Martens, Proceed. 7th Int. Zeolite Conf., Kodansha, Elsevier, 1986, p.23.
74. P.A. Jacobs, E.G. Derouane and J. Weitkamp, J. Chem. Soc. Chem. Commun. (1981) 591.
75. J.A. Martens, C. Janssens, P.J. Grobet, H.K. Beyer and P.A. Jacobs, Stud. Surf. Sci. Catal. 49A (1989) 215.
76. J.A. Martens, P.A. Jacobs and S. Cartlidge, Zeolites 9(1989) 423.
77. J.A. Martens, J. Perez-Pariente and P.A. Jacobs, Proceed. Nato Workshop on 'Chemical Reactions in Organic and Inorganic Constraint Environment' Reidel 1985, p.115.
78. J.A. Martens, M. Tielen and P.A. Jacobs, Stud. Surf. Sci. Catal. 46 (1989) 49.
79. J. Weitkamp, S. Ernst and R. Kumar, Appl. Catal. 27 (1986) 207.
80. C.W.R. Engelen, J.P. Wolthuizen and J.H.C. van Hooff, Appl. Catal. 19 (1985) 153.
81. P. Meriaudeau, G. Sapaly and C. Naccache, Stud. Surf. Sci. Catal. 49B (1989) 1423.
82. T. Inui, Stud. Surf. Sci. Catal. 44 (1988) 198.
83. T. Mole, J.R. Anderson and G. Greer, Appl. Catal. 17 (1985) 141.
84. H. Kitagawa, Y. Sendoda and Y. Ono, J. Catal. 101 (1986) 12.
85. N.S. Gnep, J.Y. Doyemet, A.M. Seco, F. Ribeiro and M. Guisnet, Appl. Catal. 35 (1987) 108.
86. N.A. Kutz, in "Heterogeneous Catalysis", Texas A&M University Press, 1984, p. 121.
87. e.g. in EPA 12,570 (1979); USP 4,300,011 (1981).
88. V.J. Frilette, W.O. Haag and R.M. Lago, J. Catal. 67 (1981) 218.
89. R.L. Gorring, J. Catal. 31 (1971) 13.

90. M. Vervecken, Y. Servotte, M. Wydoodt, L. Jacobs, J.A. Martens and P.A. Jacobs, in "Chemical Reactions in Organic and Inorganic Constrained Systems", Ed. R. Setton, D. Reidel, 1986, p. 95.
91. F.R. Ribeiro, F. Lemos, G. Perot and M. Guisnet, in "Chemical Reactions in Organic and Inorganic Constrained Systems", Ed. R. Setton, D. Reidel, 1986, p. 141.
92. W.O. Haag, in "Heterogeneous Catalysis", Texas A&M University Press, 1984, p. 93.
93. R.J. Quan, L.A. Green, S.A. Tabak and F.J. Krambeck, AIChE Nat. Meeting, New Orleans, La, 1986.
94. W.O. Haag, R.M. Lago and P.G. Rodewald, J. Molec. Catal. 17 (1982) 161.
95. W.O. Haag, Proceed. Int. Conf. Zeolites, Reno, Ed. A. Bisio and D.H. Olson, Butterworth, p.466.
96. C.D. Chang, ACS Symp. Ser. 368 (1988) 596.
97. C.D. Chang, W.H. Land and R.L. Smith, J. Catal. 56 (1979) 169.
98. C.D. Chang, C.T.W. Chu and R.F. Socha, J. Catal. 86 (1984) 289.
99. W.W. Kaeding and S.A. Butler, J. Catal. 61 (1980) 155.
100. U. Dettmeier, E.I. Leupold, H. Litterer, H. Baltes, W. Herzog and A. Wunder, Erdol und Kohle 36(8) (1983) 365.
101. P.G. Rodewald, USP 4,145,315.
102. F.A. Wunder and F.I. Leupold, Angew. Chemie 92 (1980) 125.
103. S.W. Kaiser, USP 4,524,234 (1980).
104. C.T.W. Chu and C.D. Chang, J. Catal. 86 (1984) 297.
105. M. Richter, W. Fiebig, H.G. Lischke, and G. Ohlmann, Zeolites, 9 (1989) 238.
106. D.H. Olson and W.O. Haag, ACS Symp. Ser. 248 (1984) 275.
107. A. Corma, V. Fornez, J. Perez-Pariente, E. Sastre, J.A. Martens and P.A. Jacobs, ACS Symp. Ser. 368 (1988) 555.
108. J.A. Martens, J. Perez-Pariente, E. Sastre, A. Corma and P.A. Jacobs, Appl. Catal. 45 (1988) 85.
109. J. Dewing, J. Molec. Catal. 27 (1984) 25.
110. P.A. Jacobs and J.A. Martens, Pure Appl. Chem. 58(10) (1986) 1329.
111. M. Guisnet and N.S. Gnep, in "Zeolites, Science and Technology", Eds. F.R. Ribeiro, A.E. Rodriguez, L.O. Rollmann and C. Naccache, Nijhoff, Den Haag, 1984, p. 571.
112. R. Kumar, G.N. Rao and P. Ratnasamy, Stud. Surf. Sci. Catal. 49B (1989) 1141.
113. J.A. Rabo, R.J. Pellet, P.K. Coughlin and E.S. Shamsoun, Stud. Surf. Sci. Catal. 46 (1989) 1.

Chapter 13

PREPARATION OF ZEOLITIC CATALYSTS

Herman W.Kouwenhoven(1) and Bas de Kroes(2).
(1)Technisch Chemisches Laboratorium, ETH-Zentrum, 8092 Zuerich,
Switzerland.
(2)AKZO Chemicals B.V., Research Centre Amsterdam, P.O.Box 15,
1000 AA Amsterdam, The Netherlands.

1.INTRODUCTION

Applications of zeolitic catalysts have been established in many
conversion reactions. The nature of the reactions varies
considerably and the scale of operations covers a wide range.
Highly sophisticated catalyst preparation procedures were
developed for the large scale applications in the petrochemical
industries. The considerable economic incentive for obtaining the
optimal yield structure justifies fine tuning of catalysts for
each particular application in hydrocarbon processing. Two recent
review papers (ref.1) and (ref.2) mention a large number of
potential further applications of zeolitic catalysts in chemical
reactions. Although the number of existing applications in fine
chemical reactions is presently small, this area has a large
scope and is a challenging field of applied research in catalyst
preparation.
Some characteristic properties which make zeolites so versatile
as catalyst component are:
   * Size and shape of the zeolite crystallites in many
instances depend on the actual synthesis conditions and are
variables in catalyst design.
   * Compared to the amorphous silicas and aluminas used
traditionally in catalyst preparation, the internal surface area
of zeolites is high and thermally very stable.
   * The pores are well defined and of molecular dimensions,the
actual pore size being determined by the crystallographic zeolite
structure.
   * The accessibility of a zeolite structure may be subtly
changed by ion-exchange and/or postimpregnation with certain
salts, (pore size engineering).
   * The non lattice atoms can be removed by ion-exchange

techniques.

* The non oxygen atoms in the lattice have a tetrahedral coordination (T-atoms).

* The T-atoms in the lattice are in most structures highly accessible, their nature determines the polarity of the surface and its catalytic activity.

* The chemical composition of the lattice is dependent on the synthesis conditions. For many structures the Si/Al ratio may be varied within wide limits and for some structures also the nature of the T-atoms is a variable.

* The chemical composition of the lattice may be changed by the application of the various techniques for dealumination- and T-site substitution reactions (secondary synthesis).

* The catalytic activity of the surface of a zeolite can be modified by a treatment with a reactive chemical and the modification may be limited to the outer surface of the crystals by the choice of a reagent, which due to its size, cannot reach the internal surface.

* Catalyst components having a special effect on the activity of the final composite material such as a hydrogenation/dehydrogenation function can be introduced via ion-exchange with a suitable precursor compound of the required element(s).

The large number of variables which may be addressed during the investigation of the catalytic activity of a specific zeolite for a certain type of reaction prohibits a detailed discussion of every aspect of the preparation of zeolitic catalysts in this article. Rather we will discuss the following steps in the zeolite catalyst preparation procedure:

Zeolite activation

T-site substitution

Metal emplacement

Catalyst shaping

and refer to the relevant literature for detailed discussions of each particular aspect. We will discuss in more detail the preparation of zeolitic cracking catalysts for oil conversion, of zeolitic titanium silicates,used as catalysts for phenol hydroxylation and of noble metal loaded zeolites for various applications.

2. CATALYST PREPARATION PROCEDURE

2.1 ZEOLITE ACTIVATION

2.1.1 Removal of the template

In many cases the zeolite sample intended for use as a catalyst component will be an as-synthesized product, containing template molecules which have to be removed before further processing.

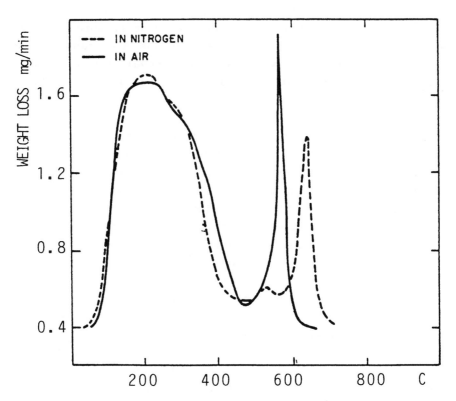

Fig.1. Differential thermogravimetric analysis in air and in nitrogen of as synthesized zeolite Omega.

Since these molecules are mostly strongly sorbed in the zeolite lattice and are often located in positions having a limited accessibility, their removal by air calcination is usually a strongly activated process. Compared to the "free" template molecules their counterparts occluded in the zeolite lattice appear to be stabilized and require a much higher temperature for complete oxidation. Moreover the oxidation process is strongly exothermic and e.g. the adiabatic temperature rise for 1%w CH_2 in a zeolite is about 600 K! Accordingly it is advisable to study the thermal decomposition both in N_2 and in O_2/N_2 mixtures using thermal analysis in order to establish the reaction pattern (see Fig.1).

In many cases it is advantageous to first thermally decompose template molecules in an inert atmosphere and to remove remaining carbonaceous residues by subsequent oxidation. Although most zeolites are thermally quite stable, a staged calcination procedure as outlined above may also prevent unwanted T-atom migration out of the zeolite lattice due to stabilization reactions during the oxidative template removal. A more detailed discussion of the template removal procedure is given in (ref.3).

2.1.2 Ion-exchange for preparation of solid acids

Removal of Na atoms by ion-exchange techniques will usually follow the template removal step. Ion-exchange of zeolites is a research subject in its own right, a recent review is presented in another chapter of this book by Townsend. For the preparation of H-zeolites the most effective and gentle method is via one or more ammonium-exchange treatments. The ammonium-exchanged zeolite is subsequently calcined to convert it to the H-form. On a large scale calcination is preferentially done in an oxygen containing atmosphere. On a small scale a staged procedure using N_2 followed by a N_2/air mixture after reaching a temperature of about 720 K is advantageous. Also for this step, a study of the reaction by thermal analysis is very useful in order to be aware of the exothermic effects which may accompany the removal of ammonia (ref.4). Removal of Na from zeolites with an internal surface consisting of various types of cages differing in accessibility, usually requires repeated ion-exchange steps combined with a calcination. It is assumed that as a result of a calcination Na atoms, which were unaccessible for ion-exchange, are redistributed over the zeolite surface and made accessible. At the same time solid state reactions occur in the zeolite and alumina is removed from the lattice. This phenomenon, which is referred to as stabilization, is discussed later in some detail in relation to the manufacture of the zeolite component for catalytic cracking catalysts. In a broader context stabilization is reviewed in (ref. 5).

For some zeolites it also possible to directly exchange Na- for H ions by a treatment with an acid. The zeolite should be resistant to acids, a general property of zeolites having a Si/Al ratio of 5 or higher.In most cases acid leaching, a more or less inescapable side effect of acid treatment, will remove aluminum from the lattice.For mordenite this results in a maximum in activity for acid catalyzed reactions at a Si/Al ratio of about 18, Table 1. Acid leaching of mordenite is discussed in detail in (ref. 6).

TABLE 1
Catalytic activity of Pt on H-mordenite for
n-pentane isomerisation. Effect of Si/Al ratio.

Na, wt %	Si/Al ratio	Relat.activity
0.03	10	100
0.02	17	135
0.02	25	84

Conditions: 523 K, 30 bar,WHSV:1 h^{-1},
molar ratio H_2/C_5:2.5. 0.5%w Pt on mordenite.

Protonic sites are required in a number of important applications of zeolitic catalysts. The overall catalytic activity depends both on number and quality of the acid sites. The total number of acid sites is equal to the total number of Al atoms in T-sites.It appears that for each particular zeolite structure the acidity per site reaches its maximum value for a specific Si/Al ratio.This ratio is equal to or higher than that required for a surface configuration in which each Al atom is isolated and has no next nearest T-sites occupied by Al atoms. For zeolites having a Si/Al ratio lower than about 9, the strength of the acid sites will not be a constant per site. The width of the curve representing the acidity distribution is related to the Si/Al ratio and the ordering of the trivalent atoms in the structure.It appears that the nature of the trivalent atom determines the acid strength of the site. This is shown by the NH_3 desorption temperature given in Table 2 for the MFI structure containing the T-atoms: B, Fe, Ga, or Al.
The relation between the number of aluminum atoms in the lattice

502

and the activity in an acid catalysed reaction of various samples of ZSM-5 is shown Fig. 2. The relations between acid strength and Si/Al ratio for various zeolite structures are further discussed in (ref.7).

The catalytic activity is strongly dependent on the degree

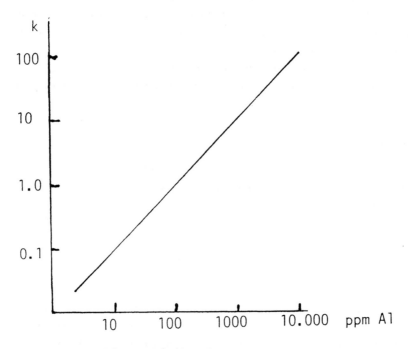

Fig.2. Hexane cracking activity, k, as a
function of Al content in H-ZSM-5 (ref.30).

TABLE 2
Nature of T-atom, acidity and activity of MFI silicates.

T Atom	NH₃ Desorption (ref.11)		C₄ Cracking (ref.12)		Rearrangement of 2-phenyl-propanal to 1-phenylpropanone 573 K (ref.13)			
	Si/T	T Max (K)	Si/T	k^* 773 K	Si/T	WHSV (h^{-1})	Conv.	Sel.
B	170	423			94	0.8	63	97
Fe	196	603			36	2.0	98	95
Ga	396	623	138	7.1				
Al	140	653	140	45.5				

*)k=ml/g/min

of Na removal, Table 3 (ref.8). The deactivating effect of a

TABLE 3
Catalytic activity of Pt-H-zeolite Y for n-pentane
hydro-isomerization. Effect of sodium content.

Crystallinity X-Ray %	Na$_2$O % w	TEMP. FOR 30% Conversion, K
90	2.02	578
80	0.27	573
80	0.02	523

Conditions: WHSV: 1 h^{-1}; 30 bar; H$_2$/C$_5$ molar ratio:2.5.
0.5%w Pt on zeolite Y.

small residual Na content on acidic activity is similar to that of small amounts of adsorbed NH$_3$. Deactivation of Brönsted acidity by sodium and ammonium ions is further discussed in recent publications (ref.9) and (ref.10). For maximum Brönsted acid activity deep removal of sodium and/or other bases is essential.

2.1.3   Introduction of other elements

Ion-exchange is frequently applied for the introduction of other elements in zeolites. Well known examples are rare earth elements in zeolite X and Y; Ca in zeolite A and Y; Cu in zeolite Y etc. Most cationic elements may be introduced in this way by direct ion-exchange. Upon further activation of these elements by reduction or metal compound formation, the zeolite is (partially) converted into the H-form. In case acidic activity is unwanted the acid sites have to be neutralised before the material can be used in the envisaged application.

An alternative method for the introduction of other elements is impregnation with a solution using the so called "dry impregnation" technique. In this method the volume of the solution containing the required weight of salt is equal to the pore volume of the zeolite to be impregnated. The available pore volume of the zeolite must be measured with the same solvent as used in preparing the solution. It is preferable to apply nitrates or salts of organic acids such as acetates, since these are easily decomposed by heating.

Impregnation of ZSM-5 with a Mg salt was applied to subtly change the effective pore diameter of ZSM-5 and increase its selectivity for p-xylene formation in toluene disproportionation (ref.14).

Ionic alkali-metal clusters such as (Na$_4$)$^{3+}$ are formed in a zeolite matrix by impregnation with NaN, and are applied in reduction reactions and in base catalyzed reactions (ref.15). The nature of these clusters has been investigated by spectroscopic techniques and it was confirmed that $(Me_4)^{3+}$ units are formed (ref.15a). Formation of $(Na_4)^{3+}$ clusters in zeolite X and Y by reaction at ambient conditions with a solution of a Li-alkyl or Li-aryl compound was recently reported in (ref.16).

2.2 T-SITE SUBSTITUTION REACTIONS

We have mentioned earlier that the framework composition of a zeolite can be modified by steaming at high temperature (>730 K) or by acid extraction. In both cases alumina is removed while the zeolite retains its crystallinity. The vacancies created in the lattice by steaming are generally repaired by SiO_2 migration.As a result of the substitution of Al by Si the unit-cell size will decrease, because the atomic radius of Si is smaller than that of Al. The relationship between unit-cell size and Si/Al ratio for zeolite Y is the subject of many reports, see for instance (ref.17).

High temperature steaming is widely applied on an industrial scale, the process is, however, rather difficult and requires strict process control for reproducible results. On the other hand acid leaching, which is also industrially applied, is easier to reproduce on a large scale. Acid leaching, however, will leave lattice positions vacant which were formerly filled by Al. Accordingly the stability of the lattice is, in many cases affected, unless the structural damage is repaired.

Several additional methods have been developed to change the T-site population in an existing zeolite such as reaction with aqueous $(NH_4)_2SiF_6$ (ref.18); reaction with an anhydrous halide in boiling CCl_4 (ref.19) and gasphase reaction with anhydrous halides such as $SiCl_4$ at high temperatures (ref.20). The products of these reactions are referred to as being ultrastabilized, provided that the aim of the reaction was to increase the Si/Al ratio of the zeolite. The reactions with aqueous $(NH_4)_2SiF_6$ and $SiCl_4$ in CCl_4 show that well-crystallised zeolite Y is surprisingly active in Al-removing reactions at temperatures <373 K. A detailed discussion which includes the introduction of other elements is presented in (ref.21). It is interesting that also the reverse reaction occurs i.e. the introduction of Al atoms in a dealuminated zeolite or a high silica material. An early report of this reaction was in (ref.22) stating that extrusion of a B-silicate having the ZSM-5 structure using an alumina binder, resulted in a final material containing lattice alumina and having a much higher catalytic activity. (Re)insertion of aluminum atoms in the faujasite framework has recently been reported in (ref.23). The framework composition was followed by solid state MASNMR measurements which showed clearly that Al reinsertion is possible from hot, basic aluminate solutions. The introduction of Ti by reaction of acid extracted ZSM-5 with gaseous TiCl₄ will be discussed later in this article in relation to the preparation of catalysts for the hydroxylation of phenol. T-site substitution reactions change the nature of the zeolite surface chemistry appreciably, their application has been compared to a secondary synthesis step (ref.23).

2.2.1 <u>T-site substitution and catalyst deactivation</u>

In many applications of zeolitic catalysts the conversion level is not a constant at constant reaction conditions. Rather, a steadily decreasing conversion is observed due to catalyst aging. In general, aging of a catalyst may be due to various deactivating surface reactions and one of the most frequently observed mechanisms is that of a surface gradually being covered by a layer of polymeric species. An example of a deactivating catalyst is shown in Fig. 3.

In order to increase the active life of the catalyst, measures may be taken depending on the mechanism of the deactivating reaction. In Fig.3 it is for example shown that introduction of platinum in the zeolite or the the application of a higher hydrogen partial pressure both result in a lower deactivation rate. The conversion however is inversely related to the hydrogen partial pressure and the activity level at which the rate of deactivation is close to zero is very low and makes application of a high hydrogen pressure not practicable.

When the deactivation is due to poremouth plugging as a result of coke formation on the outer-surface of the zeolite particles, a possible remedy is the selective removal of the

506

active sites on the outer surface. Many examples of such a measure have been reported in the literature, a recent example is the deactivation of the outer surface of ZSM-5 by reaction

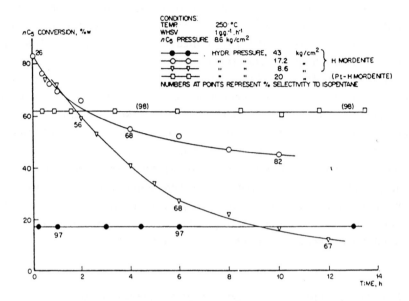

Fig.3. n-Pentane isomerisation activity of mordenite.
        Catalyst deactivation.

with $SiCl_4$ (ref.24). The outer surface deactivation is due to the removal of protons by the substitution of Al- by Si-atoms.

Deactivation by formation of carbonaceous residues on the inner surface of the zeolite is mostly due to condensation reactions of reactive intermediates. The rate of this reaction may be lowered by increasing the Si/Al ratio of the zeolite and thus increasing the separation of the protonic sites. A comparison of the performance of various samples of the zeolite differing in Si/Al ratio is applied to optimize the catalyst composition. Preferentially this is studied on samples obtained in direct synthesis, possible amongst others for ZSM-5 and zeolite Beta. In other cases secondary synthesis may be applied to dealuminate the zeolite.

Active hydroxyl groups on the inner and outer surface may also be deactivated by reaction with alkylsilyl compounds in the following type of reaction :

$$(C_nH_{(2n+1)})_3SiH + HO-Surface \rightarrow (C_nH_{2n+1})_3Si-O-Surface + H_2$$

The reaction product has a much lower thermal stability than an untreated zeolite and is easily oxidized.

In most other instances use of an hydrogen atmosphere and introduction of an hydrogenation function will appreciably lower the rate of deactivation and result in a more stable conversion level.

2.3 METAL EMPLACEMENT

Bifunctional catalysts are applied in many petrochemical and oil conversion processes and their preparation is well known. Two types of methods are used for the introduction of active metals, the first is impregnation, the second ion-exchange. The former method is for example applied in the manufacture of hydrocracking catalysts, consisting mostly of zeolite Y and a $Ni/Mo/S$ or $Ni/W/S$ hydrogenation function. Many patents describing hydrocracking catalysts have been published. According to this literature the general method for hydrocracking catalyst manufacture is to commingle a solution of salts of the metals which form the hydrogenation function, the activated zeolitic cracking component and the amorphous binder in a mixer-muller. The plasticized product is shaped by extrusion and calcined at a temperature of about 800 K. During calcination the salts are decomposed, forming oxides supported on the inner-surface, the zeolite is converted into the catalytically active form and the extrudates develop sufficient physical strength for the envisaged application, see among the many other patents (refs. 25 and 26). The finished catalyst is loaded in the reactor and in situ presulfided using e.g. a mixture of H_2 and H_2S at reaction temperature. During this treatment the metal oxides are converted into the complex sulfide phase, which forms the hydrogenation component of the hydrocracking catalyst. Although the reported manufacturing procedures appear not to be sophisticated, the finished catalysts have a high performance under severe reaction conditions. Developments in the area of unconventional activation techniques by manipulation of the framework composition of zeolites have appreciably contributed to the emerging new generation of hydrocracking catalysts.

Noble metal containing catalysts are usually prepared by ion-exchange of metalammine complexes with the zeolite. Because of the cost of noble metal, the metal loading should be low and the exposed metal surface should be as high as possible. This means that the metal must be well dispersed and have an even distribution over the available zeolite surface. The type of metal complex, the conditions of ion-exchange, the calcination procedure and the reduction conditions each have a strong effect on the dispersion of the metal in the finished catalyst. The subject is discussed in detail in recent papers (refs. 27, 28 and 29), which also review preceeding literature. We will present some additional points during our later discussion of zeolitic noble metal catalysts.

2.4 EFFECT OF CRYSTALLITE SIZE

For many zeolites the size and shape of the crystallites are dependent on the synthesis conditions. Since reactions in zeolites are frequently diffusion limited, crystallite site has a strong effect on the activity and selectivity of the catalyst. This aspect is dealt with in the chapter by Post in this book and in (ref.30). In relation to the preparation of zeolitic catalysts one must be aware of the fact that crystallite size is important for catalyst performance and that its effects should be considered in catalyst development.

2.5 CATALYST SHAPING

2.5.1 Granules, particle size <0.1 mm

Granulated catalysts consisting of particles smaller than 0.1 mm are usually manufactured by spray-drying of a slurry, which contains all the active components of the final catalyst and a binder. This unit operation cannot be scaled down to the usual laboratory type of batch size, because the dimensions of the equipment determine, amongst other properties, to a large extent the maximum size of the particles which can be made. In general the zeolite component of a spray-dried catalyst is in an activated form and the spray-dried product is ready for delivery. It is usually applied after an in situ thermal activation. Apart from its catalytic activity the physical properties of spray dried powders are important parameters and are mainly determined by the binder system used in the manufacturing process.

Properties of catalysts are measured using standardized methods, some of these were developed in the USA under the auspices of the ASTM. Other analytical methods are agreed between manufacturer and user. Some of the more general properties which are frequently measured to establish whether the product meets the agreed specifications, are collected in Table 4.

2.5.2 Catalyst pellets

Various methods are applied for the preparation of catalyst pellets, see Table 5 and zeolitic catalysts do not require specific methods. By far the most frequently applied method in commercial catalyst manufacturing is extrusion, using an hydraulic binder such as clay, alumina, silica, or a combination of these. Although the technique is straightforward, manufacturing of dense extrudates requires special equipment. Simple equipment is available for the laboratory extrusion of experimental catalysts. It is however difficult to prepare representative batches of catalyst on a laboratory scale as far as catalyst texture and strength are concerned. These aspects have a large effect on catalyst performance and may even influence the design of the reactor and the configuration of the

TABLE 4
Specified properties of shaped catalysts.

Property	Powder	Extrudates (granules)
Specific surface area, m^2/g	+	+
Pore size distribution	+	+
Pore volume ml/g	+	+
Compacted bulk density	+	+
Particle density	+	+
Skeletal density	+	+
Particle size distribution	+	+
Average length mm, (n particles)		+
Average diameter mm, (n particles)		+
Maximum/minimum diameter, mm		+
Maximum/minimum length, mm		+
Side crushing strength,N/mm (n particles)		+
Attrition %w	+	+
Chemical composition	+	+
Loss on ignition at ...K	+	+
Zeolite content %w	+	+
Zeolite unit cell size	+	+
Zeolite crystallinity	+	+
Metal dispersion/distribution	+	+
Manufacturing method	+	+
Performance test	+	+

catalyst bed(s). In many cases the extrusion mixture will contain all the components of the finished product. The extrusion mixture is plasticized by intensive mixing and the resulting dough is shaped using mostly an auger extruder. The finished catalyst is obtained after drying and calcination of the "green" extrudates. It is advantageous that the calcined product is the end product. Further operations in which shaped particles are contacted with aqueous solutions are usually complicating and costly. The finished catalyst must be conform a number of chemical and physical specifications. In an ideal case the specifications agreed upon between manufacturer and user describe the material sufficiently accurate to obviate extensive catalyst testing. A number of catalyst properties, which are often quoted in specifications are given in Table 4.

TABLE 5
Methods for the manufacture of shaped zeolites (> 0.1 mm).

Procedure	Binder	Remarks
Agglomeration	Clay	Use in separation processes
Extrusion	Alumina	Use in catalysis and most other processes
Compaction	–	May damage zeolite.
Hydrothermal conversion of shaped precursor	–	Formerly applied for mordenite synthesis, method deserves revival

2.6 ANALYSIS

A large number of variables influences the final performance of zeolitic catalysts. Accordingly it is of utmost importance that both the starting materials and the products of the various catalyst preparation steps are fully characterized and documented. Useful analytical techniques are described in the chapter by van Hooff and Roelofsen in this book. Moreover whenever commercial samples are available their use is strongly advised since these materials are generally prepared on a large scale. Use of commercial samples may alleviate the appreciable reproducibility problems, which are usually faced during research on the application of zeolitic catalysts.

3   FLUIDIZED CATALYTIC CRACKING (FCC) CATALYSTS

3.1   FCC PROCESS,GENERAL DATA

The FCC process is the most important refinery process for the production of gasoline components by the conversion of various feedstocks, such as vacuum gasoil and heavier fractions, into lighter products via cracking reactions. A typical product composition is given in Table 6. Valuable products are: LPG, gasoline   and   light   cycle   oil;   fuel   gas   and   residue   are byproducts and coke on catalyst represents a loss which is removed by coke burning in the regenerator.

FCC is a continuous regenerative process, a process scheme for a riser cracker is given in Fig.4. The preheated feed is injected at the bottom of the riser reactor and mixed thoroughly with the

TABLE 6
Typical Product Composition FCC Process
Feedstock:Vacuum gasoil + Residue

Component	% w
Fuel gas (H_2;C_1;C_2) and LPG (C_3;$C_3^=$;C_4;$C_4^=$)	10-15
Gasoline	40-50
Light cycle oil	15-20
Residue	15-25
Carbon on catalyst	1-3

freshly regenerated, hot catalyst. The catalyst to oil weight-ratio in the mixture is 5 to 6 and its temperature is about 773 K. The residence time in the riser reactor is about 3 seconds and after this very short reaction time the catalyst is completely deactivated by coke deposition. Deactivated catalyst and hydrocarbons are separated in the steam stripper vessel. Coke on catalyst is subsequently removed by burning in air at 1030 K in the regenerator reactor.The residence time in the regenerator is about 10 min. The regenerated catalyst is mixed with some fresh catalyst and transported to the reactor.

An average FCC unit has a catalyst inventory of about 200 tons and its throughput is about 22000 barrels/day = 3.300 ton/day. Because of the extreme reaction conditions the FCC catalyst must

512

have a high attrition resistance and a very high physical strength, moreover its thermal stability must be very high in order to withstand the regeneration process. Some typical data on the process conditions are collected in Table 7.

A review of the distribution of the around 330 FCC units over the various areas is given in Table 8. About 50% of the installed FCC capacity is located in the USA and Canada, the largest gasoline market. World wide catalyst consumption is about 1000 ton/day or 350.000 ton/year.

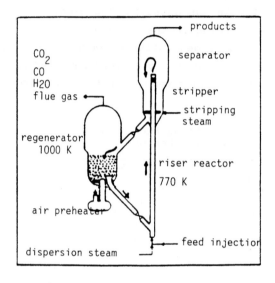

Fig.4. Riser cracker.

TABLE 7
Average FCC unit
Typical data

Catalyst inventory	200 ton
Fresh catalyst rate	2-3 t/d
Feed rate	3300 t/d
Cat/oil ratio w/w	5-6
Linear velocity in riser reactor	10 m/sec
Temp in riser	800 K
Temp of feed	530 K
Temp. in regenerator	1000 K

TABLE 8
Number of FCC units

USA/Canada	176
Latin America	39
Asia/Pacific	37
Europe	54
Middle East/Africa	17
Eastern Europe	13
Total	336

3.2 FCC CATALYST PRODUCTION CAPACITY,MAIN CATALYST PRODUCERS

Worldwide the installed production capacity for FCC catalyst is about 550.000 ton/year, see Table 9. The appreciable overcapacity results in a very competitive situation amongst the producers. The accompanying total capacity for Y-zeolites is about 100.000 ton/year, since the average percentage of zeolite in the FCC catalyst is about 20 % weight. The main FCC catalyst producers and the location of their plants are listed in Table 10.

TABLE 9
Worldwide production capacity
FCC catalysts,kton/year

USA	350
Europe	100
Japan,Brazil,China	100

TABLE 10
Main catalyst manufacturers

Company	Plants
Grace	USA/Europe
Engelhardt	USA/Europe
Akzo	USA/Europe/Brazil
Filtrol	USA
UOP	USA/Europe
Crosfield	Europe
Shokubai-Kasei	Japan

3.3 FUTURE DEVELOPMENTS IN FCC

The very competitive market situation and economical and environmental requirements in FCC have led to many catalyst improvements in recent years. Even further improvements in product performance have to be substantiated in order to meet future demands anticipated for the FCC process. Expected economic demands are:
* Improved gasoline yield.
* Increased gasoline octane number.
* Increased yield by lower gas and coke formation.
* Expanded boiling range of feedstock (resid cracking).
* Improved metal (Ni,V) tolerance of FCC catalyst.
* Decreased catalyst losses by dust emissions.
* Decreased SO_2 emissions.

These goals are most probably reached at lowest cost by tailoring the FCC catalyst rather than by adaptation of the FCC process. In most of the areas for process improvement mentioned above appreciable progress has been made in recent years due to

514

improvements in catalyst performance.

3.4   COMPOSITION OF THE FCC CATALYST

3.4.1   MATRIX EFFECTS
The two main components of a FCC catalyst are the active zeolite and the matrix material, the latter acts as a continuous binder phase between the zeolite crystallites. As such the matrix determines the physical properties of the catalyst. In recent years the function of the matrix has been expanded and includes now catalytic activity to crack large molecules into fragments small enough to enter the zeolite pore system. In addition the properties of the matrix include activity to trap V, Ni and alkali compounds, which may be present in residual fractions. Structural damage to the zeolite function by the uptake of Ni and V and loss in acidity by Na poisoning are thus prevented. Components in the matrix precursor mix are alumina, silica-alumina gel and pretreated clays.

3.4.2   ZEOLITE COMPONENT
The main zeolite component in modern FCC catalysts is based on zeolite Y, Fig 5. The zeolite has a pore opening of about 0.8 nm, it is activated by ion-exchange with NH$_4$ ions or three-valent rare earth ions. Acid sites are generated by calcination of the ion-exchanged zeolite and these acid sites are active in the catalytic cracking process.

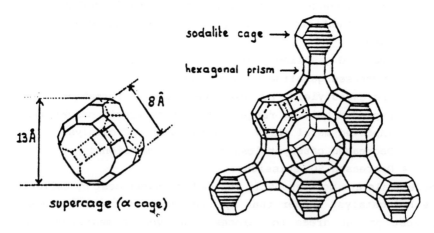

Fig. 5. Faujasite structure.

3.4.2.1 <u>Acid site configuration</u>

Important factors related to the acid site configuration are
* The number of acid sites.
* The strength of the acid sites.
* The distribution of the acid sites in the lattice.
The acid site configuration determines to a large extent the
ratio of the various chemical reactions occurring during the
cracking process. In this respect an important factor is the
degree of isolation of the acid sites, illustrated in Fig. 6.

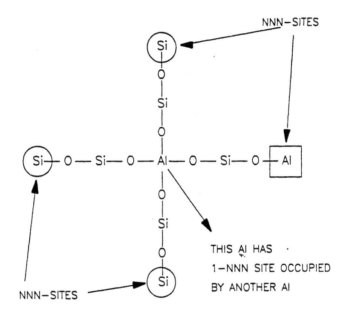

Fig. 6. Acid-site configuration. For highest degree
of Al isolation all NNN sites to be occupied by Si.

In this discussion we call Al sites without an Al atom on a NNN
position 0-NNN sites. Reactions of o-xylene depend strongly on
the acid site configuration, this is illustrated in Fig. 7.
The ratio between the rate of isomerisation and the rate of
disproportionation depends on the degree of isolation of the Al
atoms in the zeolite lattice (average number of Al atoms on NNN
sites/unit cell).

The effects of the acid site configuration on catalyst
performance are an active field of research. For FCC catalysts
the situation is complicated due to the fact that their

516

configuration changes during the useful life in the FCC unit. The
change in active site configuration may be studied in

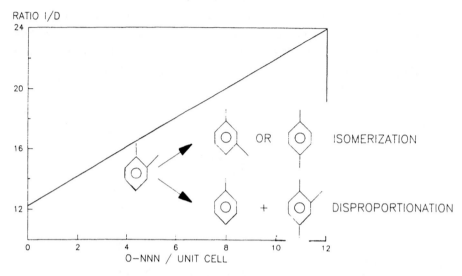

Fig.7.o-Xylene conversion over zeolite H-Y. Effect of acid-site
    configuration on isomerization/disproportionation ratio.

the laboratory by the application of simulated deactivation
procedures. It appears that both the rate of deactivation and its
mechanism are dependent on the quality of the zeolite used in the
manufacture of the catalyst and on the hydrothermal stability of
the zeolite component. Changes in acid site configuration occur
as a result of exposure to high temperature in the presence of
steam, conditions which are common in a FCC unit. Al is removed
from T-sites and forms extra (ref.32) (non)framework alumina as
a separate phase. The vacated T-sites are partly filled by silica
from other parts of the lattice (Fig.8) some structural damage
will however always remain. Recent investigations have shown
that extra framework alumina is also catalytically active. During
the alumina migration a transition state occurs where alumina is
partly bound to the zeolite structure. It appears that this
transition state alumina has a strong effect on the catalytic
activity (refs 33, 34, 35, 36).

3.4.3 MANUFACTURE OF Y ZEOLITE COMPONENT

It has been stated earlier that both the quality and the activation procedure of the zeolite used in the manufacture of the FCC catalyst have a strong influence on its performance. The activated zeolite is one of the raw materials in the FCC catalyst manufacture.

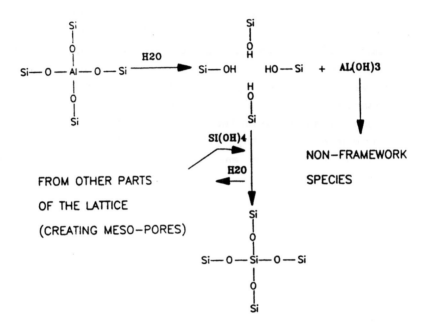

Fig. 8. Solid state reactions in zeolite Y during steaming.

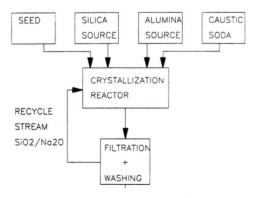

Fig. 9. Industrial synthesis of zeolite Y.

518

3.4.3.1 <u>Crystallization of Na-Y</u>

A simplified block diagram of the crystallization process is presented in Fig. 9. A silica source, an alumina source and caustic soda are mixed in water and sometimes a seeding mixture is added in order to accelerate the crystallization process.
Upon mixing the ingredients an amorphous silica-alumina gel is formed. The crystallization process is carried out 373 K and lasts 10-50 h, depending on the desired Si/Al ratio in the final product. The mechanism of the crystallization of zeolite Y from the amorphous gel is not yet well understood. Moreover it appears that zeolite Y is one out of the crystalline structures which can be formed at the reaction conditions (temp. about 375 K). Fig. 10 shows that for example the composition of the reaction mixture has a strong influence on the end product. This illustrates that the homogeneity of the gel is an important factor. An inhomogeneous gel may easily result in locally different crystalline structures.

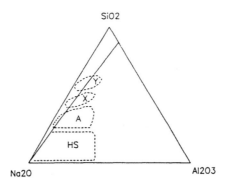

Fig. 10. Synthesis of zeolite Y. Product composition depends on
molar ratio of the components.(HS is hydroxysodalite)

These impurities accelerate the recrystallization of an eventually formed zeolite Y into other phases, see Fig.11. Scale up of laboratory recipes is accordingly an important step in the commercialization of zeolites. Additional problems arising during scale up of zeolite synthesis are discussed below.

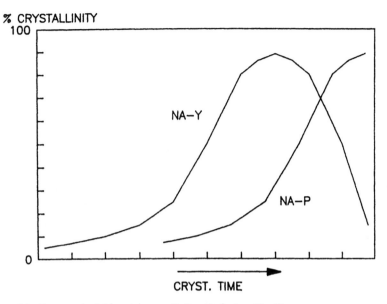

Fig. 11.Recrystallization of Na-Y into Na-P.

The crystallization process consists of two stages as is illustrated in Fig.12. It is generally accepted that during the induction period small nuclei are formed, which upon reaching a

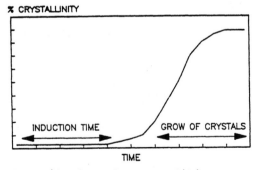

Fig. 12.Product purity depends on conditions
during induction period.

critical concentration show an accelerated growth to the final
crystals (ref.37). Practical experience shows that for zeolite
Y the induction period represents a very delicate process stage
which can easily be disturbed by agitation or in general by shear
forces. Seeding of the synthesis mixture alleviates some of the
above mentioned problems. It results in an end product of higher
purity, which however contains very small crystals. A common
problem of zeolite Y slurries is presented by their pseudoplastic
behaviour. A settled zeolite Y slurry will form a massive and
very hard sediment that is extremely difficult to handle. On the
other hand a "solid" filtercake is transformed into a milky
liquid under the influence of shear. This property of zeolite Y
slurries has important consequences for the lay-out of a
production unit.

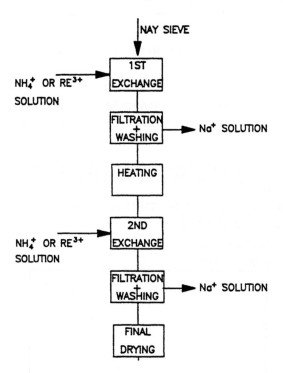

Fig.13. Ion-exchange of zeolite Y for sodium removal. In each
step about 70% of the sodium is removed.(RE:Rare earth)

3.4.3.2 Ion-exchange and zeolite activation
During this process step NaY is converted into the catalytically

active material. A simplified block diagram of the activation procedure is given in Fig. 13.

About 70% of the Na ions are replaced by NH_4 or rare earth (RE) ions during the first ion-exchange and an intermediate heating step is applied to convert the zeolite into the H-form. The temperature of the intermediate calcination is usually between 823 and 1050 K, the presence of steam during this process step is essential. During heating Na is redistributed over the zeolite surface and is thus made available for further ion-exchange to about 70% replacement by NH_4. After the second ion-exchange the Na_2O content is less than 1 %w. During the calcination some Al is removed from the zeolite lattice, the product is usually described as stabilized Y. This product has a higher Si/Al ratio in its framework than the NaY raw material from which it was made and contains some non framework alumina (NFA). The nature of NFA and its role in the catalytic activity of the various forms of stabilized Y sieve materials is a widely investigated research subject. The conditions of the stabilization process determine the acid site configuration of the activated zeolite. Important parameters are :

* The Si/Al ratio of the raw Na-Y,
* The Na_2O content before the heating step,
* Temperature, residence time and H_2O partial pressure during the heating step,
* The Na_2O content after the second ion-exchange step.

With the exception of the heating step zeolite activation does not present difficult technological problems. Control of process conditions during calcination in large scale equipment is the most difficult step in scale up to commercial operation.

The latest trend in FCC catalyst manufacture is to subject the zeolite component to further activation procedures. Such a process step may consist of a repeated intermediate heating step followed by an ion-exchange. More Na is removed and the product has an higher Si/Al ratio in its framework. Alternatively T-site substitution reactions may be applied on the Na-Y material using the procedure described in (ref.13).

3.4.4 Manufacture of the composite FCC catalyst

The last step in the manufacture of FCC catalyst is its shaping into a fluidizable powder. To this end an aqueous slurry

is prepared containing the matrix precursors and the activated
zeolite component. The slurry is spray dried in a large
continuous unit and the product is shipped to the consumers. Some
typical properties are collected in Table 11.

TABLE 11
FCC catalyst,typical properties.

Pore Volume,H_2O,ml/g	0.20-0.35
Surface area,m^2/g	250-350
App. bulk density,g/ml	0.7-0.8
Zeolite content,%w	20-30

Particle size distribution	
μm	%w
< 149	97
< 105	80
< 80	65
< 40	20
< 20	2

4   PREPARATION OF Ti CATALYSTS FOR HYDROXYLATION OF PHENOL

Titanium silicates having the MFI structure (TS-1) have been
reported by Enichem and affiliated companies in recent patents
and contributions in the open literature (refs. 38, 39). These
materials have interesting properties as selective catalysts in
oxidation- and epoxidation reactions using aqueous H_2O_2 as an
oxidant. TS-1 is prepared by direct synthesis (ref. 40) and has
a unit cell volume which indicates that Ti is incorporated in the
framework. Additional evidence for framework Ti is the presence
of an IR band at about 960 cm^{-1}. Spectroscopic characterization
of TS-1 by IR and reflectance UV-VIS is discussed in (ref.41).
The lowest Si/Ti ratio reported by Enichem for TS-1 is about 39
and the synthesis of TS-1 as given in their publications is very
similar to that of ZSM-5. The most important aspect is
prevention of the formation of a separate TiO_2 phase in the
precursor SiO_2/TiO_2 gel. This is effected by coprecipitating TiO_2
and SiO_2 at a low temperature and hydrothermal synthesis in the
absence of alkali. Free TiO_2 in the gel will not be incorporated
into the silicalite lattice during hydrothermal synthesis. A TS-1
type catalyst containing free TiO_2 lacks selectivity, stability
and activity. This aspect of the synthesis of TS-1 is discussed
in more detail in (ref.42).

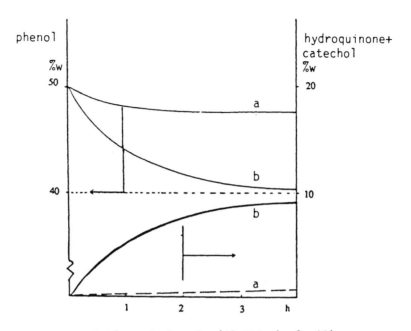

Fig. 14. Hydroxylation of phenol with H_2O_2.(ref. 42).
a= TS-1 + added free TiO_2; b= TS-1 and Ti-ZSM-5;conditions: 353
K, 20 g phenol; 1g catalyst; 15.6g methanol;4 ml 36% H_2O_2.

Removal of the template from the as synthesized TS-1 is a
critical step in catalyst preparation. In the patents a final
temperature of 823 K is mentioned. Results reported in (ref.42)
show that after calcination at this temperature free TiO_2 is
found in a material which was originally pure TS-1. Direct
synthesis of Ti-silicalite has also been reported in (ref.43),the
lowest Si/Ti ratio reported here is 16.4.

Ti silicates may also be prepared by a T-site substitution
reaction at elevated temperature of $TiCl_4$ and a zeolite. This
reaction is described for acid extracted ZSM-5 having a Si/Al
ratio 2000 (ref.42), at temperatures between 673 and 773 K. The
reactions have two interesting aspects, the first is that such
a high Si/Al ratio can be obtained by extracting ZSM-5 (Si/Al=50)
three times with one molar HCl at 353 K. The second is that the
uptake of Ti is limited to about one Ti per unit-cell. The
product (Ti-silicalite) appears to be similar to TS-1. It has an
increased unit-cell size, shows an IR band at 960 cm^{-1} and at the
same Ti content equals the performance of TS-1 as a catalyst in
the hydroxylation of phenol with H_2O_2. The two types of material

appear to be similar, both have however to be more fully characterized in order to define differences, if any. A comparison of the performance of TS-1 and Ti silicalite is presented in Fig. 14, which additionally clearly demonstrates the deleterious effect of free TiO_2 on catalyst performance.

5 ZEOLITE SUPPORTED NOBLE METAL CATALYSTS

Scaling up laboratory recipes for the manufacture of highly dispersed, zeolite supported noble metal catalysts is a challenging task in their commercialization. Depending on the zeolite component, three types of catalysts may be considered. The first type is represented by the bifunctional catalysts containing an acidic zeolite function(ref.27). The second type is represented by monofunctional catalysts, containing a neutral (ref.28) zeolite function. The third type is represented by monofunctional catalysts containing an hydrophobic zeolite function (ref.29).

Zeolite supported noble metal catalysts are best prepared by ion-exchange using an aqueous solution of an cationic metal ammine complex, such as $Pt(NH_3)_4X_2$. In general the metal complex has a strong interaction with the zeolite. As a result the metal will be deposited near the outer surface of the crystals, forming a thin layer of high metal concentration. This may be remedied by the addition of a competing cation, such as NH_4 and the application of a long equilibration time. The ion-exchange procedure is in some detail discussed for the various catalysts in the references given above and the effect of a competing cation is specifically mentioned in (refs.29 and 44). Using Al free silicalite as support hydrophobic Pt catalysts were prepared for D_2/H_2O exchange. It was found that the ion-exchange capacity of silicalite is strongly dependent on the pH of the exchanging solution. At a pH of about 9 a Pt load of 8%w could be prepared, whereas at pH=4 the uptake capacity was about 0.8%w. A summary of this preparation is presented in Table 12. In commercial catalyst manufacture the application of competitive ion-exchange is technically feasible. Costwise its application presents some problems however since the metal uptake by the zeolite is not complete and the noble metal accordingly has to be recovered from the spent ion-exchange solution. Calcination of the noble metal ion-exchanged zeolite presents in general some problems in large scale operations. It has however been firmly established that using an Al-containing zeolitic support, an oxidizing atmosphere is a prerequisite for obtaining a well dispersed metal function in the finished catalyst.

TABLE 12
Preparation of Pt on silicalite by ion-exchange (ref.29)

Silicalite	Commercial,type 115 from UCC
Metal emplacement	Cation-exchange with $Pt(NH_3)_4Cl_2$
pH ion-exchange	4 ->Pt about 0.9%w
	9.5 ->Pt about 8%w(addition NH_4OH)
Drying	400 K in air
"Calcination"	573 K in vacuo
Reduction	573 K in H_2

Remarks
Commercial material attractive
At high pH high silica zeolites show ion-exchange capacity in excess of number non Si framework atoms.
Pt dispersion dependent on ion-exchange conditions:
Low pH -> poor dispersion (low H/Pt)
High pH -> good dispersion (high H/Pt)

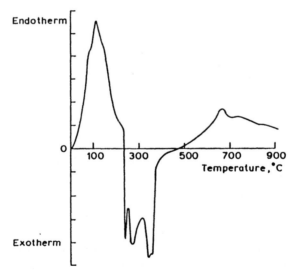

Fig.15. Thermal analysis of Pt loaded NH_4-zeolite Y in air. Strong exothermic effect.

A careful study by thermal analysis of the reactions occurring during calcination will be required in order to develop an optimal procedure. Particularly for materials prepared by

526

competitive ion-exchange in the presence of an ammonium salt,
thermal analysis will yield valuable information. Results of such
a study for Pt(NH$_3$)$_4$-NH$_4$-zeolite Y are shown in Fig.15. Comparison
with TPD data of NH$_4$-zeolite Y in Fig.16 (ref.45) shows that
unless proper measures are taken, the calcination may easily
result in extensive thermal damage due to its highly exothermic
character.

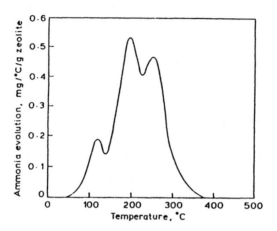

Fig.16. NH$_3$ TPD of NH$_4$-zeolite Y.(ref. 45)

The adiabatic temperature rise due to the burning of one %wt of
NH$_3$ is about 250 K! Judicious choice of a temperature programme
for the calcination in some cases may be sufficient to prevent
temperature excursions. In other cases it may be necessary to
develop an alternative preparation procedure to circumvent
problems during calcination.

Reduction of the calcined material is generally simple and
the reaction occurs at relatively low temperature. Care should
however be taken that the partial pressure of water is low since
well dispersed noble metals may sinter in mixtures of H$_2$O and H$_2$
at temperatures above 720 K. Finishing of a silicalite-based
catalyst prepared by competitive ion-exchange in the presence
of NH$_3$ is possible without an intermediate calcination (ref.29).
The presence of chloride ions during reduction of a Pt(NH$_3$)$_4$
loaded silicalite appears to have a detrimental effect on the Pt
dispersion of the finished catalyst.

6   CONCLUSION

In the preparation and commercial manufacture of zeolitic catalysts results of fundamental studies in zeolite chemistry are very often quickly applied. Specially in large applications such as FCC and hydrocracking even a small improvement in catalyst selectivity often has a considerable economic value. Because of the strong research effort in the area of zeolites, new catalysts having an improved performance are produced at a high rate. This lively interaction between fundamental and applied science makes research into the use of zeolitic catalysts very interesting.

For additional reading excellent reviews are found in (refs. 30 and 46). A review of relevant patents is given (ref.47).

REFERENCES

1   W.Hölderich,M.Hesse and F.Näumann,Angew.Chemie 100,(1988), 232

2   H.van Bekkum and H.W.Kouwenhoven, Recl.Trav.Chim.Pays Bas 108,(1989),283

3   M.Soulard, S.Bilger, H.Kessler and J.L.Guth, Zeolites, 7, (1987),463.

4   C.Dimitrov,Z.Popova,S.Mladenov,K,-H.Steinberg and H.Siegel in D.Kallo and Kh.H.Minachev (eds.),"Catalysis on Zeolites" Akad.Kiado Budapest 1988, p. 135

5   H.Bosaceck and V.Patzelova in ref 4 ,p. 169.

6   B.L.Meyers,T.H.Fleisch,G.J.Ray,J.T.Miller and J.B.Hall J.Catal. 110, (1988), 82.

7   J.D.Barthomeuf, Materials,Chemistry and Physics,17,(1987), 49.

8   H.W.Kouwenhoven in W.M.Meier and J.B.Uytterhoeven (eds), "Molecular Sieves", Adv.in Chem.Series 121, Washington, 1973, p.529.

9   W.K.Hall, J.Engelhardt and G.A.Sill in P.A.Jacobs and R. A. van Santen (eds), "Zeolites,Facts, Figures,Future", Elsevier,Amsterdam 1989,Stud.Surf.Sci.Catal.49, p. 1253.

10   P.O.Fritz and J.H.Lunsford, J.Catal. 118, (1989), 85.

11   C.T.W.Chu and C.D.Chang, J.Phys.Chem. 89, (1985), 1569.

12   D.K.Simmons, R.Szostak, P.K.Agrawal and T.L.Thomas, J.Catal.106,(1987),287.

13   W.F.Hoelderich, Pure & Appl.Chem. 58, (1986), 1383.

14   N.Y.Chen, W.W.Keading and F.G.Dwyer, J.Am.Chem.Soc. 101, (1979), 6783.

15   L.R.M.Martens,P.J.Grobet and P.A.Jacobs, Nature,315,(1985), 568.

15a R.E.H.Breuer,E.de Boer and G.Geismar, Zeolites 9,(1989),336.

16   K.B.Yoon and J.K.Kochi, J.Chem.Soc.Chem.Commun. 1988, 510.

528

17   M.W.Anderson and J.Klinowski, J.Chem.Soc.Far.Trans.I 82,
      (1986),1449.
18   G.W.Skeels and D.W.Breck in D.Olson and A. Bisio(eds),
      Proc.Sixth Int.Zeol.Conf.,1983 Butterworths,U.K.,1983, p.
      87.
19   L.V.C.Rees and E.F.T.Lee, PCT, WO 88/01254 ,(1988).
20   H.K.Beyer, I.M.Belenykaja, F.Hange,M.Tielen,P.J.Grobet
      and P.A.Jacobs, J.Chem.Soc.Far.Trans.I, 81,(1985),2889.
21   P.Fejes, I.Kirisi, I.Hannus and Gy.Schöbel in ref 4,
      p. 205.
22   C.T.W.Chu,J.Catal.93,(1985),451.
23   H.Hamdan,B.Sulikowski and J.Klinowski,J.Phys.Chem 93,(1989),
      350.
24   J.R.Anderson, Y.F.Chang and A.E.Hughes,Catal.Lett.2,(1989),
      279.
25   P.Dufresne and C.Marcilly, EPA 0.162.733,(1985).
26   A.Hoek,T.Huizinga and I.M.Maxwell, EPA 0.247.679(1988).
27   M.Guerin,C.Kappenstein,F.Alvarez,G.Gianetto and M.Guisnet
      Appl.Catal.45,(1988), 325.
28   S.T.Homeyer,W.M.H.Sachtler in ref 9,p. 975.
29   H.A.Rangwala,J.A.Szymura,S.E.Wanke and F.D.Otto, Can.Journ.
      Chem.Eng.66,(1988),843.
30   W.O.Haag and N.Y.Chen, in L.L.Hegedus (ed)"Catalyst
      Design,Progress and Perspectives", J.Wiley & Sons
      1987, p. 163.
31   E.van Broekhoven and H. Wijngaards, Proc. Ketjen Catalyst
      Symposium 1988, F 8,p. 1.
32   J.Scherzer in " Catalytic Materials" ACS Washington, 1984,
      p. 157.
33   L.Kubelkova, S.Beran, A.Malecka and V.M.Mastikhin,
      Zeolites, 9, (1989), 12.
34   G.Garralon, A.Corma and V.Fornes, Zeolites, 9,(1989),84.
35   E.H.van Broekhoven, S.Daamen, R.G.Smeink, H.Wijngaards and
      J.Nieman in ref.9, p. 1291.
36   R.M.Barrer,Zeolites,1, (1981),130.
37   P.A.Jacobs andJ.A.Martens,"Synthesis of High Silica
      Aluminosilicate Zeolites, Stud Surf.Sci.Catal. 33,
      Elsevier,Amsterdam 1987, p. 80.
38   G.Perego, G.Bellussi, C.Cormo, M.Taramasso, F.Buonomo,
      A.Esposito in Y.Murakami,A.Ijima and J.W.Ward (eds),
      Proc. Seventh Int.Conf., Stud.Surf.Sci.Catal.28,
      Elsevier,Amsterdam 1986, p. 12.
39   B.Notari in P.J.Grobet,W.J.Mortier,E.F.Vansant and
      G.Schulz-Ekloff (eds.), "Innovation in Zeolite
      Materials Science" Elsevier, Amsterdam, 1988,p. 413.
40   see for example M.Taramasso,G.Perego and B.Notari,   D.E.
      30.47.798,1986.
41   M.R.Boccuti,K.M.Rao,A.Zecchina and G.Leofanti in "Structure
      and Reactivity of Surfaces", C.Morterra,A.Zecchina,G.Costa
      (eds.)Stud.Surf.Sci.Catal. Elsevier,Amsterdam 1989, p. 133.
42   B.Kraushaar-Czarnetzki,Characterization and Modification
      of Zeolites and Related Materials,Thesis Eindhoven, 1989.
      B.Kraushaar-Czarnetzki and J.C.van Hooff
      Cat.Lett.1,(1988),81;Cat.Lett.2,(1989),43.
43   M.C.Chen e.a. in "Catalysis 1987" ,J.W.Ward (ed.),Stud.Surf.
      Sci.Catal.38,Elsevier,Amsterdam 1988, p. 253.
44   H.W.Kouwenhoven and H.J.A.van Helden Ger.Offen 1.816.822
      (1968)
45   R.J.Mikovski and J.F.Marshall,J.Catal. 44,(1976),170.

46   R.Szostak."Molecular Sieves", Van Nostrand Reinhold, New
      York, 1989.
47   P.Michiels, "Molecular Sieve Catalysts", EPO Applied
      Technology Series, vol. 9, Pergamon Press, Oxford, 1987.

Chapter 14

COKE FORMATION ON ZEOLITES

H.G. KARGE

Fritz-Haber-Institut der Max-Planck-Gesellschaft, Faradayweg 4-6, 1000 Berlin 33 West (FRG)

ABSTRACT

A number of techniques and methods suitable for investigation of coke formation on acidic zeolite catalysts is presented and their application illustrated by pertinent results. Such tools for studying coked zeolite catalysts are: various spectroscopic techniques (IR, NMR, ESR, UV-VIS); gravimetry (TGA, also combined with GC and MS) for the determination of the amount of coke deposits and their H/C ratio; extraction of the coked zeolite and chemical analysis (GC, MS) of the extracts; adsorption measurements. Particular attention is paid to the problem of the nature of coke components and, in this context, some emphasis is laid on the distinction between different types of coke. Important phenomena related to coke lay-down such as the effect of acidity, shape selectivity, mechanism and kinetics and location of coke deposition are also discussed.

INTRODUCTION

Coke formation is the most frequent cause of catalyst deactivation in acid-catalyzed hydrocarbon reactions. Therefore, a great number of studies had been carried out already on coke formation on the classic cracking catalysts, i.e. amorphous silica/alumina which was employed before the advent of zeolite catalysts [1]. Also, coke formation was investigated on so-called bifunctional catalysts which exhibit, besides a hydrogenation function, i.e. usually a metal component, an acidic function as well. Examples are platinum, palladium, or nickel supported by acidic alumina or silica/alumina [2]. Similarly, coke formation on desulfurization and denitrogenation catalysts was studied [3].

It is interesting to note, however, that in the first decade of zeolite catalysis little attention was paid to the phenomenon of coke formation, with the exception of a few studies such as the pioneering work by Eberly et al. [4-5]. It was only in the late seventies that some papers appeared, dealing with coke formation on zeolites [6-9]. Since then we have to notice an ever increasing interest in this topic, on the aspects of both industrial application and fundamental research on zeolite catalysis.

The economical point of view of coke formation in industrial hydrocarbon processing cannot be overestimated: Even though the activity of a coked zeolite catalyst may be recovered by coke combustion, regeneration is frequently incomplete and, in any case, requires additional investment.

According to a current view of coke formation this is usually visualized as the build-up of very bulky, polyaromatic systems with even graphite-like structure [7, 10]. However, this view might be too narrow in many cases of deactivation of zeolite catalysts caused by deposition of carbon-containing residues during hydrocarbon reactions. Frequently, such reactions are carried out at relatively low temperatures where the exclusive or preferential formation of polyaromatic residues is unlikely. Such reactions are, for instance, the alkylation of benzene by olefins [11], or the alkylation of isobutane by butenes [12-13], over acidic zeolite catalysts. Nevertheless, catalyst deactivation, sometimes even a very rapid one, is observed in such reactions as well. It must be ascribed, then, to carbonaceous deposits which are olefinic or to some extent even paraffinic in nature.

Therefore, a more general definition of coke, underlying tentatively the present study, is the following one: *"Coke consists of carbonaceous deposits which are deficient in hydrogen compared with the coke-forming reactant molecule(s)"*.

However, not only the nature of those deposits called "coke" is a matter of debate and the subject of many investigations. There are numerous problems related to coke and coke formation on acidic zeolites: very important questions concern the effects of reaction temperature and pressure, time on stream, reactant composition, catalyst properties (such as structure, acidity and binder), location and kinetics of coke deposition. Even though a large body of data and results has been accumulated on coke formation during the last decade, many questions are still open and controversially discussed. Therefore, the aim of the present review is not, primarily, to report on general results of research in the field of coke formation. Rather, particular attention is paid to the methods and techniques which were developed and employed to tackle the above-mentioned aspects and problems related to coke on zeolite catalysts.

The great variety of techniques comprises, inter alia, gravimetric or thermogravimetric analysis(TGA); mass spectrometry (MS); coke combustion, combined with gas chromatography (GC), TGA or MS; spectroscopic methods like infrared (IR), electron spin resonance (ESR), nuclear magnetic resonance (NMR, e.g. ^{1}H NMR, ^{13}C NMR, ^{129}Xe NMR) with and without high resolution techniques (magic angle spinning or MAS, cross polarisation or CP, high field decoupling), ultra violet and visible spectroscopy (UV-VIS), photoelectron spectroscopy (PES). Besides spectroscopic methods, which usually do not require destruction of the coke/zeolite system in order to liberate coke-constituting species, also treatment of the coked catalysts with mineral acids, followed by extraction and chemical analysis, was employed. Adsorption measurements with various adsorbates on the fresh and coked samples were carried out especially to elucidate the locus of coke deposition, i.e. internal pore structure vs. external surface of the zeolite crystallites.

This enumeration of tools for coke investigation is certainly not exhaustive, and new methods are likely to appear in the near future. However, the above-mentioned methods and techniques seem to be the most important and frequently employed ones. Thus, they will be briefly presented in the following paragraphs and their application illustrated by suitable examples and significant results. In conclusion, an attempt will be made to summarize some aspects of coke formation which seem to be, at the present state of the art, of general validity.

METHODS OF COKE INVESTIGATION AND RESULTS

1. Determination of the H/C ratio

The H/C ratio of carbonaceous deposits is an important parameter in characterizing the coke, even though the composition of coke may be complex and comprises constituents with different hydrogen content. The determination of the H/C ratio (via oxidation and analysis of the H_2O and CO_2 formed) is, in the case of zeolites, by no means trivial. The main source of error results from H_2O adsorption from the ambient atmosphere. Surprisingly, this is scarcely discussed in the literature, and frequently H/C ratios of coke deposits on zeolite catalysts are reported without any indication of the method of determination.

To solve the principal problem one additional measurement is necessary. Thus, Entermann et al. [14] monitored the oxygen consumption in a recirculation apparatus during coke combustion.

Weitkamp et al. [15-16], developed another procedure which minimizes the experimental error due to adsorbed or structural water of the coked zeolite catalyst and, thus, provides reliable results. The principle of this method can be visualized from the scheme of Figure 1 (for experimental details see Refs. 15 and 16). The procedure starts

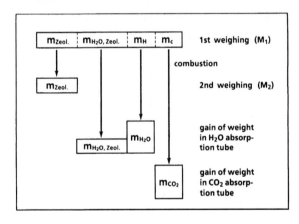

Fig. 1. Scheme for the determination of the H/C ratio in coked zeolite catalysts [15, 16]

with weighing of the coked zeolite containing the mass of the zeolite matrix, m_{zeol}, the mass of adsorbed and structural water, $m_{H_2O, zeol}$, and the masses of hydrogen, m_H, and carbon, m_c, of the coke. The data obtained by this first weighing are $M_1 = m_{zeol} + m_{H_2O, zeol} + m_c + m_H$. After combustion, the increase in mass of the absorption tubes for H_2O (filled with magnesium perchlorate) and for CO_2 (filled with sodium asbestos) is determined, i.e. $(m_{H_2O, zeol} + m_{H_2O})$ and m_{CO_2}. Finally, the weight of the dry zeolitic material, generally converted to the amorphous state, is determined via "back weighing". This step is, in fact, the crucial one. However, if readsorption of water is carefully excluded this step provides $M_2 = m_{zeol}$. The evaluation of the H/C ratio follows from equ. (1) through (7), where equ. (7) contains only the measured data, viz. M_1, M_2, m_{H_2O} and m_{CO_2} (atomic weights rounded).

534

$$\Delta M = M_1 - M_2 = m_{H_2O, \text{Zeol.}} + m_c + m_H \qquad (1)$$

$$m_{CO_2} = \frac{44}{12} m_c \qquad (2)$$

$$m_C = \frac{12}{44} m_{CO_2} \qquad (3)$$

$$m_{H_2O} = m_{H_2O, \text{Zeol.}} + 9\,m_H \qquad (4)$$

$$m_{H_2O} - \Delta M = 8m_H - m_c = 8\,m_H - \frac{12}{44} m_{CO_2} \qquad (5)$$

$$m_H = \frac{1}{8} \left\{ m_{H_2O} - \Delta M + \frac{12}{44} m_{CO_2} \right\} \qquad (6)$$

$$H/C = \frac{m_H}{1/12\,m_c} = \frac{3}{2} \frac{m_{H_2O} - \Delta M + \frac{12}{44} m_{CO_2}}{12/44\,m_{CO_2}} \qquad (7)$$

Weitkamp et al. checked the procedure with several test hydrocarbons and determined the H/C ratio of coke formed upon reaction of various hydrocarbons (n-heptane, 2,2,4-trimethylpentane, 1-hexene, methylcyclohexene, toluene) over LaNaY zeolite.

   More recently, Karge et al. [17-18] combined TGA (using a temperature-programmed microbalance, Perkin Elmer, TGS 2) and GC. The scheme of the experimental set-up is depicted in Figure 2. Prior to combustion of the coke in a stream of oxygen (if necessary

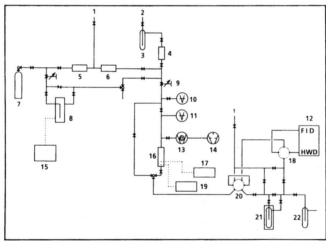

Fig. 2. Scheme of the experimental set-up for the determination of the H/C ratio in coked zeolite catalysts

1: He; 2: synthetic air, 3: molecular sieve trap; 4: gas drying, 5, 6: gas cleaning; 7: feed; 8: saturator; 9: fine adjustment valve, 10: baratron; 11: ionization gauge; 12: GC; 13: turbo molecular pump; 14: roughing pump; 15: thermostat; 16: microbalance; 17: temperature controller; 18: four-way valve, 19: microbalance controller; 20: gas sampling valve; 21: cooling trap; 22: bubbler.

in contact with CuO as a catalyst), adsorbed water and desorbable hydrocarbons are removed from the coked zeolite in a stream of dried N_2 at e.g. 475 K. The desorbed compounds are trapped at 100 K. When the desorption from the catalyst is completed, the trap is rapidly heated to 500 K and the trapped compounds purged into the GC for analy sis. Similarly, the products of the subsequent coke combustion (H_2O, CO_2) are trapped and analyzed. Tests with measured amounts of adsorbed water and with suitable hydrocarbons (e.g. coronene, H/C = 0.5) confirmed the reliability of the TGA/GC technique. The weight loss measured during desorption or combustion is compared with the weight of the desorbed materials and combustion products evaluated from the GC analysis. Not only the H/C ratio of coked samples obtained from catalytic runs in separated reactors could be determined with the help of the apparatus and procedure described, but the coke formation could be carried out in the microbalance as well, when a feed stream was passed over the catalyst pretreated inside the balance. In this case, the amount of deposited carbonaceous materials was monitored via the weight gain. Subsequently, the H/C ratio was determined as outlined. In this way, the H/C ratio of coke deposits formed for instance through reaction of ethylene or methanol over H-ZSM-5 catalysts was measured as a function of reaction temperature and time on stream. Examples of results are presented in Figures 3 and 4.

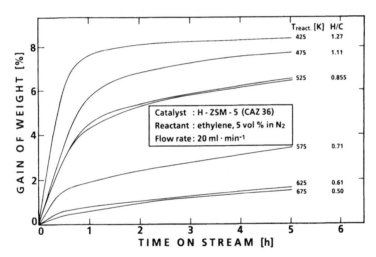

Fig. 3. TGA measurements of coke formation upon reaction of ethylene as a function of temperature.

Figure 3 demonstrates the weight gain of a H-ZSM-5 catalyst (lot label: CAZ 36) due to coke deposition during the reaction of ethylene at reaction temperatures (T[react]) increasing from 425 to 675 K. Also, the H/C ratios after 6 h on stream are indicated. Two periods of coke formation are observed: a rapid initial deposition of carbonaceous materials (during the first 30 to 60 minutes) and a subsequent period of much slower coking. The amount of coke deposited after a given time (e.g. 5 h) decreases with increasing reaction temperature. However, the lower the reaction temperature the shorter the time on stream after which the weight gain levels off. The H/C ratio significantly decreases with increasing T[react].

The situation is even more complicated because the H/C ratio decreases not only with increasing reaction temperature as indicated in Figure 3 but also with time on stream.

536

This is shown for one particular reaction temperature in Figure 4. The "aging" of coke which is demonstrated by Figure 4 depends, in turn, on the reaction temperature and may affect the type of up-take curves, such as depicted in Figure 3, in a rather complex manner.

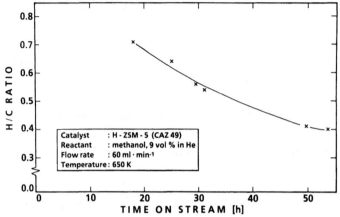

Fig. 4. H/C ratio of coke as a function of time on stream.

Table 1 summarizes results of coke formation obtained via TGA/GC for two different ZSM-5 catalysts (lot labels CAZ 36 and CAZ 49) when subjected to a stream of (i) ethylene/nitrogen and (ii) methanol/helium at various reaction temperatures. The main differences between the two catalysts were as follows: CAZ 36 possessed a higher number of acidic OH groups per gram, lower acidity strength and larger crystallite sizes ($\emptyset \approx 4$ μ) compared to CAZ 49 ($\emptyset \approx 0.4$ μ) (see also Ref. 19).

Table 1   Coke formation on two H-ZSM-5 catalysts

| Temperature | CAZ 36 | | | | CAZ 49 | | | |
| | ethylene | | methanol | | ethylene | | methanol | |
[K]	wt%	H/C	wt%	H/C	wt%	H/C	wt%	H/C
425	8.3	1.27			7.2	1.81		
475	7.8	1.11			5.8	1.50		
525	6.6	0.86			4.8	1.38		
575	3.6	0.71	2.9	1.26	2.5	1.35	1.0	1.52
625	1.6	0.61	2.4	1.12	1.1	1.12	0.9	1.42
675	1.5	0.50	2.0	0.83	0.5	1.02	0.8	1.32

Feed: 5 vol% ethylene in N_2 or 9 vol% methanol in He; flow rate: 20 ml · min^{-1} in a thermobalance, TGS 2; time on stream: 6 h

Results similar to those reported here were arrived at by other authors: Magnoux et al. [20], Guisnet et al. [21] and Schulz et al. [22] who also observed a decrease of the H/C ratio of coke as a function of reaction temperature, *viz.* with coke formed upon propene transformation over ultrastable H-Y catalysts (USH-Y) [20] and upon methanol conversion over H-ZSM-5, RE-Y, or H-MOR [22]. The GC measurements of the H/C ratio by Magnoux et al. [20] refer to the so-called soluble coke. Also, aging of the carbonaceous deposits, i.e. declining H/C ratio with time on stream was found by Schulz et al. [22]. Finally, it was reported in several cases that the rate of coke deposition was particularly high during the initial period of the coke forming reaction. Thus, Neuber et al. [23] found rapid deactivation in a first stage of the reaction due to coke formed by rapid cracking of 1-methylnaphthalene over H-Y and ZSM-20 at 573 K. Similar results were reported by Solinas et al. [24] on coke deposition during methylnaphthalene isomerization over H-Y, whereas Schulz et al. [22] measured a high initial deactivation upon methanol conversion over RE-Y only for a high reaction temperature (753 K).

2. Infrared Spectroscopic Investigations of Coke Formation

IR spectroscopy was one of the first techniques employed to investigate the nature of carbonaceous deposits laid down on zeolite catalysts during hydrocarbon reactions. Eberly Jr. [5] observed upon reaction of 1-hexene over H-Y the appearance of a prominent IR band in the CH deformation region around 1600 cm^{-1}. He ascribed this band to highly unsaturated carbonaceous residues, most likely polyolefins, as was suggested by comparison with spectra typical of such species which had been reported earlier [25-26].

Reaction of olefins over acidic mordenite catalysts (H-MOR, Be-MOR) was studied via IR by Karge et al. [27]. At low temperatures (about 350 K to 375 K) instantaneous polymerization upon contact of the olefin with the zeolite surface was observed. IR bands of saturated hydrocarbons (ν_{as} [CH$_3$] = 2960 cm^{-1}; ν_{as} [CH$_2$] = 2930 cm^{-1}; ν_s [CH$_3$] = 2860 cm^{-1}) predominated in the spectra (compare also Refs. 6 and 28) which were obtained in an IR cell for static measurements [29]. On successive heating to higher temperatures in high vacuum the coke band (around 1585 cm^{-1}) appeared. However, no CH bands at wavenumbers higher than 3000 cm^{-1}, which would be indicative of aromatics, were observed. Simultaneous MS measurements of the species desorbed into the high vacuum indicated fragments [CH]$_n$ with n = 2, 3 . . . 7 [6]. Conversion measurements conducted separately in a microflow reactor but under similar conditions revealed parallel formation of coke (polyolefins?), gaseous olefinic polymers and paraffins, indicating significant hydrogen transfer. Both IR and conversion experiments with catalysts pretreated at various temperatures in high vacuum suggested that complete removal of the acidic OH groups, i.e. dehydroxylation as evidenced by IR, resulted in a loss of any olefin conversion including coke formation [6]. The dehydroxylated catalysts exhibited high densities of so-called Lewis sites [30] but they regained the capability for coke formation only when rehydroxylation occurred either due to traces of water present in the feed or formed by oxidation of the hydrocarbons; oxidation may be caused by oxygen from the zeolite lattice and/or oxygen strongly adsorbed on the catalysts. Thus, the presence of Lewis sites, as indicated by IR after pyridine adsorption, provided no activity for generation of coke; rather, the presence of Brønsted sites was the indispensable requirement

538

for this process to occur (see also Refs. [12, 31]. This is at variance with results reported by Weeks et al. [28] who claimed that polymerization of 1-butene took place on dehydroxylation sites of deammoniated H-Y.

IR spectroscopy has not only the advantage of enabling us to investigate coke formation without destruction of the zeolite matrix containing the residues (*vide infra*, section 7). It can also be successfully employed for studying coke formation *in situ*, i.e. observing the deposition of carbonaceous materials on the working catalyst. This requires the use of a suitable IR flow reactor cell which may be connected to a device (MS, GC) for analysis of the product stream. Several types of such cells are described in the literature, see for instance Refs. [32-35].

Eisenbach and Gallei [7] studied the formation of coke upon conversion of n-hexane and n-hexene over acidic faujasite-type catalysts. They found that the presence of olefins remarkably favoured coke formation, as indicated by the growth of the coke band. These authors, however, ascribed the coke band around 1600 cm^{-1} to graphite-like deposits.

Systematic studies of *in-situ* IR and conversion measurements in order to follow the formation of coke on working zeolite catalysts were conducted by Karge et al. [36-37] and Lange et al. [38]. IR cells, complementary equipment and procedure used by these authors are described in detail in Ref. 33. Some of the results obtained are reviewed in the following paragraphs.

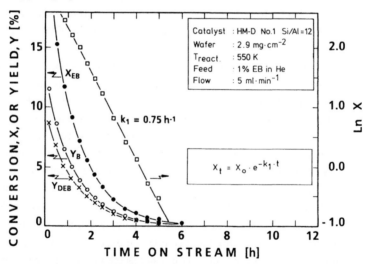

Fig. 5. Decrease in activity of an acidic mordenite upon coke deposition during conversion of ethylbenzene as a function of time on stream [36].

Figure 5 shows the decrease in activity of a slightly dealuminated hydrogen mordenite catalyst during conversion of ethylbenzene (EB) under conditions where, besides the main reaction, *viz.* disproportionation to benzene (B) and diethylbenzenes (DEB's), dealkylation occurs to a small extent. The olefin formed gives rise to coke formation, which results in deactivation of the catalyst. Deactivation, measured via the decrease in ethylbenzene conversion, X, with time on stream, t, follows a first order relationship, *viz.*

$X_t = X_o \cdot e^{-k_1 \cdot t}$, where X_o is the initial conversion and k_1 the rate constant; k_1 charac-terizes the different deactivation behaviour of various mordenites with systematically varied Si/Al ratios [36].

Spectra obtained from the working catalyst during ethylbenzene conversion are dis-played in Figure 6. The most interesting feature of the set of spectra obtained in the region of stretching modes of acidic OH groups (active Brønsted sites) around 3600 cm^{-1} is the constancy of their integrated absorbance (dashed areas); the deviations observed are within the limit of experimental error. This means that the active sites were not poisoned or consumed even though the activity decreases (similar to the behaviour shown in Figure 5) and, consequently, the coke band around 1600 cm^{-1} (see middle part of Figure 6) ceased to grow.

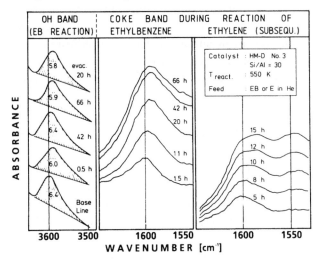

Fig. 6. In-situ IR spectra of a working hydrogen mor-denite catalyst during con-version of ethylbenzene [36]

However, not only did the intensity of the band of acidic OH groups remain unaffect-ed during the ethylbenzene reaction but these sites were also still active. This was con-firmed when, after 66 h under a stream of ethylbenzene in helium, the feed stream was changed to ethylene in helium. Immediately, a conversion of ethylene was measured. Prior to scanning the spectra under the ethylene/helium stream a new base line was established or, in other words, the last coke band spectrum of the ethylbenzene reaction (obtained after 66 h) was subtracted from all the following ones measured after the change of the feed. The result is demonstrated by the right hand part of Figure 6: the coke band started again to grow.

In conclusion, deactivation of the mordenite catalyst during the reaction of ethylben-zene was caused by deposition of coke blocking the active sites for the relatively large ethylbenzene molecules whereas the smaller ethylene molecules still had access to the active centres. Similar observations were made with H-ZSM-5 catalysts.

However, Bibby et al. [39] in their study of the formation of coke on and its removal from H-ZSM-5 MTG catalysts claimed that the acidic OH groups (band at 3610 cm^{-1})

were almost completely eliminated with increasing build-up of carbonaceous deposits up to about 5 wt%. Subsequently, a reinforced intensity of the 3610 cm^{-1} band was observed at ca. 7.5 wt% followed by another fall at 10 wt%. Simultaneously, the band of the non-acidic OH groups (3740 cm^{-1}) continuously declined. These features are difficult to understand and may originate, to a certain extent, from spectroscopic difficulties with the heavily coked samples.

In contrast, the OH groups of H-Y were significantly more affected by coke deposition upon olefin reaction [37]. Eisenbach and Gallei [6] reported analogous behaviour of Ca-Y during n-hexane conversion over Pt/Ca-Y and even claimed irreversible consumption of OH groups due to coke formation.

The intensity of the coke band around 1600 cm^{-1} proved to be a suitable measure for the amount of coke deposited [36, 40-42], even though the nature of the coke might be rather complex. In this context it is worth noting that the position of the so-called coke band around 1600 cm^{-1} in all *in-situ* IR experiments of coke formation shifted to lower wavenumbers when the time on stream was extended. This may indicate that the structure of the main components of the coke changed with time on stream, likely in the direction of more unsaturated compounds, e.g. aromatic species [5]. This corresponds to the aging of the coke (compare section 1).

A plot of the integrated absorbance of this band vs. the amount of coke deposited, measured for instance by a balance, provides usually straight lines [40-41]. Sometimes, however, a distinct change in the slope of such plots is observed indicating a marked alteration in the nature of the coke due to aging [*vide supra*, Figure 4 and Ref. 43].

Analogous *in-situ* IR studies as described above with ethylbenzene were carried out by using an ethylene/helium stream from the beginning of the experiment [37-38]. Results are shown in Figures 7 and 8.

Upon contact with the olefin at low temperatures (300 - 450 K) the intensity of the band of acidic OH groups at 3605 cm^{-1} decreased. Simultaneously a broad band around 3500 cm^{-1} appeared indicating a shift of a part of the 3605 cm^{-1} band to lower wavenumbers due to hydrogen bonding with adsorbate species (Figure 7). The overall absorbance was almost constant.

In the region of the CH stretching modes again the bands of paraffinic species developed (ν_{as} [CH$_3$] = 2958 cm^{-1}; ν_{as} [CH$_2$] = 2926 cm^{-1}; ν_s [CH$_3$] = 2872 cm^{-1}; ν_s[CH$_2$] = 2856 cm^{-1}). The region between 1500 and 1300 cm^{-1} (Figure 8) exhibited the corresponding CH deformation bands, *viz.* δ_{as} [CH$_3$] = 1485 cm^{-1}; δ_{as}[CH$_2$] = 1468 cm^{-1}; δ_s[CH$_3$] = 1382 cm^{-1} and δ_s[CH$_2$] = 1359 cm^{-1}.

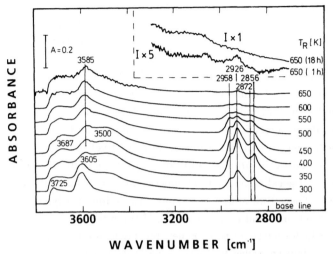

Fig. 7. IR spectra of carbonaceous deposits formed by reaction of ethylene over hydrogen mordenite [38].

When the reaction temperature was increased above 500 K, the situation significantly changed: the band of acidic OH's was essentially regenerated; the CH bands were generally weakened whereby the ratio CH_3/CH_2 seemed to increase, indicating substantial cracking and branching of the carbonaceous residues on the surface. Finally, at relatively high reaction temperatures and/or extended time on stream, a weak band around 3080 cm^{-1} (Figure 7, insert) and bands around 1600 cm^{-1} and 1540 cm^{-1} were observed

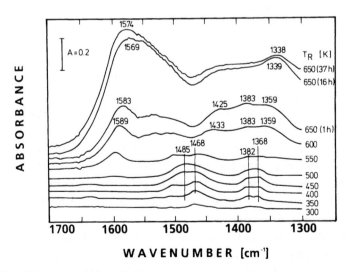

Fig. 8. IR difference spectra of carbonaceous deposits formed by reaction of ethylene over hydrogen mordenite [38].

(Figure 8). The first band was most likely indicative of CH stretching modes of aromatics while the 1600 cm[-1] band ("coke" band) could be ascribed to polyalkenes and/or aromatic species. The band at 1540 cm[-1] was also found by Fetting et al. [41] under similar conditions and had been assigned to alkylnaphthalenes. However, it may be also typical of polyphenylene structures [44]. After prolonged time on stream at high temperatures (650 K), both bands around 1600 and 1540 cm[-1] collapsed into one broad band with its centre at about 1570 cm[-1]. Similar experiments and, in general, coincident results were reported in Refs. [40, [45] and [46]. Interestingly, Haber et al. [46] reported formation of aromatic deposits upon reaction of olefins such as 1-butene and 1-decene over H,Na-Y at relatively low temperatures, *viz.* 398 and 363 K, respectively.

The above-described *in-situ* IR results suggest discrimination between at least two types of coke, viz. a so-called coke I (low temperature coke, "soft coke", "white coke") constituted mainly of paraffinic and, to a lesser extent, olefinic or polyolefinic species (weak band around 1600 cm[-1]) and a so-called coke II (high-temperature coke, "hard coke", "black coke") which consists of polyalkenes, but predominantly of alkylaromatics and polyaromatics (intense band below 1600 cm[-1], band above 3000 cm[-1]).

3. Nuclear Magnetic Spin Resonance Spectroscopy

^{1}H-NMR spectroscopy has been successfully employed for chemical analysis of hydrocarbon extracts from coked zeolites (*vide infra*, section 7). Also, proton magnetic resonance spectroscopy provided valuable elucidation of problems related to the location of the carbonaceous deposits on coked zeolites (*vide infra*, section 13). The new technique of high resolution NMR, in particular MAS NMR of [13]C [47], turned out to be a powerful tool for the investigation of coke and coke precursors. This method was elegantly applied, for instance by Derouane et al. [48], van den Berg et al. [49], Neuber et al. [23] and Maixner et al. [50] when investigating coke formation upon reaction of alkenes over hydrogen mordenite or H-ZSM-5 and conversion of polynuclear aromatics. In the study of Neuber et al. [23] samples of H-Beta, H-ZSM-5 and H-Y were deactivated during conversion of 1-methylnaphthalene at 453 K. Their CP/MAS ^{13}C NMR spectra were obtained, in fact, after contact with the ambient atmosphere and showed prominent signals at 19 ppm and 127 ppm vs. TMS. These signals are typical of methyl groups attached to the aromatic ring and to carbon from the aromatic nuclei, respectively [51]. Increase of the temperature of reaction over H-Beta to 673 K resulted in a slight decrease of the CH_3 intensity relative to that of the 127 ppm signal due to a small extent of dealkylation. Thus, the coke structure formed upon 1-methylnaphthalene conversion over H-Beta, H-ZSM-5 and H-Y was interpreted by the authors to be partially demethylated aromatics with a limited number of aromatic rings.

^{13}C-MAS NMR spectra can be much improved if feed hydrocarbons for the coking reaction are used which are enriched in [13]C. In Figure 9 sets of spectra are displayed which were obtained from H-MOR catalysts activated at 700 K and subsequently coked in an atmosphere of 1 kPa of 1,2 - [13]C - C_2H_4 [38]. The series of experiments started with a sample contacted with ethylene at 300 K. Prior to each further heating step the

1,2 - ^{13}C-C$_2$H$_4$ atmosphere was renewed, since previous measurements [52] had proven that this procedure closely simulates on-stream conditions. However, immediately before carrying out the MAS NMR measurements the samples had to be contacted with the ambient atmosphere, *viz.* when they were placed into the rotor tube. For comparison, CP measurements were therefore conducted without MAS under exclusion of ambient atmosphere. Even though the quality of the spectra was lower than in the MAS case no substantial difference with respect to the main signals was encountered.

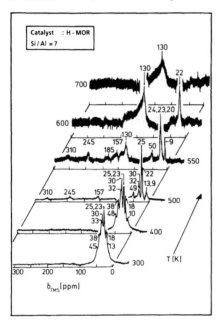

Fig. 9. ^{13}C-MAS NMR spectra of carbonaceous deposits formed by reaction of 90% ^{13}C-enriched ethylene over hydrogen mordenite [38].

At low temperatures, i.e. up to 500 K, signals were observed which gave evidence for the formation of branched and possibly also linear alkanes. The carbons of paraffinic chains, CH$_3$ - CH$_2$ - (CH$_2$)$_n$ - , may contribute to the NMR lines observed at 13, 25, and 30 to 33 ppm (Figure 9). A large fraction of branched chains was indeed indicated by the signal around 40 ppm and an intensity ratio for the lines at 13 and 25 ppm lower than 1 : 1 [47]. Interestingly, a line was observed at 48 to 50 ppm which was possibly due to formation of alkoxide, i.e. species such as CH - CH$_2$ - O - zeolite. The oligomeric deposit remained nearly unchanged upon heating to 450 K. Upon heating up to 550 K, the carbonaceous species isomerized and subsequently cracked whereby smaller molecules were formed. This is derived from the change in the spectra between 0 and 40 ppm, where the lines around 18 and between 30 and 40 ppm significantly decreased, leaving lines at 20, 23, 24 and 50 ppm and, concomitantly, causing the appearance of lines at higher shifts. At high coking temperatures (700 K) the signal of highly unsaturated species (130 ppm) absolutely predominated.

Around 500 K, various carbocations developed which gave rise to well-defined NMR lines at 130, 157, 185, 245 and 310 ppm. The line at 130 ppm is ascribed to highly unsaturated species, most probably aromatics; that at 157 ppm together with the 245 ppm line is likely to indicate acyclic and cyclic carbocations. The line at 185 ppm, which appears after heating the deposit to 550 K, together with the line of "high-temperature" coke species at 130 ppm, could be indicative of protonated benzene, alkylbenzenes or polynuclear aromatics [47, 53-54]. The protonated carbon of the benzene ring would then be responsible for the signal around 50 ppm.

The existence of polyenic carbocations is not clearly supported by the present results, since this would require not only the appearance of the 130 ppm line but also the presence of a line in the range between 80 and 100 ppm [55]. The occurrence of uncharged polyenes, however, is not excluded (compare section 5).

Thus, the results of these ^{13}C MAS NMR investigations are in good agreement with the discrimination of two types of coke as derived from IR results (compare previous section as well as sections 4 and 5).

More recently, Anderson et al. [56] developed a very promising technique permitting ^{13}C MAS NMR studies of coke formation under conditions which are even closer to those of real catalysis. They kept a series of small tubes, containing the catalyst H-ZSM-5, under the feed stream (methanol) at reaction temperature (650 K) and followed the conversion of methanol to gasoline by on-line GC. At certain intervals the tubes with used catalyst samples were sealed off. The sealed tubes could be investigated without intermittent contact of the coked catalyst samples with the ambient atmosphere. Thus, the authors were able to follow the build-up of coke precursors and coke by ^{13}C MAS NMR as a function of time on stream and other reaction parameters. Up to about 30 hydrocarbon species held in the pore system of the working catalyst were identified. Surprisingly, also an intense signal was observed which was ascribable to CO.

CP ^{13}C MAS NMR spectroscopy was also applied for the investigation of coke formation upon conversion of n-hexane (not enriched in ^{13}C) over H-ZSM-5 catalysts at 813 K. The results showed formation of aromatic ($\delta = 128$ ppm) and, to a lesser extent, most likely olefinic carbonaceous residues [57]. Finally, the same technique was used by Andersen et al. in their work on retained and desorbed products from reaction of 1-hexene over H-ZSM-5 [58].

4. Electron Spin Resonance Spectroscopy

It is a general experience that upon coke formation on zeolites free radicals appear. These can be investigated by ESR. The experiments can be carried out in X or Q band (8.2 to 12.4 or 33 to 50 GHz, respectively; Ref. [59]), using static EPR cells of fused silica [52] or even tiny flow reactors for *in-situ* measurements [60-61]. Leith [62] and Kucherov and Slinkin [63-66] were the first to publish interesting ESR spectra obtained after contact of activated acidic zeolites with olefins under static conditions.

The particular features of the ESR signals provide information about the type of coke formed. Thus, Leith [62] and Kucherov and Slinkin [63-64] observed a well-resolved 15-line ESR signal with a hyperfine splitting constant of a = 0.8 mT and a g-value of 2.0028 when they adsorbed propylene on hydrogen mordenite. The multiplicity of lines proved that the olefin must have oligomerised and, indeed, the observed spectrum could be ascribed to oligomeric olefinic or allylic species [62-64]. In similar experiments by Lange et al. [52] this signal was well reproduced. At higher reaction temperatures (350-450 K) under static conditions but with renewed ethylene atmosphere prior to each heating step, an ESR spectrum developed with seven hyperfine lines ($a = 1.6$ mT) and a g-value of 2.0025. This spectrum was almost identical to that obtained from γ-irradiated glassy allyl chloride [67] and similar to the spectrum shown in Figure 11 for under-stream conditions at 430-470 K [61]. At even higher temperatures (above about 500 K), however, the hyperfine splitting of the spectrum vanished and the signals collapsed into a single line with a g-value close to that of free electrons, indicating species with a highly delocalized π-electron system.

When the numbers of spins per gram of coked catalyst were plotted as a function of reaction temperature with all the other conditions unchanged, curves resulted as shown in Figure 10. This figure demonstrates the clearly separated two regions of two types of coke, _viz._ the oligomeric olefinic/allylic deposits (below 500 K, multiple line spectra, $a = 1.6$ mT) vs. highly unsaturated hydrocarbons (above ca. 550 K, single line spectrum, $g = 2.0023$). The spin concentration of the latter coke species steeply increased with temperature.

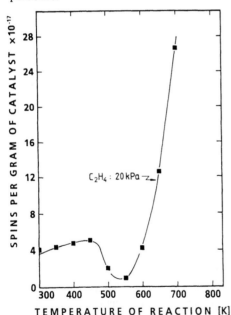

Fig. 10. Intensity of the EPR spectra (spin density) after reaction of ethylene over hydrogen mordenite as a function of temperature; ethylene atmosphere renewed after each heating step [52].

Similar results were obtained by in-situ measurements in an ESR flow reactor [61], as is illustrated by Figure 11, where a stream of 8 vol% ethylene in helium was passed at successively higher reaction temperatures over a catalyst of hydrogen mordenite which was activated at 700 K in a flow (15 ml · min^{-1}) of dry air (1 h) and nitrogen (1 h) prior to reaction.

The number of radicals typical of the low-temperature coke (multiple-line ESR spectra) decreased with decreasing Al content of the zeolite, i.e. decreasing number of acidic sites (Brønsted sites) which are required for formation of oligomeric species. This is demonstrated by Figure 12. In the series used, the density of Lewis sites, which might be responsible for the stabilization of these radicals, decreased in the same sequence as the density of the Brønsted sites. However, the effect of the zeolite acidity (type, number and strength of the acidic sites) of dealuminated mordenites used in the experiments of Ref. [52] on the formation of "high-temperature coke" (single line ESR spectra) was much more complex and requires further investigation.

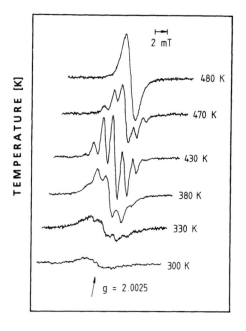

Fig. 11. In-situ EPR spectra of radicals formed during coke formation upon ethylene reaction over hydrogen mordenite [61].

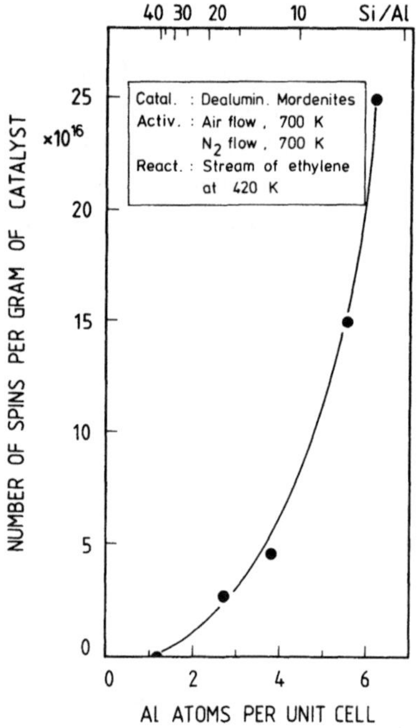

Fig. 12. Density of the radicals (spins) formed during ethylene reaction over hydrogen mordenites as a function of their Al content (acidity) [61].

Even though only a small fraction of the carbon atoms of the coke deposits corresponded to radicals (about 0.1 to 1%), the ESR results can be regarded as representative for the formation of carbonaceous residues on the zeolite catalysts. This is suggested by the results presented in Figure 13. A close (linear) correlation was found between the concentration of radicals of high-temperature coke (determined via ESR) and the total amount of coke formed on hydrogen mordenite at 573 K (determined via TGA). Interestingly, the maximum deposition of coke was 12 wt-%, irrelevant of what sample of the series of dealuminated mordenites (compare Figure12) was employed for the ethylene reaction. This seems to suggest that under the conditions of this reaction the maximum deposit of coke is related to the primary pore structure (*vide infra*, section 13). This differs from the results obtained when the coke-forming reaction was the dealkylation of the more bulky ethylbenzene (compare Ref. 36). In this case deactivation was caused by site blockage; the ultimate loading with coke depended on the dealkylation activity and decreased with decreasing Al content (number of active Brønsted sites).

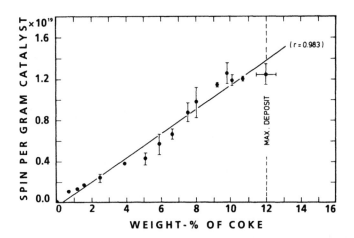

Fig. 13. Correlation between the number of radicals and the amount of high-temperature coke deposited during the reaction of ethylene over hydrogen mordenite at 573 K [61].

5. Ultraviolet-visible Spectroscopy

Even though zeolite samples usually exhibit low transparency and, moreover, the transmittance is strongly reduced upon coke deposition, coking of zeolites can also be studied by UV-VIS spectroscopy in transmittance mode. This would even enable us to carry out spectroscopic investigations with the very same samples in other spectroscopic ranges, e.g. in IR, ESR etc. The techniques are described in detail elsewhere [68-69]. However, measurements in reflection mode are also possible [70]. This is particularly noteworthy, since it may happen that the spectroscopic transmittance of the samples is indeed too low.

UV-VIS spectroscopic studies related to coke formation were carried out by only a few authors [19, 70-72]. It turned out that UV-VIS spectroscopy is rather sensitive and capable of providing valuable information about intermediates, coke precursors and final coke structures.

As an example, a set of difference spectra (i.e. with the background spectrum of the zeolite matrix having been subtracted) is depicted in Figure 14. These spectra were obtained under static conditions with a H-ZSM-5 sample (total Si/Al ratio 33.5). The transparent wafer (12.6 mg · cm^{-2}) was pretreated in high vacuum (10^{-5} Pa, 10 h) and subsequently loaded at room temperature with methanol (0.5 mmol per g zeolite).

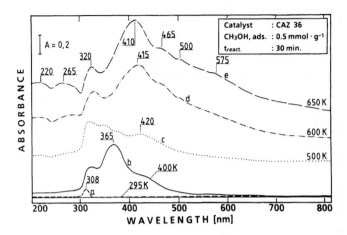

Fig. 14. UV-visible difference spectra of a methanol-loaded H-ZSM-5 catalyst (CAZ 36) after heating at successively higher temperatures [19].

Immediately after contact with methanol a small band at 308 nm appeared (difference spectrum a); absorbance in this range (around 310 nm) is usually observed after interaction of olefins with acidic adsorbents and originates from π-allylic carbocations [73-75]. Subsequent stepwise heating to 400, 500, 600 and 650 K of the adsorbate/catalyst system in a closed UV-VIS cell led to the spectra b, c, d and e, respectively, which exhibit a number of bands typical of intermediates and carbonaceous residues.

At low temperature, e.g. 400 K, the most prominent band is that around 365 nm. UV bands in the range 360-385 nm are ascribable to polyenylic cations such as $[CH_2 \cdots C(Me) \cdots CH \cdots CH \cdots CH_2]^+$, see Refs. [71-73, 76]. The broad band around 320 nm may still contain some contributions of π-allylic cations but mainly arises from cyclohexenyl cations (315 nm), since such cyclization products of olefin interaction on acidic catalysts are indeed indicated by UV bands in the range of 275-330 nm [70, 75]. The shoulder at 420 nm is most probably due to more bulky species such as diphenylcarbenium ions [77] or polyalkylaromatics and condensed aromatic compounds [70, 73] which, to a small extent, may even form at low temperatures.

Signals of such species, however, become predominant at higher reaction temperatures at the expense of the polyenic species which are weakened and, finally, vanish upon the transformation of "low-temperature coke" into "high-temperature coke". Pertinent assignments of the bands which appear at higher reaction temperatures are as follows: a pair of bands (320 and, more intense, 410-420 nm) is indicative of bulky aromatic species (diphenyl or polyphenyl carbenium ions, polyalkylaromatics, condensed aromatic ring systems, see Refs. 70, 77). Evidence of formation of such species at high temperatures (600-700 K) is also provided by bands at 500 and 575 nm [70, 77]. Bands observed at 220, 265 and 465 nm may be ascribed to dienes, cyclohexadiene and/or benzene and cations of substituted benzenes, respectively.

Similar experiments with olefins (ethylene, propylene, butene, butadiene, hexatriene etc.) resulted in spectral features which were, at least to a large extent, essentially the same as described above [78]. However, the fact that in UV-VIS studies polyenic cations were observed is at some variance with the ^{13}C MAS NMR results (see section 2.3), but this may indicate that the UV-VIS technique is very sensitive and only a small fraction of the polyalkenes need to be present as carbocations in order to be identified in the 360-385 nm region.

In agreement with the preceding considerations of the IR, NMR and ESR results, the UV-VIS investigations illustrate again that two types of coke should be distinguished (coke I and coke II, see section 4); at low temperatures, olefinic/polyolefinic constituents predominate whereas at higher temperatures the main components of the carbonaceous deposits are bulky aromatic and polyaromatic species.

However, it could be shown that the preferential occurrence of one type of coke or the other depends not only on the reaction temperature but on properties of the catalysts as well. Thus, for instance, small crystallites of H-ZSM-5 (average diameter 0.4 µm) with fewer but more strongly acidic sites than those used in the experiments of Figure 14 favoured the formation of coke I even at a relatively high reaction temperature [19]. This is illustrated by Figure 15 comparing the spectra of two H-ZSM-5 catalysts which differ in their properties as outlined above.

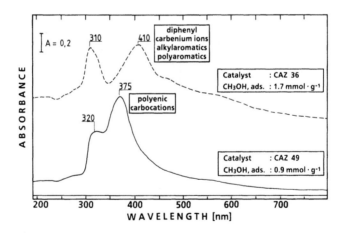

Fig. 15. UV-visible difference spectra of two different H-ZSM-5 catalysts (CAZ 36, CAZ 49) loaded with methanol and heated at 650 K.

Differences in behaviour with respect to formation of carbonaceous deposits, as demonstrated by the two examples of Figure 15, may have significant consequences in the catalytic performance with respect to the desired reaction (e.g. methanol-to-gasoline conversion) as well as for the catalyst regeneration [19, 78].

550

6. Photoelectron Spectroscopy

Only recently, photoelectron spectroscopy, i.e. X-ray photoelectron spectroscopy (XPS) and Auger electron spectroscopy (AES), was applied in studying coke formation [79, 80]. Determination of the C/Si ratio as a function of coke loading was used to distinguish between deposition in the interior pore system and on the external surface of the zeolite crystallites (Ref. [79], see section 13). Fleisch et al. [80] claimed that their investigations by core electron energy loss spectroscopy (CEELS) enabled them to distinguish clearly between coke compounds with sp^3 and sp^2 hybridization. They reported good agreement of the results obtained by this technique and CP MAS NMR measurements.

The interpretation of XPS spectra of coked zeolite catalysts, however, is difficult. Assignments of signals might be erroneous because of severe conductivity problems; the disturbance caused by charging effects depends on the degree and topology of the surface coverage with coke. Nevertheless, photoelectron spectroscopy is a promising and attractive method for coking studies, all the more as modern spectrometer devices permit essentially *in-situ* measurements, i.e. investigation of the coked surface without any intermittent contact with ambient atmosphere [81].

Auger Electron Spectroscopy (AES) enables us to avoid disturbances of the above-mentioned type because of the high electron energy used. Figure 16 shows an example of a carbon Auger electron spectrum [82]. It was obtained from an H-ZSM-5 (CAZ 49)

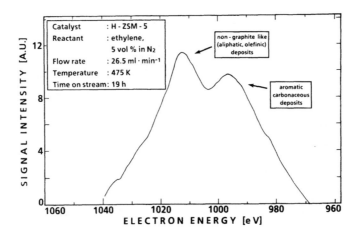

Fig. 16. Carbon Auger spectrum of an H-ZSM-5 catalyst (CAZ 49) after coking in an ethylene stream at 475 K.

sample which was covered with coke after 19 h exposure to a stream of 5 vol% ethylene in N_2 (total pressure 100 kPa) at 475 K. The spectrum exhibits two broad peaks, *viz.* at 995 eV and around 1010 eV. The former peak is ascribable to highly aromatic carbonaceous deposits ("carbon black" type, see Ref. [83]). These species are probably very bulky and, therefore, their presence is restricted to the external surface of the crystallites. The high energy peak around 1010 eV is indicative of deposits richer in hydrogen and aliphatic or olefinic in character. However, a distinction between sp^3 and sp^2 hybridization is not possible. Although quantitative estimations about the relative abundance of the two types of coke constituents cannot be derived from these results, the appearance of both sufficiently resolved signals in the spectrum of Figure 16 is in qualitative agreement with the spectroscopic results described in sections 2 through 5.

A sample, which was coked in a feed stream of 12 vol% methanol in He at 650 K for 19 h, provided AES results similar to those shown in Figure 16 [82].

Spectra obtained by XPS from the same H-ZSM-5 samples (coked with ethylene or methanol) still exhibited the Si and Al signals at 103.2 eV and 75.0 eV, respectively. However, the peak areas of the coked samples were reduced by about 85 % compared to the corresponding peak areas of the fresh catalysts. Since the depth of observation of the XPS method is restricted to a few monolayers, a coverage of the external surface of the zeolite grains with a thick homogeneous layer of coke would result in a complete suppression of the Si and Al signals. Therefore, one has to conclude that the external zeolite surface was not completely covered by a homogeneous coke layer; rather, a topologically heterogeneous coverage is suggested.

This has been confirmed by more recent work using a significantly improved technique which allows even for "in-situ" experiments [81]. Results obtained so far refer to coking of an H-ZSM-5 catalyst (activated in high-vacuum at 675 K prior to reaction) under the stream of diluted ethylene (5 vol% in He) at 650 K. In view of well-established XPS data for carbon-containing species [84] the results may be interpreted as follows:

(i)   coke deposition in the form of a thin (1 to 2 monolayers), complete and conductive film of carbonaceous materials starts to occur already in the very first stage of the experiment (0.4 to 0.8 wt% coke, obtained by coking with ethylene);

(ii)   upon longer exposure to the feed stream thicker carbon-containing islands grow on top of the first layer, exhibiting electrical charging effects and rendering the coke deposition topologically heterogeneous;

(iii)   in agreement with earlier results (IR, NMR, ESR, UV-VIS, AES), two types of carbonaceous deposits can be distinguished, *viz.* paraffinic or olefinic, hydrogen-rich ones (corresponding to coke of type I) and residues similar to "carbon black" but still non-graphite species (corresponding to coke of type II).

Photoelectron spectroscopy will certainly allow for further valuable and very detailed *in-situ* investigations of the nature of coke and the mechanism of its formation.

7. Extraction of coked zeolite catalysts

Venuto et al. [85] were the first to dissolve coked zeolite catalysts and analyse the extracts of the solution by GC. Since they used RE-X catalysts, a relatively mild acid treatment was sufficient to destroy the zeolite matrix. Coke forming reactions were, *inter alia*, alkylation of benzene by olefin (at 485 K) and olefin polymerisation. After destruction of the coked zeolites (H/C ratios of coke about 1.3) the acidic suspensions were extracted with CHCl$_3$. The black tar extracted from the RE-X catalysts (deactivated during the alkylation of benzene by olefin) was a complex mixture of high-molecular-weight aromatic compounds (highly alkylated benzenes, polyalkylnaphthalenes and higher condensed polycycles). Interestingly, essentially the same species were found trapped in the pore system of the zeolite catalyst when ethylene alone was passed over RE-X at 485 K. This suggests that, at least to a great extent, the reactions of ethylene and intermediates formed therefrom are responsible for deactivation of the RE-X catalyst employed for the alkylation of benzene. Becker et al. [11] arrived at the same conclusion when they studied the deactivation of hydrogen mordenite during alkylation of benzene with ethylene or propylene. However, Walsh and Rollmann [86] reported on an interesting effect of the feed composition on coke formation. They used radiotracer methods and found that the coke was mainly produced from the aromatics when a mixture of paraffins and aromatics was passed over Y- and mordenite-type catalysts at 635 K.

According to Venuto et al. [31, 85], formation of polynuclear aromatics inside the zeolite pore structure is due to a complex series of side reactions of ethylene and/or polyalkylbenzenes, including polymerisation, hydrogen transfer and dehydrocyclisations.

The method of acid treatment and extraction of coked zeolite catalysts was further developed and extensively used by Guisnet and co-workers in a series of systematic studies on coke formation [20-21, 87-92].

The techniques used by these authors have been described in Refs. [20, 87]. In a first step the coke components, trapped in the channels and cavities of a zeolite pore system, are liberated by dissolution of the aluminosilicate matrix in hydrofluoric acid (40%) at room temperature. Subsequently the soluble components are extracted by CH_2Cl_2 as a solvent; in some cases "insoluble coke" remains in the form of black particles (*vide infra*). The extracts are analysed by GC, HPLC, ^{1}H-NMR and MS.

A serious problem related to the extraction method, however, is whether or not changes in the original chemical nature of the coke occur upon the rigorous treatment with mineral acids. Guisnet and co-workers [20, 87] have carried out a series of tests which seem to confirm that the procedure employed does not generate any artifacts. They loaded inert carriers (SiO_2) with, for instance, phenanthrene or 1-tetradecene and submitted these systems to the same treatment as was applied to coked catalysts. No chemical modification of the hydrocarbon test materials was observed. However, it is still not certain that this also holds for highly reactive species (e.g., polyenic compounds etc.) when in contact with the active surface of the catalyst. Concerted experiments, using spectroscopic tools and the extraction method, are being carried out to clarify this point [93].

Guisnet's group studied coke formation upon propene conversion over ultrastabilized H-Y, (US H-Y), H-MOR and H-ZSM-5 [20], and on n-heptane cracking over H-Y, hydrogen mordenite (H-MOR), H-ZSM-5, hydrogen offretite (H-OFF) and hydrogen erionite (H-ERI) [21, 88-92].

As an illustration, we describe some results obtained by Magnoux and Guisnet when investigating the nature of coke deposits and the mode of deactivation on n-heptane cracking over H-ERI [90]. The coke loading was varied through the variation of the time on stream.

The authors found that at low coke loadings the coke was "non-polyaromatic"; the aromaticity increased with severity of coke deposition. Evidence for this general trend was provided, *inter alia*, by ^{1}H-NMR (Figure 17). Particularly with high coke loadings

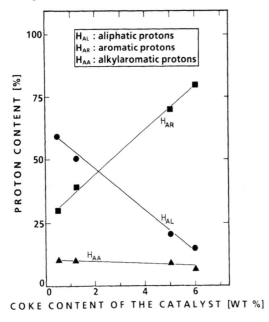

Fig. 17. ^{1}H NMR analysis of the soluble fraction of coke (formed upon n-heptane reaction over H-ERI); details see text [90]

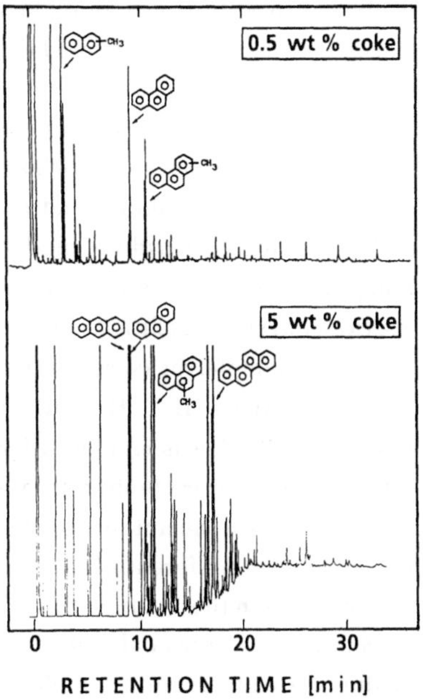

Fig. 18. Gas chromatographic analysis of extracts of coke formed upon n-heptane reaction over H-ERI [90]; details see text.

(5 to 6 wt%), polyaromatic compounds in the form of small black particles (H/C < 0.5) were detected. The "soluble coke", however, was a rather complex mixture. Depending on the coke loading, i.e. time on stream (aging), a number of main families of compounds constituting the carbonaceous deposits were deduced from the GC and MS results. The upper part of Figure 18 shows the main compounds forming the carbonaceous residues after a short time on stream or low coke loading (0.5 wt%). Some of the main families of coke constituents determined at higher loadings are indicated in the gas chromatogram in the lower part of Figure 18. At even higher coke content (6 wt%), accumulated during n-heptane cracking over H-ERI, the composition of the soluble coke was less complex; it was mainly comprised of anthracene, phenanthrene and chrysene. Similar results were obtained for coke formed over H-ZSM-5 and H-MOR.

The method of extraction was also applied by Andersen et al. [58] in their study on retained and desorbed products which occur upon reaction of 1-hexene over H-ZSM-5 catalysts. Their attempt to derive a detailed scheme of the hydrocarbon chemistry, involved in coke formation during this reaction, is based to a large extent on the results obtained by the extraction technique. Evidence was provided of a key role of cyclopentadienes as important intermediates in the route to coke formation.

8. Adsorption measurements

Comparison between the adsorption capacity of fresh and coked zeolite catalyst samples is frequently used to characterize the deposition of carbonaceous materials in the void volume of the zeolite structure. More specifically, adsorption measurements are used to clarify whether the coke is predominantly laid down in the zeolite pores or on the outer surface.

There is a relatively large number of studies of this type [see, e.g., Refs. 20, 39, 89-90, 92-97]. Adsorbate species, which are very often used in this context, are nitrogen, water, ammonia, trimethylamine and small paraffin molecules. However, also more bulky molecules such as pyridine, methyldiisopropylamine, substituted benzenes, methylene blue, which had no access to the particular internal volume, were used in order to probe the external surface of the zeolite crystallites [96-97].

As an example, measurements by van Hooff et al. [94] were used to compare different zeolite structures (H-Y, H-MOR, H-ZSM-5) with respect to pore filling. Cracking of n-hexane at 573 K was employed for coke formation. The void volumes of the fresh and coked catalysts were probed by n-butane adsorption. Results are presented in Table 2. From these data the authors concluded that in the case of H-Y and H-MOR coke deposition proceeded preferentially inside the pore structure whereas this occurred only to a minor extent in the channels of H-ZSM-5.

Similarly, Karge et al. [98] determined the BET analogous surface area of a series of dealuminated mordenite samples prior and subsequent to coking by ethylene (m[cat] : 0.015 g activated at 675 K in flowing N_2 (100 ml · min^{-1}); feed: 5 vol% ethylene in nitrogen; $\dot{v}$: 16 ml·min^{-1}; T[react] : 650 K; time on stream (TOS) : 100 h). The measurements were carried out with an "Omnisorb 360" apparatus [99] or with a "Quantasorb" (Quanta Chrome Comp., N.Y.). The uncoked samples were degassed at 675 K, the coked samples at 625 K prior to the measurements of the surface area. The results showed that 75-90% of the BET analogous internal surface was lost upon a coke loading of about 12 wt%.

Table 2. Integral conversion, coke deposition and decrease of pore-volume after 300 min reaction at 573 K for the 3 different zeolite catalysts [94]

Catalyst	Integral Conversion gC / gCat.	Coke Deposition wt%	Coke Selectivity %	Pore Volume (PV) ml/g		
				initial	final	ΔPV
H-ZSM-5	1.82	1.93	1.1	0.145	0.134	0.011
H-Mord.	0.34	4.75	14	0.099	0.022	0.077
H-Y	0.22	2.59	12	0.141	0.084	0.057

Bibby et al. [96] used basic probe molecules (ammonia, pyridine) for monitoring the change of the adsorption capacity of H-ZSM-5 catalysts caused by coke deposition. From their results they simultaneously derived conclusions concerning the available acidic centres before and after coke loading (see also section 11) and the locus of coke deposition (section 13).

Hydrocarbon molecules with different effective molecular diameters (n-hexane, 3-methylpentane) were employed by Guisnet and co-workers [92] in order to determine not only the change in adsorption capacity as a function of coke deposition but also to elegantly discriminate between the various localizations of the deposited coke components in, for instance, the different cavities of hydrogen offretite (see section 13).

9. Temperature-programmed oxidation

Temperature-programmed oxidation should be a suitable tool to discriminate between different types of coke deposited on a catalyst. It has been, *inter alia*, successfully applied to coke investigation on bifunctional catalysts. Although bifunctional zeolite catalysts are usually employed under hydrogenating conditions which remove, to a large extent, polymerisable, unsaturated coke-forming species, coke formation is frequently encountered with this type of catalyst as well. One example was already mentioned (see section 2) as studied *via* in-situ IR by Eisenbach and Gallei [7].

Barbier [2], in fact, investigated coke formation on reforming catalysts which used Al_2O_3 as a support (Pt/Al_2O_3). He studied the coke deposits formed upon conversion of cyclopentane/cyclopentene mixtures and was able to discriminate, *via* temperature-programmed combustion of the deposits, between coke laid down on the metal and coke located on the support. Analogous investigations should be possible with coked bifunctional catalysts containing acidic zeolites as a support.

PARTICULAR PHENOMENA RELATED TO COKE FORMATION

10. Selectivity of coke formation

The idea that coke formation in zeolites is a shape-selective reaction was first advanced by Rollmann [8] and Rollmann and Walsh [9]. In their studies they systematically investigated the correlation between structure properties, sorptive behaviour, catalytic selectivities and coking (aging) tendencies of a series of acidic zeolites (erionite, ferrierite, TMA-offretite, mordenite, zeolite L, zeolite Y and H-ZSM-4). It was shown that those catalysts (ferrierite, erionite) which exhibited a high (reactant) selectivity in the conversion of a five-component feed (2,3-dimethylbutane, 3-methylpentane, n-hexane, benzene, toluene) were also characterised by low coke yields and low aging rates. By contrast, the large pore zeolites (H-MOR, H-L, H-Y, H-ZSM-4) did not selectively convert the feed mixture and, correspondingly, suffered from severe coking and aging (see Table 3). TMA-offretite represented a unique case in the group of zeolites investigated. It contains both 12- and 8- ring channel systems, both accessible by n-hexane. However, only the 12-ring channels are capable of accommodating the larger cyclohexane or benzene molecules. This intermediate position of TMA-offretite is reflected in its selectivity and coking behaviour.

556

Table 3.   Coke Selectivities[a] [8]

Catalyst	Selectivity	
	Coke[b]	Normalized[c]
Erionite	0.14	0.16
Ferrierite	0.03	0.04
TMA offretite	0.7	0.8
Mordenite	0.3	0.4
Zeolite L	0.4	0.5
ZSM-4	0.4	0.5
Zeolite Y	2.2	2.9

[a] 5-Component feed; 3.3 days on stream; all catalysts treated under a standard cycle with WHSV = 3 h[-1]; T = 589-811 K; p = 1.4 · 10^6 Pa; H_2/HC = 3; [b] grams/100 g conversion, observed. [c] grams/100 g conversion, normalized to 50% conversion using first-order approximation.

Already Rollmann [8] suggested in his work that the formation of more bulky structures such as cycloparaffins ("coke precursors") is sterically hindered by the restrictive pore systems of reactant or product shape-selective catalysts, e.g. ferrierite and erionite. In a subsequent paper Rollmann and Walsh [9] confirmed their view of shape selectivity in coke formation (Figure 19) and also included the investigation of the new prominent ZSM-5 catalyst with its remarkably low propensity to form coke. Furthermore, the concept of geometrical restrictions for the formation of spaceous transition states of coke building species ("transition state selectivity") was discussed [10].

However, a modified hypothesis for the low-coking tendency of H-ZSM-5 in comparison with H-Y and H-MOR was advanced by Schulz et al. [100]. These authors assumed that only the first stage of formation of carbonaceous deposits is governed by shape-selective properties of the surrounding of the acidic sites inside the pore structure: on H-ZSM-5 this, indeed, results in low initial deposition of carbonaceous residues. The further growth of coke, however, does not proceed, according to Schulz et al. [100], via a carbenium ion mechanism but via direct methylation of the carbonaceous deposits (by the feed, e.g. olefin or methanol) followed by elimination of methane. For stoichiometric reasons at least one carbon atom is added to the coke when one CH_4 molecule is evolved. Thus, evolution of CH_4 is seen as a suitable indicator for growth of coke (compare also Ref. [101]).

In view of this concept one would expect more or less pronounced induction periods for coke formation on shape--selective catalysts such as H-ZSM-5. But this is not the case, at least not always, and the mechanism and kinetics of coke formation seem to depend on various factors (see, e.g. Figure 8 in Ref. 19). Therefore, more work is needed, employing

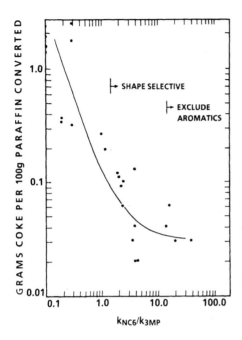

Fig. 19. Coke yield vs. shape selectivity of paraffin conversion over acid zeolite catalysts [8]; the shape selectivity is measured through the ratio of the first order rate constants, k_{NC6}/k_{3MP}, for disappearance of the respective paraffin isomers (n-hexane, 3-methylpentane) at 700 K.

various techniques, to check the interesting hypothesis by Schulz et al. In any case it is obvious that the effect of shape selectivity with respect to coke formation has to be considered primarily in the case of the so-called type II coke (see section 5) and does not play a major role, if any, with type I coke (low-temperature or white coke).

11. Effect of acidity

The role of acidity (nature of acidic sites, i.e. Brønsted vs. Lewis centres, number and strength of sites) was already mentioned several times in preceding sections. However, the relationship between coke formation and acidity is far from evident, and it seems too early to make an attempt of proposing general correlations between both phenomena.

As mentioned in section 4 (Figure 12), the yield of low-temperature coke continuously decreased with the Al content of dealuminated mordenites, whereas the effect on high-temperature coke, although existent and measurable, depended in a rather complex manner on the Si/Al ratio [61]. The coking behaviour of a series of mordenites with systematically varied (framework) Si/Al ratio during hydroisomerization of n-hexane was studied by Haas et al. [102]. These authors found that synthetic mordenites steamed at 725 K or strongly dealuminated by leaching with hydrochloric acid (Si/Al up to 39) exhibited improved activity, stability and regeneration properties. Also, Kubelková et al. [45] did observe an effect of acidity (strong Brønsted sites and electron accepting centres, i.e. cationic Al species) on coke formation when ethylene was reacted over ZSM-5 or modified Y-type catalysts. Temperature-programmed desorption of ammonia from fresh and coked catalysts seemed to indicate that the most acidic sites were preferentially poisoned by coke.

In contrast, Rollmann [8] did not observe any influence of the modulus SiO_2/Al_2O_3 on the coking behaviour of two pairs of differently dealuminated hydrogen mordenite and H-Y catalysts. Similarly, Schulz et al. [100] claimed that the acidic sites are required only for the initial slow stage of formation of coke precursors in the methanol-to-gasoline reaction at 650 and 750 K and that the acid sites are not involved in further coke build-up.

On the other hand, Bibby et al. [103] reported that in fact the overall deposition of coke during methanol conversion over H-ZSM-5 was higher the lower the Si/Al ratio, whereas the initial coke formation was not related to the aluminium content. Furthermore, it was deduced from ammonia and pyridine adsorption measurements on H-ZSM-5 samples that during the initial period of coke deposition (up to about 4 to 5 wt% of coke during methanol reaction at 650 K) a rapid loss of acid Brønsted sites occurred. This was followed by a much slower decrease in the number of acidic OH groups. However, after a total coke deposition of about 15 wt% the number of acidic sites, indicated by the bases, dropped to zero and the activity of the catalysts had completely vanished. This is at some variance with the observations by Karge et al. [36] who found via *in-situ* IR and conversion measurements that the number of acidic OH groups in H-MOR and H-ZSM-5 remained essentially unaffected in spite of severe coking and loss of activity. This was explained by deactivation through blocking rather than consumption or poisoning of sites.

Although there seems to be noticeable disagreement about the role of the number (and sometimes also the strength) of Brønsted sites with respect to coke formation over acidic zeolites, most authors agree that the presence of acidic OH groups is the indispensible requirement for coking to occur (at least in the initial stage of the process). Only a few studies seem to suggest a predominant effect of Lewis sites on the propensity of the zeolite to form coke [28]. In general, however, these investigations have not excluded that Brønsted centres were present on the catalyst surface; they may form, for instance, through a (limited) rehydroxylation in which Lewis sites are involved [104].

12. Mechanism and Kinetics

The current view of the pathways of coke formation is presented in the scheme of Figure 20. Many intermediates and precursors of coke, suggested by this scheme, were identified (see sections 2-7) as well as constituents of the "white" coke or type-I coke, generally formed at lower temperatures (sections 2-6). It has been shown that similar species occur during deposition of carbonaceous materials, regardless of whether the feed contains paraffins, olefins or methanol. However, the actual network of the reactions depends most probably on the specific situation (catalyst, feed, temperature etc.) and has not yet been experimentally established and unambiguously evaluated in a particular case.

The kinetics of coke deposition is frequently described by the, in fact, empirical Voorhies equation [105]

$$A = k_c \cdot t^{n_c} \, , \quad n_c \geq 0 \tag{8}$$

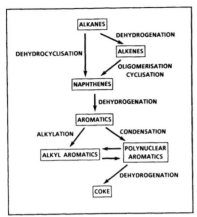

Fig. 20. Current view about coke formation on zeolites.

where A = amount of coke deposited [wt%], k_c = constant [wt% · min$^{-n_c}$], t = time on stream [min], n_c = parameter. Voorhies obtained a value of n_c close to 0.5 and, therefore, suggested that coke deposition is related to diffusion, the coke itself being the diffusion barrier. However, subsequent studies of a number of workers (compare Ref. [1]) have shown that values of n_c occurred which were significantly higher than 0.5. Moreover, it turned out that both k_c and n_c were affected by the nature of the feed as well as by reaction conditions, e.g. space velocity. More elaborate relations for deactivation and coke deposition are, for instance, provided in Refs. [106-108].

On the basis of the *in-situ* IR and conversion measurements upon ethylbenzene reaction over dealuminated mordenites (presented in section 2) it was found that both aging (activity decrease) and coke deposition could be approximated by a relationship of first order in time [36, 61] (Figure 21).

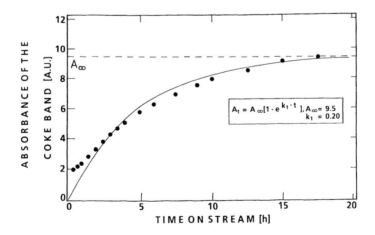

Fig. 21. Coke formation (measured via the absorbance of the coke band) upon ethylbenzene reaction over hydrogen mordenite (Si/Al = 12) as a function of time on stream; first order relationship.

560

The results allowed elimination of the variable t and presentation of the loss of activity as a function of coke deposited (Figure 22). The slopes of the straight lines in Figure 22 provide a measure for the sensitivity of the catalyst to coke formation: the steeper the straight line the lower the amount of coke required to effect a given decrease in conversion.

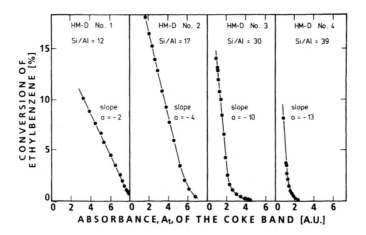

Fig. 22. Sensitivity of hydrogen mordenites (with various Si/Al ratios) to coke formation upon conversion of ethylbenzene (see text) [36].

Another approach, relating the coke yield not simply to the time on stream but taking into account the different times of exposure of the catalyst, in order to obtain a distinct state of deactivation, was advanced by Schulz et al. [22,100]. These authors defined a normalized time on stream (NTOS) which is the real time on stream divided by the time elapsed until unconverted feed (methanol) was observed due to deactivtion of the catalyst. This facilitated the comparison of various catalysts with respect to their catalytic performance.

13. Localization of Coke Deposition

An interesting and important problem related to coking zeolite catalysts is the localization of the coke deposits. Usually the question arises whether the coke is preferentially formed inside the zeolite pore system or on the external crystallite surface. In some cases, even more detailed information is required when various types of cavities of a zeolite structure are possible loci of coke deposition.

In Figures 23 and 24 the coke deposition upon the reaction of a relatively small coke-forming molecule (ethylene) is compared for three catalysts, viz. H-Y, H-MOR and H-ZSM-5. The measurements both via the absorbance of the IR band around 1600 cm^{-1} (coke band, see section 2) and via the weight gain (TGA) seem to suggest that the final

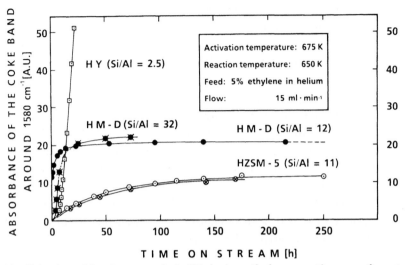

Fig. 23. Coke deposition (measured via IR) during ethylene reaction over three types of zeolites (H-Y, HM-D, H-ZSM-5) with different void volumes [37].

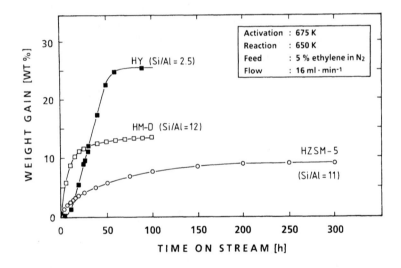

Fig. 24. Coke deposition (measured via TGA) during ethylene reaction over three types of zeolites (H-Y, HM-D, H-ZSM-5) with different void volumes [37].

coke levels in Figures 23 and 24 (26 wt%, 12 wt% and 8 wt%) reflect the sequence of the void volumes of H-Y, H-MOR and H-ZSM-5: 0.38 (supercages only), 0.20 or 0.13 (total or main channel volume), and 0.16 $cm^3 \cdot g^{-1}$, respectively. The final coke level of 8 wt% in the case of H-ZSM-5 (see Figure 24) is in remarkably good agreement with the value found by Sexton et al. [79] for internal coke formation, i.e. channel filling (*vide infra*). The IR results (Figure 23) show indeed the same sequence (and a good reproducibility) of the results as the TGA measurements; in fact, for H-Y the final coke level was not

562

observed due to the very low transmittance of a heavily coked H-Y sample. If one assumes an average density of $1.5 \text{ g} \cdot \text{cm}^{-3}$ of the coke, the above figures for mordenite correspond to a 50% filling of the total void volume [109] $(0.20 \text{ cm}^3 \cdot \text{g}^{-1})$ or a 75% filling of the main channel system $(0.13 \text{ cm}^3 \cdot \text{g}^{-1})$. Thus, it is likely that, under the reaction conditions applied, a significant fraction of the amount of the final coke loading is accommodated inside the pore system.

On the basis of his data, Rollmann [8] arrived at the same conclusion. The situation is similar to that observed with H-Y and H-ZSM-5. Moreover, the decrease of the internal adsorption capacity measured for water and nitrogen (see section 8) after coke formation also suggests that at least a large fraction of the coke was deposited in the internal pore system of H-MOR and H-ZSM-5.

Similarly, Bibby et al. [39] derived from their ammonia and pyridine adsorption studies on fresh and coked H-ZSM-5 samples that at least the initial period of coke formation occurred in the interior of the zeolite crystallites where the acidic active sites are located.

In their XPS work on coking of H-ZSM-5 under a stream of methanol in nitrogen at 643 K, Sexton et al. [79] also derived from the C/Si ratios that up to 8 wt% coke the filling of the pores predominated. The C/Si ratio increased linearly with the amount of coke deposited. Formation of coke on the external surface occurred at higher loadings and was indicated by an exponential increase of the C/Si ratio (Figure 25). The results were in reasonable agreement with the theory advanced by these authors.

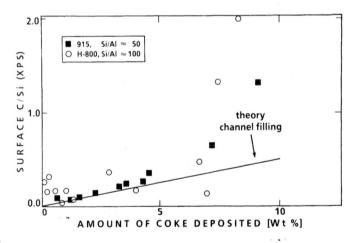

Fig. 25. C/Si ratio of coke deposited from a stream of methanol in nitrogen at 645 K on H-ZSM-5 catalysts [79].

Already Rollmann [8] pointed out that even in the structure of narrow- or medium-sized channel systems such as TMA-erionite (or ZSM-5) bulky coke species may form if the intersections of the channels provide sufficient space to accommodate these residues or, if necessary, the transition states through which they form.

However, in many cases H-ZSM-5 catalyst samples bear more than 20 wt% coke after complete deactivation in, for instance, a stream of methanol. In such events it is obvious that not all of the coke could be deposited in the internal pore system. Rather, a great part of the coke must be located on the external surface and fill the intercrystalline volume.

As an example, reference is made to the work of Bibby et al. [103]. For a first period of coking these authors observed indeed that the sorption capacity or the void volume of their H-ZSM-5 catalysts decreased linearly with the amount of coke deposited during methanol reaction at 635 K. They concluded that internal coke formation occurred intimately associated with the active sites. Internal deposition was confirmed by the X-ray diffraction (XRD) patterns which showed similar changes of the lattice parameters as were observed upon occlusion of large organic molecules or bulky ions. However, it was also observed that the maximum amount of coke deposited (up to 26.7%) exceeded the measured decrease of the void volume, indicating formation of external coke as well. In a subsequent study, Bibby et al. [96] confirmed their results via adsorption experiments with trimethylamine (TMA) and ethyldiisopropylamine (EDA). TMA had access to the pore volume of H-ZSM-5 and probed the total number of acidic sites whereas EDA provided information only about the external surface. Again, the authors arrrived at the conclusion that the reduction of the sorption capacity was caused by simple pore filling.

Andersen et al. [58] showed that, after deacidification of the external surface of H ZSM-5 by interaction with $SiCl_4$, the fraction of higher aromatics decreased when 1-hexene was reacted over this catalyst. Hence, in the case of the non-modified H-ZSM-5 those bulky species (naphthalenes, phenanthrenes, anthracenes) seem to form preferentially outside the pore system. According to Andersen et al., coke constituents initially formed in the interior of a (non-modified) H-ZSM-5 crystal might, at higher temperatures (T[react] $\geq$ 600 K) migrate to the external zeolite surface where, at acidic sites, conversion to more bulky, involatile residues is possible.

Applying transmission electron microscopy (TEM) in combination with electron energy loss spectroscopy (EELS), Gallezot et al. [110] were able to locate external coke deposits formed upon n-hexane cracking over ultra-stabilized H-Y, H-OFF and H-ZSM-5. Their techniques provided also information about the structure of the coke constituents, *viz.* coronene- or pentacene-like species.

An elegant technique to reliably discriminate between coke deposited in the intracrystalline and the intercrystalline volume was developed by Bülow et al. [111]. They carried out ^{1}H-NMR measurements on H-ZSM-5 samples coked by n-hexane cracking at 775 K or mesitylene reaction at 800 K. By means of NMR pulsed field gradient techniques the intracrystalline self-diffusion coefficient in solid samples, D_i, was measured for methane and propane. NMR tracer desorption experiments provided data about the fraction γ of molecules which are able to leave the crystallite during the observation time; γ is related to the effective diffusion coefficient D_d. While D_i is affected only by carbonaceous deposits inside the pore system and not by external coke layers, D_d is sensitive to both types of coke localization.

From the evaluation of D_i and D_d Bülow et al. [111-112] were able to conclude that (i) upon n-hexane cracking two stages of coke deposition occur: during the initial period coke is laid down in the intracrystalline channel network, during the final periods coke is mainly formed near the external surface; (ii) coke formed from mesitylene is, even at the beginning of the reaction, predominantly deposited close to the external crystallite surface.

Finally, another method should be mentioned which appears to be a suitable tool for determination of the coke distribution, $viz.$ the ^{129}Xe NMR technique [113]. This technique rendered possible, for instance, measurements of the free volume of a H-Y zeolite accessible for Xe atoms before and past coking upon reaction of n-hexane or propylene. Also, it allowed for the discrimination between internally and externally deposited residues.

Frequently, the sizes of channels and cavities of a particular zeolite architecture are not uniform. As an example offretite was already mentioned (section 10). This structure contains two types of pores, $viz.$ linear cylindrical channels (12-membered rings, diameter > 0.67 nm) and gmelinite cages (8-membered rings, diameter 0.45-0.50 nm) which interconnect the channels. Guisnet and co-workers [92] studied the localization of coke deposits formed upon n-heptane cracking on hydrogen offretite. Combining their determination of the coke composition (conducted analogously to the example of section 7) and adsorption measurements using four differently sized adsorbates (i.e. nitrogen, n-hexane, NC6, and 3-methylpentane, 3MP) these authors were able to localize the various coke species related to their period of formation. (i) In the initial stage, i.e. at low coking, small compounds are formed which are trapped in the gmelinite cages. This reduces the total adsorption capacity, e.g. for N_2, but does not hinder NC6 or 3MP from entering the pores. (ii) At higher coke loadings (0.8 through 4.5 wt%), the coke comprises polyaromatic molecules overflowing into the large channels. They limit or block access to the large channels, in particular for 3MP. (iii) For even greater coke content (> 4.5 wt%) very bulky polyaromatic compounds form in the pores close to the external surface preventing also N6C from penetrating the large channels and the gmelinite cages.

Systematic and detailed adsorption experiments such as described by Guisnet and co-workers [90, 92], employing adsorbates with suitably selected sizes, shapes and adsorptive properties, promise to provide better information as to where the deposition of carbonaceous residues occurs when coke is formed on a working zeolite catalyst. Most likely, such measurements can be advantageously combined with spectroscopic investigations. In-$situ$ studies, i.e. adsorption intermittent with periods of reaction but without contact to the ambient, would be desirable.

SUMMARY AND CONCLUDING REMARKS

It has been shown that there exists a variety of techniques which are suitable for investigating coke formation during hydrocarbon reactions over acidic zeolite catalysts and related phenomena. In many cases they were indeed successfully applied. It appears from the large body of results accumulated in this field so far that some general conclusions may be derived. For instance, the following features seem to be well established.

(i) At least two classes of coke may be distinguished by several criteria, *viz.* coke I (white or soft coke, preferentially formed at low reaction temperatures) and coke II (black or hard coke, mainly occurring at higher temperatures, e.g. at T[react.] > 500 K), even though the borderline between the two types of coke may not always be very sharp.

(ii) The so-called coke II, to which much more attention usually has been paid, is constituted mainly of polyaromatics, frequently of condensed ring systems.

(iii) Not only higher reaction temperatures but also extended time on stream ("aging of coke") favour the formation of more bulky, polyaromatic coke components with a low H/C ratio.

(iv) Coke formation is a shape-selective process and is significantly influenced by the architecture of the particular zeolite catalyst.

However, there are other aspects of coke formation, whose investigation seems to lead to conflicting or, at least, ambiguous results. Examples are the role of acidity, i.e. nature, number and strength of acidic sites on coke formation, and the locus of coke deposition, in particular with narrow- or medium-pore-size zeolites. However, the effect of acidity, the location of coke deposition, kinetics and mechanism of coke formation etc. most probably depend on a number of specific conditions of the process under consideration. Therefore, even more systematic and detailed studies may be required to obtain a consistent and satisfying description of these phenomena.

NOTE (added after preparation of the manuscript)

Very recently, a review article by Guisnet et al. [114] appeared in which the authors, after a brief overview of the pertinent literature, present an excellent and concise report of their studies on coking and deactivation of zeolites and the effect of zeolitic pore structures. It is based, inter alia, on Refs. [20-21] and Refs. [87-92]. This article may be very helpful for a reader entering the field of coke formation on zeolite catalysts.

ACKNOWLEDGEMENTS

The author thanks those colleagues and co-workers who contributed to this review. Particularly, he thanks Professor J. Weitkamp (University of Stuttgart) for helpful discussion and criticism. Financial support by the Bundesminister für Forschung und Technologie (BMFT), Projects 03C 111 1 and 03C 231 1, is gratefully acknowledged. The author gratefully acknowledges permission from Elsevier Sci. Publ., Academic Press and Butterworth Publ. to use published materials as cited in the text.

REFERENCES

[1]   E.E. Wolf and F. Alfani, Catal. Rev.-Sci. Eng. 24 (1982) 329-371.

[2]   J. Barbier, in: "Catalyst Deactivation 1987", Proc. 4th Int. Symp., Antwerp,
       Belgium, Sept. 29-Okt. 1, 1987 (B. Delmon and G.F. Froment, Eds.), Elsevier,
       Amsterdam 1987; Studies Surf. Sci. Catalysis 34 (1987) pp.1-19.

[3]   A. Arteaga, J.L.G. Fierro, F. Delannay and B. Delmon, Appl. Catal. 26 (1986)
       227-249.

[4]   P.E. Eberly, Jr., C.N. Kimberlin, Jr., W.H. Miller and H.V. Drushell, Ind. Eng.
       Chem., Process Des. Dev. 5 (1966) 193-198.

[5]   P.E. Eberly, Jr., J. Phys. Chem. 71 (1967) 1717-1722.

[6]   H.G. Karge and J. Ladebeck, Proc. Symp. Zeolites, Szeged, Hungary, Sept. 11-14
       1978 (P. Fejes, Ed.); Acta Universitatis Szegediensis, Acta Physica et Chemica,
       Nova Series 24 (1978) pp. 161-167.

[7]   D. Eisenbach and E. Gallei, J. Catal. 56 (1979) 377-389.

[8]   L.D. Rollmann, J. Catal. 47 (1977) 113-121.

[9]   L.D. Rollmann and D.E. Walsh, J. Catal. 56 (1979) 139-140.

[10]   S. Csicsery, in: "Shape Selective Catalysis, Route to Chemical Fuels", Proc. of a
       Symp., ACS Meeting, Seattle, USA, March 20-25, 1983, pp. 116-141.

[11]   K.A. Becker, H.G. Karge and D. Streubel, J.Catal. 28 (1973) 403-413.

[12]   J. Weitkamp, Proc. 5th Int. Conference on Zeolites, Naples, Italy, June 2-6, 1980
       (L.C.V. Rees, Ed.) Heyden, London, 1980, pp. 858-865.

[13]   J. Weitkamp, in: "Catalysis by Zeolites", Proc. Int. Symp., Ecully (Lyon), France,
       Sept. 9-11, 1980 (B. Imelik et al., Eds.) Elsevier, Amsterdam, 1980; Studies Surf.
       Sci. Catalysis 5 (1980) 65-75.

[14]   W. Entermann and H.C.E. van Leuven, Anal. Chem. 44 (1972) 589-590.

[15]   J. Weitkamp, in: "Innovation in Zeolite Materials Science", Proc. Int. Symp.,
       Nieuwpoort (Belgium), Sept. 13-17, 1989 (P.J. Grobet et al., Eds.) Elsevier,
       Amsterdam, 1988; Studies Surf. Sci. Catalysis 37 (1988) pp. 515-534.

[16]   C. Chen, Diploma Thesis, Engler-Bunte-Institut der Universität Karlsruhe,
       1983.

[17]   H.G. Karge, H. Darmstadt, R. Amberg, Z. Zheng and M. Łaniecki, Appl. Catal.,
       submitted for publication.

[18]   H. Darmstadt, Diploma Thesis, Fritz-Haber-Institut der Max-Planck-
       Gesellschaft / Technische Hochschule Darmstadt, 1988.

[19]   H.G. Karge, M. Łaniecki, M. Ziołek, G. Onyestyák, A. Kiss, P. Kleinschmit and
       M. Siray, in: "Zeolites: Facts, Figures, Future", Proc. 8th Int. Conf. Zeolites,
       Amsterdam, The Netherlands, July 9-14, 1989 (P.A. Jacobs and R. van Santen,
       Eds.) Elsevier, Amsterdam, 1989; Studies Surf. Sci. Catalysis 49 (1989) pp.
       1327-1337.

[20]   P. Magnoux, P. Roger, C. Canaff, V. Fouche, N.S. Gnep and M. Guisnet, in:
       "Catalyst Deactivation 1987", Proc. 4th Int. Symp., Antwerp, Belgium, Sept. 29-
       Okt. 1, 1987 (B. Delmon and G.F. Froment, Eds.), Elsevier, Amsterdam 1987;
       Studies Surf. Sci. Catalysis 34 (1987) pp. 317-330.

[21]   M. Guisnet, P. Magnoux and C. Canaff, in: "New Developments in Zeolite
       Science and Technology", Proc. 7th Int. Zeolite Conf., Tokyo, Japan, August
       17-22, 1986 (Y. Murakami, A. Iijima, and J.W. Ward, Eds.) Kodansha, Tokyo and
       Elsevier, Amsterdam, 1986, pp. 701-707.

[22] H. Schulz, W. Böhringer, W. Baumgartner and Zhao Siwei, in: "Catalyst Deactivation 1987", Proc. 4th Int. Symp., Antwerp, Belgium, Sept. 29-Okt. 1, 1987 (B. Delmon and G.F. Froment, Eds.), Elsevier, Amsterdam 1987; Studies Surf. Sci. Catalysis 34 (1987) pp. 479-492.

[23] M. Neuber, S. Ernst, H. Geerts, P.J. Grobet, P.A. Jacobs, G.T. Kokotailo and J. Weitkamp, in: "Catalyst Deactivation 1987", Proc. 4th Int. Symp., Antwerp, Belgium, Sept. 29-Okt. 1, 1987 (B. Delmon and G.F. Froment, Eds.), Elsevier, Amsterdam 1987; Studies Surf. Sci. Catalysis 34 (1987) pp. 567-577.

[24] V. Solinas, R. Monaci, E. Rombi and L. Forni, in: "Catalyst Deactivation 1987", Proc. 4th Int. Symp., Antwerp, Belgium, Sept. 29-Okt. 1, 1987 (B. Delmon and G.F. Froment, Eds.), Elsevier, Amsterdam 1987; Studies Surf. Sci. Catalysis 34 (1987) pp. 493-500.

[25] R.N. Jones and C. Sandorfy, in: "Chemical Applications of Spectroscopy", Interscience Publishers, Inc., New York, N.Y. 1956.

[26] E.R. Blout, M. Fields and R. Karplus, J. Am. Chem. Soc. 70 (1948) 194-198.

[27] H.G. Karge, Proc. 4th Int. Zeolite Conf., (Molecular Sieves II), Chicago, Ill., USA, April 18-22, 1977 (J. Katzer, Ed.); Am. Chem. Soc., Washington D.C.; ACS Symp. Series 40 (1977) pp. 584-595.

[28] T.J. Weeks, Jr. and A.P. Bolton, Proc. 3rd Int. Conf. Molecular Sieves, Zürich, Switzerland, Sept. 3-7, 1973; Recent Progress Reports (J.B. Uytterhoeven, Ed.) Leuven University Press, Leuven, 1973. pp. 426-431.

[29] H.G. Karge, Z. physik. Chem. [N.F.] 122 (1980) 103-116.

[30] H.G. Karge, Z. physik. Chem. [N.F.] 76 (1971) 133-153.

[31] P.B. Venuto and P.S. Landis, Advances in Catalysis 18 (1968) 259-371.

[32] H.G. Karge and K. Klose, Z. physik. Chem. [N.F.] 83 (1973) 92-99.

[33] H.G. Karge, W. Abke, E.P. Boldingh and M. Łaniecki, Proc. 9th Iberoamerican. Symp. Catalysis, Lisbon, Portugal, July 16-21, 1984 (M.F. Portela, Ed.) Jorge Fernandes, Lisbon, 1984, pp. 582-593.

[34] E. Gallei and E. Schadow, Rev. Sci. Instrum. 45 (1974) 1504-1506.

[35] W.C. Hecker, A. Bell, R.F. Hicks, C.S. Kellner and B.J. Savatsky, J. Catal. 71 (1981) 216-218.

[36] H.G. Karge and E.P. Boldingh, Catalysis Today 3 (1988) 53-63.

[37] H.G. Karge and E.P. Boldingh, Catalysis Today 3 (1988) 379-386.

[38] J.-P. Lange, A. Gutsze, J. Allgeier and H.G. Karge, Appl. Catal. 45 (1988) 345-356.

[39] D.M. Bibby, G.D. McLellan and R.F. Howe, in: "Catalyst Deactivation 1987", Proc. 4th Int. Symp., Antwerp, Belgium, Sept. 29-Okt. 1, 1987 (B. Delmon and G.F. Froment, Eds.), Elsevier, Amsterdam 1987; Studies Surf. Sci. Catalysis 34 (1987) pp. 651-658.

[40] P. Kredel, Dissertation, Technische Hochschule Darmstadt, 1988.

[41] F. Fetting, E. Gallei and P. Kredel, Ger. Chem. Eng. 7 (1984) 32-38.

[42] F. Fetting, E. Gallei and P. Kredel, Chem.-Ing.-Tech. 54 (1982) 606-607.

[43] H.G. Karge and M. Łaniecki, to be published.

[44] DMS Arbeitsatlas der Infrarot-Spektroskopie, Verlag Chemie, Weinheim, 1972.

[45] L. Kubelková, J. Nováková, M. Tupá and Z. Tvarůzková, Proc. Int. Symp. Zeolite Catalysis, Siófok, Hungary, May 13-16, 1985 (P. Fejes and D. Kalló, Eds.) (Acta Physica et Chemica Szegediensis) Petöfi Nyomada, Kecskemét, 1985, pp. 649-657.

568

[46]   J. Haber, J. Komorek and T. Romotowski, Proc. Int. Symp. Zeolite Catalysis, Siófok, Hungary, May 13-16, 1985 (P. Fejes and D. Kalló, Eds.) (Acta Physica et Chemica Szegediensis) Petöfi Nyomada, Kecskemét, 1985, pp. 671-679.

[47]   E. Breitmeier and W. Völter, "^{13}C-NMR Spectroscopy, Methods and Application in Organic Chemistry", Wiley Interscience, New York, 1972, pp. 221-223.

[48]   E.G. Derouane, J.P. Gilson and J.B. Nagy, Zeolites 2 (1982) 42.

[49]   J.P. van den Berg, J.P. Wolthuizen, A.D.H. Claque, G.R. Hays, R. Huis and J.H.C. van Hooff, J. Catal. 80 (1983) 130-144.

[50]   S. Maixner, C.Y. Chen, P.J. Grobet, P.A. Jacobs and J. Weitkamp, in: "New Developments in Zeolite Science and Technology", Proc. 7th Int. Zeolite Conf., Tokyo, Japan, August 17-22, 1986 (Y. Murakami, A. Iijima, and J.W. Ward, Eds. Kodansha, Tokyo and Elsevier, Amsterdam, 1986, pp. 693-700.

[51]   J.B. Stothers: "Carbon-13 NMR Spectroscopy" in "Organic Chemistry", - A Serie: of Monographs, Vol. 24, Academic Press, New York, London 1972, pp. 90-100.

[52]   J.-P. Lange, A. Gutsze and H.G. Karge, J. Catal. 114 (1988) 136-143.

[53]   G.A. Olah, J.S. Staral, G. Asencio, G. Liang, D.A. Forsyth and G.D. Mateescu, J. Am. Chem. Soc. 100 (1973) 6299-6308.

[54]   G.L. Nelson and E.A. Williams, Progr. Phys. Org. Chem. 12 (1976) 229-342.

[55]   R.N. Young, Progr. Nucl. Magn. Reson. Spectrosc. 12 (1979) 261-286.

[56]   M.W. Anderson and J. Klinowski, in: "Recent Advances in Zeolite Science"; Proc of the Annual Meeting of the British Zeolite Association (BZA), Cambridge, UK, 1989, (J. Klinowski and P.J. Barrie, Eds.) Studies Surf. Sci. and Catalysis, Elsevier, Amsterdam, 52 (1989), pp. 91-112.

[57]   D. Neuhaus, Diploma Thesis, Karl-Marx-Universität Leipzig, Sektion Physik, 1989.

[58]   J.R. Andersen, Y.-F. Chang and R.J. Western, J. Catal. 118 (1989) 466-482.

[59]   J.E. Wertz and J.R. Bolton. "Electron Spin Resonance in Elementary Theory and Practical Applications", McGraw Hill, New York, N.Y., 1972.

[60]   H.G. Karge, E.P. Boldingh, J.-P. Lange and A. Gutsze, Proc. Int. Symp. Zeolite Catalysis, Siófok, Hungary, May 13-16, 1985 (P. Fejes and D. Kalló, Eds.) (Acta Physica et Chemica Szegediensis) Petöfi Nyomada, Kecskemét, 1985, pp. 639-647.

[61]   H.G. Karge, J.-P. Lange, A. Gutsze and M. Łaniecki, J. Catal. 114 (1988) 144-152.

[62]   I.R. Leith, J. Chem. Soc. Chem. Commun. (1972) 1282-1283.

[63]   A.V. Kucherov and A.A. Slinkin, Kinet. Katal. 23 (1982) 1172; Engl. Translat.: 997-1003.

[64]   A.V. Kucherov and A.A. Slinkin, Kinet. Katal. 24 (1983) 947;. Engl. Translat.: 804-810.

[65]   A.V. Kucherov and A.A. Slinkin, in: "Structure and Reactivity of Modified Zeolites"; Proc. Int. Conf., Prague, ČSSR, (P.A. Jacobs et al., Eds.) Elsevier, Amsterdam 1984; Studies Surf. Sci. Catalysis 18 (1984). pp. 77-84.

[66]   A.V. Kucherov, A.A. Slinkin, D.A. Kondratyev, T.N. Bodarenko, A.M. Rubinstein and Kh.M. Minachev, J. Mol. Catal. 37 (1986) 107-115.

[67]   P.B. Ascough and H.E. Evans, Trans. Faraday Soc. 60 (1964) 801-808.

[68]   H.G. Karge, S. Trevizan de Suarez and I.G. Dalla Lana, J. Phys. Chem. 88 (1984) 1782-1784.

[69]   H.G. Karge, M. Ziołek and M. Łaniecki, J. Catal. 109 (1988) 252-262.

[70]   J.C. Vedrine, P. Dejaifve and E.D. Garbowski, in: "Catalysis by Zeolites", Proc.
        Int. Symp., Ecully (Lyon), France, Sept. 9-11, 1980 (B. Imelik et al., Eds.)
        Elsevier, Amsterdam 1980; Studies in Surface Sci. Catalysis 5 (1980) pp. 29-37.

[71]   H. Förster, S. Franke and J. Seebode, J. Chem. Soc. Faraday Trans. I 79 (1983)
        373-382.

[72]   M. Łaniecki and H.G. Karge, Proc. 6th Int. Symp. Heterogeneous Catalysis,
        Sofia, Bulgaria, July 13-17, 1987 (D. Shopov et al., Eds.) Publ. House Bulgarian
        Academy of Science, Sofia 1987, pp. 129-134.

[73]   E.D. Garbowski and H. Praliaud, J. Chim. Phys. Phys. Chim.-Biol. 76 (1979)
        687-692.

[74]   E.D. Garbowski, J.-P. Candy and M. Primet, J. Chem. Soc. Faraday Trans. I 79
        (1983) 835-844.

[75]   G.N. Asmolov, O.V. Krylov, O.V. Bragin and D.B. Furman, Kinet. Katal. 18
        (1977) 730; Engl. Translat. 18 (1977) 609-613.

[76]   T.S. Sørensen, J. Am. Chem. Soc. 87 (1965) 5075-5084.

[77]   G.A. Olah, C.U. Pittmann, Jr., R. Waack and M. Doran, J. Am. Chem. Soc. 88
        (1966) 1488-1495.

[78]   H.G. Karge, M. Ziołek and H. Darmstadt, in preparation.

[79]   B.A. Sexton, A.E. Hughes and D.M. Bibby, J. Catal. 109 (1988) 126-131.

[80]   Th.H. Fleisch, G.W. Zajac, B.L. Meyers, G.J. Ray and J.T. Miller, in "Catalysis:
        Theory to Practice", Proc. 9th Int. Congr. Catalysis, Calgary, Canada, June 26-
        July 1, 1988 (M.J. Phillips and M. Ternan, Eds.) The Chemical Institute of
        Canada, Ottawa, Ontario, 1988, pp. 483-490.

[81]   M. Muhler, R. Schlögl and G. Ertl, Surface Sci. 189 (1987) 69-79.

[82]   R. Schlögl, H.G. Karge, H. Darmstadt and M. Wesemann, in preparation.

[83]   E. Tegler, G. Wiech and H. Faessler, J. Atomic Mol. Phys. 14 (1981) 1273-1282.

[84]   R. Schlögl, Surface Sci. 190 (1987) 861-872.

[85]   P.B. Venuto and L.A. Hamilton, I & EC Product Research and Development 6
        (1967) 190-192.

[86]   D.E. Walsh and D. Rollmann, J. Catal. 49 (1977) 369-375.

[87]   M. Guisnet, P. Magnoux and C. Canaff, in: "Chemical Reactions in Organic and
        Inorganic Constrained Systems" (C.R. Setton, Ed.) Nato Series C 165, Reidel,
        Dordrecht, 1986, pp. 131-140.

[88]   P. Magnoux, P. Cartraud, S. Mignard and M. Guisnet, J. Catal. 106 (1987)
        235-241.

[89]   P. Magnoux, P. Cartraud, S. Mignard and M. Guisnet, J. Catal. 106 (1987)
        242-250.

[90]   P. Magnoux and M. Guisnet, Zeolites 9 (1989) 329-335.

[91]   P. Magnoux, M. Guisnet, S. Mignard and P. Cartraud, J. Catal. 117
        (1989)495-502.

[92]   S. Mignard, P. Cartraud, P. Magnoux and M. Guisnet, J. Catal. 117 (1989)
        503-511.

[93]   H.G. Karge and H. Darmstadt, in preparation.

[94]   J.G. Post and J.H.C. van Hooff, in: "New Developments in Zeolite Science and
        Technology"; 7th Int. Zeolite Conf., Tokyo, Japan, August 17-22, 1986 (Japan
        Assoc. of Zeolite, Ed.) Tokyo, 1986, Preprints of Poster Papers, Paper 1D-20, pp.
        249-250.

[95]   D.M. Bibby, in: "New Developments in Zeolite Science and Technology"; 7th Int.
        Zeolite Conf., Tokyo, Japan, August 17-22, 1986 (Japan Assoc. of Zeolite, Ed.)
        Tokyo, 1986, Preprints of Poster Papers, Paper 1D-11, pp. 231-232.

570

[96]   D.M. Bibby and C.G. Pope, J. Catal. 116 (1988) 407-414.
[97]   G.P. Handreck and T.D. Smith, J. Chem. Soc. Faraday Trans. I 84 (1989) 645-654.
[98]   H.G. Karge, R. Fiedorow, J. Schütze and R. Schlögl, unpublished results.
[99]   W.J.M. Pieters and A.F. Venero, in: "Catalysis on the Energy Scene", Québec, Canada, Sept. 30-Oct. 3, 1984 (S. Kaliaguine and A. Mahay, Eds.) Elsevier, Amsterdam 1984; Studies Surface Sci. Catalysis 19 (1984) pp. 155-163.
[100]   H. Schulz, W. Böhringer, W. Baumgärtner and Zhao Siwei, in: "New Developments in Zeolite Science and Technology"; Proc. 7th Int. Zeolite Conf., Tokyo, Japan, August 17-22, 1986 (Y. Murakami, A. Iijima and J.W. Ward, Eds.) Kodansha, Tokyo and Elsevier, Amsterdam, 1986, pp. 915-922.
[101]   A. Corma and B.W. Wojciechowski, Catal. Rev.-Sci. Eng. 27 (1985) 29-150.
[102]   J. Haas, F. Fetting and L. Gubicza, Proc. Int. Symp. Zeolite Catalysis, Siófok, Hungary, May 13-16, 1985 (P. Fejes and D. Kalló, Eds.) (Acta Physica et Chemica Szegediensis) Petöfi Nyomada, Kecskemét, 1985, pp. 659-669.
[103]   D.M. Bibby, N.B. Milestone, J.E. Patterson and L.P. Aldridge, J. Catal. 97 (1986) 493-502.
[104]   H.G. Karge, Z. phys. Chem. [N.F.] 99 (1975) 241-254.
[105]   A. Voorhies, Ind. Eng. Chem. 37 (1945) 318-322.
[106]   G.F. Froment and K.B. Bischoff, Chem. Eng. Sci. 16 (1961) 189-201.
[107]   G.F. Froment and K.B. Bischoff, Chem. Eng. Sci. 17 (1962) 105-114.
[108]   M.B. Ajinka, W.H. Ray and G.F. Froment, Ind. Eng. Chem., Process Des. Dev. 13 (1974) 107-112.
[109]   D.W. Breck and R.W. Grose, Advances in Chemistry Ser. 121 (1973) 319-329.
[110]   P. Gallezot, C. Leclerq, M. Guisnet and P. Magnoux, J. Catal. 114 (1988) 100-111
[111]   M. Bülow, J. Caro, J. Völter and J. Kärger, in: "Catalyst Deactivation 1987", Proc. 4th Int. Symp., Antwerp, Belgium, Sept. 29-Okt. 1, 1987 (B. Delmont and G.F. Froment, Eds.) Elsevier, Amsterdam 1987; Studies Surf. Sci. Catalysis 34 (1987) pp. 343-354.
[112]   J. Kärger, H. Pfeifer and W. Heink, Proc. 6th Int. Zeolite Conf., Reno, USA, July 10-15, 1983 (D. Olson and A. Bisio, Eds.) Butterworths, Guildford, Surrey, UK, 1984, pp. 184-200.
[113]   T. Ito, J.L. Bonardet, J. Fraissard, J.B. Nagy, C. André, Z. Gabelica and E.G. Derouane, Appl. Catal. 43 (1988) L5-L11.
[114]   M. Guisnet and P. Magnoux, Appl. Catal. 54 (1989) 1-27.

Chapter 15

HYDROCARBON PROCESSING WITH ZEOLITES

I.E. MAXWELL AND W.H.J. STORK

KONINKLIJKE/SHELL—LABORATORIUM, AMSTERDAM (Shell Research B.V.) Badhuisweg 3,
1031 CM Amsterdam, The Netherlands

ABSTRACT
    This chapter on hydrocarbon processing with zeolites covers both existing
and new catalytic applications of zeolites in oil refining and gas conversion.
By way of introduction some structural trends related to these industries are
discussed to provide some background in which to relate the current and future
developments. Further,the rather unique properties and limitations of zeolites
are discussed in order to put the applications in hydrocarbon processing into
perspective.
    Some more evolutionary type developments are covered where the introduction
and continued improvement of zeolite catalysts has had a major impact on
existing process technologies. Examples of such technologies include catalytic
cracking, hydrocracking, paraffin isomerization and olefin oligomerization.
    New emerging processes in which the application of zeolitic catalysts has
led to new process concepts are also included. The unique shape selective
properties of zeolites are shown to play a dominant role in this developing
field of applications. Some examples of such emerging process technologies are
catalytic dewaxing, methanol—to—gasoline (MTG) and LPG—to—aromatics (CYCLAR)
conversion.
    Finally, some general trends are discussed in terms of zeolite catalysis and
how these might be expected to have a further impact on hydrocarbon process
technology in the future. Significant opportunities are believed to exist for
further developments related to existing processes. For example, the use of
zeolitic co—catalysts to modify the performance of generic catalytic cracking
catalysts is expected to grow in the future. The discovery and application of
new synthetic molecular sieves leading to new processes also holds promise for
the future. In addition, the advances in understanding which are being achieved
by the recent emphasis on theoretical studies should potentially reduce the
present highly empirical approach of the catalytic chemist.

1. INTRODUCTION

1.1. Background

    Zeolites are currently being used industrially on a very large scale. Table

1 an gives an estimate of the zeolite production in 1986, from which it can be

seen that the total production in that year had exceeded some 500 000 tonnes.

The greater part of this production was intended for ion—exchange applications

in detergents, using zeolite A, but large amounts had also been produced for

adsorption (drying; physical separations of hydrocarbons) and catalytic appli-

cations. In terms of volume most of the zeolitic catalysts are used in oil

refining, where especially zeolite Y is employed on a large scale in catalytic

cracking; applications in gas and chemical processes, although growing, are at present still limited.

TABLE 1

Industrial applications of zeolites (data for 1986)

Ion exchange	>400 000	tonnes/annum
Adsorption	50 000	"
Catalysis	65 000	"
Petrochemical	350	"
Refining, speciality	2 500	"
Refining, Y-sieve	62 000	"

In this survey the applications of zeolites in oil refining and gas-conversion-processes will be reviewed. Before embarking on a detailed discussion, however, we will first briefly summarize the characteristics that make zeolites so valuable in their catalytic applications.

Next the basics of oil refining will be discussed, i.e. the conversion and upgrading of the various oil fractions into transportation fuels boiling in the gasoline and middle-distillate range. The important applications of zeolites in oil refining will then be discussed, with the various processes being reviewed in more detail. This will be followed by a discussion of the new emerging gas-conversion processes which make use of zeolite catalysts. After a brief introduction to the basics of gas conversion, the processes for the conversion of methanol, synthesis gas, light paraffins and olefins will be dealt with separately.

This chapter does not pretend to provide a comprehensive review of all the studies that have been carried out on each topic, but is rather intended to present a selection of perhaps the most relevant papers to provide the reader with a general impression of the "state of the art".

1.2. <u>Zeolite properties</u>

On the basis of their present use in oil refining, various important aspects of zeolites can be identified. It is, furthermore, noteworthy that many crucial catalytic properties can often be tailored to suit a particular application. These properties can be summarized as follows:

(i) Zeolites especially the silica-rich materials, often exhibit a high thermal and hydrothermal stability silica-rich materials. This allows their use as catalysts at high temperatures, and their frequent oxidative regeneration even in the presence of some steam.

(ii) Zeolites are crystalline microporous solids, allowing a relatively high concentration of reactant molecules inside the zeolite cages; thus, in the presence of a zeolite the reaction will proceed as if it were carried out at higher pressure compared to most other catalytic solid surfaces. This so-called cage effect has been nicely illustrated by Rabo (ref. 1), as shown in Table 2 for n-hexane cracking over KY.

TABLE 2

Zeolites: non-acid catalysis n-hexane over KY, 500 °C (from ref. 1)

$$n\text{-}C_6H_{14} \longrightarrow \rightleftharpoons \rightarrow C_3H_7^{\cdot} \begin{array}{c} \xrightarrow{a2} C_2H_4 + CH_3^{\cdot} \\ \xrightarrow[a_3]{} C_3H_6 + R^{\cdot} \end{array}$$

	a_2	a_3 (%)
Thermal	44	11
KY	18	39

KY results in: − 5x higher cracking
− different selectivity
(less C_1, C_2)
KY favours bimolecular reactions

(iii) Zeolites can be synthesized with a wealth of different structures. Thus, one can distinguish between zeolites with pore openings consisting of, for example 8-, 10-, 12- and, recently, 18-membered rings, as discussed in detail in the chapters by Flanigen, Van Koningsveld, and Jansen and Wilson. Moreover, the pore system can be in one, two or three dimensions, it can contain pores of different sizes, and consist of pipes and/or cages, etc. The fact that the pore openings are of very much the same dimensions as hydrocarbon molecules, i.e. from about 0.3 to 1.2 nm, is the very characteristic feature that forms the basis of the catalytic shape-selective behaviour of zeolites.

(iv) Zeolites and related materials can nowadays be prepared with a wide variety of chemical compositions. Originally, zeolites were aluminosilicates with an Si/Al atomic ratio between 1 and 5; with the advent of templated synthesis and controlled modification procedures (e.g. ultrastabilization) higher ratios (up to infinity) have become attainable, such as in ZSM-5.

(v) Isomorphous replacement of especially Al (e.g. by Ga, Fe ,B ,Ti) in zeolites has rapidly become more important. Further, some five years ago researchers at UOP discovered the new family of molecular sieves based on aluminophosphates, where two Si atoms in a zeolite had been replaced by one Al and one P atom (ref. 2). Currently, also SAPOs, MeAPOs, MeAPSOs, etc. are available, where in an AlPO structure further elements have been isomorphously introduced. Finally, the composition range has been further expanded to include iron phosphates etc.

(vi) Zeolites generally have ion-exchange properties which can be used to introduce metals in a highly dispersed manner, but also – through ammonium-ion exchange followed by calcination – to introduce acidity. The acidity of zeolites is, in general, very high, being about 1000 times higher than that of amorphous silica-alumina, and is directly related to the framework aluminium content, as can be seen from Fig. 1 (refs. 3,4,5). While the concentration of

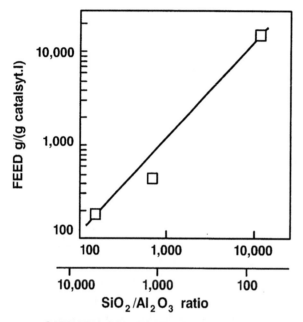

CATALYTIC CONVERSION (ACTIVITY PER Al SITE)
IS THE SAME FOR ANY SILICA/ALUMINA RATIO WHEN
THE FEED RATE IS ADJUSTED PROPORTIONATELY

Fig. 1. Zeolites: activity in acid catalysis

the acid sites can be adjusted through the framework Al content, the intrinsic acidity of the sites can be modified by replacing Al by Ga, Fe, and B ,for example, with the acidity decreasing in this order, as is illustrated in Fig. 2 (refs. 6,7).

(vii) The near-molecular pore dimensions of zeolites result in a wide range of diffusivities (spanning some 10 orders of magnitude), which impart the rather unique shape-selective properties to these materials. At least three types of shape-selective catalytic behaviour have been identified to date; they include:

    – reactant selectivity
    – product selectivity
    – restricted transition state selectivity

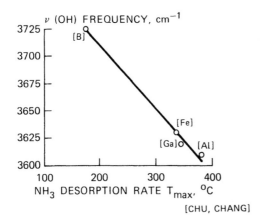

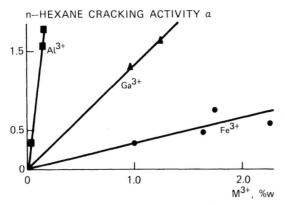

Fig. 2. Zeolites: acidity and isomorphous substitution (from refs. 6, 7)

These different forms of shape selectivity are illustrated in Fig. 3 (refs. 8,9).

The applications of zeolites in oil refining and the newly emerging gas-conversion processes will be discussed with emphasis on the special features of these rather unique materials as compared to conventional catalyst systems.

2. OIL-CONVERSION PROCESSES

2.1. Basics of oil refining

The refining of crude oil is a major industry in which worldwide some 40 million barrels (one barrel contains 167 litres) of crude oil are being processed daily in numerous refineries. The basic principles involved in oil refining are always similar but the trend is towards increasing complexity.

Crude oil is first split into various fractions in a primary atmospheric distillation step. In this way the raw materials for the main oil products are

576

REACTANT SELECTIVITY

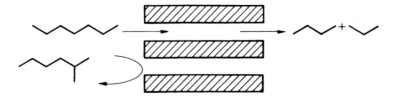

PRODUCT SELECTIVITY

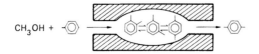

RESTRICTED TRANSITION–STATE SELECTIVITY

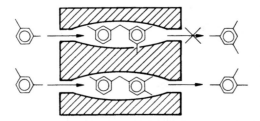

Fig. 3. Different types of shape selectivity (from ref. 8)

recovered, which include naphtha (transportation fuel and petrochemical feedstock; boiling range: C_5 to about 180 °C) to produce gasoline, and middle
distillates to produce kerosine (aviation fuel; boiling range: about 130 to
300 °C), diesel/gas oil (transportation and domestic fuel; 150–370 °C). In
addition, base oils for lubricants can be recovered (Fig. 4). These fractions
are the primary oil products (i.e. transportation fuels and naphtha for chemical feedstock), but many of these primary products need further purification
and upgrading before they can be used as final products.

In fact, the further upgrading of these primary oil products is becoming
increasingly important as environmental legislation becomes more stringent.
Sulphur, for example, is removed from oil fractions by means of catalytic
hydrodesulphurization (HDS) using Co(Ni)/Mo/alumina catalysts. Combustion properties are important for the transportation fuels which must also be upgraded
to meet product quality requirements. Thus, the C_5–C_6 tops fraction may need to

577

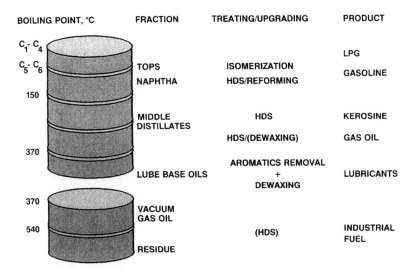

Fig. 4. A simple oil refinery (basics)

be isomerized to branched paraffins to increase the octane number, while the octane of the naphtha fraction may be increased by aromatization in the catalytic reforming process. For gas oils the wax content may impede the cold flow properties and thus require a dewaxing step which in a modern refinery may be catalytic.

This simple (so-called "hydroskimming") refinery produces the required boiling point fractions, but utilizes in this way only part of the crude oil. The heavier fractions of the barrel (boiling above about 370 °C) can only be used as low-value industrial fuel. The more modern so-called "complex refinery" has additional processing capabilities (see Fig. 5) which enable increased

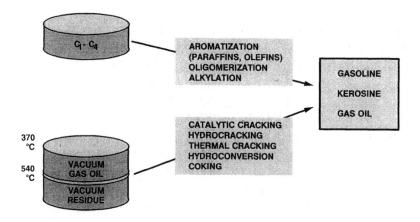

Fig. 5. Conversion options in a complex refinery

amounts of transportation fuels to be recovered by (partial) conversion of this heavy residue fraction of the barrel of oil. This is often achieved by first separating the residue in a vacuum distillation tower into fractions boiling below and above 540 °C, and then converting the lighter fraction (vacuum gas oil) into gasoline and middle-distillate products in a catalytic cracking unit or in a hydrocracking unit, for example.

The remaining heavy vacuum residue can be converted into more valuable lighter fractions (as shown in Fig. 5) by means of thermal processes (i.e. vis-breaking or coking) or catalytically in the presence (i.e. hydroconversion) or absence of hydrogen (residue cat. cracking). These two options are fundamentally different in the way they achieve the necessary increase in hydrogen to carbon ratio of the products, i.e. by adding hydrogen or by removing carbon (i.e. as coke).

The overall refinery yield of transportation fuels can be further increased by converting the light gases associated with the crude oil or those produced in the refinery into higher molecular weight hydrocarbons. These "synthesis" processes include alkylation of propene/butene/i-butane mixtures to yield high-octane gasoline-range products and oligomerization of olefins. These various processing options can be translated into many different refinery schemes, and it usually depends on many factors, such as local crude supply, desired product slate, product quality requirements and desired flexibility, etc., as to which scheme will be optimal in a particular situation.

A discussion on a modern oil-refining complex is not complete without considering the petrochemical industry, since this uses significant amounts of oil products as feedstock. Basically, the chemical industry requires as building blocks components such as synthesis gas, olefins and aromatics (benzene, toluene and xylenes) (BTX). These are usually manufactured from naphtha, which is converted to olefins and aromatics in a thermal cracking unit; currently, there is a trend to use also heavier oil fractions as feedstock for this unit. Synthesis gas can be manufactured from naphtha in a steam reformer, but also from coal or heavy oil residues by gasification. Aromatics (BTX) can also be recovered from the high-octane product from a catalytic reformer (see above). Given the increasing amounts of oil products required by the petrochemical industry as feedstock, these outlets can significantly influence the optimal refinery scheme.

Now that the various refining processes and options have been outlined, it is of interest to see where zeolite catalysts are being applied. Fig. 6 shows the various crude oil boiling fractions, which are now considered as feedstocks, and indicates for which processes zeolite catalysts are either already being used commercially or are likely to be applied in the near future. It is encouraging from the viewpoint of the zeolite chemist to observe that the list

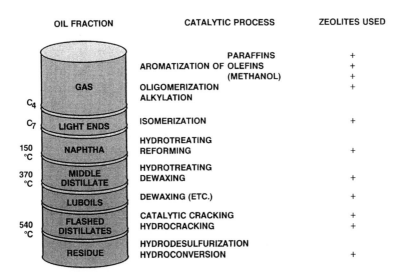

OIL FRACTION   CATALYTIC PROCESS   ZEOLITES USED

	PARAFFINS	+
	AROMATIZATION OF OLEFINS	+
	(METHANOL)	+
GAS	OLIGOMERIZATION	+
	ALKYLATION	
C_4		
C_7 LIGHT ENDS	ISOMERIZATION	+
150 °C NAPHTHA	HYDROTREATING REFORMING	+
370 °C MIDDLE DISTILLATE	HYDROTREATING DEWAXING	+
LUBOILS	DEWAXING (ETC.)	+
540 °C FLASHED DISTILLATES	CATALYTIC CRACKING HYDROCRACKING	+ +
RESIDUE	HYDRODESULFURIZATION HYDROCONVERSION	+

Fig. 6. Zeolite catalysis in oil refining

of applications where zeolitic catalysts are competitive and are often much improved compared to other catalyst systems appears to be continuously growing.

In the following sections of this chapter the various applications of zeolites in oil-refining processes will be discussed in more detail.

2.2. Paraffin hydroisomerization

Environmental considerations have brought about a rapid phase-out of the lead addition to gasoline, which has been further stimulated by the fact that automobile exhaust catalysts are rapidly poisoned by lead additives. Thus, unleaded gasoline is rapidly penetrating transportation fuel markets on a global scale and it is predicted, for example, that by 1990 in Western Germany, The Netherlands, and Scandinavia, unleaded gasoline will have about 40 % of the market share. Obviously, further increases in unleaded gasoline can also be expected in the future until lead is ultimately removed as a component in gasoline. Since lead was added to the gasoline to increase the octane number, other means to achieve this must be employed.

In Table 3 the various hydrocarbon components of a typical US gasoline pool are listed together with their octane numbers (ref. 10). Gasoline octane can, for example, be boosted by increasing the catalytic reformer severity (i.e. removing low-octane paraffins), by increasing paraffin alkylation, adding MTBE (methyl tertiary butyl ether) and by replacing straight run tops (C_5–C_6 paraffins) with isomerized products. Such paraffin isomerization can bring about a substantial increase in the octane number, as can be seen from Table 4. For example, n-pentane and especially n-hexane have much lower octane numbers than their corresponding branched isomers.

TABLE 3

Typical composition of U.S. gasoline pool (1987) (from ref. 10)

	%vol	Octane range
Butanes	6	91 – 93
Lt. straight run	4.5	55 – 75
Cat. Crackate	36	84 – 89
Hydrocrackate	3	85 – 87
Coker	1	60 – 70
Alkylate	11	90 – 94
Reformate	34	86 – 96
MTBE	1	106 –110
Isomerate	3.5	80 – 90

TABLE 4

Paraffin isomerization

Research octane numbers of C_5 and C_6 paraffins

	RON–O
Pentanes	
n–Pentane	62
Methylbutane	93
2,2-dimethylpropane	83
Hexanes	
n–Hexane	29
2–Methylpentane	78
3–Methylpentane	76
2,2–Dimethylbutane	92
2,3–Dimethylbutane	104

This skeletal isomerization conversion is an equilibrium–limited reaction, as shown in Fig. 7, with a low reaction temperature favouring isomerization and thus high–octane products. A very low reaction temperature of some 150 °C can be employed by using chlorided alumina type catalysts containing a noble metal such as platinum. However, these catalyst systems suffer from the disadvantage of being quite sensitive to trace impurities in the feed such as water. By contrast, zeolite–based catalysts which operate at somewhat higher temperatures (i.e. 250 °C) are more robust and can withstand low levels of impurities such as sulphur and water in the feedstock. Such catalysts were first developed for the Shell Hysomer process using platinum/mordenite as the zeolite component. At these reaction temperatures the equilibrium is such that some normal paraffins are not converted to branched products, as can be seen from Fig. 7, and it can therefore be attractive to combine the Hysomer isomerization process with the ISOSIV iso/normal separation process, developed by UOP.

In the ISOSIV process the normal paraffins are selectively absorbed by zeolite 5A, while the branched isomers cannot enter the small pores of the molecular sieve. The effluent of the Hysomer process can thus be separated in the ISOSIV step and the normals recycled back to the catalytic section and so complete conversion to iso-paraffins is achieved. This is the basis of the joint Shell/UCC process, TIP (Total Isomerization Process), as shown in Fig. 8, which results in an increase of some 9-10 octane (RON-O) points compared to the single-step Hysomer process (ref. 10). The Hysomer and TIP processes are being extensively applied; although the first unit was started up in 1969, the real increase in capacity was in the eighties as a result of the growth in octane demand. More recent process improvements of the TIP process have resulted in a 10-15 % cost reduction (10).

Hydroisomerization is a typical acid catalysed reaction, and the mechanism is believed to be as shown in Fig. 9. In this mechanism carbenium ions are formed through protonation of alkenes; such ions can then rearrange, and desorb as isomers, or be cracked through beta scission. The intermediate in the rearrangement is assumed to be a protonated cyclopropane species, as shown in Fig. 9, which would also explain why direct C_4 isomerization does not occur in these systems. This mechanism also explains the bifunctional requirements (i.e. hydrogenation and acidic functions) of the catalyst; nevertheless, it should be stressed that also bare H-mordenite has a significant isomerization activity (ref. 11), as shown in Fig. 10. However, to obtain optimal selectivity and stability the bifunctional catalyst is much to be preferred.

Weitkamp has published (ref. 12) results on the cracking and isomerization over a bifunctional zeolite catalyst for alkanes of various chain lengths, as shown in Fig. 11. Higher molecular weight alkanes are much more reactive than lower ones and clearly the isomerization proceeds at lower temperatures than the hydrocracking. While C_6 paraffins can be isomerized rather selectively, this selectivity decreases with increasing chain length. These facts explain why paraffin isomerization so far has concentrated on C_5-C_6: a feedstock with a wide boiling range easily leads to cracking of the heavier components and only isomerization of the lighter ones. Further, isomerization of the heavier alkanes is also less selective.

Clearly, considerable effort has been devoted to the development of high-performance hydroisomerization catalysts by industrial laboratories. In general terms the mordenite has to be converted from the Na form ex-synthesis into the active highly acidic form, which involves not only ion exchange of Na for H,

582

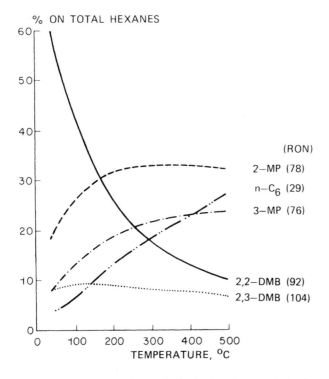

Fig. 7. Thermodynamic equilibria for hexane hydroisomerization

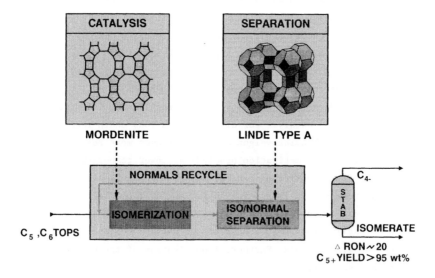

Fig. 8. Paraffin Total Isomerization Process (TIP)

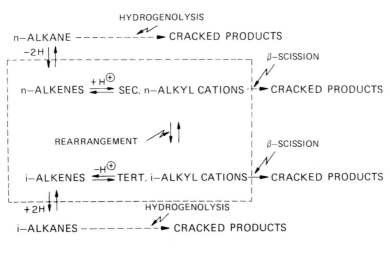

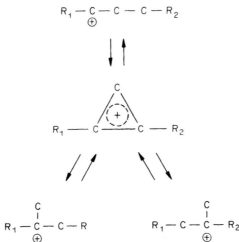

Fig. 9. a) Mechanism for n—alkane hydroisomerization
       b) Protonated cyclopropane intermediate

584

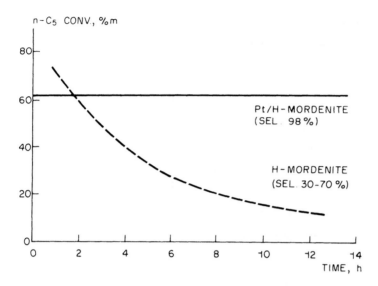

Fig. 10. Influence of metal function in pentane hydroisomerization (from ref. 11)

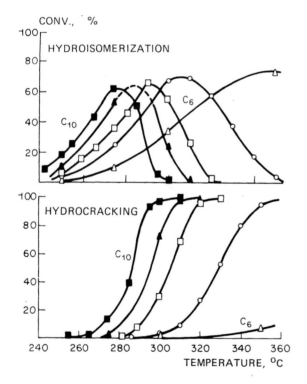

(Pt/Ca—ZEOLITE Y)

Fig. 11. Comparison of hydroisomerization and hydrocracking for C_6–C_{10} paraffins (from ref. 12).

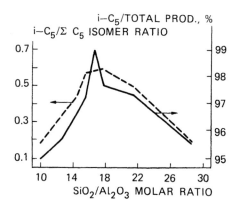

Fig. 12. Effect of mordenite SiO_2/Al_2O_3 ratio for pentane hydroisomerization
(from ref. 13)

but also an adjustment of the silica/alumina ratio. Fig. 12 gives an example
of the influence of the Si/Al ratio on activity and selectivity (ref. 13);
solid-state NMR has proved to be a powerful technique to monitor the zeolite
changes during such treatments (ref. 14). Another step which has been shown
(ref. 15) to influence performance is the calcination, as shown in Fig. 13.

Research on improved isomerization catalysts continues, as is demonstrated
by the growing patent literature, and indeed Fig. 14 shows that significant
potential benefits may be obtained from catalysts with improved activity, as
has been calculated, for example, using a hydroisomerization process model
developed by Shell (ref. 10).

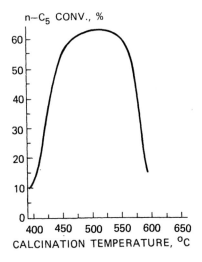

Fig. 13. Effect of mordenite calcination temperature on pentane hydroisomerization activity (from ref. 15)

586

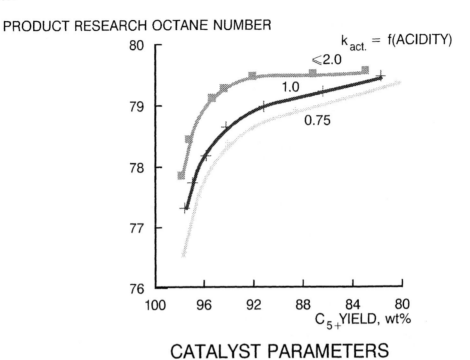

Fig. 14. Hydroisomerization catalyst modelling (from ref. 10)

2.3. Gasoline-upgrading processes

Two different types of gasoline-upgrading process can be distinguished which utilize zeolite-based catalysts. The first type involves a shape-selective cracking mechanism and the second type makes use of shape-selective aromatization reactions.

2.3.1. Shape-selective cracking. This process makes use of the shape-selective properties of medium-pore zeolites such as ZSM-5 to preferentially crack paraffinic components in gasoline to light gaseous hydrocarbons. An octane increase is achieved since the most reactive n-paraffin molecules have the lowest research octane number and vice versa (ref. 16) as is demonstrated for ZSM-5 in Fig. 15.

The first commercial process of this type was the so-called selectoforming process from Mobil (ref. 17), which was introduced in the mid-1960s and later improved with the M-forming process (ref. 16), whereby the more shape-selective ZSM-5 zeolite was introduced as the catalyst. Gulf researchers, too, (ref. 18) were active in this field and reported selective upgrading of reformates and straight-run gasolines using the medium-pore one-dimensional zeolite ferrierite.

RELATIVE CRACKING RATES

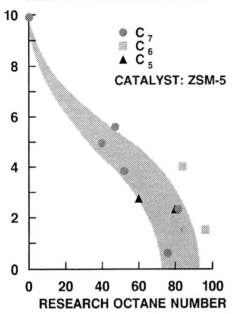

Fig. 15. Cracking rates vs octane number: ZSM–5 catalyst (from ref. 16)

The hydroprocessing conditions typically applied for gasoline upgrading via shape–selective cracking are mild pressures (30 – 50 bar) and temperatures in the range 300 – 350 °C. The commercial success of these processes has, however, been rather limited due to the relatively high loss in gasoline yield compared to the octane gain. In addition, the saturated by–product light gases have, in general, a relatively low value.

More recently, this basic concept has, however, been applied in catalytic cracking technology where a shape–selective zeolite cracking component (e.g. ZSM–5) is introduced as a co–catalyst to increase gasoline octane. This will be discussed in more detail in the section dealing with catalytic cracking.

2.3.2. Aromatization reactions. An interesting more recent development has been the application of zeolites to gasoline upgrading by means of selective aromatization reactions. Aromatization is an important catalytic reaction in a modern refinery/petrochemical complex to produce both high–octane gasoline and benzene/toluene/xylenes (BTX). The technologies currently applied are Fluid Catalytic Cracking (FCC) and Catalytic Reforming.

A new approach which has been developed by Chevron workers (ref. 19) makes use of the fact that conventional reforming catalysts have relatively poor aromatization selectivities for C_6 and C_7 aliphatic components, as shown in

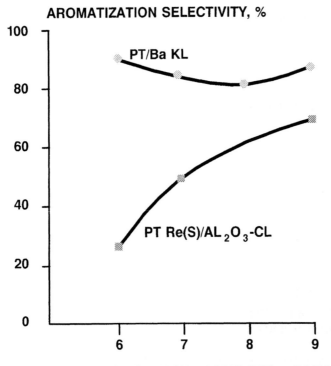

AROMATIZATION SELECTIVITY, %

NUMBER OF CARBON ATOMS PER n-PARAFFIN MOLECULE

Fig. 16. Catalytic reforming with zeolite L (from ref. 19)

Fig. 16. By contrast, a novel catalyst based on zeolite L has been shown to exhibit high selectivities towards aromatics formation including low-molecular-weight paraffins such as hexanes and heptanes (see Fig. 16). The catalyst is composed of highly dispersed platinum clusters in barium-exchanged potassium zeolite L. A novel feature of the catalyst is the lack of acidity, with the aromatization properties being attributed to a combination of the platinum clusters and the shape-selective properties of the zeolite.

On the basis of this new catalyst system Chevron has developed a new process called AROMAX (ref. 20), which is primarily intended to be applied for aromatization of the lighter feedstocks and is thus complementary to conventional catalytic reforming. A major drawback of the catalyst system, however, appears to be the high sensitivity to sulphur contaminants. This generally requires an extra feedstock pretreatment step, which may have limited the application of this process route to date.

Nevertheless, new process routes to achieve gasoline octane upgrading with high selectivity are expected to be of interest in the future as the lead phase-out in gasoline continues on a global scale.

2.4. Catalytic dewaxing

The presence of waxy molecules, which essentially consist of normal paraffins, is often a problem in terms of product quality for gas oils and lubricating base oils. In particular, the cold-flow properties of these oil products are sensitive to even quite low concentrations of waxy components and therefore processing is required to selectively remove these n-paraffin molecules.

If the wax is selectively removed from a gas oil, for example, a heavier fraction may be included in this product, which can often be attractive economically. The chemistry of lubricating base oils is more complex: basically, an ideal lube oil consists of saturated molecules with some ring structures and slightly branched paraffin chains which do not easily crystallize. Thus, from the straight-run oil fractions the aromatic components are removed by furfural extraction or by hydroprocessing, and then the wax (highly linear paraffin molecules) is selectively removed. This is conventionally carried out by solvent extraction, but nowadays catalytic dewaxing is becoming increasingly popular. It has been estimated that in 1990 the world-wide catalytic dewaxing capacity for gas oils will amount to some 75 000 barrels per stream day, and for lubricating base oils to 60 000 bpsd (i.e. 10 % of the total dewaxing capacity).

The catalytic process is carried out in trickle-flow operation over a bifunctional shape-selective zeolite catalyst, under a hydrogen flow. The hydrogenation function (often noble metals, though also Ni, and Ni/W have been described) improves the catalyst stability. Depending on the type of feedstock (i.e. hydrotreated or not) the catalyst life can vary from weeks to years; regeneration is often achieved through hot hydrogen stripping. The technology is at present dominated by Mobil: more than 70 % of the catalytic dewaxing units are based on Mobil zeolite catalysts and process technology. This is the result of a large research effort, which is also reflected in their extensive patent literature on this subject. However, most of the major oil companies are active in this field.

In catalytic dewaxing one uses the shape-selective properties of a zeolite to selectively hydrocrack the normal paraffins to lighter products. This shape selectivity is illustrated in Fig. 17, taken from (ref. 21), for ZSM-5: clearly the cracking rate decreases with increasing degree of branching. The selective removal of the normal paraffins in gas oil dewaxing is nicely illustrated in Fig. 18, taken from (ref. 22), where the sharp peaks of the normal paraffins in the gas chromatograph of the feed are clearly absent in the product after catalytic dewaxing. Fig. 19 shows that with increasing depth of dewaxing and thus decreasing pour point of a lube oil, the alkanes are selectively converted (ref. 23). The medium-pore-size (10 ring) zeolites appear to be particularly

590

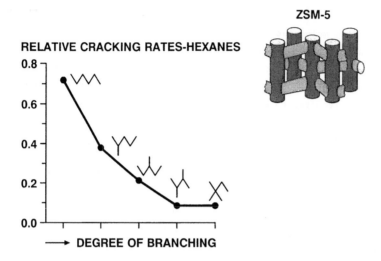

Fig. 17. Shape selective catalytic dewaxing (from ref. 21)

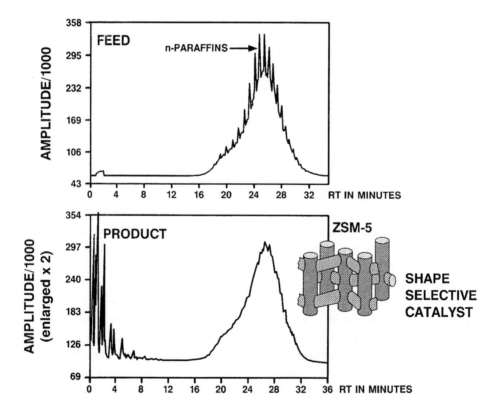

Fig. 18. Shape selective catalytic dewaxing (from ref. 22)

FEEDSTOCK,%

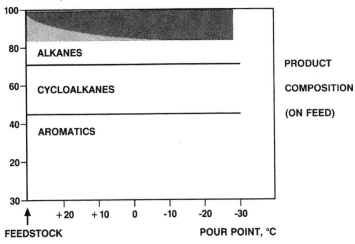

Fig. 19. Molecular composition of lube oil as a function of pour point (from ref. 23)

suitable to obtain a high selectivity, as is demonstrated in the comparison between ZSM-5 and mordenite (ref. 23), (the first generation BP catalyst), as shown in Table 5.

TABLE 5

Dewaxing with ZSM-5 vs. mordenite (from ref. 23)

Catalyst		Feed	ZSM-5		Pt/mordenite	
Conversion,	%w	0	20	23	27	36
Product pour point,	°C	+35	−12	−26	−12	−26
Viscosity index		−	94.3	89.4	89.6	77.6

However, also within the medium pore size zeolites, optimization is possible, as is illustrated in Table 6, where ZSM-5 and ZSM-23 are compared (ref. 23). ZSM-23, which has a somewhat smaller pore size, gives a higher lube oil

TABLE 6

Dewaxing over ZSM-5 and ZSM-23 (from ref. 23)

		ZSM-5	ZSM-23
Pore size,	nm	0.54 x 0.56	0.45 x 0.56
Conversion,	%wt	15	11
Product pour point,	°C	−12	−12
Viscosity index		101.0	108.7

592

yield, of a higher quality (viscosity index, VI) than ZSM-5 at equal product pour point. Apart from zeolite structure, also crystallite size and aluminum content are important variables to further optimize catalyst performance. As shown in Fig. 20, (ref. 4) there is a significant advantage in activity and stability for the smaller particles, while Fig. 21, (ref. 5) demonstrates that the cracking activity can also be optimized by controlling the aluminium content and steaming conditions.

Normally, in catalytic dewaxing the wax molecules are hydrocracked, often to C_3-C_4, and the selectivity differences are thus due to the ability of the zeolite to crack wax while leaving the valuable lube oil molecules unchanged. An alternative way to improve selectivity would be to isomerize the wax (to lube oil components) instead of cracking it. Clearly, for this purpose zeolites with somewhat larger pore sizes must be used, and Mobil has filed a number of patents for this approach using zeolite beta. Table 7, taken from (ref. 26), illustrates that such a catalyst may indeed give significantly higher oil yields than a ZSM-5 based one.

A further recent development is partial dewaxing by isomerization followed by selective dewaxing through cracking ("cascade two-stage dewaxing"). Another interesting development is the use of SAPO-11 by Chevron; although the SAPO-based catalyst is some 70 °C less active than ZSM-5, it does give higher yields of lube oil, with improved product properties as reflected in a higher VI. According to Chevron, dewaxing using this new catalyst is accomplished by both isomerization and cracking, thus leading to the improved performance.

TABLE 7

(From ref. 26)

	Isomerization of raw light gas oil		
	Example No.		
	8 (Pt/Beta)	9 (Ni/ZSM-5)	10 (Zn/Pd/ZSM-5)
Reaction pressure kPa	6996	5272	6996
Temperature, °C	402	368	385
LHSV	1	2	2
Products percent:			
C_{1-4}	2.3	8.6	15.9
C_5-165 °C	16.1	11.4	19.8
165 °C +	81.6	79.1	64.3
Total liquid product pour point, °C	-53	-34	-54

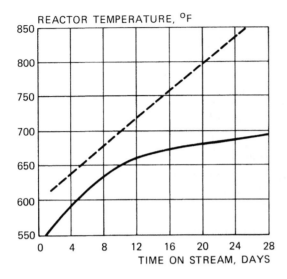

Fig. 20. Influence of ZSM-5 crystallite size (from ref. 24)

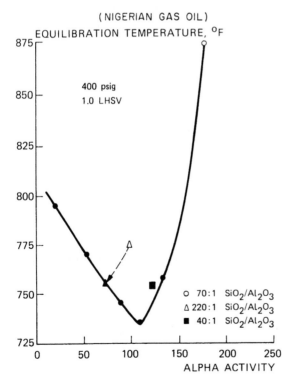

Fig. 21. Effect of alpha activity (adjusted through steaming) on catalytic dewaxing activity (from ref. 25)

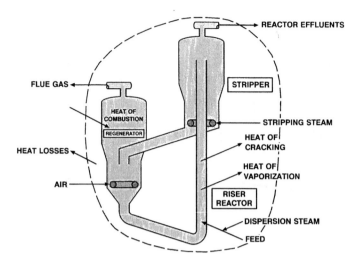

Fig. 22. Simplified overall FCC configuration and heat balance

2.5. Catalytic cracking

Catalytic cracking is a moving-fluidized-bed process for the conversion of
vacuum distillates and residues into olefinic gases, high-octane gasolines and
diesel oil fractions over an acidic catalyst (Fig. 22). Fluid catalytic cracking has been a principal factor in refining since the late 1940s; a prime
reason for the durability and perhaps the most significant characteristic of
the cracking process is the flexibility in terms of the variety of feedstocks
from a wide range of crudes which can be processed (refs. 27-29). This is
accomplished by cracking the vaporized feed over a solid acid catalyst (i.e.
zeolite Y) at a temperature between 470 and 540 $^{\circ}$C and a pressure between 0.5
and 1.5 kg/cm^2 gauge in a riser reactor (1-3 s contact time). The catalyst
promotes a mechanism of catalyst action which can to a large degree be
explained by carbonium-ion chemistry (ref. 28).

After catalyst/vapour separation at the end of the riser, the catalyst is
transported to the regeneration section where about 1 %w carbon on catalyst and
a small amount of hydrogen present as hydrocarbon on the catalyst is burnt off.
The reactions in the regenerator are exothermic, and produce temperatures of
about 700 $^{\circ}$C and a 20 % steam environment. The hot catalyst circulates between
the reactor and regenerator, acting as a heat carrier. Conventional FCC using
vacuum distillate feeds operates in heat-balanced operation, where the heat
released in the regenerator is sufficient to vapourize and crack the feed.

However, the present trend towards heavier-residue-containing feedstocks
results in high coke makes, which can cause excessive regenerator temperatures
in heat-balanced operation, leading to, for example, hydrothermal catalyst

deactivation and low catalyst circulation rates (i.e. cat/oil ratios). To solve these problems new developments such as regenerator catalyst cooling have been introduced to remove the excess heat and give increased operational flexibility (ref. 27). Hence, modern residue FCC units with this configuration work in non—heat—balanced operation.

2.5.1. <u>Zeolite Y catalysts</u>. The introduction of zeolite Y base catalysts in the mid—1960s has had a dramatic impact on the performance of FCC units. This is illustrated in Fig. 23, where the pronounced increase in gasoline yield is evident. Further, the high activity and low coke make of zeolite catalysts enabled the reactor technology to advance from dense fluidized beds to

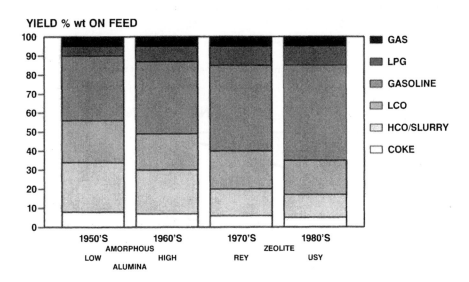

Fig. 23. FCC catalyst development history (from ref. 1)

short—contact—time risers with an associated improvement in performance (see Fig. 24). However, the recent shift towards residue—containing feedstocks and the need for increased gasoline octane (due to the removal of lead additives for environmental reasons) are posing major challenges to FCC catalyst manufacturers.

The first generation of zeolite FCC catalysts (ref. 28) involved the use of acidic rare—earth—ion exchanged zeolite Y (i.e. REY zeolites). These catalysts were not only much more active for cracking reactions than the previous amorphous systems but they also influenced the composition of the gasoline. In fact, as shown in Fig. 25, gasoline octane (i.e. both MON and RON) was significantly reduced. This can be attributed to an increased rate of hydrogen transfer reactions between olefins and naphthenes giving paraffins and aromatics. It

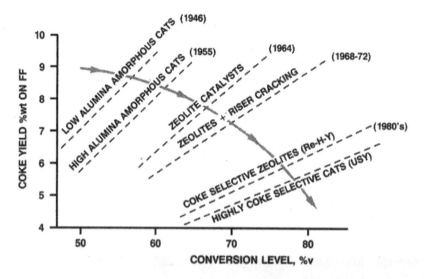

Fig. 24. Trends in catalytic cracking catalyst performance (from ref. 1)

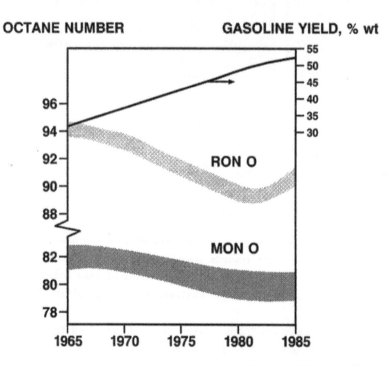

Fig. 25. FCC gasoline octane and yield development history (from ref. 1)

TABLE 8

Catalytic Cracking:

Comparison of FCC gasoline compostion as a function of catalyst
type (from ref. 32)

Conditions: reaction temperature = 520 $^{\circ}$C, conversion = 67 %,
contact time = 1.2 s

Catalyst		Rare-earth Y	USY	Partial rare-earth USY
Composition C_5-C_{10}				
Fraction of gasoline, %w				
Saturates	normal	2.1	2.0	2.3
	branched	21.5	14.7	16.4
Olefins	normal	12.1	11.6	13.1
	branched	19.4	23.1	23.6
Cyclic	saturates	11.5	11.8	9.7
	olefins	6.6	12.0	7.8
Aromatics		26.8	24.8	27.1
Total olefins		38.1	46.7	44.5
Calculated				
RON		89	94	94
MON		78	79	82

is this increased paraffin content of the gasoline which is not sufficiently
compensated for by the aromatics formation that results in decreased gasoline
octane.

This problem has led to the introduction of so-called ultrastable zeolite Y
(i.e. USY zeolites) materials (refs. 30,31), which exhibit lower unit cell
parameters and thus reduced acid site density. Since hydrogen transfer is
bimolecular this is suppressed in the lower acid site density zeolite Y based
catalysts. This effect is illustrated in Table 8, which shows the FCC gasoline
composition as a function of zeolite type (ref. 32). The enhanced olefin make
due to hydrogen-transfer suppression is also shown to result in an increase in
gasoline octane, particularly RON. An additional benefit of the USY zeolites
has been reduced coke make, which has enabled the processing of heavier feedstocks with retention of heat-balanced operation. These benefits led to a quite
dramatic shift away from REY- to USY-based catalysts as shown, for example, in
Fig. 26 for North America.

598

CATALYST MARKET SHARE, %

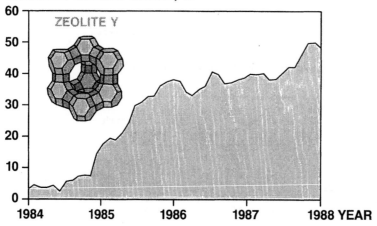

Fig. 26. Application of ultrastable zeolite Y in catalytic cracking in
North America (from ref. 32)

However, more recently some of the shortcomings of USY-type catalysts have
been recognised and these include both low activity and insufficient gasoline
MON. This has led to the hybrid-type catalysts whereby a balance is achieved
between the REY and USY zeolites in order to optimize the advantages of both
types of catalyst. As shown in Table 8, these partial rare-earth USY-type
catalysts exhibit a better balance in terms of hydrogen-transfer reactions
whereby a significant improvement is realized in terms of gasoline MON quality
(ref. 32). In addition, these new-generation catalysts make use of more modern
techniques to manufacture USY, which resulting in a more nearly defect-free
zeolite Y structure and an associated improvement in performance, particularly
in terms of hydrothermal stability.

Despite these catalyst developments FCC gasoline quality remains a problem
in the modern refinery complex, which is further exacerbated by the poorer
quality of feedstocks which are being processed. This gasoline octane problem
is reflected in the so-called "octane dip", which is shown in Fig. 27, where
the MON and RON show a marked minimum in the mid-boiling range (i.e. C_8 molecular weight).This can be largely attributed to the presence of low octane saturated cyclic components in this carbon number range (ref. 33), as shown in Fig.
28. Thus, there is scope for further improvement of FCC catalysts, which will
require a good understanding of zeolite structure/catalytic performance
relationships.

2.5.2. <u>ZSM-5 additives</u>. Several petroleum companies, viz. Exxon, Mobil and Gulf
(ref. 34), have studied shape-selective zeolite catalytic processes for improving the octane quality of reformate and gasoline streams. The basic concept is

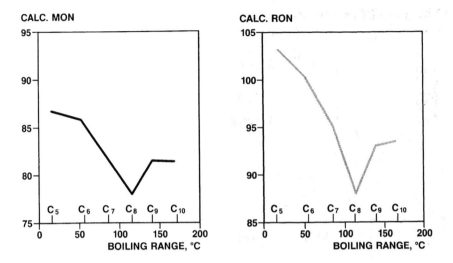

Fig. 27. FCC gasoline octane as a function of boiling range

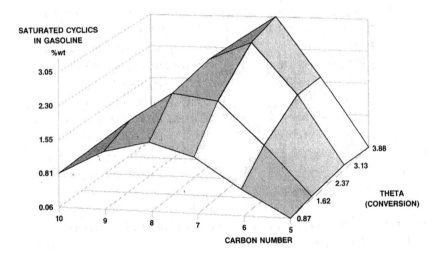

Fig. 28. FCC gasoline composition as a function of conversion (from ref. 33)

similar to catalytic dewaxing which involves the application of reactant shape selectivity by which n—paraffin molecules are preferentially removed by cracking to light (gaseous) hydrocarbons. The octane increase is achieved because the most reactive n—paraffin molecules have the lowest research octane number and vice versa for a medium—pore—size pentasil zeolite, such as ZSM—5.

In FCC it has been used primarily in the form of a high—concentration, separate particle additive; however, it has also been successfully employed as a composite, i.e. a catalyst containing ZSM—5 and a faujasite cracking component in the same particle (ref. 35). As discussed above, the purpose of ZSM—5 is to selectively centre–crack n–paraffins to primarily propylene and butylene as by–products, both of which are suitable for use as alkylation feedstock. Hence, if alkylation capacity is available, the alkylate yield more than compensates for the cracked gasoline loss, resulting in a net increase in gasoline plus alkylate.

Up to the present, the ZSM–5 additive has been mainly used in conjunction with REY catalysts, where the gasoline fraction is rich in n–paraffins. More recent studies (36), where ZSM–5 has been applied with USY–type FCC catalysts, have shown that low–octane components of the gasoline are also in this case preferentially cracked out to C_3 and C_4 olefins but with an associated loss in gasoline yield (see Fig. 29). A corresponding increase in (calculated) RON and

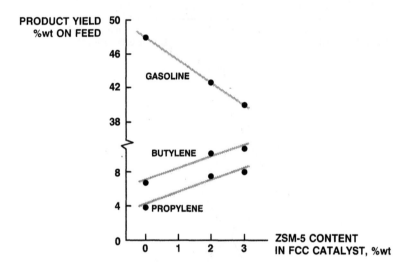

Fig. 29. Product yields: ZSM–5 FCC co–catalyst (from ref. 36)

MON is observed, as shown in Fig. 30. An interesting finding of this study (ref. 36) was that under FCC conditions no preferential cracking of straight–chain relative to branched components occurred, as had been expected on the basis of catalytic dewaxing technology. The octane improvement was thus shown to be primarily due to an increase in concentration of aromatics in FCC gasoline (Table 9). The observed cracking of branched paraffins likely reflects the reduced shape selectivity of the ZSM–5 zeolite at the relatively high operating temperatures applied in FCC (i.e. 480–520 $^\circ$C) compared to catalytic dewaxing (300–350 $^\circ$C).

This relatively new concept of using zeolite co–catalysts to modify the performance of a generic FCC catalyst system (e.g.zeolite Y) can, in principle, significantly increase the product flexibilty of the catalytic cracking unit and thus further developments in this area can be expected in the future.

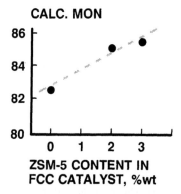

Fig. 30. Calculated motor octane number: ZSM-5 FCC co-catalyst (from ref. 36)

TABLE 9

Catalytic Cracking:

Gasoline composition with ZSM-5 FCC co-catalyst

(from ref. 36)

Added ZSM-5,		%wt	0	2	3
Conversion,		%wt	68.5	71.1	70.3
Gasoline yield,		%wt	48.2	42.5	39.8
Gasoline,		%wt			
Saturates	normal		2.0	1.7	1.8
	branched		17.9	14.8	14.0
Olefins	normal		12.7	10.1	8.3
	branched		26.3	26.5	28.0
Cyclics	saturated		5.5	8.1	7.5
	olefinic		4.9	2.7	2.4
Aromatics			30.6	36.1	37.8
Calculated					
RON			95.9	98.7	100.1
MON			82.7	85.2	86.2

2.5.3. <u>Residue catalysts</u>. With the continued growth of residue processing in FCC units increasing attention is being given to improving the performance of catalysts for this type of operation. Some catalyst aspects of importance in this respect are:

- coke make
- bottoms conversion

    - metals resistance

    - gas make

Both the zeolite and the matrix play an important role in the optimal performance of a residue catalyst (ref. 37). Some properties of the zeolite structure which are considered to be important are :

    - mesopore structure for bottoms conversion

    - unit cell parameter for coke make

    - defect structure for hydrothermal stability

    - resistance of zeolite structure to vanadium attack

The resistance to performance decline due to nickel and vanadium contamination is one of the most challenging problems in the development of modern residue FCC catalysts. Approaches to this problem include the use of nickel passivation agents (e.g. antimony or bismuth) to suppress the undesirable dehydrogenation reactions which are catalysed by nickel (ref. 38). For vanadium metal traps (e.g. basic alkaline earth oxides or metals such as strontium) are often applied which act as scavengers and thus prevent the vanadium from destroying the zeolite structure. However, although many of these techniques are effective they are generally limited to relatively low levels of contamination. This has led to the development of catalyst rejuvenation techniques which are claimed not only to remove metals but also to substantially restore catalyst activity (refs. 39,40).

Many challenging problems remain to further improve the performance of FCC catalysts as the requirements in terms of both feedstock (i.e. poorer quality) and products (i.e.better quality) become ever more demanding. It is vital to the future of the FCC process that these challenges are met and catalysts and in particular zeolites will play an important role in these developments.

2.6. Hydrocracking

Hydrocracking is an oil conversion process of growing importance due to a number of structural trends in the refining and petrochemicals industries. These include a shift towards an increased demand for middle-distillate product (i.e. kerosine, diesel and gas oil) relative to gasoline which is particularly evident in the rapidly developing countries in, for example, the Pacific Basin and the Indian continent (Fig. 31).

Another trend which is evident on a global scale is the increasing environmental legislation. Hydrocracking which inherently produces oil products which have low concentrations of heteroatoms (i.e. sulphur and organic nitrogen) and aromatic components is, from an environmental viewpoint, an attractive processing route.

At least three types of hydrocracking process configurations can be distinguished. The simplest so-called single-stage configuration is that in which

THOUSAND BARRELS/DAY

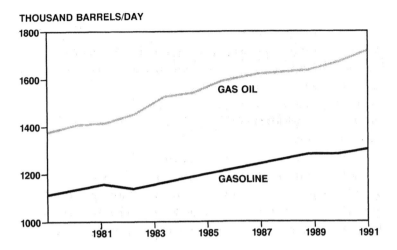

Fig. 31. Demand for gas oil and gasoline in the pacific basin

both the hydrotreating and the hydrocracking reactions are carried out in a
single reactor (Fig. 32). Alternatively, the hydrotreating and hydrocracking
steps may be performed in separate reactors. The most conventional, so—called
two—stage configuration consists of two reactors with interstage product
removal (Fig. 33), whereby gaseous by—products such as H_2S and NH_3 are not
carried through to the second—stage reactor. A more modern and cost—effective
process is the series—flow configuration, which has no such interstage product
separation (Fig. 33) and thus requires more robust second—stage catalyst
systems such as those based on zeolites.

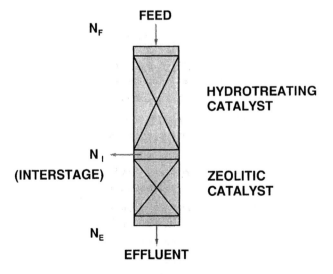

Fig. 32. Stacked bed reactor configuration

604

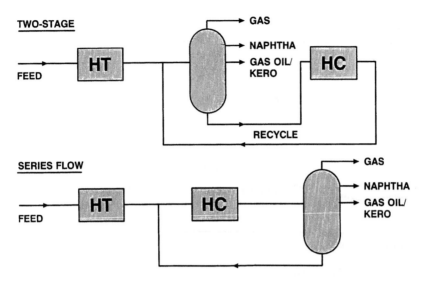

Fig. 33. Hydrocracking process configurations

A typical feature of the hydrocracking process is the product yield flexibility, as is demonstrated in Fig. 34, whereby the primary product can be, for example, naphtha, kerosine or gas oil. The process conditions and, perhaps more importantly, the catalyst type applied can be used to shift the product yield as required (ref. 41).

2.6.1. <u>Zeolite-based catalysts</u>. Although the first commercial hydrocrackers made extensive use of amorphous inorganic oxides such as aluminosilicates as catalysts it was soon realised that zeolites with their well-defined pore structure offered substantial advantages (ref. 42). One of the major advantages of zeolite catalysts is their markedly reduced coking tendency, which results in significantly improved catalyst stability. This is particularly important when heavy feedstocks are processed where high rates of catalyst deactivation can become a major process constraint. The application of zeolite catalysts, for example, within the Shell Group (ref. 42) has been an important factor in achieving a significant increase in feedstock heaviness and associated economic benefits (Fig. 35).

Another important characteristic of zeolite catalysts is their uniform and relatively high concentration of Bronsted acid sites which results in high levels of hydrocracking activity. This is reflected in lower initial reactor temperatures and, together with reduced rates of coke deactivation, results in enhanced catalyst life. Typical performance data for amorphous aluminosilicate and zeolite catalysts are shown in Fig. 36 where the benefits of the zeolite catalyst in terms of both initial activity and stability are apparent.

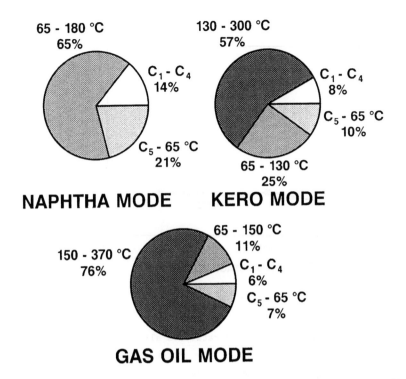

Fig. 34. Hydrocracking process product flexibility (from ref. 42)

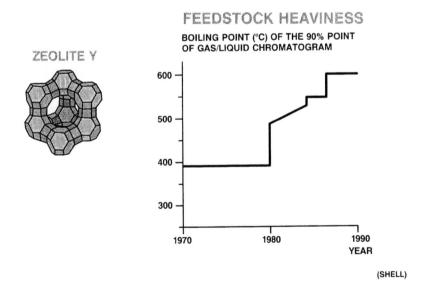

Fig. 35. Hydrocracking application of zeolite Y (from ref. 42)

606

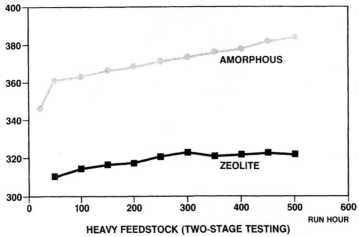

Fig. 36. Hydrocracking catalysts (from ref. 44)

As regards selectivity, the zeolite based catalysts also exhibit a different behaviour compared to amorphous systems (ref. 43). First-generation zeolitic hydrocracking catalysts are generally more naphtha- and thus less middle-distillate-selective than their amorphous counterparts. In North America this has not been a disadvantage since in this region hydrocrackers are normally operated in a naphtha-selective mode. However, as previously mentioned, in the developing countries where there is a high demand for middle-distillate product, naphtha-selective zeolite catalysts are less applicable.

Another disadvantage of first-generation zeolite Y catalysts has been their relatively poor stablility with respect to selectivity with time on stream. This performance deterioration, which results in increased gas make, is particularly apparent when operating with heavy feedstocks and recycle to extinction (refs. 43,44).

This has led, in recent years, to a significant research effort by a number of companies (refs. 41,45-47) into the development of new zeolite-based catalysts which are more middle-distillate-selective. Studies at Shell (ref. 44), for example, have shown that the product selectivity of zeolite Y-based hydrocracking catalysts can be profoundly influenced by varying the unit cell dimension (see Fig. 37). This parameter can be regarded as a measure of the catalyst acid site density. First-generation zeolite Y-based catalysts typically exhibit high acid site densities. As shown in Fig. 37 it has been found that by decreasing this unit cell parameter and thereby lowering the acid site density the selectivity towards middle-distillate products can be markedly improved without any significant activity penalty. Further, this shift towards

KEROSINE SELECTIVITY

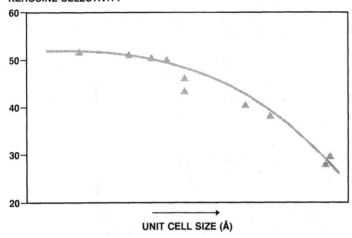

UNIT CELL SIZE (Å)

Fig. 37. Zeolitic hydrocracking catalysts (from ref. 44)

heavier products also results in reduced gas make and thus increased liquid
yields.

It should be noted, however, that in addition to reduced acid site density
lower—unit—cell zeolite Y materials are known to possess other characteristics
such as a meso—pore stucture and extra—lattice alumina (ref. 47), which most
likely also contribute to the improved catalyst performance. In fact, low—unit—
cell zeolite Y—based catalysts have been extensively studied for catalytic
cracking and are currently widely applied to achieve reduced coke make and
improved gasoline octane (ref. 31). Although it is tempting to draw analogies
it should be emphasized, however, that due to the markedly different catalytic
chemistry involved direct comparisons between catalyst properties and performance for these two processes should be treated with caution.

In addition, the stability with respect to selectivity during catalyst
ageing is also markedly improved for low unit—cell—zeolite Y catalysts. This
has been demonstrated in the laboratory by means of an accelerated ageing test
(Fig. 38), where the low— and high—unit—cell zeolite Y—based catalysts are
compared under the same process conditions. These improved performance features
of low—unit—cell zeolite Y—based hydrocracking catalysts have now also been
demonstrated in commercial practice under recycle conditions (ref. 44).

2.6.2. <u>Single—stage stacked—bed configuration</u>. In recent years the performance
of single—stage hydrocracking configurations has been substantially improved by
the introduction of so—called stacked beds (ref. 48). As shown in Fig. 32, this
involves combining both hydrotreating and zeolitic catalysts within the same
reactor configuration. The major advantage of such stacked—bed systems is that

608

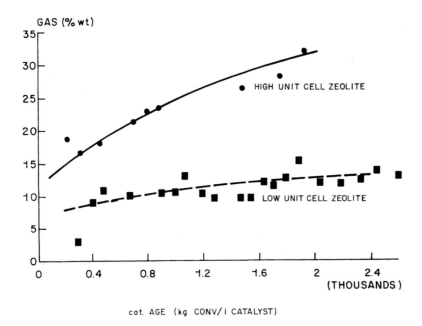

Fig. 38. Zeolitic hydrocracking catalysts: accelerated ageing test (from ref. 44)

the conversion per pass and/or the space time yield can be significantly increased. It is the high cracking activity of zeolites in the presence of relatively high concentrations (100 – 20 ppm) of organic nitrogen compounds which makes these types of catalyst particularly useful for this application. The very substantial activity gains that can be typically achieved by the application of zeolite catalysts in such stacked beds are shown in Table 10. Further, the activity advantages of the stacked-bed systems increase with conversion, which can be attributed to the apparent activation energies of these dual catalyst systems being higher than that of the single hydrotreating catalyst (ref. 48).

However, one of the potential limitations of these stacked beds in the past has been the somewhat reduced selectivity to middle-distillate products caused by the introduction of the zeolite component. The recent development of more highly middle-distillate-selective zeolite catalysts has now substantially reduced this disadvantage with retention of the activity gain (ref. 48), as is demonstrated in Fig. 39 and Table 10.

The above discussion pertains to the relatively high pressures (e.g. 100–150 bar pressure) which prevail in conventional hydrocracking. However, the stacked-bed systems using zeolite catalysts can equally well be applied (ref. 49) under mild hydrocracking conditions, which are typified by lower operating pressures (e.g. 50–80 bar pressure).

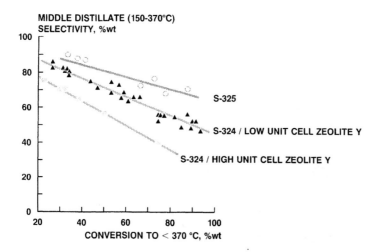

Fig. 39. Middle distillate selectivity: hydrocracking catalyst stacked beds (from ref. 48)

TABLE 10

Hydrocracking Process:

Comparison product of selectivities for single/stacked

beds (from ref. 48)

Catalyst	S324		S324/LOW UCZ		S324/HIGH UCZ
Conversion to <370 °C, %wt	50	70	50	70	70
Gas	4	5	3	4	5
Naphtha/gasoline	23	28	25	45	56
Kerosine	21	24	33	25	20
Gas oil	52	43	39	26	19
Activity T req, °C	T	T+14	T−35	T−30	T−32

high unit cell zeolite: HIGH UCZ
low unit cell zeolite : LOW UCZ

2.6.3. <u>Series–flow/two–stage configurations</u>. New–generation zeolite catalysts have been successfully applied in conventional two–stage and series–flow modes of operation (ref. 44). In particular, as previously mentioned, the high activity and nitrogen resistance of zeolite catalysts have made the lower–cost series–low mode of operation a technically feasible option which is likely to grow in importance.

610

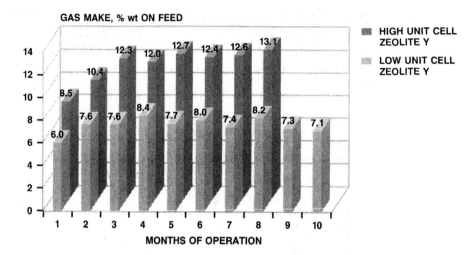

Fig. 40. Zeolitic hydrocracking catalysts: commercial performance (series flow) (from ref. 44)

An example of the markedly improved performance experienced in commercial series—flow operation of new—generation zeolitic hydrocracking catalysts is shown in Fig. 40. Clearly, the new low—unit—cell zeolite Y—based catalyst results in a marked reduction in gas make compared to the previously applied high—unit—cell zeolite—based anologue. Laboratory data (ref. 44) indicate that this improved perfomance is particularly enhanced when processing heavier feedstocks. Since this heavy feedstock trend is expected to continue it is likely that these types of zeolite catalysts will find increasing application in hydrocrackers in the future.

Further, in view of the increasing importance of hydroprocessing routes and the significant advances that have already been made in applying zeolite—based hydrocracking catalysts it is most likely that this field of research will receive considerable attention in the future.

2.6.4. The Paragon process. Another interesting recent zeolite—catalyst—related development in the field of hydrocracking is the so—called Paragon process developed by Chevron (ref. 50). This new process concept couples a conventional high—pressure hydrocracker with a low—pressure step, which makes use of a shape—selective zeolite catalyst such as ZSM—5 (Fig. 41). In this last reactor normal and slightly branched paraffinic hydrocarbons are selectively cracked to olefins in the C_3–C_8 range (Fig. 42), which can thus significantly increase the gasoline yield and octane quality compared to a conventional hydrocracker. However, although the concept is elegant the process does suffer from a number of disadvantages such as the fact that the organic nitrogen level of the feed to the shape—selective zeolite catalyst may not exceed 1 ppm to avoid rapid catalyst deactivation. Further, the combination of high— and low—pressure reactors within a process is expensive in terms of compression—decompression costs. These disadvantages would appear, at least to date, to have retarded any commercial developments.

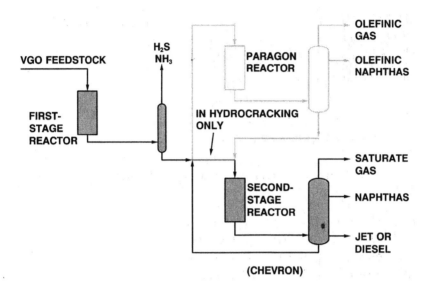

Fig. 41. The Paragon process (from ref. 50)

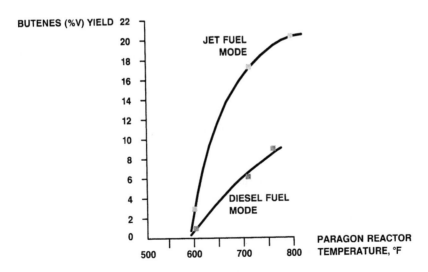

Fig. 42. Paragon process olefin yields (from ref. 50)

2.7. <u>Residue hydroprocessing</u>

As described in the Introduction, atmospheric and vacuum—distilled residues, i.e. the oil fractions in which the heaviest molecules of a crude oil have been concentrated, are in themselves not very attractive products. The various contaminants in a crude oil, such as sulphur and nitrogen compounds, metals such as nickel and vanadium, and large aromatic molecules (asphaltenes) which are easily transformed into coke, are concentrated in these fractions. The highest concentrations of these contaminants are found, of course, in the vacuum residue. Further, the properties of various residues depend strongly on the crude origin, some examples of which are given in Table 11.

TABLE 11

Typical properties of heavy oil fractions

Feedstock		Arabian Heavy Vacuum Distillate	Arabian Heavy Atmospheric Residue	Arabian Heavy Vacuum Residue	Maya Vacuum Residue
Yield on crude,	%wt	19	54	35	40
Sulfur,	%wt	2.9	4.5	5.4	5.3
Vanadium,	ppmw	0.5	97	150	635
Nickel,	ppmw	0.7	31	48	127
C$_5$ asphaltenes,	%wt*	–	12	20	34
H/C ratio		1.70	1.55	1.44	1.37

*The material which flocculates when mixing the residue with n—pentane is called pentane (C$_5$) asphaltenes.

The traditional outlet of residual oil has been industrial and marine heavy fuels but with the increasingly stringent environmental standards these residues could no longer be blended as such into these fuels but had to be desulphurized first. Thus, in the seventies a large number of residue hydrodesulfurization units were built (particularly in Japan), reducing the sulphur content in the residue by some 80—90 %.

However, these low—sulphur residual fuels remain less valuable than gasoline or middle—distillate oil products and thus there existed an economic driving force not only to desulphurize in these units but also to achieve (by a proper choice of catalysts and conditions) an ever—increasing conversion of residue into distillates. This is illustrated for example in Fig. 43, where the conversion in Shell residue units is plotted versus time (ref. 51). Thus, catalysts that gave increased conversion to more valuable lighter products were developed.

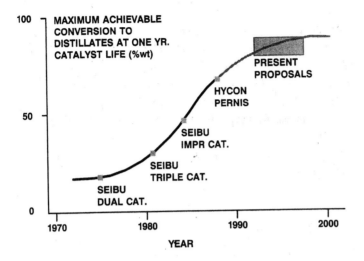

Fig. 43. Residue hydroprocessing conversion development in Shell (from ref. 51)

Given the tendency in distillate hydrocracking to process increasingly heavier feedstocks it is not surprising that also hydrocracking type catalysts, including zeolitic catalysts were studied. Obviously the residual feedstocks, with their high contaminant levels and large refractory molecules, are more difficult to process than the distillate feedstocks discussed earlier. On the other hand, even modest levels of conversion are economically attractive so that complete conversion is by no means necessary.

In Japan, with its large interest in residue hydroprocessing, a national project was started by the formation of RAROP, the Research Association for Residual Oil Processing, in which oil and chemical companies cooperated with major government funding. Although, they certainly are not the only group active in this field, they appear to have taken the lead in the development of zeolitic catalysts for this purpose and have presented several publications on this subject (refs. 52-56).

It would appear that the starting point of these investigations has been a search for a catalyst that is active for cracking in the presence of H_2S, as could be judged from toluene disproportionation and cumene cracking experiments. Several metal-loaded ultrastable Y-sieves were studied, from which an Fe-loaded USY, prepared by treatment of USY with ferric nitrate at pH=1.6-2.1 turned out to be particularly active (refs. 52,53). A detailed study was carried out to characterize this catalyst, from which it was concluded that ferric complexes will be bonded to framework vacancies created by acid-induced dealumination (Fig. 44); these iron complexes then apparently grow into small iron oxide particles of about 1 nm size due to the pH increase accompanying the zeolite dealumination step (ref. 52). The acidic properties seem to result from

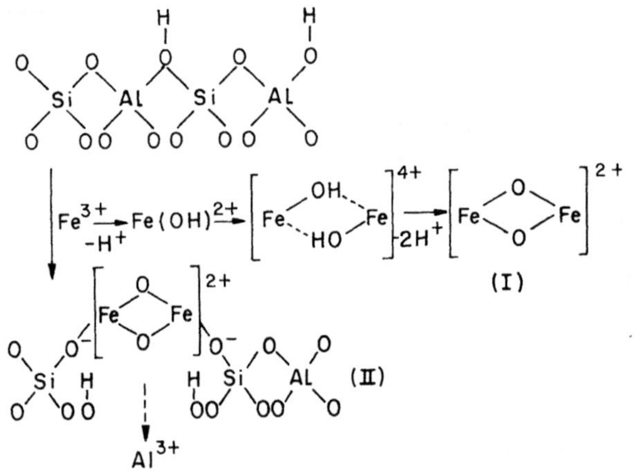

614

Fig. 44. Ferric complex induced zeolite framework dealumination (from refs. 52, 53)

the adsorption of H_2S on the supported Fe species.

The Fe-containing zeolite was incorporated in a catalyst by extrusion with alumina and impregnation with cobalt and molybdenum. This catalyst was then tested with a Kuwait atmospheric residue in trickle-flow operation at 13.5 MPa, space velocity 0.3 h^{-1}, and 400-410 °C. Surprisingly, the catalyst exhibits a cracking activity which increases at 410 °C during the first 1500 hours; this is attributed to a graphitization of the carbonaceous species deposited on the catalyst. Ultimately, the catalyst gives about 65 %w conversion to material boiling above 343 °C, with a high selectivity to naphtha, under the quoted conditions (ref. 54). No data have been published that indicate whether the lighter residue fractions are converted preferentially, as one would expect if the interior of the zeolite crystals still plays a dominant role in the conversion reactions. If, for example, the outer surface of the zeolites would be completely coked, rendering the inside inaccessible, no such selectivity would be expected. One may wonder whether the iron particles have another role, apart from inducing acidity, and are involved in reducing the coke-forming tendency. Itoh (ref. 57) has published a study on clay catalysts that have been promoted with iron, in which the iron appears to halve the coke formation in what is essentially a thermal cracking regime.

It appears that the zeolite catalysts ("R-HYC") have already been used commercially, in combination with residue HDS catalysts, for some years in the hydroprocessing of atmospheric residues, operating under the conditions as indicated above, and in line with the experience in distillate hydrocracking at relatively high temperatures. Thus, it would appear that there is indeed quite some scope to apply zeolite-based catalysts to upgrade even the heaviest fractions of the crude oil barrel.

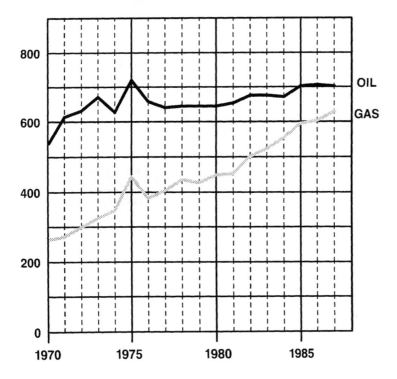

BILLION BARRELS/BARRELS OIL EQUIVALENT

Fig. 45. Evolution of proven world oil and gas reserves

3. GAS-CONVERSION PROCESSES

3.1. <u>Basics of gas conversion</u>

The evolution of proven world oil and gas reserves (Fig. 45) shows a dramatic increase in the latter and a levelling off the former. This trend is expected to continue, which will result in a shift towards the use of gas as a feedstock for the manufacture of transportation fuels and even petrochemicals.
Recognition of this situation has led to increased global interest in new
emerging technologies which can efficiently convert natural gas into liquids
and higher added-value-products. The incentive to produce liquids from gas also
stems from the desire to employ such process technology directly at remote
natural gas field sites to save on transportation costs. This new technology
could thus potentially compete very favourably with pipelines and LNG tankers
which are the present prime means of gas transportation to consumers.

Some examples of such process routes are shown schematically in Fig. 46,
where both methane and LPG can be employed as feedstocks. It is evident from
this diagram that some routes involve direct conversion from natural gas to
liquids (e.g. ARCO, (ref. 58) direct oxidative coupling) and others make use of

616

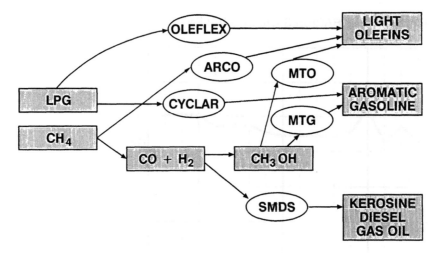

Fig. 46. Catalytic gas upgrading processes

synthesis gas (e.g. Shell Middle Distillate Synthesis; (ref. 59)) and methanol
(e.g. Mobil Methanol to Gasoline, (ref. 60)) as intermediates. Several of these
process routes make use of zeolite-based catalysts, which will be discussed
individually.

3.2. Conversion of methanol to olefins

The Mobil process (refs. 59) for the conversion of methanol to olefins (MTO)
can be considered as an intermediate step of their methanol-to-gasoline (MTG)
process (ref. 60). As shown in Fig. 47 methanol is initially converted into

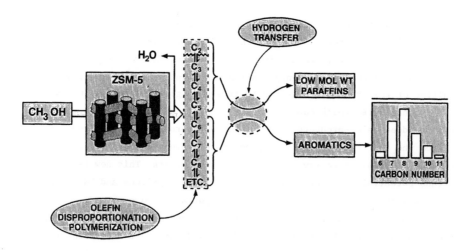

Fig. 47. Methanol to gasoline reaction mechanism (from ref. 61)

olefins over ZSM-5 followed by oligomerization reactions and hydrogen transfer, which lead to the selective production of gasoline-range aromatics. The product distribution of methanol conversion over ZSM-5 is shown as a function of reciprocal space velocity (contact time) is shown in Fig. 48. From this figure it can be seen that dimethyl ether is also an important intermediate product. By a judicious choice of process conditions (e.g. temperature, space velocity, pressure and methanol partial pressure) the product distribution can be maximized towards the production of light olefins (refs. 61,62).

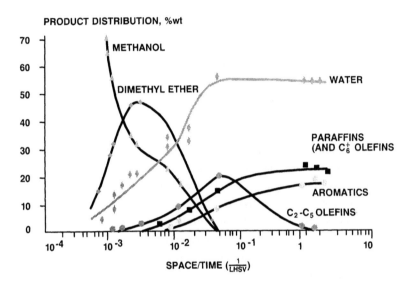

Fig. 48. Methanol to hydrocarbons reaction path: HZSM-5 (371 $^{\circ}$C) (from ref. 61)

A particularly interesting feature of this zeolitic catalytic chemistry is the C-C bond coupling step which is effected starting from a C_1 hydrocarbon (methanol). Although there have been extensive mechanistic studies (refs. 63-66) the detailed chemistry of this remarkable catalysis is by no means as yet fully understood. The shape-selective properties of the ZSM-5 zeolite structure play an important role in the catalytic performance. For example, the relatively high selectivity to lower olefins (see Table 12), which is achieved by suppression of cyclization and hydrogen-transfer reactions, can be attributed not only to an appropriate choice of process conditions but also to transition-state shape selectivity. The relatively low coke make (see Table 12) is also a reflection of the shape selective behaviour of this catalyst system.

618

TABLE 12

Methanol conversion to olefins (from ref. 67)

Number of zeolites oxygen atoms/ring	"Small pore" 8	"Medium pore" 10
Typical catalysts	erionite zeolite T chabazite ZK-5 ZSM-34 SAPO-34	ZSM-5 modified ZSM-5
Typical yields, %wt		
$C_2^=$	25-55	5-25
$C_3^=$	20-50	20-40
Total olefins	50-95	50-80
C_5+	1-15	25-60
Coke	0.5-5	less than 0.5

A variety of "small"- and "medium"-pore zeolites other than ZSM-5 have also been studied (ref. 67) for methanol conversion, as shown in Table 12. Interestingly, the "small"-pore zeolites such as ZSM-34 and SAPO-34 give higher yields of lower olefins and, in particular, ethylene. This can likely be attributed to a greater degree of shape selectivity and thus further suppression of consecutive reactions. However, this is not entirely consistent with the higher coke makes which apparently occur when using these smaller pore materials.

UOP have recently claimed (ref. 68) that SAPO-34 is a highly selective catalyst for converting methanol to ethylene and propylene. Selectivities to these lower olefins of up to 85 % are claimed at 100 % methanol conversion (see Table 13) but detailed process conditions were not published. This does,

TABLE 13

Methanol Conversion to Olefins with "medium pore"
silicoaluminophosphate zeolite, SAPO 34
(Union Carbide, from ref. 68)

Methanol conversion	100	100
Product yields, %wt		
Ethylene	61.1	43.0
Propylene	27.4	41.8
Butenes	5.4	10.8
Pentenes	—	—
Total olefins	93.9	95.6

however, provide an interesting example of potential new catalytic applications using the AlPO–derived molecular sieve materials.

There have also been quite some studies on the modification of various zeolites to enhance the selectivity towards lower olefins production from methanol. The techniques employed include cation exchange, pore modification by silica deposition, dealumination and phosphorous impregnation. All these studies point to the importance of both acidity and pore structure in determining optimal catalytic performance.

Although the methanol to lower olefin yields obtained to date are remarkably high (e.g. using SAPO–34) the current world methanol prices are such that this technology cannot compete with conventional steam cracking using oil–derived products as feedstocks. However, if in the future the efficiency of methanol manufacturing would be significantly improved and natural gas becomes the preferred feedstock then this zeolite/molecular sieve–based process route may become commercially attractive.

3.3. Conversion of methanol to gasoline

In the Mobil methanol–to–gasoline (MTG) process, the reactions of the MTO process are carried one step further. The olefins are thus intermediates that react further to form aromatics and lower paraffins (Fig. 48). Historically, of course, the MTG process was first, developed in the seventies and early eighties at the peak of the energy crisis and high oil prices. The technology was initially developed for fixed–bed reactors but this was later followed by the demonstration of fluid–bed operation on a pilot-plant scale, first for MTG (1984) and then for MTO (1985).

As previously mentioned, the basic catalytic chemistry is related to that of the MTO process but with some important differences. The effect of zeolite structure is shown in Table 14, where the reaction products of methanol passed over various zeolites at 370 °C are shown (ref. 69). The small-pore zeolites cannot easily accommodate aromatics and therefore yield mainly lower olefins, while the medium-pore zeolites yield primarily aromatics and lower paraffins (a combination required by the overall H/C mass balance) and some non-aromatic C_5+ product, depending on the conversion level. Wide-pore zeolites, however, give high concentrations of heavy aromatics, as shown also in Fig. 48, from (ref. 69).

Over the medium-pore zeolites the largest aromatic molecule produced is essentially durene (1,2,4,5-tetramethylbenzene) and although the wide pore sieves permit further alkylation they also exhibit a high selectivity to coke formation, which results in a short catalyst life (Table 15). Table 14 and Fig. 49 illustrate that profound differences in product selectivities are found for zeolites with various pore sizes. Other subtleties that have been studied include the effect of Al content, impregnation with phosphorus, addition of Zn, Mn, etc. (ref. 69). Such catalyst variations can be used to better separate the MTO and the MTG modes of operation. Alternatively, this can be achieved through reduction of the methanol partial pressure by means of steam, for example, or by increasing the temperature and space velocity, which both favour olefins selectivity.

TABLE 14

Methanol conversion to hydrocarbons over various zeolites (370 $^{\circ}$C, 1 atm, 1 LHSV) (from ref. 69)

	Hydrocarbon distribution (%wt) in				
	Erionite	ZSM-5	ZSM-11	ZSM-4	Mordenite
C_1	5.5	1.0	0.1	8.5	4.5
$C_2^=$	0.4	0.6	0.1	1.8	0.3
C_2	36.3	0.5	0.4	11.2	11.0
$C_3^=$	1.8	16.2	6.0	19.1	5.9
C_3	39.1	1.0	2.4	8.7	15.7
$C_4^=$	5.7	24.2	25.0	8.8	13.8
C_4	9.0	1.3	5.0	3.2	9.8
C_5+ aliphatic	2.2	14.0	32.7	4.8	18.6
A_6	–	1.7	0.8	0.1	0.4
A_7	–	10.5	5.3	0.5	0.9
A_8	–	18.0	12.4	1.3	1.0
A_9	–	7.5	8.4	2.2	1.0
A_{10}	–	3.3	1.5	3.2	2.0
A_{11}+	–	0.2	–	26.6	15.1

TABLE 15

Dimethyl ether conversion: effect of zeolite type (from ref. 67)

Catalyst	Residual Activity, %	Coke, %wt	Coke/hydro-carbons x 10^3
HZSM-5	90	1.5	0.3
HY	8	9.3	40
H-Erionite/Chabazite	0.15	8.6	90
H-Mordenite	0.15	6.8	200

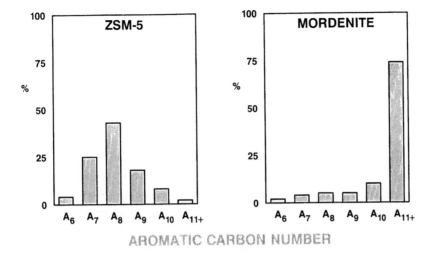

Fig. 49. Methanol to gasoline: effect of zeolite pore size (from ref. 69)

In the scale—up to commercial operation two factors are of particular importance, namely catalyst deactivation and heat production. Despite the relatively
high stability of the ZSM-5 catalyst, deactivation in the MTG operation is pronounced. The catalyst is progressively coked, and the total catalyst inventory
is completely deactivated after about one month under these conditions. The
heat production in the MTG reaction is high, leading to an adiabatic temperature rise of some 650 °C (ref. 61).

Two process setups have been developed to cope with this heat release. In
the older fixed—bed system, the methanol is first dehydrated to dimethyl ether,
releasing 20 % of the heat, and is then in a second stage reacted further to
gasoline. In the second stage a recycle of light gases is applied, not only to
dissipate the heat but also to increase the gasoline selectivity. The commercial MTG plant in New Zealand, with a capacity of 14500 bbl/d of gasoline,
started up in 1985/86, and has the setup described above. The second stage is
carried out in five reactors in parallel (four for the reaction and one for
regeneration). The New Zealand complex further contains a product treatment
step, to reduce the durene content of the gasoline to 2 % by weight.

Fluid bed operation was an obvious alternative to the multi—reactor fixed-
bed operation. This has been developed in a 100 b/d demonstration unit in
Germany, Wesseling, which ran in the MTG mode of operation during 1984. The
temperature in such a fluidized bed is, of course, uniform and part of the
catalyst is continuously withdrawn for regeneration. By varying the amount of
this stream, the average catalyst activity can be controlled. The fluid—bed
performance has considerable advantages over the fixed bed in terms of thermal

efficiency, gasoline yield and octane quality (refs. 67,69).

Finally, in the context of the methanol-to-gasoline route, the combination of MTO plus MOGD (Mobil olefins to gasoline and diesel, which is discussed in more detail in the following section) should be mentioned, which according to Mobil (ref. 67) has the advantage of offering a high degree of product flexibility.

3.4. Conversion of synthesis gas to gasoline

In New Zealand the MTG plant is combined with a methane steam reformer to produce synthesis gas and a methanol synthesis plant to overall produce gasoline from natural gas (ref. 70). With such a combination of processes one attempts to improve the economics through a close integration of the various steps. In the Topsoe integrated gasoline synthesis process (TIGAS) the methanol synthesis and the MTG reaction are integrated without methanol being isolated as an intermediate product. A multifunctional catalyst was developed which permitted production of oxygenates (methanol and DME) from synthesis gas at moderate pressures, which thus improves the process integration. The process has been developed in an industrial pilot plant. Compared to the Mobil MTG version the hydrogenation of the olefinic gases over the oxygenate catalyst is a disadvantage (ref. 71).

A more rigorous approach towards integration is to directly convert synthesis gas into aromatic gasoline in a single step. This necessitates combining the CO hydrogenation function and the aromatization function into a single catalyst, possibly also including a CO shift function. From the patent literature it would appear that this approach has indeed been extensively studied. Mobil, for example, has worked on ZnO/Cr_2O_3-ZSM 5 (ref. 72) catalyst systems, and has also studied combinations of Fischer-Tropsch functions and ZSM-5. Shell also has quite an extensive patent coverage on direct gasoline synthesis, and both BP and UOP would appear to have been active in this area. As mentioned above, apart from the routes involving methanol-type intermediates, Fischer-Tropsch-based routes will convert synthesis gas into distillate-range products. The Shell SMDS (Shell Middle Distillate Synthesis) process is an example of such a route. However, these process routes do not typically make use of zeolitic catalysts.

3.5. Conversion of light paraffins to aromatics

The direct conversion of LPG into aromatic components is a potentially interesting route to produce both high-octane gasoline components and benzene/toluene/xylenes as petrochemicals. BP and UOP are jointly developing (refs. 73-75) a new process (CYCLAR) to convert C_3-C_5 paraffins and olefins into aromatics-rich products. This process makes use of a gallium-loaded ZSM-5 type

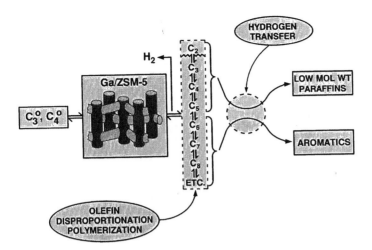

Fig. 50. LPG to aromatics: CYCLAR process (UOP/BP)

catalyst, as shown in Fig. 50. The mechanism involves the initial dehydrogenation of paraffins to form lower olefins and hydrogen, followed by olefin oligomerization and cyclisation reactions in which hydrogen transfer is likely to play an important role. In many respects this mechanistic scheme resembles the one used to describe the MTO and MTG processes (see Fig. 47).

There have been quite some studies (ref. 76) aimed at finding a substitute for gallium as the dehydrogenation/cyclization catalyst component. Although zinc silicates, for example, have been shown to be active for aromatization the zinc component is unfortunately too volatile under the relatively high process temperatures employed. Platinum has also been shown to further enhance the aromatization activity (ref. 76) of, for example, the zinc silicate catalyst system.

A typical product yield from the CYCLAR process is shown in Table 16, from which it can be seen that the aromatic product is very rich in benzene and toluene. This particular distribution of aromatics is significantly different from that obtained in the continuous platforming process. Further, the product distributions are similar for propane and butane feedstocks. The resultant octanes (RON and MON) of the gasoline fractions are high, thus reflecting the selectivity towards aromatic products (see Table 16).

The process conditions employed include relatively high temperatures (475-575 °C) and low pressures (2-10 bara), which thermodynamically favour the formation of aromatics. The rate of catalyst deactivation is such that moving-bed reactor technology has to be employed with continuous catalyst regeneration, similar to that applied in the low-pressure continuous catalytic reforming (CCR) process. A commercial demonstration unit for the CYCLAR process will shortly come into operation at the BP Grangemouth refinery.

TABLE 16

LPG to aromatics (CYCLAR): Typical products yields
(from ref. 75)

Overall composition, %wt		Aromatics composition, %wt	
Total aromatics	64	Benzene	31
Fuel gas	30	Toluene	41
Hydrogen	6	Xylenes/ethylbenzene	17
		C_9+	11
	100		100

Reactor temperature : 475–575 $^{\circ}$C
Pressure   : 2 – 10 bara
Catalyst contact time: 5 – 20 s
LHSV   : 2–4
Catalyst   : Ga/ZSM-5
Feedstock   : propane

In the short term, current feedstock/product price structures may somewhat
limit the commercial potential of LPG-to-aromatics conversion technology. However, in niche situations where the LPG price is low, hydrogen is in short
supply and aromatics are required for petrochemicals this new zeolite-catalyst-
based process technology may well be an attractive option.

3.6. Olefin oligomerization

Although olefin oligomerization has been known and used to, for example,
upgrade catalytic cracker off-gas streams in refineries for some time (e.g. the
CATCON polygasoline process (ref. 77) developed by UOP) there has been renewed
interest in this type of process based on improvements obtained by the application of zeolite catalysts. In particular, both Mobil (refs. 78–80) and more
recently Shell (ref. 81) have developed zeolite-catalyst-based processes (i.e.
MOGD and SPGK, respectively) in recent years. Perhaps the main advantage of the
new zeolite catalysts is the yield flexibility, whereby via an appropriate
adjustment of process conditions a switch in primary product between gasoline
and middle distillates is possible.

The MOGD process makes use of ZSM-5 zeolite as the shape-selective catalyst
system, which when operated at relatively low temperatures produces mainly
middle-distillate product via olefin-oligomerization reactions. By increasing
the operating temperature and adjusting the fractionation conditions, gasoline
can be produced as the primary product (Table 17). This process has been
applied on a semi-commercial scale at the Mobil refinery complex in Paulsboro
(ref. 80).

TABLE 17

Olefin Oligomerization: Distillate and Gasoline
modes for the Shell Polygasoline Kero Process
(SPGK) (from ref. 81)

	Max distillate mode	Gasoline mode
C_1-C_2	–	–
C_3	1	4
C_4	2	5
C_5-165 °C gasoline	18	–
165 °C+ distillate	79	–
C_5-204 °C gasoline	–	84
204 °C+ distillate	–	7

$C_3^=/C_4^=$ feed

The SPGK process has only recently been announced (ref. 81) and it is based
on a modified zeolite catalyst system. Both gasoline and middle–distillate
products can be produced as primary products by appropriate adjustment of the
operating conditions (Fig. 51). The attractiveness of olefin oligomerization to
produce transportation fuels depends very much on the individual situation in
each refinery. In modern large, fully integrated refinery complexes this type
of technology can be exploited to produce, for example, high–quality middle
distillate products. However, this particular utilization of lower olefins also

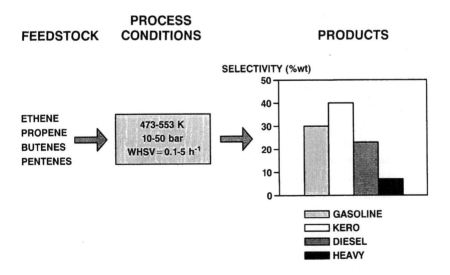

Fig. 51. Olefin oligomerization: Shell SPGK process (from ref. 82)

has to compete with alternative applications, e.g. as feedstocks to petrochemical processes, which in some cases may be more attractive. An advantage of the
zeolite-based olefin oligomerization routes is that normally only minimal
olefin-feed purification is required.

Future developments in this field could possibly be towards applications
related to the manufacture of petrochemicals rather than that of transportation
fuels given the relatively high prices of lower olefins. In this regard there
may be many new opportunities to develop more sophisticated non-acid zeolite
catalysts (ref. 82) rather than the present predominantly acid-catalysed
systems.

4. FUTURE DEVELOPMENTS

As already indicated in the Introduction and as will by now be evident from
the foregoing discussion, zeolites are playing an increasingly important role
in the modern complex oil refinery. Looking to the future two parallel developments can be expected.

Firstly, improved zeolite catalysts will continue to be developed for
existing applications. This will likely be more evident for the larger-volume
and thus the more important conversion processes such as catalytic cracking and
hydrocracking. There is undoubtedly still considerable scope to further modify
the existing zeolite catalysts and/or introduce new zeolite systems. For
example, in catalytic cracking the introduction of co-catalyst systems based on
shape-selective zeolites is an interesting new development, which could
conceivably gain in importance in the future as a means of increasing the
product flexibility of this process. In fact, even modest improvements in
catalyst performance can have a significant effect on the economics of these
large-scale processes. In the hydrocracking process the zeolite catalysts are
enabling heavier feedstocks to be processed, a trend which is expected to
continue in the future. Further, the lower-cost-series-flow and single-stage/
stacked-bed configurations which exploit zeolite catalysts will likely be the
preferred choice for hydrocracking in new projects in the future. The challenge
in this field is to develop zeolite-based catalysts that are sufficiently
flexible to cover the complete spectrum of oil product requirements.

Secondly, it is expected that new zeolite-catalyst-based processes will
continue to be introduced as the demands for deeper conversion and improved
product quality continue unabated. New processes which produce "synthetic"
products whereby lighter components are converted into higher-molecular-weight
products will almost certainly play an increasing role in the refinery of the
future. In this regard it is also conceivable that natural gas will be increasingly used as a feedstock for reasons of both availability and environmental
friendliness. Thus, both direct and indirect methane-conversion processes will

become important and will not necessarily need to wait for substantially higher oil prices in order to be exploited, as has been thought in the past. Some of these new gas-conversion processes are likely to make use of zeolite catalysts.

There are indications that at the bottom of the barrel zeolite-based catalysts are already being commercially used for residue hydroprocessing. The challenge for this application is to combine the relatively small-pore zeolites with larger-pore amorphous components in order to achieve the desired high conversion of high-molecular-weight molecules. In this field the recent synthesis of the first 18-membered ring aluminophosphate Fig. 52; (ref. 83) molecular sieve (VPI-5) represents an exciting development, which could potentially lead to a new family of heavy-feedstock-processing catalysts.

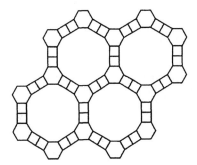

FIRST SYNTHETIC 18-MEMBERED RING (1.2 nm) MOLECULAR SIEVE

Fig. 52. Aluminophosphate-based molecular sieves (from ref. 83)

Theoretical studies on zeolites have received a great deal of attention in recent years. The progress made in computing, molecular graphics and computational chemistry has provided a strong stimulus in this field. Although these developments are still very much in their infancy it is to be expected that the insight gained from these theoretical studies will be sufficiently advanced to assist in understanding the catalytic processes taking place within the micropores of a zeolite. Although a full rationalisation may well be many years hence given the complexity of catalytic chemistry, any progress made in this field could provide another stimulus in the future, allowing modest steps to be made towards "catalyst design". This more advanced level of understanding would potentially enable the catalytic chemist to design a zeolite catalyst system that yields the desired products from a preferred feedstock and would thus reduce the arduous trial-and-error process in the laboratory. However, for the time being many new catalysts and zeolite-based processes will continue to rely heavily on empiricism and serendipity.

628

5. ACKNOWLEDGEMENTS

The authors would like to express their sincere gratitude to their many colleagues for the stimulating discussions on the subject of zeolite catalysis. Particular thanks are due to the following people in relation to the present review article: Drs. J. van den Berg, J. Biswas, A. Esener, J. Gosselink, A. Hoek, T. Huizinga, A. Klazinga, M. Manton, J. Minderhoud, G-J. den Otter, M. Post and H. Schaper.

REFERENCES

1 J. Rabo, in: Zeolite Science and Technology, F.A. Ribeiro, A.E. Robriques, L.D. Rollman and C. Naccache (Eds.), Martinus Nijhoff, The Hague, 1984, p. 291.
2 E.M. Flanigen, R.L. Patton and S.T. Wilson, Structural, Synthetic and Physicochemical Concepts in Aluminophosphate Based Molecular Sieves, in: P.J. Grobert, W.J. Mortier, E.F. Vansant and G. Schulz-Ekloff, (Eds.) Innovation in Zeolite Materials Science, Proc. Nieuwpoort Symposium, (1987), Elsevier, 1988, pp. 13-29.
3 J.N. Miale, N.Y. Chen and P.B. Weisz, J. Catal. 6, (1966) 278-287.
4 W.O. Haag, R.M. Lago and P.B. Weisz, Nature, 309, (1984) 589-591.
5 P.B. Weisz, Chem Tech, 1987, p. 368.
6 C.T.W. Chu and C.D. Chang, J.Phys.Chem., 89, (1985) 1569-1571.
7 M.F.M. Post, T. Huizinga, C.A. Emeis, J.M. Nanne and W.H.J. Stork, An Infrared and Catalytic Study of Isomorphous Substitution in Pentasil Zeolites, in: H.G. Karge and J. Weitkamp (Eds.), Zeolites as Catalysts, Sorbents and Detergent Builders: Applications and Innovations, Proc. Wurzburg (1988) Symposium, Elsevier, 1989, pp. 365-377.
8 S.M. Csicsery, Zeolites, 4 (1984), 202.
9 I.E. Maxwell, J. Inclusion Phenom., 4 (1986) 1-29.
10 G.J. den Otter, G.V. Tonks, F.H.H. Khouw and I.E. Maxwell, Shell/UOP Isomerization Process for Octane Upgrading of Light Ends, to be published, Erdoel, Erdgas, Kohle.
11 H.W. Kouwenhoven, Molecular Sieves, ACS Advances in Chemistry Series 121, 1973, pp. 529-539.
12 J. Weitkamp, Hydrocracking and Hydrotreating, in: J.W. Ward (Ed.), ACS Symposium Series 20, American Chemical Society, 1975, pp. 1-27.
13 P.B. Koradia, J.R. Kiovsky and M.Y. Asim, J. Catal., 66 (1980) 290-293.
14 G.R. Hays, W.A. van Erp, N.C.M. Alma, P.A. Couperus, R. Huis and A.E. Wilson, Zeolites, 4 (1984) 377-383.
15 J.A. Gray and J.T. Cobb, J. Catal., 36 (1975) 125-141.
16 P.B.Weisz, Pure Appl.Chem., 52 (1980) 2091-2103.
17 S.D. Burd, Jr. and J. Mazuik, Hydrocarbon Process., 51 (1972) 97-102.
18 J.P. Grannetti and A.J. Perrotta, Ind. Eng. Chem. Process Des. Dev., 14 (1975) 86-92.
19 T.R. Hughes, W.C. Buss, P.W. Tamm and R.L. Jacobson, in: Y. Murakami, A. Lijima and J.W. Ward (Eds.), 7th Int. Zeolites Conf. Proc., Kodansha/Elsevier, Tokyo, 1986, pp. 725-732.
20 D.V. Law, P.W. Tamm and C.M. Detz, Energy Progress, Vol.7, N4, December 1987, pp. 215-222.
21 K.W. Smith, W.C. Starr and N.Y. Chen, Oil Gas J. 78 (1980) May 26, 75-84.
22 N.Y. Chen, R.L. Gorring, H.R. Ireland and T.R. Stein, Oil Gas J. 75 (1977) June 6, 165-170.
23 J.G. Bendoraitis, A.W. Chester, F.G. Dwyer and W.E. Garwood, in Y. Murakami, A. Lijima and J.W. Ward (Eds.), 7th Int. Zeolites Conf. Proc., Kodansha/Elsevier, Tokyo, 1986, pp. 669-675.
24 US patent 3,968,024 (1976).

25 GB patent 2,027,742 (1980).

26 US patent 4,419,220 (1977).

27 J.E. Naber, P.H. Barnes and M. Akbar, The Shell Residue Fluid Catalytic Cracking Process, Japan Petroleum Refining Conference, Tokyo, October 19–21, 1988.

28 B.W. Wojciechowski and A. Corma, Catalytic Cracking, Marcel Dekker, New York, 1986.

29 L. Upson, I. Dalin and W.R. Wichers, Heat balance – The Key to Cat. Cracking, 3rd Katalistiks FCC Symposium, Amsterdam, May 26–27, 1982.

30 B. de Kroes, C.J. Groenenboom and P. O'Connor, New Zeolites in FCC, Ketjen Catalysts Symposium 1986, Scheveningen, May 25–28, 1986.

31 L.A. Pine, P.K. Maher and W.A. Wachter, J. Catal., 85, (1984) 466.

32 J. Biswas and I.E. Maxwell, The Relationship Between Generic Catalyst Types and FCC Performance, to be published, Applied Catalysis.

33 J. Biswas and I.E. Maxwell, Octane Enhancement Induced By Zeolites in Fluid Catalytic Cracking, Paper 8th International Zeolite Conference, Amsterdam, July 10–14, 1989

34 J.M. Maselli and A.W. Peter, Catalytic Reviews Science Engineering, 26, (1984) 525–554.

35 H.F. Henz, V.M. de Marco Meniconi and J.M. Fuoco, Petrobas Experience with Octane Enhancement in Resid Cat. Cracking, Ketjen Catalyst Symposium '86, Scheveningen 1986.

36 J. Biswas and I.E. Maxwell, Octane Enhancement in Fluid Catalytic Cracking. Part I, to be published, Applied Catalysis

37 J. Biswas and I.E. Maxwell, Recent Process and Catalyst-Related Developments in Fluid Catalytic Cracking, submitted to Applied Catalysis.

38 A.R. English and D.C. Kowalzyk, Oil Gas J. 82 (1984) July 16, 127.

39 F.J. Elvin, NPRA Meeting March 1986, Paper AM–86–41.

40 F.J. Elvin, NPRA Meeting March 1987, Paper AM–87–44.

41 J.W. Ward, Design and Preparation of Hydrocracking Catalysts, G. Poncelet and P.A. Jacobs (Eds.), Preparation of Catalysts III, Elsevier Science Pub., Amsterdam, 1983.

42 I.E. Maxwell, Catalysis Today, 1 (1987) 385–413.

43 T. Yan, Ind. Eng. Chem. Process Des. Dev., 22 (1983) 154–160.

44 I.E. Maxwell, T. Huizinga, A.A. Esener, A. Hoek, F. van der Meerakker, W.H.J. Stork and O. Sy, submitted to Oil Gas J.

45 M.E. Reno, B.L. Schaefer, R.T. Penning and B.M. Wood, 1987 NPRA Annual Meeting, Paper AM–87–60.

46 P.J. Nat, 1988 NPRA Annual Meeting, Paper AM–88–75.

47 J. Scherzer, in: Catalytic Materials; Relations between Structure and Reactivity, ACS Symposium Series 248, T.E. Whyte et al. (Eds.), 1984, p. 157.

48 A.A. Esener and I.E. Maxwell, "Studies in Surface Science and Catalysis, Hydrotreating Catalysts", (Eds.) M.L. Occelli and R.G. Anthony, vol. 50, pp. 263–271.

49 J.W. Gosselink, A. van de Paverd and W.H.J. Stork, Mild Hydrocracking, Optimization of Multiple Catalyst Systems for Increased VGO Conversion, Catalysts in Petroleum Refining Conference, Kuwait, 1989.

50 D.J. O'Rear, Ind. Eng. Chem. Res. 26, (1987), 2337–2344.

51 J.M. Oelderik, S.T. Sie and D. Bode, Applied Catalysis, 47 (1989) 1–24.

52 S. Hikada, A. Ino, K. Nita, Y. Maeda, K. Morinaga, and N. Yamazoe, Modification of Y Zeolite by Ferric Nitrate Solution Proc. 7th Int. Zeolite Conf., Tokyo, 1986, pp. 329–336.

53 S. Hikada, A. Ino, S. Nakai, H. Shimakawa, T. Mibuchi and K. Nita, Acidic Properties of FeHY-zeolites in the Presence of H_2S, paper H–9 of the Ketjen Catalysts Symposium, Amsterdam, 1986

54 K. Nita, S. Nakai, S. Hidaka, T. Mibuchi, H. Shimakawa, Ken-Ichi and K. Inamura, Deactivation and Reactivation Behaviour of Zeolite Hydrocracking Catalyst for Residual Oil, p. 501 in: B. Delmon and G.F. Froment, (Eds.) Catalyst Deactivation, Elsevier Science Publishers, 1987.

630

55 H. Sue, The Idemitsu (Kosan Co. Ltd.) Hydrocracking Process produces
   Distillate from Residual Oils, 3rd UNITAR Heavy Crude Tarsands Int. Conf.,
   Long Beach, California, August 1985.
56 H. Sue and M. Fujita, Idemitsu (Kosan Co.) R–HYC (residuum hydrocracking)
   Catalysts for Resid Conversion, 51st API Refining Dep. Mid–year Meeting
   San Diego, California, December 1986.
57 T. Itoh, New Hydrocracking Catalyst for Heavy Oil Upgrading, Symposium on
   Advances in Resid Upgrading, Division of Petroleum Chemistry, ACS Denver
   Meeting, April 5–10, 1987.
58 N.W. Green and R.V. Ramanathan, Conversion of Natural Gas to Transport
   Fuels, AICHE Spring Meeting, March 6–10, New Orleans, 1988, Paper 70B.
59 M.J. van de Burgt, C.J. van Leeuwen, J.J. Dell'Amico and S.T. Sie, The Shell
   Middle Distillate Synthesis Process, paper T11, at the Methane Conversion
   Symposium, University of Auckland, New Zealand, 27 April – 1 May, 1987.
60 H.H. Gierlich, W. Dolkmeyer, A. Avidan and N. Thiagarajan, Chem.–Ing.–Tech.
   58, Heft 3 (1986), MS 1462–1486.
61 C.D. Chang, Catal. Rev.–Sci. Eng., 25 (1) (1983) 1–118.
62 C.D. Chang, Catal. Rev.–Sci. Eng., 26 (334) (1984) 323–345.
63 J.P. van den Berg, J.P. Wolthuizen and J.H.C. van Hooff, Proc. of the Fifth
   Conference on Zeolites, L.V.C. Rees (Eds.) Heyden, London, 1980, p. 649.
64 T. Mole, J. Catal., 84 (1983) 423.
65 G.A. Olah, H. Doggwieler, J.D. Felberg, S. Frohlich, M.J. Grdina,
   R. Karples, T. Keumi, S. Inaba, W.M. Ip, K. Lammertsma, G. Salem and
   D.C. Tabor, J. Am. Chem. Soc., 106 (1984) 2143.
67 A.A. Avidan, Gasoline and Distillate from Methanol, Methane Conversion
   Symposium, University of Auckland, New Zealand, 27 April – 1 May, 1987.
68 J.M.O. Lewis, paper at the Dewitt Symposium, Houston (1988).
69 C.D. Chang and A.J. Silvestri, ChemTech, October 1987, 624–631.
70 Oil Gas J. 83 (34) August 1985, p. 38–39.
71 J. Topp–Jorgensen and J.R. Rostrup–Nielsen, Oil Gas J. 84 (1986), May 19,
   1986.
72 C.D. Chang, H.W. Lang and R.L. Smith, J. Catal., 56 (1979) 169–173.
73 Pet. Times, 91, Aug. 1987, 2213–2214.
74 R.E.Bolan, PEP Review N85–3–3, SRI International 1987.
75 US Patent 4,528,412 (1985).
76 T. Inui, Y. Makino, F. Okazumi and A. Miyamoto, in: P.J. Grobet, W.J.
   Mortier, E.F. Vansant and G. Schulz–Ekloff (Eds.), Innovation in Zeolite
   Materials Science, Proc. Int. Sympos., Nieuwpoort (Belgium), Elsevier, 1988,
   pp. 487–494.
77 V.N. Ipatieff and G. Egloff, Oil Gas J., 33 (52) 1934, 31.
78 W.E. Garwood, Prepr, Am. Chem. Soc. Div. Pet. Chem., 27 (2) (1982) 573.
79 US Patent 3,960,778 (1974).
80 S.A. Tabak and J.H. Beech, Mobil Olefin to Gasoline and Distillate (MOGD)
   Process, paper presented at the AICHE Spring National Meeting, Houston,
   April 2–6, 1989.
81 J.P. van den Berg, K.H.W. Roebschlaeger and I.E. Maxwell, The Shell
   Polygasoline and Kero (SPGK) Process, paper presented at the 11th North
   American Meeting of the Catalysis Society, Dearborn, USA, May 7–11, 1989.
82 I.E. Maxwell, Advances in Catalysis, 31, (1982), 1–76.
83 M.E. Davies, C. Saldarriaga, C. Montes, J. Garces, C. Crowder, Nature, 331,
   (1988) 698.

Chapter 16

"ZEOLITES IN ORGANIC SYNTHESES"

W.F. Hölderich

BASF Aktiengesellschaft, Ammoniaklaboratorium, D-6700 Ludwigshafen, Germany

and H. van Bekkum

University of Technology Delft, Laboratory of Organic Chemistry,
Julianalaan 136, 2628 BL Delft, The Netherlands

1. INTRODUCTION

Present-day main applications of synthetic zeolites include the use as cation exchangers in detergent formulations, as selective adsorbents (molecular sieving) in drying and separations, and as catalysts in the process industry. In the latter area by far the greatest use is in catalytic cracking, other important areas being hydroisomerization and aromatics processing (cf. Chapter 15). In a not too far future methanol-to-lower-olefins processes might constitute another important application.

Compared to the successful use of zeolites in hydrocarbon processing, their use in the synthesis of organic intermediates and fine chemicals is in a relatively early state of development. Two reasons for this are
- many of the target organic systems to be synthesized are too bulky to be built in or to desorb from the zeolite pore systems
- the average synthetic organic chemist is not acquainted with zeolites and their potential, other than their use as drying agent.
One of the first to draw attention to the broad potential of the combination of acidity and shape selectivity offered by zeolites in organic reactions was Venuto (1). The tendency to utilize this potential for specific, highly selective syntheses in the field of intermediates and fine chemicals has continued steadily in recent years as reviewed by the present authors (2, 3). Additional valuable reviews are given in ref. (4).

Zeolites and molecular sieves may be modified in many ways; they can be tuned over a wide range of acidity and basicity, many cations can be introduced by ion exchange and isomorphous substitution is possible also allowing build-in of isolated redox centers in the lattice. Moreover metal crystallites and metal complexes can be entrapped within the microporous environment. Altogether rich opportunities are provided of designing and employing heterogeneous catalysts that are tailored to suit the reactions desired. In this respect the experience gained in petrochemistry offers valuable guidance.

This chapter contains:
- a short discussion on general aspects of zeolites which are of relevance to their use in organic syntheses;
- a paragraph on non-catalytic uses of zeolites in organic preparative procedures;
- a section on zeolite catalysis in organic syntheses with focus on the following areas
  - o Isomerization reactions
  - o Electrophilic substitution of arenes
  - o Oxidation reactions
  - o Cyclization reactions including heterocyclic ring formation
  - o Nucleophilic substitution and addition
  - o Zeolite-catalyzed two- and multi-step syntheses

2. GENERAL ASPECTS

2.1. Effects of zeolite structure

As mentioned in earlier chapters a distinction is made between large-, medium- and small-pore zeolites. The entrance to molecules of these three categories is limited by rings of 8, 10 or 12 oxygens, respectively, as shown in Fig. 1. For more detailed geometric information on a number of important zeolites, see the Appendix. The crystallographic accessibilities given in the Appendix may be increased by about 10% to obtain the limiting sizes of entering reactants and desorbing product molecules.

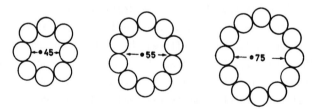

Fig. 1. Rings of oxygen atoms as present in zeolites A, ZSM-5 and Faujasite (X/Y); diameters in nm.

Table 1 gives some further examples of these three classes, also indicating whether the pore system is one-dimensional (parallel channels, no interconnections, code I), or three-dimensional (interconnected channels or cages, code III). Furthermore the Si/Al ratio or minimum Si/Al ratio is given.

Table 1. Zeolites and molecular sieves classified according to accessibility of and connections in pore system.

Small pore	Si/Al	Dim.	Medium pore	Si/Al	Dim.
A	1	III	ZSM-5	≥ 10	III
Chabazite	2	III	Silicalite	∞	III
Erionite	3	III	Boralite[a]	∞	III
			TS-1[b]	∞	III
			ZSM-11	≥ 10	III
			Ferrierite	≥ 5	III
			Theta-1/Nu-10/		
			ZSM-22	≥ 10	I
			AlPO-11[c]	0	I
			SAPO-11	< 0.2	I

	Large pore	Si/Al	Dim.
	Faujasite/X/Y	≥ 1	III
	SAPO-37	< 0.2	III
	Beta	≥ 5	III
	Mordenite	≥ 5	II
	Offretite	≥ 3.5	I
	L	≥ 3	I
	Omega	≥ 2.6	I
	ZSM-12	≥ 11	I
	AlPO-5[c]	0	I
	SAPO-5	< 0.2	I

[a] ZSM-5 (MFI) structure with aluminum isomorphously replaced by borium, $Si/B \geq 20$.

[b] Ibid. but aluminum replaced by titanium, $Si/Ti \geq 30$.

[c] $Al/P = 1$.

In zeolites which are closely related with respect to composition (Si/M range, M = Al, Ga, Fe, B, Ti) and building units, quite different channel arrays can exist. Thus, in ZSM-5 and ZSM-11, zig-zag and/or orthogonal connections are present, respectively, and also intergrowths of these endpoints exist, whereas ZSM-22 and ZSM-12 can be visualized as a parallel bundle of macaroni.

634

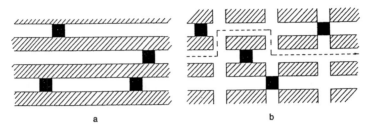

Fig. 2. Coke-formation in noninterconnecting channel systems (a) and in three-dimensional systems (b).

Such structural differences influence the effective lifetime of zeolitic catalysts. Catalyst deactivation takes place largely as a result of coke deposition (with a spectrum ranging from polycyclic aromatics, especially at higher temperatures, to long-chain, wax-like paraffinic deposits, particularly at lower temperatures). In general, zeolites with a three-dimensional structure will deactivate less rapidly than zeolites with a one-dimensional framework. In three-dimensional zeolites a channel which is blocked at one point can still be utilized for the catalytic reaction via the interconnections, as schematically shown in Fig. 2.

Furthermore it seems that the larger the connecting cages or crossings are, the larger the ease of coke-formation and deactivation is. Thus X- and Y-zeolites with supercages with a diameter of 1.3 nm generally deactivate faster than medium pore pentasil zeolites with crossings of about 0.9 nm diameter (4). Another important variable in coke formation presumably is the Si/Al ratio which determines the number and nature of the acidic sites. See Chapter 14 for a full account of coke formation.

Zeolites can display different types of shape selectivity (5-12) in organic reactions. *Reactant* selectivity allows only part of the reacting molecules to pass through the catalyst pores. In *product* selectivity, only compounds with the correct dimensions can diffuse out of the pores. Only those reactions can take place of which the transition state fits within the internal zeolite pores; this may lead to *restricted transition state* selectivity. For a detailed discussion the reader is referred to Chapter 12.

It may be noted that restricted transition state selectivity belongs to a family of *reaction* selectivities having in common that the surrounding zeolite affects the relative importance of parallel and/or consecutive reactions. Of particular importance in organic synthesis often is the so-called restricted growth type selectivity in which the formation of higher addition, substitution or oligomerization products is suppressed due to the dimensions of these products.

Shape selectivity can only operate when the reaction occurs within the zeolite pore framework. There are, however, cases in which it is uncertain whether reaction takes place on both the inner and the outer surfaces or, on the other hand, only on the outer surface of the zeolite. Furthermore, examples are known in which the molecular dimensions would certainly permit reaction within the zeolite framework, but where the course of the reaction is apparently not determined by shape selectivity but instead by other parameters such as thermodynamic considerations.

These phenomena are discussed in detail in a recent paper (13) of one of the authors.

2.2. Modification of zeolites

The properties of zeolites and thus their catalytic behaviour can be varied within certain limits by modifying the zeolites either during or after the actual synthesis (cf. Chapter 5).

Variations in synthesis formulation and in synthesis conditions allow
- to synthesize a particular type of zeolite;
- to establish a certain Si/Al ratio, often taking into account some limits (cf. Table 1) pertaining to the zeolite in question;
- to introduce other T-atoms (e.g. B, Ga, Fe, Ti), again often a maximum build-in percentage of such T-atoms applies (cf. Table 1);
- to synthesize relatively large or small crystals of the zeolite under consideration. The crystal size has a direct bearing on the inner/outer surface area ratio and on the average length of the intracrystalline diffusion paths of realting molecules. Generally, activity decreases (14) with increasing crystal size, whereas selectivity increases provided a homogeneous Al distribution exists;
- to modify the morphology of the zeolite crystals. Systems having parallel channels, such as zeolite L or mordenite, show the fastest crystal growth in the direction of the channels. By manipulation of the growth rates, morphologies may be obtained (15, 16) which represent more efficient catalysts.

2.2.1. Introduction of Brönsted and Lewis acidity

The number of counter ions in zeolites, which is directly related to the content of trivalent T-atoms, can be adjusted by building-in different ratios of tetravalent (Si) and trivalent T-atoms.

In general, zeolite synthesis yields the neutral sodium form. By means of ion-exchange with NH_4-salts and subsequent calcination one obtains the proton i.e. the acidic form of the zeolites.

Factors affecting the acid strength *per site* include the Si/T ratio and the type of T-atom.

The acidity of the well-studied pentasil zeolites is known to fall according to the T-atom sequence Al > Ga > Fe >> B. In this way the acidity spectrum ranges from the mildly acidic borosilicates to the strong acidic sites in alumino- and gallosilicates, with high Si/T ratio. As a consequence, any organic reaction subject to proton catalysis can, in principle, also be catalyzed by a zeolite. Some advantages over conventional Brönsted catalysts will be listed in a subsequent paragraph.

In some cases the conversion into the proton form may also be carried out by direct ion-exchange with an aqueous mineral acid (HCl, HNO_3). The strong zeolite ZSM-5 is an example. For many other zeolites a treatment with mineral acid will also modify the properties of the zeolite, e.g. by dissolving T-atoms out of the framework (aq. HCl) or by lowering the acidity (aq. H_3PO_4) and at the same time increasing the number of acidic centers, or by increasing the acidity by HF-treatment.

Ion-exchange from aqueous solution also allows the introduction of many types of cations, e.g. transition metals, lanthanides into zeolites. In several cases existing knowledge of exchange equilibria allows one to achieve a desired cation loading in a zeolite. Again, any type of homogeneously cation-catalyzed reaction can, in principle, be catalyzed by a zeolite carrying this cation.

The zeolite lattice, acting as a macro-ligand, often offers the cation several oxygen surroundings, which can impose restrictions to the cation's coordination chemistry and its catalytic activity. For instance in zeolites X and Y (see Fig. 3 the cation positions range from locked-in and inaccessible for reactants (positions S_I and $S_{I'}$) to the active and mobile S_{III} position in the supercages. Recent careful studies (18) provide a detailed picture of the ion-exchange of La^{III} and Ce^{III} at various temperatures in the zeolites Na-Y and Na-X. It appears that at ca. 80 °C La^{III} and Ce^{III} enter the sodalite units and hexagonal prisms. The latter offer cation positions having a high-oxygen coordination. In order to be able to penetrate the 6-ring windows of the sodalite unit and the hexagonal prism, the La and Ce water coordination has to be strongly adapted. Obviously, it is important to know the distribution of the active cations and their mobility under reaction conditions.

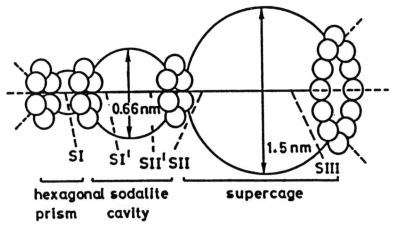

Fig. 3. Cation positions in the faujasite lattice (adapted from ref. 17).

<u>2.2.2. Basic zeolite catalysts</u>

The main emphasis of zeolite catalysis has been upon reactions catalyzed by strong or weak acids. If, however, complete ion-exchange with alkali metal ions such as K, Rb or Cs is carried out, it is possible not only to neutralize the Brönsted acid centers, but also to prepare weakly basic zeolites. This is particularly true if these neutral zeolites are then impregnated with alkali hydroxides.

For some examples the reader is referred to a paper of Corma et al. (19) reporting on condensations of benzaldehyde with e.g. diethyl malonate over alkali metal exchanged X- and Y-zeolites. The catalytic acitivity was found to increase with decreasing Si/Al ratio and with increasing size of the alkali cation. The ractions are assumed to involve ionization of the activated C-H bond followed by attack of the negatively charged reactant on the carbonyl group of benzaldehyde.

In another recent study on base-catalysis by alkali-modified X- and Y-zeolites (20) the conversion of isopropanol (towards propene on acid sites, towards acetone on basic sites) is taken as a probe reaction.

In another approach Martens et al. (21) developed sodium cluster-containing X-, Y- and L-zeolites by thermal decomposition of adsorbed sodium azide. These systems were able to catalyze 2-butene isomerization and aldol condensations.

For a review on basicity and basic catalytic properties of zeolites we refer to ref. (21a) and (2b).

2.2.3. Introduction of metal (0) and of metal complexes

Metal ion or metal-complex exchange, followed by a proper reduction procedure, leads to metal(0)-reduced metallic particles, e.g. Pt, Pd, inside the zeolite. Such metal entities may activate C-H bonds and may serve as catalyst in (de)hydrogenation, aromatization and oxidation reactions. They also exert a stabilizing effect in several reactions and catalyze oxidative reactivation.

Generally the noble metal is introduced into the zeolite by ion exchange with an aqueous solution of a cationic metal-ammine complex ($Pt(NH_3)_4^{2+}$ or $Pd(NH_3)_4^{2+}$) while performing the ion-exchange in competition with NH_4-ions (cf. Chapter 13). The subsequent calcination (NH_3-removal) and reduction (H_2) has been studied in great detail by Sachtler et al. (22) for the systems Pt-in-Y and Pd-in-Y. The location (supercage or sodalite cage, cf. Fig. 3) of the naked noble metal ion prior to reduction has a large effect on the ease of reduction and the degree of association after reduction.

Small metal complexes can be exchanged like the above-mentioned noble metal-amine complexes or adsorbed into zeolites; larger complexes, sized up to the diameter of cages or channel crossings, can be constructed inside the zeolite. Various complexes ranging from metal carbonyl complexes to phthalocyanines in - mainly Y - zeolites have been prepared and characterized. Recent illustrative work (23) on metal carbonyl complexes involves adsorption of $Ru(CO)_5$ (diam. .63 Å) into zeolite NaY, transformation in the zeolite to $Ru_3(CO)_{12}$ (diam. .92 Å) and subsequently towards a Ru hydrogenation catalyst. The entrapment of metal phthalocyanines into zeolites (222) also implies the heterogenization of homogeneous catalysts. Here the zeolite imposes factors leading to very interesting regioselectivity and reactant selectivity in the field of oxidation (Section 6.6). Some reports (24) mention the catalytic use of zeolite-encapsulated chiral complexes, constructed by adding chiral ligands to metal ions inside the zeolite. For a review on metal complexes in zeolites, see ref. (17).

2.2.4. High temperature treatments

The calcination of zeolites serves in the first place to create the H-form by decomposition of the ammonium-form or by removing organic molecules incorporated into the zeolite framework during the synthesis. In the second place, at temperatures exceeding 400 °C, there is increased dehydroxylation which leads to transformation of Brönsted acid sites into Lewis acid centers. Since this means that the catalytic properties are affected, it is necessary to carry out the tempering under carefully controlled conditions.

Calcination also serves regeneration of deactivated (coked) zeolitic catalysts.

The tempering of zeolites in the presence of water is a well-known method for preparing thermally stable zeolites with a long active lifetime. This so-called steaming brings about dealumination (alumina migrates out of the zeolite framework into the cages) with partial curing of the lattice by insertion of Si. The products obtained in this way are known as ultrastable zeolites e.g. US-Y.

In another dealumination procedure, in which the zeolite is treated with $SiCl_4$ in the gas phase Al is replaced by Si without developing defect sites in the framework. Liquid phase dealumination techniques include treatment with $(NH_4)_2SiF_6$ (24) and the use of a solution of $SiCl_4$ in carbon tetrachloride (25).

2.2.5. Tuning accessibility of zeolites

As has already been mentioned in an earlier chapter, fine tuning of the effective pore width of zeolites, subsequent to their synthesis, can be achieved by ion-exchange with cations of different sizes.

A further possibility is by means of CVD (chemical vapour deposition) methods, in which Si- or Ge-alkoxides are deposited at the mouth of the zeolite pore and then subjected to heat treatment. In this way the pore opening is artificially narrowed (26).

By reacting zeolites with diborane or silane Vansant et al. (26a) were able to alter the adsorption properties of zeolites such as H-mordenite profoundly.

2.2.6. Inertization of external surface of zeolites

The active centers located in the outer surface of zeolites do not display any shape selective behaviour. This negative effect is particularly marked in the case of small crystallites.

Accordingly the aim is to reduce the Al content on the external surface as far as possible. This can be achieved either by lowering drastically the Al concentration in the reaction solution towards the end of the zeolite synthesis resulting in the formation of an SiO_2 coating (27-29) or alternatively, by neutralizing or poisoning the acid centers of the outer surface subsequent to the synthesis. Bulky nitrogen bases such as 4-methylquinoline (30, 91) or silanes bearing bulky substituents such as triphenylchlorosilane (31, 32) are suitable for this purpose.

2.3. Shaping zeolitic catalysts. Binders

With few exceptions zeolites in the pure state, in particular the Si-rich organophilic types, are difficult to formulate into extrudates, tablets or microspheres for use in fluidized beds. It is therefore necessary to add a binder material for this purpose.

Suitable binders are various aluminum oxides, preferably boehmite, amorphous aluminosilicates, finely divided SiO_2, mixtures of SiO_2 and Al_2O_3, TiO_2, ZrO_2, and clays. After molding, the extrudates or tablets are for example dried at 110 °C/16 h and calcined at 500 °C/16 h. It is also possible to obtain advantageous catalysts by molding the isolated zeolite immediately after drying and subjecting it to calcination only after molding.

When extruding binder-free or peptization aids are used, being for example ethylcellulose, stearic acid, potato starch, formic acid, oxalic acid, acetic acid, nitric acid, ammonia, amines, silica esters and graphite or mixtures thereof (32a). Part of these aids have the ability to form chemically new interfaces between particles, others work physically as glue. Upon calcination most of these materials will be removed. In this way any side-reactions caused by the binder material are avoided.

When zeolite-catalyzed reactions are conducted in the liquid phase use can of course also be made of the pure zeolite crystals denoted as zeolite powder.

2.4. Use of zeolites in the liquid phase

As mentioned in a preceding paragraph some high silica zeolites (Si/Al > 10) are stable in aqueous acid medium at moderate temperatures. Generally low silica zeolites (A, X, Y) are not stable under conditions of low pH, and care should be taken when such zeolites are used in a low pH slurry technique as catalyst or adsorbens. The same applies when acid-unstable zeolites are used in reactions producing inorganic acid, e.g. aromatic halogenation, or in formulations containing strongly Al-coordinating ligands.

It is recommended in case some doubt arises as to the stability of the zeolite under the conditions applied, to check the crystallinity of the spent zeolite by X-ray analysis and/or adsorption capacity.

Finally it should be recalled that in liquid phase zeolite-catalyzed reactions the solvent should be carefully chosen. Because of its own adsorption the solvent has an effect on the reactant concentrations inside the zeolite. Rate and selectivity can be profoundly influenced in this way (33).

2.5. Some general remarks

The broad range of techniques which can be employed during the zeolite synthesis as well as the modifications which can be made at a subsequent stage afford an extremely large number of possibilities for optimizing the catalyst for a particular reaction. This is known as catalyst design or catalyst tailoring. Each individual measure which is adopted can influence activity, selectivity, stability and life time of the zeolite catalysts.

Zeolite catalysis is obviously best served by a balanced strength of adsorption of reactants together with just weak adsorption of products. When relatively strongly adsorbing product molecules are involved in liquid-phase batch experiments, progressive deactivation will occur, whereas in gas-phase flow work, pore filling with reactants and products will reach a steady state.

It should be mentioned in this connection that, in the micro-domains of the zeolite crystals, the rate and selectivity of a given reaction are governed by the local "concentrations" of reactants and catalytic sites whereas the products, as observed outside the zeolite - in the bulk gas or liquid - also reflect the diffusion rates of reactants and products, the probability of consecutive reactions and also contributions of the outer zeolite surface - if not inertizised - and of the binder material when present.

2.6. Advantages of zeolites

The easily reproducible production of well defined zeolite surfaces and of catalyst prepared from them are important advantages.

Major applications of zeolites are in reactions catalyzed by proton acids and Lewis acids, where the change from a homogeneous to a heterogeneous procedure brings advantages in respect of easy separation and disposal of the catalyst, avoidance of corrosion etc. In this regard, their shape selectivity often has an advantageous effect on the composition of the product.

As carriers for active components, zeolites make it possible to increase the activity and stability of the catalyst towards steam and high temperature, e.g. by doping with rare earths. Doping with suitable metals enables hydrogenation and oxidation reactions to be carried out. Bifunctional zeolitic systems can be easily designed and prepared. In all these cases the shape selectivity of the support is an additional factor controlling the reaction (2d).

The thermal stability of the zeolites permits them to be used above 150 $^\circ$C. They are therefore advantageous for reactions in which the thermodynamic equilibrium requires high temperatures. At the same time carrying out the reaction at a high temperature level makes it possible to recover process heat effectively.

Zeolite catalysts can be regenerated readily with air merely by burning off the coke, which is frequently responsible for their deactivation; after this treatment they generally regain their initial activity.

3. NON-CATALYTIC USE OF ZEOLITES IN ORGANIC SYNTHESIS

Application of zeolites in organic synthesis includes non-catalytic and catalytic uses. The first category comprises
- drying and purification of reactants and media;
- separation of products;
- application as re-usable reagent;
- use as reactant disperser and as slow release carrier;
- use as reactant concentrator;
- use as scavenger, e.g. in shifting equilibria.
Some examples may illustrate these methods.

3.1. Drying and purification of reactants and media

Important variables in the use of zeolites as selective adsorbents (34) are the pore size (accessibility), the Si/Al ratio and the counterions. Zeolites with low Si/Al (type A, type X, Si/Al = 1) are highly hydrophilic due to the high cation content. Zeolites with high Si/Al ratio (e.g. silicalite-1, ZSM-5 with Si/Al $\to \infty$) are organophilic and will selectively adsorb organic compounds such as ethanol, 1-butanol and phenol from an aqueous solution.

Drying of liquids and gases constitutes one of the early and well-established uses of zeolites. Especially the hydrophilic zeolite A (NaA, KA) is used for this purpose. Alternatively, the more acid-stable small-pore zeolite chabazite can be used. At low water pressures or concentrations zeolites are superior to conventional sorbents like silica. The high affinity for water is coupled with a fairly high saturation capacity (~ 25 wt %).

When drying organic liquids size and nature of the organic compound have of course a bearing on the drying efficiency; the more hydrophobic the compound, the deeper the drying. Table 2 lists some examples (35). For drying of small molecules, e.g. acetonitrile, zeolite KA is the desiccant of choice.

Table 2. Drying of solvents (ref. 35) with zeolite NaA (5% wt/v) at 30 °C.

Solvent	Residual water, ppm
Toluene	.01
Benzene	.03
Dichloromethane	.07
Diethyl ether	2
1,4-Dioxane	13
Tetrahydrofuran	28

In drying the enzyme invertase zeolite NaA proved to be somewhat more effective than the classical drying agent P_2O_5.

Zeolites can be useful in keeping reaction formulations dry *during* the reaction. Thus zeolite NaA proved to be (36) an essential component for selectivity and activity in the stereoselective Sharpless epoxidation of allylic alcohols (Fig. 4). Primarily the role of the zeolite appears to be the protection of the Ti catalyst from traces of water present or formed in the reaction medium.

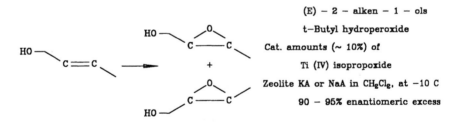

Fig. 4. Zeolite A as drying agent in enantioselective epoxidation (ref. 36).

Another chiral Ti-catalyst serves the addition of 1,1-disubstituted olefins to methyl glyoxylate (37). Again the presence of zeolite NaA is reported to be required for obtaining high stereoselectivity. Its precise role is not clear, as yet.

Finally we mention in this paragraph the use of zeolite NaX for the selective adsorption (38) of hydroperoxide impurities from ether solvents or alkene reactants. The zeolite is more effective than the conventional alumina

644

adsorbens. When using a Co^{II}-exchanged zeolite X decomposition of peroxide impurities can be achieved.

3.2. Separation of products

Zeolites are used widely as selective adsorbents in a range of large scale (39) as well as small scale separation processes.

Separations may be a result of differences in size and shape of the molecules to be separated or may be due to different affinity of the zeolite towards the compounds to be separated.

Table 3 lists some known separations between or within classes of organic compounds and indicates the separation principle and the overall class of zeolites applied.

Table 3. Adsorptive separations by zeolites.

Class(es) of compounds	Separation due to	
	size	affinity
Linear/branched aliphatics	small pore zeolite	
Functionalized aliphatics		large pore
Disubstituted aromatics (o,m,p)	medium pore	large pore
Trisubstituted aromatics	medium pore 1,2,4 vs 1,2,3/1,3,5	large pore
Tetrasubstituted aromatics disubstituted napthalenes	large pore	
Monosaccharides		large pore
Oligo- vs monosaccharides	large pore	

Well-established separations *by shape* - a form of exclusion chromatography - include for instance the separation of linear and branched alkanes over CaA, and the separation of 1,2,4-trisubstituted benzenes from their larger 1,2,3- and 1,3,5-substituted isomers over medium pore zeolites (e.g. ZSM-5, AlPO-11).

Separations by size on large pore (X, Y) zeolites are amongst others the separation of 2,6- and (the larger) 2,7-disubstituted naphthalenes and separation of some 1,2,4,5-tetrasubstituted benzenes from their larger 1,2,4,6-isomers.

The recent discovery of the super large pore molecular sieve VPI-5 (an AlPO with an 18-member pore opening) by Davis et al. (40) will substantially enlarge the potential of zeolites and molecular sieves in separation of organic compounds.

In some well-known separations *on affinity* suitably exchanged and/or modified X- and Y-zeolites are applied to separate
- linear alkenes from linear alkanes
- isomeric disubstituted benzenes
- isomeric trisubstituted aromatics.

In addition to application in separating regio-isomers zeolites may become useful in the separation of stereoisomers. In the carbohydrate field diastereomeric monosaccharides have been separated over X- and Y-zeolites with cation variation over the alkali and alkaline earth metals (41). For instance glucose/fructose separation is achieved over CaY or CaX offering an alternative for the commonly used sulfonated polystyrene resin (in the Ca-form).

Mannose/glucose separation can be performed over BaX or BaY. Differences in the strength of the cation-carbohydrate complexes are assumed to play an important role in these separation processes.

Disaccharides are to be considered - in view of their dimensions - as boundary cases for adsorption into the Y pore system. Consequently the temperature may be an important variable (42).

So far no examples are known of separation of enantiomers over zeolites. This would require a chiral zeolite or a zeolite loaded in a stable and homogeneous way with a chiral adsorbate.

An example of selective adsorption of organics from an aqueous medium by a zeolite is the room temperature adsorption of caffeine from aqueous solutions using dealuminated, steam-treated zeolite Y, having a Si/Al ratio of ≥ 13. The adsorbed caffeine may be recovered by ethanol extraction. It is claimed that the coffee flavour components are not affected by this extraction (43).

Substantial research efforts are presently devoted to the design and development of zeolite-based membranes in which zeolites - embedded into an organic or inorganic matrix - govern the selective passage of molecules.

3.3. Application as re-usable reagent

Zeolites may serve as a carrier for oxidizing cations (Fig. 5). Thus $Ce^{IV}Y$ can be applied (44) to oxidation reactions, e.g. pinacol cleavage, and the zeolitic reagent can be subsequently regenerated by high-temperature air oxidation. In this way, some conventional oxidations by metal compounds might be turned into clean procedures, provided that metal-ion extraction does not occur during the process.

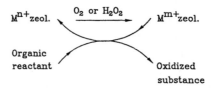

Fig. 5. Regenerable zeolite oxidant.

The high cation-exchange capacity of zeolites has been applied in the use of Ag-exchanged zeolites A, X, and Y to promote (45) the coupling of protected monosaccharides. In view of the size of the reactants, the reaction has to take place at the outer zeolite surface (Fig. 6), which means that particle size and surface roughness are important factors. The precise way of action and fate of the Ag^{I} ions have to be settled yet.

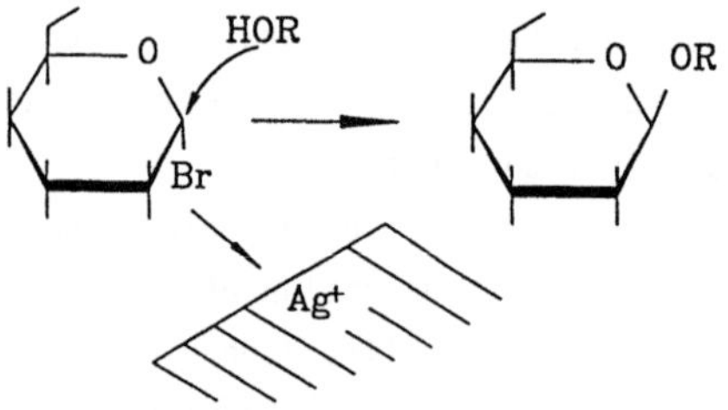

Fig. 6. Carbohydrate coupling by Ag^{I} zeolite.

3.4. Use as reagent moderater and as selective reaction host

Chemicals can be adsorbed in zeolites either from the gas phase or by using solutions in appropriate solvents, up to complete pore volume saturation if required. The charged zeolite can be used in slow-release techniques - e.g. in reactions which are difficult to control as to regioselectivity and consecutive reactions. An example is the use of bromine-loaded zeolite CaA for para-selective bromination of aniline and the toluidines (46).

Interesting effects of microporous solids (present in near-stoichiometric amount) on yield and regioselectivity of aromatic halogenation and nitration by activated reagents were recently reported by Smith (47). Thus para-selective chlorination of monosubstituted benzenes is obtained (Fig. 7) using t-butyl hypochlorite and zeolite NaHX in acetonitrile. The enhanced p-selectivity is ascribed to steric factors (shape selectivity).

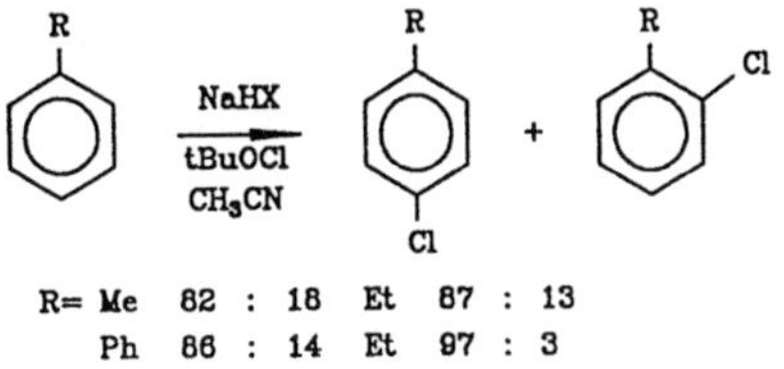

R= Me 82 : 18   Et 87 : 13
   Ph 86 : 14   Et 97 : 3

Fig. 7. Chlorination of aromatics (2.5 mmole) with tBuOCl (2.5 mmole) on 1.5 g of zeolite in acetonitrile (10 ml).

Using benzoyl nitrate onto Al^{III}-exchanged H-mordenite p,p'-dinitration of terphenyl (1,4-diphenylbenzene) was achieved (47).

Another interesting example of a zeolite acting as a host lattice for reacting molecules is the photochlorination of n-alkanes, adsorbed into a ZSM-5 type zeolite (48). The selectivity for primary C-H conversion of dodecane, S_t, relative to internal chlorination varies from 1.2 to 7.8 compared to $S_t = 0.4$ for photochlorination of liquid dodecane. The selectivity for terminal chlorination reaches a maximum at loadings of 1-2% which corresponds to 10-20% filling of the zeolite pore system. At these low loadings the alkane might prefer the straight channels of the zeolite (cf. Fig. 2), and the chlorine atoms might (mainly) arrive through the sinusoidal channels.

648

3.5. Zeolites as reactant concentrator

The principle of physical catalysis was elegantly demonstrated by Dessau (49) who studied the cyclodimerization of butadiene to yield vinylcyclohexene over several large-pore Na zeolites and other porous materials. It appeared that Na zeolites, having pores > 0.55 nm and a carbon molecular sieve with pores of 1.3 nm, showed catalytic activity. Smaller pore zeolites, such as ZSM-5 and an amorphous silica with pores >> 1.5 nm, showed hardly any catalytic activity for this reaction. The porous material (zeolite, carbon molecular sieve) concentrates the reactant for the bimolecular reaction and the effect is comparable to the results obtained under high pressure conditions.

3.6. Use as scavenger, e.g. in shifting equilibria

The ability of (hydrophilic) zeolites to selectively adsorb water from a reaction mixture was already mentioned in Section 3.1. Zeolites of the A type are also known to be efficient hydrochloric and hydrobromic acid scavengers which can be of use in organic syntheses.

The yield of common organic reactions is often limited by a relatively low equilibrium conversion. Such equilibria, e.g. acetalization, esterification, enamine formation, transacetalization and transesterification, can be brought to essential completion by selective adsorption of the small side-product (water, methanol) into a suitable zeolite.

For Brönsted-acid-catalyzed water-forming reactions - acetalization, esterification - an obvious combination would seem, at first sight, to be a conventional homogeneous proton catalyst together with zeolite KA. Problems may arise, however, from cation exchange leading to catalyst deactivation. Preferred combinations (cf. Fig. 8) consist of KA together with a solid acid catalyst: a sulfonic acid resin, silica-alumina, or a large-pore zeolite such as HY or rare earth (RE)-Y (50). In the last method (II), the reactants should fit in the Y zeolite, which moreover should be chosen with a relatively high Si/Al ratio, i.e. relatively hydrophobic in order to prevent water accumulation in the catalyst. Also a high Si/Al mordenite has been studied (51) as esterification catalyst.

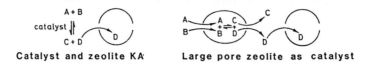

Catalyst and zeolite KA   Large pore zeolite as catalyst

Fig. 8. Shifting equilibria by selective side-product removal.

In another water-producing reaction, the enamine synthesis, an amine and a carbonyl compound are involved and the preferred catalyst is expected to be of the Lewis acid type. In this case, a good adsorbent/catalyst combination is zeolite KA and alumina (52).

In another approach in order to deal with partial equilibrium conversion a zeolite is used to adsorb selectively the desired *product* after which non-converted reactants (and side products) are recycled. As an example we mention the preparation (53) of 1-O-octyl-β-D-glucoside by selective adsorption onto zeolite NaY from a reaction mixture of an enzymatic synthesis starting from glucose and 1-octanol.

4. ZEOLITE-CATALYZED ISOMERIZATIONS

4.1. General remarks

Zeolites have a substantial potential for catalysis of isomerization reactions. In some industrial isomerization processes relatively large quantities of non-regenerable catalyst (H_2SO_4, $AlCl_3$) are used and zeolite catalysis might lead to clean technology. An interesting option for several products is the combination isomerization/separation: a zeolite-catalyzed isomerization is followed by a chromatographic separation over zeolites and recycle of the unwanted isomers, thus avoiding organic waste products. Established integrated processes of this type include the TIP Process for total isomerization of alkanes (see Chapter 15) and several processes providing di- or trisubstituted benzenes, of which some examples will be given.

In isomerization reactions of aliphatic compounds, a distinction must be drawn between rearrangements of the carbon skeletons, in which C-C single bonds are broken and formed, and double bond isomerization reactions, in which the carbon skeleton remains unchanged. The shape selectivity of the zeolites is intended to cause the reactions to take the desired path.

4.2. Double bond isomerization of olefins

Double bond isomerizations can be performed using acidic catalysts via a carbenium ion mechanism or using basic catalysts via a carbanion mechanism at the allylic proton (54, 55). The advantage of the basic over the acidic catalysts for this reaction is that cracking and skeletal isomerization are suppressed.

One of the most extensively investigated isomerization reaction is the

650

conversion of butenes; in the steam cracking of naphtha, after butadiene and isobutene have been removed, a C_4-fraction containing a varying proportion of 2-butene to 1-butene is obtained and this ratio can be changed if desired by isomerization. For this isomerization, as well as for the isomerization of longer-chain olefins, acid or basic zeolites can be applied and this is subject of numerous publications (56).

For example, the reaction can be controlled to give the desired isomer by adjusting the acidity of the catalyst and the temperature. At 175-200 °C, using an acid boron pentasil, 88% of the 1-butene is converted into 2-butene in a *trans/cis* ratio of 5:3. Na-boron zeolite is suitable for isomerization in the reverse direction; at 450 °C *trans*-2-butene is converted in a yield of 60% into 1-butene and *cis*-2-butene in a ratio of 7:5 (57-60).

4.3. Double bond isomerization of olefins with functional groups

In the isomerization of 2-alkylacroleins into 2-methyl-2-alkenals according to reaction (1) a boron pentasil zeolite converts the terminal double bond without reduction in selectivity through skeleton isomerization or reactions at the aldehyde group. Thus Ce-boron zeolite is an effective heterogeneous catalyst which, at 300 °C, gives an 82% selectivity of tiglaldehyde (reaction 1), R = CH_3) at a conversion of 20% (61); conventional catalysts such as S-poisoned Pd catalyst in the presence of H_2 exhibit the disadvantage of considerable formation of hydrogenation products.

$$R-CH_2-\overset{\overset{\displaystyle CH_2}{\|}}{C}-CHO \longrightarrow RC\overset{\overset{\displaystyle CH_3}{|}}{\underset{H}{=}}C-CHO \qquad (1)$$

R= alkyl, aryl, arylalkyl

The isomerization of pentenoic acid esters is possible over noble metal or transition metal doped Y (62, 63) and pentasil zeolites (64, 65) without attack or cleavage of the ester group. For example (64) using a pentasil-type borosilicate impregnated with 0.8 wt % Pd methyl 3-pentenoate at 180 °C and WHSV = 1,4 h^{-1} is converted in a product mixture of 86.9% 3-, 8.0% 4-, 3.7% trans-2-pentenoate.

If the olefin is substituted by an acetoxy group, the pentasil zeolites are not only capable of isomerizing the double bond, but they also produce a rearrangement through a simultaneous shift of the functional group (66). Thus 1,4-diacetoxybutene - see reaction (2) - is converted into 3,4-diacetoxybutene on boron and aluminum pentasils at 300 °C, at a conversion of over 50%. The

selectivity is 50-60%. This transformation is reversible; if 3,4-diacetoxy-
butene is passed over boron zeolite at 200 °C, the 1,4-product is obtained with
a selectivity of 80%. Hitherto, allyl rearrangements of this type have only
been carried out in the liquid phase using SeO_2 or Pt or Pd halides in the
presence of O_2. The use of pentasil zeolites in the gas phase offers an
alternative to this.

$$CH_2 - CH = CH - CH_2 \rightleftharpoons CH_2 = CH - CH - CH_2 \qquad (2)$$
$$\quad\; | \qquad\qquad\qquad | \qquad\qquad\qquad | \quad\; |$$
$$\;\; OAc \qquad\qquad\quad OAc \qquad\qquad\quad OAC\; OAc$$

The aim of selective double bond isomerization is only attained to a limited
extent, since, inter alia, mixtures of isomers are formed. The target product
can, however, be obtained by removing and recycling the undesired isomers.

4.4. Skeletal isomerization of saturated hydrocarbons

In the chemistry of intermediates, sometimes a selective rearrangement of the
carbon skeleton is required. Skeletal isomerization reactions of alkanes take
place under surprisingly mild conditions over strongly acidic zeolites. Thus,
n-butane is converted into isobutane at 200 °C on H-Y- of La-Y-zeolite (67).
Bifunctional Pt-H-mordenite at 250 °C is applied for alkane isomerization in
the TIP process (Chapter 15); for a discussion of the reaction mechanism
including the role of Pt in reducing coke formation and cracking, see ref. 68.
Alkanes containing only one side methyl group are formed as the primary
products when long-chain alkanes are isomerized using bifunctional catalysts.
Thus, n-undecane is isomerized on Pt/Ce-Y-zeolite in a stream of hydrogen at
275 °C at conversions of up to 40% into a mixture of isoundecanes, without
cracking processes manifesting themselves (69). Reactions of this type can also
be used to characterize shape-selective zeolites (see Chapter 12).

An interesting skeletal rearrangement is the conversion of
tetrahydrodicyclopentadiene into adamantane, reported by Honna et al. (70, 2d)
to take place over rare earth exchanged Y-zeolites in a H_2/HCl atmosphere at
250 °C. Initially, an *endo-exo* isomerization takes place. A stable catalyst is
obtained by adding a hydrogenation component (Pt-Re-Co). Selectivities up to
50% have been observed.

On the basis of a mechanistic study, including H-D exchange experiments,
Lau and Maier conclude (71) that strongly acidic zeolite sites first form
pentacoordinate carbonium ions, which lose H_2 upon transformation into
carbenium ions. A sequence of 1,2-shifts then leads to adamantane.

Intermolecular hydride and proton transfer are assumed to occur.

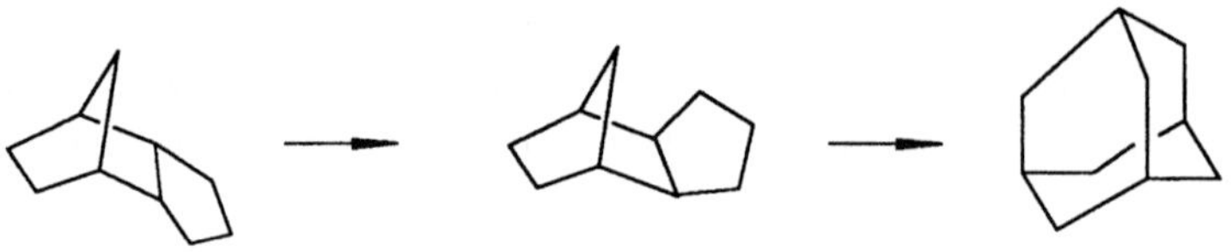

(3)

RE—Y,H$_2$/HCl , 250   °C

Several workers have reported on the use of zeolites in the isomerization of
terpene hydrocarbons. Here, proton-acid and non-proton catalysis generally
leads to quite different product mixtures. For the thermal isomerization of
β-pinene to myrcene (reaction (4)), which is of industrial interest, a maximum
selectivity of 83% was reported (72) using a zeolite coded TSZ-642.

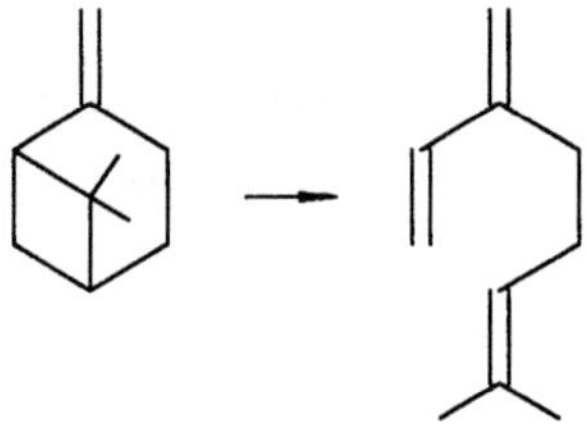

(4)

4.5. Isomerization of functionalized saturated systems

It is already evident from the allyl rearrangement of diacetoxybutenes (Section
4.3) that the skeleton of hydrocarbons containing carbonyl groups can be
isomerized by pentasil zeolites without the carbonyl group being attacked.
Further examples of rearrangement reactions will be given in this capture.

4.5.1. Aldehyde-ketone rearrangement

$$R_1R_2R_3C - CHO \longrightarrow R_1R_2CH - \overset{O}{\overset{\|}{C}} - R_3 \qquad (5)$$

Preparation of ketones from aldehydes would be an attractive route, as the
latter are easily accessible via oxo synthesis, for example. This type of
isomerization is known, e.g. with tin-, molybdenum- and copper-containing mixed

oxide catalysts or using cerium oxide on alumina. Disadvantages of these processes are the low selectivity obtained with satisfactory yields and the fact that steam is needed in order to get high selectivity and long life time. Because of this, unsymmetric, substituted ketones are generally prepared in industry by the condensation of various organic acids followed by decarboxylation. The unavoidable formation of symmetrically substituted ketones and carbon dioxide are disadvantages in this process.

It has now been found by Hölderich et al. at BASF that ketones can be obtained advantageously by isomerization of aldehydes using pentasil zeolites (73). A plus point of this method is that the addition of steam is not necessary. The results are summarized in Table 4. As to the priority of the migrating group it is found that the methyl group migrates exclusively if phenyl, benzyl and methyl groups are present. This could be caused by the high shape selectivity of the pentasil zeolites.

Table 4. Aldehyde-ketone rearrangement (73) over pentasil zeolites.

Feed	Zeolite	Condit.[d]	Product	Conv.	Selec.
2-Methyl-propanal	boron[a]	400 °C 2 h^{-1}	Butan-2-one	42%	90%
Pival-aldehyde	boron[a]	360 °C 2 h^{-1}	3-Methyl-butan-2-one	92%	85%[e]
2-Phenyl-propanal	boron[a]	400 °C 0.8 h^{-1}	1-Phenyl-propan-2-one	63%	97%
2-Phenyl-propanal	iron[b]	400 °C 2 h^{-1}	1-Phenyl-propan-2-one	98%	95%
2-Phenyl-propanal	iron[c]	400 °C 2 h^{-1}	1-Phenyl-propan-2-one	100%	87%
3-Phenyl-2-methyl-propanal	boron[a]	400 °C 2 h^{-1}	4-Phenyl-butan-2-one	59%	85%

[a] SiO_2/B_2O_3 = 47, without binder.
[b] SiO_2/Fe_2O_3 = 18, without binder.
[c] Extruded with boehmite as a binder in the ratio 60:40 wt %.
[d] Under isothermal conditions in a tube reactor; WHSV h^{-1} = g feed/g catalyst and hour.
[e] Plus 12% isoprene.

The iron-zeolite is more active than the boron-zeolite, as is shown by the conversion of 2-phenyl-propanal. This tendency is found e.g. in the conversion of methanol to olefins too (74). The iron-zeolite formed into extrudates with boehmite as a binder is less selective than the pure zeolite catalysts. Using the alumo-zeolite ZSM-5 only low selectivity is obtained at high conversion rate even when steam is used (75), showing that the zeolite's Brönsted acidity is not a major factor in this reaction.

4.5.2. Pinacol rearrangement

The pinacol rearrangement is an acid-catalyzed dehydration of 1,2-diols leading to carbonyl compounds, reaction (6). In a few cases unsaturated alcohol and conjugated dienes are obtained too. This applies regardless of the use of a homogeneous or a heterogeneous catalyst. Both reactions of the 1,2-diols take place on zeolites.

$$
\begin{array}{ccc}
\underset{\underset{OH}{|}}{\overset{\overset{H_3C}{|}}{H_3C-C}} - \underset{\underset{OH}{|}}{\overset{\overset{CH_3}{|}}{C}} - CH_3 & \longrightarrow & \underset{\underset{CH_3}{|}}{\overset{\overset{H_3C}{|}}{H_3C-C}} - \overset{\overset{O}{\|}}{C} - CH_3
\end{array}
\qquad (6)
$$

When using zeolites of the faujasite and pentasil type (76) yields between 41 and 83% are achieved at 105 °C, the best performance being obtained with H-Y zeolite.

Within the faujasite group it was found (77) that differences in Brönsted acidity (promoting the pinacol arrangement) and in basicity (promoting β-H elimination) result in differences in selectivity for ketone and diene formation, respectively. If the temperatures are raised to 400 °C or higher, dienes are preferably formed.

1,2-Diols containing a primary OH group are rearranged, after dehydration, to give aldehydes. For example, phenylglycols, which can contain a wide variety of substituents in the aromatic nucleus, are converted (reaction (7)) into phenylacetaldehydes on aluminum and boron pentasils at 250-300 °C. The selectivity at complete conversion is between 88 and 96% (78). In this reaction, the zeolites are superior to the amorphous metal-doped or undoped aluminum silicates which give yields of 50-86%. Like phenylglycol itself, ethers and esters of the latter can be isomerized in high yields (91-96%) (78). These reactions are only of academic interest and will probably not be used in industry, since the 1,2-diols are usually prepared from the corresponding epoxides, which undergo rearrangement by zeolites to give such aldehydes too (cf. Section 4.5.3).

$$Ph-\overset{\overset{\displaystyle OH}{|}}{CH}-CH_2OH \quad \longrightarrow \quad Ph-CH_2-C\overset{\displaystyle O}{\underset{\displaystyle H}{\diagdown}} \qquad (7)$$

4.5.3. Rearrangement of epoxides

Activation of epoxides for ring opening reactions can be achieved by Brönsted acid catalysts (addition of a proton to the epoxide oxygen) as well as by Lewis acid catalysts (coordination of the epoxide oxygen to a multivalent cation). As to the regioselectivity of the ring opening electronic as well as steric factors can play a role. These general considerations apply to the opening of the epoxide ring to give aldehydes using heterogeneous catalysts, such as zeolites, in the liquid or gas phase. Indeed, zeolites in their proton form (H-ZSM-5) as well as the essentially neutral Ti-silicalite (TS-1) are found to catalyze this type of reactions. The reaction of styrene oxide or alkyl- or alkoxy-substituted styrene oxides using titanium zeolites in acetone or methanol at 30-100 °C results in phenylacetaldehydes, according to reaction (8) with selectivities higher than 90% at conversions between 90 and 100% (79). In a subsequent patent (80) the Enichem workers report on the direct conversion of vinylbenzene compounds to phenylacetaldehydes over TS-1 using aqueous H_2O_2 in e.g. methanol as solvent. Selectivities are over 90%, the epoxide is assumed to be the intermediate.

$$(8)$$

R = alkyl, aryl, arylalkyl, halogen, haloalkyl, alkoxy, alkylthio

It is also possible to prepare phenylacetaldehydes containing a very wide variety of substituents on zeolites in the gas phase, without using solvents. Particularly attractive features of the acidic boron-, iron- and aluminum-pentasils are their high service lives and yields (> 90%) (78, 81).

The high regioselectivity observed in the isomerization of styrene oxides towards phenylacetaldehydes may be related to the stabilization of the developing α-cation by the adjacent phenyl ring. In a similar way the selective conversion of phenylglycol into phenylacetaldehyde can be understood.

In zeolite-catalyzed isomerization of aliphatic epoxides the observed regioselectivities are relatively low. Also rearrangements have been reported. Thus, 2-methyl-2,3-epoxybutane is converted at 150 °C on pentasil type zeolites to 51.6% methyl isopropyl ketone, 40.4% pivalaldehyde and 7.4% isoprene (82). The pivalaldehyde results from a pinacol-type methyl shift in the intermediate ring-opened carbenium ion.

Several materials, including some zeolites (83) have been tested as catalysts for the isomerization of α-pinene oxide towards the fragrance intermediate camphorene aldehyde (reaction (9)).

$$\tag{9}$$

It appears (84) that certain Lewis acid cataysts are less active but more selective than strong Brönsted acids in this isomerization. Careful design will be required to develop a zeolite catalyst with a better performance than the presently used $ZnCl_2$-catalyst.

4.5.4. Rearrangement of cyclic acetals

Another example of C-O bond cleavage over zeolites is the conversion of cyclic acetals (85) leading to aldehydes rather than to olefinic compounds. Acetals of neopentylglycol are isomerized, particularly on pentasil zeoliltes, at 250-400 °C to neoalkanals with > 90% selectivity at 50-80% conversion (equation (10)).

$$R_2 - \overset{\displaystyle R_1}{\underset{\displaystyle H}{C}} - O - \overset{\displaystyle R_3}{\underset{\displaystyle H}{C}} - \overset{\displaystyle R_4}{\underset{\displaystyle R_5}{C}} - C \overset{O}{\underset{H}{\big/\!\!\big/}} \tag{10}$$

R^1, R^2, R^4, R^5 = H, alkyl, alkenyl, aryl, arylalkyl, alkylaryl, arylalkenyl, alkenylaryl, heterocyclic residue; R^3 = H, alkyl

The transition state and/or product diffusion shape selectivity of the zeolites leads to an increase in the yield of the linear neoalkanals. Here once again zeolites as acidic catalysts are superior to conventional catalysts such as silica gel. The rearrangement is interesting in terms of the reaction mechanism, because an intramolecular hydride shift is involved. Accordingly, in the field of zeolite catalysis hydride ion mechanism must be counted alongside mechanisms based upon carbenium ions on acidic zeolites, carbanions on basic zeolites and radical mechanism.

4.5.5. Ring enlargement reactions

Several zeolite-catalyzed ring enlargement reactions exist which are basically of the Wagner-Meerwein type, i.e. protonation of the reactant is followed by a dehydration/migration step and subsequently the intermediate enlarged C^+-containing ring system is stabilized by a final reaction step.

In the following sections the formation of O- and N-containing heterocyclic rings by such a ring enlargement will be exemplified.

4.5.5.1. Conversion of tetrahydrofuran derivatives into dihydropyran systems

Zeolites (e.g. X) are reported (86) to be better catalysts than alumina for the rearrangement of tetrahydrofurfuryl alcohol at approximately 350 °C to give 2,3-dihydro-2H-pyran. 1,2,5-Pentanetriol can also be used as a starting material for the latter compound.

It is known that alkyl-substituted 1,2,5-pentanetriols can be converted in the presence of p-toluenesulphonic acid to 2-hydroxymethyltetrahydrofurans (95% yield), which compounds can, in turn, be transformed in a second step by a gas phase reaction on Al_2O_3 to 2,3-dihydropyrans (eqn. 11). The disadvantages are two reaction steps and the poor yield of the rearrangement (50%).

658

The dehydration of 1,2,5-pentanetriol on zeolite catalysts proceeds to a large extent directly to 2,3-dihydropyran. When the reaction is carried out on a boron zeolite (H-form) at 350 ^{o}C, 73% conversion and 70% selectivity are achieved. A by-product is the reaction intermediate 2-hydroxymethyltetrahydrofuran. If the catalyst is doped, e.g. with W (3.1% by wt) both activity and selectivity are increased, and 2,3-dihydropyran is obtained with a selectivity of 85% at 100% conversion (87). So by employing a bifunctional zeolite catalyst it is possible to bring about dehydration and subsequent rearrangement in a single step.

An example of ring enlargement of lactones is the zeolite-catalyzed conversion of dihydro-5-(hydroxymethyl)-2-furanones into 3,4-dihydro-2-pyrones (88), which are intermediates, inter alia, for insecticides.

4.5.5.2. The Beckmann rearrangement

The most important industrial example of the Beckmann rearrangement is the reaction of cyclohexanone oxime to ϵ-caprolactam (eqn. 12), which is the starting material for Nylon-6. The classical synthetic route involves the oximation of cyclohexanone with hydroxylamine-sulphate and the subsequent rearrangement of the oxime in concentrated sulfuric acid. Approximately 2 t $(NH_4)_2SO_4$ per t caprolactam are inevitably obtained as co-product. Further problems encountered include handling a large amount of fuming sulfuric acid and corrosion of the apparatus. In order to eliminate these problems, attempts have been made for many years to switch from a homogeneous to a heterogeneous process and many types of catalyst systems have been tested.

As long ago as the 1960's Venuto and Landis used zeolites for this purpose (1). X- and Y-zeolites as well as mordenite in the H-form or doped with rare earth or transition metals are employed; e.g. cyclohexanone oxime (30 wt % dissolved in benzene) is converted over HY at 380 °C and WHSV = 1.2 h^{-1} to ϵ-caprolactam with 76% selectivity and 85% conversion during the first two hours. The principal by-product is 5-cyanopent-1-ene. As the reaction is continued, the overall conversion decreased to about 30% after 20 hrs with a drop to 50% selectivity for caprolactam.

Generally the rapid catalyst aging and low selectivity are major drawbacks for the use of X- and Y-zeolites and mordenite. It is not possible to avoid these disadvantages by employing the strongly acidic, hydrophobic H-ZSM-5 with Si/Al = 156 (89); a 14% solution of cyclohexanone oxime in benzene at 350 °C, 1 atm and LHSV = 1.7 h^{-1} was nearly quantitatively converted for a period of 15 h. Afterwards the conversion drops rapidly to about 40% at 21 h on stream.

As to mechanistic considerations, until recent years a Brönsted acid catalyzed reaction was assumed: protonation of the OH group followed by concerted dehydration/migration and subsequently water addition. As shown by the selective phenyl migration in (E)-acetophenone oxime (90) the group trans with respect to the OH group migrates. This is also observed in the homogeneous Beckmann rearrangement.

As a result of recent patents and papers of Sato et al. (91-93) the present focus is on zeolite catalysts with reduced acidity particularly on the outer surface and upon other weakly acidic microporous materials.

Sato et al. showed for instance that a high Si/Al ratio and inertization of the outer surface - where the reaction was shown to take place - are beneficial in the case of H-ZSM-5 catalyst. Table 5 shows the effect of external surface silanation.

Table 5. Comparison of silanated with nonsilanated H-ZSM-5 in the reaction[a] of cyclohexanone oxime to ϵ-caprolactam (ref. 93).

Catalyst	Time on stream (h)	Conversion (%)	Selectivity (%)
silanated[b]	3.3	100	95.0
	31.0	98.2	95.0
nonsilanated	3.3	100	79.7
	27.0	95.8	89.4

[a] Reaction conditions: 8 wt % solution of oxime in benzene, 350 °C, WHSV = 11.7 h^{-1}, 1 atm, CO_2 as carrier gas, oxime/CO_2/benzene = 1/5.6/18.3 mol.

[b] H-ZSM-5 with Si/Al = 1600, treated with chlorotrimethylsilane at 350 °C for 4 h.

Persuing the idea of reducing the acidity of the zeolite in order to get high selectivity and long life time of the catalyst micro-porous materials such as the weakly acidic non-zeolitic molecular sieves, e.g. the medium pore sized SAPO-11 or SAPO-41 are used for the Beckmann rearrangement (94). Using SAPO-11, a 5 wt % solution of cyclohexanone oxime in acetonitrile is converted at 350 °C, atmospheric pressure and WHSV = 10.8 h^{-1} to ϵ-caprolactam with 95% selectivity and 98% conversion.

Further improvement in service life is required to constitute a favourable alternative to the homogeneous process now practiced.

4.5.6. Benzamine rearrangement

The synthesis of substituted pyridines by rearrangement of aminobenzenes in the presence of ammonia constitutes an interesting reaction. Thus aniline was converted (95) at 510 °C over H-ZSM-5 with 52% selectivity into α-picoline (2-methylpyridine). In the absence of NH_3 diphenylamine is the main product.

In a similar way 1,3-diaminobenzenes rearrange (96) to a mixture of 2- and 4-aminopyridines (eqn. 13, R = H).

A mixture of 1,3-diaminobenzene and NH_3 (molar ratio 1:60) is converted at 350 °C and 190 bar on H-ZSM-5 to 2-amino-6-methylpyridine with 83% selectivity at 43% conversion. A comparison with silica-alumina or alumina under the same reaction conditions shows the superior properties of the zeolite over other acid catalysts not possessing zeolite structure. This is a useful new route for aminopyridines, compounds which were hitherto only available by the complicated Tschitschibabin-reaction of sodium amide with pyridine.

The mechanism of these exciting reactions has not yet been elucidated, but two possible reaction routes have been advanced (95).

4.6. Isomerization of arenes

4.6.1. Skeletal isomerization of alkylsubstituted arenes

The conversion of substituted arenes into their isomers is of industrial interest and frequently serves as a model reaction for the shape selectivity of the zeolites (32). An important example is the isomerization of xylenes which is one of the processes already carried out on a large industrial scale using zeolites as catalysts. It forms the subject of numerous publications and is in operation, e.g. in the UK in a 250000 t/year plant. In this instance the shape selectivity of the zeolites is utilized to increase the proportion of p-xylene in C_8-arene fractions.

The mechanism of xylene interconversion is believed to involve proton addition followed by 1,2-methyl shift.

Polynuclear aromatic compounds can also be isomerized using zeolites. 1-Methylnaphthalene is converted (40%) into 2-methylnaphthalene with a selectivity of 95% using H-Na-Y-zeolite at 270 °C in a stream of hydrogen, but there is a loss of activity of ca. 50% in the course of 10 h (97). For a recent study of this isomerization over H-ZSM-5 and the large pore zeoites H-ZSM-12 and HY, see ref. (98).

4.6.2. Isomerization of arenes containing functional groups

The principle of the xylene isomerization is applicable to a number of toluene derivatives carrying functional groups such as hydroxyl, amino, nitrilo and halogen. As shown, e.g. by Weigert (99,100), zeolites of the ZSM-5 type are particularly suited for such equilibrations. The highly acidic H-ZSM-5 is able to protonate the aromatic nucleus, to form the intermediate carbenium ion, also when strongly deactivating groups as cyano are present. Thus a contact time of 3 s at 500 °C suffices (99) to establish the three component equilibrium of 46% ortho-, 34% meta- and 20% para-tolunitrile over H-ZSM-5.

A zirconium-MFI-containing zeolite is able to isomerize o- and/or p-toluidine into mixtures of the o-, m- and p-isomers and is superior to the H-ZSM-5 catalyst in terms of both product yield and catalyst lifetime (101). At 430 °C o-toluidine is converted on the zirconium zeolite to an o-, m-, p-mixture in a weight ratio of 37:45:15. Aniline can be applied as a diluent (100). Conditions are less severe than applied in the benzamine-to-picoline isomerization. Also 2-ethylaniline can be isomerized.

Toluidine is produced by the hydrogenation of nitrotoluene mixtures of approximate composition 63% o-, 4% m- and 33% p-isomers. It is, however, the m-isomer which is of greatest importance as an intermediate for dyestuffs and agrochemicals. The nitration-reduction process affords a toluidine mixture containing only 4% of the m-isomer. In a subsequent isomerization process on a pentasil zeolite a toluidine mixture rich in the m-isomer is obtained. The components can be separated and the o-/p-isomers are recycled.

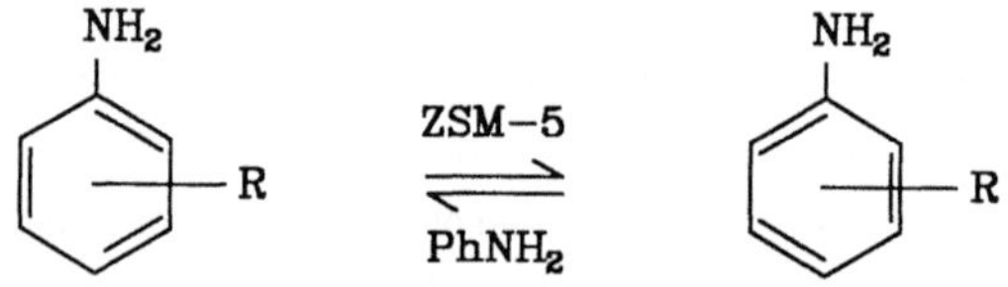

$$o,p \quad \xrightarrow[\text{PhNH}_2]{\text{ZSM-5}} \quad o,m,p \tag{14}$$

R= Me,Et

It may be noted that here the equilibrium mixture is obtained and not as with the xylenes a mixture in which the p-isomer dominates. Apparently diffusion rate differences do not prevail in the toluidine isomerization; factors are the higher reaction temperature and the fact that amines diffuse slower - presumably more jumpy due to protonation/deprotonation - through the H-ZSM-5 channels than hydrocarbons of the same size.

For a comparison of results on toluidine isomerization as well as a proposed reaction mechanism, see ref. (102).

Several, mainly patent publications, deal with isomerization (and separation) of chlorinated benzenes such as chloro- and dichlorotoluenes, dichloro- and trichlorobenzenes, over zeolites. Often integrated processes can be designed combining zeolite-catalyzed isomerization with separation over zeolites and including recycle of unwanted isomers.

The isomerization of the dichlorotoluenes may serve as an example. Upon direct chlorination 2,4- and 2,5-dichlorotoluene are the main products. The 2,6-, 2,3- and the 3,4-isomer are present in low amounts and the 3,5-isomer is absent in such mixtures. Isomerization covering all dichlorotoluenes (Fig. 9) is achieved over the large pore zeolites beta, omega and mordenite in the H-form at temperatures 300-350 $^{\circ}$C in hydrogen (103, 104). By doping the zeolite with Re, Ag or Ni a stable catalyst is obtained (105). Thus AgH-mordenite at 350 $^{\circ}$C remained completely stable over 150 h as measured by the amount of 2,6-dichlorotoluene formed (8.6%).

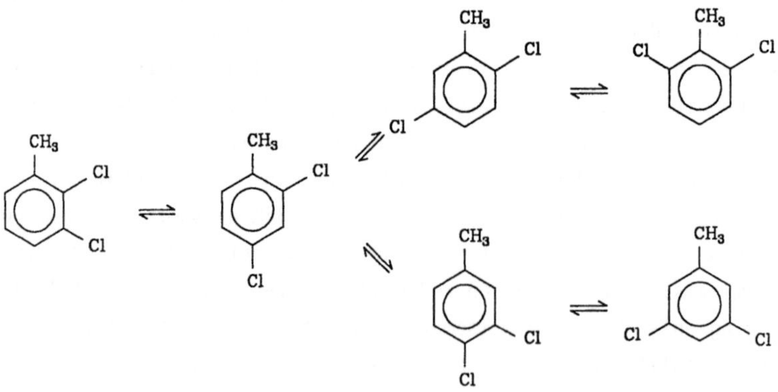

Fig. 9. Isomerization of dichlorotoluenes on a large pore zeolite.

Specific isomers can be separated from dichlorotoluene mixtures by adsorptive separation using a simulated moving bed, wherein a faujasite type zeolite is used as the adsorbent (106). The specific isomer is for example 2,6-dichloro-toluene which is the starting material for the herbicide 2,6-dichloro-benzonitrile.

When using a pentasil zeolite, Re-doped (105) or Zr-doped (106) the attainable equilibrium is limited to 2,4-, 2,5- and 3,4-dichlorotoluene (Fig. 10). The slightly larger 2,3-., 2,6- and 3,5-isomers are assumed to take part in the equilibrium at the crossings in the zeolite but are too bulky to diffuse through the channels and leave the zeolite crystal. A fine case of product selectivity.

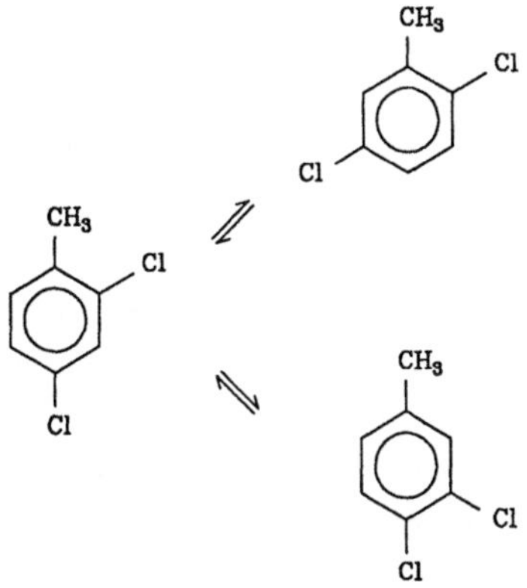

Fig. 10. Isomerization of dichlorotoluenes on a medium pore zeolite.

In accordance with the above equilibrium limitation the medium pore molecular sieve $AlPO_4$-11 is able to separate 2,6-dichlorotoluene from its 2,4- and 2,5-isomers; for the latter two compounds the sieve is accessible.

Finally it may be noted in this paragraph on chlorobenzenes that the classical isomerization catalyst, $AlCl_3$, (i) will give all six isomers, (ii) is corrosive, (iii) cannot be regenerated.

4.6.3. Isomerization of substituted heteroaromatics

Isomerization reactions of heteroaromatic compounds, e.g. alkylthiophenes and halogenothiophenes are also catalyzed by zeolites. Thus 2-methylthiophene in admixture with steam is converted at 330 °C on H-ZSM-5 to a product mixture containing 53% 2- and 42% 3-methylthiophene (107). 2-Chlorothiophene can be converted into 3-chlorothiophene (84%) with minor dechlorination (5%) taking place (reaction 15). The 2-chloro-isomer is formed preferentially in the chlorination of thiophene. Taking into account the easy separation of 2- and

3-chlorothiophene zeolite-catalyzed isomerization provides a good route to the 3-isomer. For a review on isomerization of halothiophenes, see ref. (108).

$$\text{(structure)} \rightleftharpoons \text{(structure)} \quad \left(+ \quad \text{(structure)} \right) \qquad (15)$$

4.6.4. The Fries rearrangement

The Fries rearrangement is the isomerization of phenyl esters towards o- and p-acylphenols, and is usually performed using a relatively large amount of $AlCl_3$ as a catalyst. The obvious interest in the development of a heterogeneously catalyzed process has resulted in studies on the use of cation exchangers in the H-form (Amberlite, Nafion) clay catalysts and fluorided alumina.

Two recent studies (109, 110) on the use of zeolites in the Fries reaction of phenyl acetate towards o- and p-hydroxyacetophenone do not reveal high activity and selectivity for the zeolite catalysts (cf. also Section 5.5, aromatic acylation). The addition of water improved lifetime and selectivity in the case of H-ZSM-5 (110a). This was not observed when using HY as the catalyst.

5. ELECTROPHILIC SUBSTITUTION OF ARENES

5.1. Alkylation of arenes

The liquid-phase Friedel-Crafts alkylation of arenes using Lewis-acid catalysts ($AlCl_3$, $FeCl_3$ or BF_3) is carried out on a large industrial scale, e.g. for the preparation of ethylbenzene. The use of such catalysts entails problems of corrosivity, toxicity, work-up and effluent pollution. This makes it desirable to replace homogeneous catalysis for electrophilic substitution reactions by heterogeneous catalysis. Furthermore, the gase phase also offers advantages in the field of process technology, such as utilization of the reaction heat.

The gas phase alkylation of aromatic nuclei in the presence of shape selective, acidic zeolite catalysts, the advantages of such catalysts compared with classical Friedel-Crafts reactions, the para-substitution of monoalkylated aromatics as well as the technical realization of ethylbenzene production via the Mobil-Badger process have been described extensively (e.g. 2, 111-114).

The Mobil-Badger process is an outstanding and technically proven example of the alkylation of aromatic compounds, in the USA ethylbenzene has been produced since 1980 in a 500000 t/year plant from ethylene and benzene over phosphorous-doped ZSM-5-zeolite. Ethene is completely converted to ethyl benzene, with 99% selectivity, at about 400 $^\circ$C, under approximately 20 bar, at a WHSV of 3 h^{-1} (based on ethene) and benzene/ethene mole ratio of 6 to 7. Benzene conversion is about 20% (115).

The pentasil zeolite is shape selective; the channels allow the transport of benzene, ethylbenzene and diethylbenzenes but prevent the emerging of polyalkylated products. Diethylbenzene, which is undesirable, can be converted to ethylbenzene with benzene over the same catalyst.

In respect of process technology, energy savings and environmental protection, the advantages of this heterogeneous catalyst over the conventional homogeneous catalyst are:
- alkylation at 400 $^\circ$C permits as much as 95% of the heat to be recovered
- the catalyst is non-corrosive and need not be separated off or processed
- there are no disposal problems as in the case of AlCl$_3$ hydrolysis
- and deactivated catalysts can be easily regenerated.

These advantages often also apply in the case of other electrophilic substitution reactions.

Whereas ethylation ideally takes place in the gas phase, in many alkylation reactions with longer-chain alkenes it is necessary to revert to the liquid phase; as the chain length increases, cracking reactions must be expected in the gas phase. It is then preferable to use the large-pore ZSM-12-, X- and Y-zeolites. Successful use of zeolites in the alkylation of toluene has been reported for alkenes as large as 1-hexadecene (116).

The mechanism of alkylation on H-zeolites is assumed to involve formation of carbenium ions (by addition of H$^+$ to alkenes, or by protonation and dehydration of alcohols) which attack the aromatic nucleus as depicted in reaction (16) for ethylation.

$$
\underset{\text{ZeOH}}{H_2C=CH_2} \;\rightleftharpoons\; \underset{\text{ZeO}^-}{\overset{+}{H_3CCH_2}} \;\overset{\text{ArH}}{\rightleftharpoons}\; \underset{\text{ZeO}^-}{\overset{\overset{\displaystyle H}{+/}}{Ar-CH_2CH_3}} \;\rightleftharpoons\; \underset{\text{ZeOH}}{ArCH_2CH_3} \tag{16}
$$

When subjecting a substituted aromatic system ArX to zeolite-catalyzed alkylation the initial regioselectivity will be determined by the relative stabilities of the 3 transition states ArX(R)$^+$ leading to the o-, m- and p-isomers. It will be clear that the zeolite micro domain may add electronic (adsorption) and steric factors to such factors inherently pertaining to the ArXR$^+$ system. Taking into account the small differences in the inherent stability of o- and p-ArXR for X = alkyl, OH, halogen and other o,p-directing X, such effects imposed by the zeolite may have a substantial influence on the o/p ratio.

When the reaction conditions are such (e.g. high temperature) that isomerization occurs, differences in diffusion rates may play a role and e.g. crystal size may become a factor. Summarized:

```
                                    o  Inherent electronic
                                       + steric factors
                                                            Transition state
 Ar X      ortho    X               o  Factors imposed by   selectivity
                   /                   zeolite micro-domain
  +    ←── meta  Ar ⊕
                   \                                         Product size
 R ⊕       para     R               o  Under isomerization   selectivity
                                       conditions
                                       ArXR diffusion rates
                                       crystal size effects
```

As to consecutive reactions in aromatic alkylation such compounds may - depending upon their size and the pore size of the zeolite used - show up in the product mixture or just exist in the larger cages and crossings and take part in equilibria. For a convincing direct NMR observation of the latter phenomenon in aromatic methylation, see ref. (117).

As to regioselective alkylation the focus has been on selective p-alkylation. In a comparative study (118) of the zeolites H-Y, H-mordenite and H-ZSM-5 in toluene alkylation - under assumed equilibrium conditions - just H-ZSM-5 showed pronounced p-selectivity, especially when large crystals were used with homogeneous Al-distribution.

This important feature of the pentasil zeolite may be further improved by modification, e.g. by adding Mg- and/or P-compounds. The part played by doping in increasing p-selectivity is discussed comprehensively in ref. 119-121.

Several papers and patents deal with the alkylation - with methanol or lower olefins - of oligo-aromatics. Gas phase as well as liquid phase techniques have been applied.

Thus the industrially important methylation of naphthalene has been carried out (122-127) over H-Y, H-mordenite and H-ZSM-5. Mixtures of 1- and 2-methylnaphthalene are obtained with H-ZSM-5 exhibiting the highest selectivity towards the 2-isomer (and in the consecutive reaction towards 2,6-dimethylnaphthalene).

These results raised discussions on *outer surface catalysis*. Thus Fraenkel et al. (125) assume that methylated naphthalenes do not fit into the ZSM-5 pore system and that accordingly the shape selective methylation of (Me) naphthalene over H-ZSM-5 occurs in external Brönsted acid sites. These external surface sites are located in "half" channel intersection cavities which can accommodate well the mono- and dimethylnaphthalene products. Derouane (128) supported this idea while introducing the general term "molecular nesting".

Weitkamp et al. (126) showed that the selective methylation can be explained by conventional considerations. These authors measured the adsorption of the methylnaphthalenes onto ZSM-5; at 100 °C only the 2-isomer is adsorbed and in a quantity (6.5 wt %) well exceeding the amount calculated for outer surface adsorption. It is assumed that at the reaction temperature, 400 °C, also 1-methylnaphthalene can enter the pores of ZSM-5.

at 100 °C - adsorbed   not adsorbed onto ZSM-5

Another example in this area is the selective p,p'-isopropylation of biphenyl (eqn. 17) over modified mordenites in liquid phase batch experiments. Reported selectivities towards 4,4'-diisopropylbiphenyl are 84% (at 69% conversion) over fluorided mordenite at 150 °C and 20 bar (129) and 74% (at 98% conversion) over highly dealuminated mordenite at 250 °C (130).

(17)

5.2. Alkylation of phenols and aromatic amines

Arenes containing functional gorups such as a hydroxyl or amino group can also be alkylated on zeolitic catalysts. These conversions, however, are more complex than the alkylation of alkylbenzenes, because attack at the nucleus as well as at the functional group can take place, i.e. in the case of phenol not only carbon (C-) but also oxygen (O-) alkylation is possible. Reviews on this topic have been written in which the mechanisms are discussed (131, 13). Often such reactions are carried out in the vapour phase at atmospheric pressure and at temperatures between 200 and 350 ^{0}C using pentasil and faujasite-type zeolites. A broad spectrum of products, consisting essentially of cresols, xylenols, anisoles, methylanisoles and diphenyl ether, is obtained when phenol is methylated using zeolites (eqn. 18).

$$
\begin{array}{c}
\text{OH} \\
\bigcirc \quad + \quad H_3\text{COH} \quad \longrightarrow \quad
\begin{array}{l}
\text{Anisole} \\
\text{o,p,m–Cresols} \\
\text{Methyl anisoles} \\
\text{Xylenols} \\
\text{Diphenyl ether}
\end{array}
\end{array}
\qquad (18)
$$

Both Brönsted and Lewis acid catalysis can play a role in phenol alkylation; Brönsted sites will activate the alkylating agent by protonation, Lewis acid sites, e.g. Al^{III}, will activate both phenol by coordination/deprotonation *and* the alkylating reactant by coordination. Obviously when phenol and alkylating molecule coordinate to one and the same Lewis site only O-alkylation or ortho-alkylation can (initially) occur. This is indeed observed in the methylation of phenol over alumina at 260 ^{0}C (cf. ref. (114)).

For recent work on the methylation of phenol over US-Y and a discussion on the sequence of reactions, see ref. 132 and 133.

Generally kinetic control is observed in the zeolite-catalyzed alkylation of phenol, yielding o- and p- (and 2,4- and 2,6-) isomers whereas meta-isomers, which can be accommodated by the zeolites tested, and which are the thermodynamically most stable isomers, virtually do not show up. Apparently more severe conditions are required for the o,p-to-m-isomerization.

When alkylating phenol with small alkylating agents generally the o-isomers predominate (114). The o/p-picture changes upon applying more bulky alkylating molecules. A more bulky transition state is required and this may result in zeolites with increased shape selectivity for p-products. Examples include phenol t-butylation over NaH-Y (134) and alkylation with 1-octanol (135) over RE-Y, H-mordenite and H-ZSM-12.

In the latter case an additional selectivity plays a role: the initially formed primary carbenium ion 1-C_8^+ will rapidly isomerize to the secondary 2-C_8^+ which can undergo further H-shifts towards 3-C_8^+ and 4-C_8^+ before alkylating phenol. Over RE-Y, as well as on Amberlyst-15, all three octyl structures are formed whereas with the less spacious mordenite and ZSM-12 just the 2- and 3-octyl systems are observed (see Fig. 11).

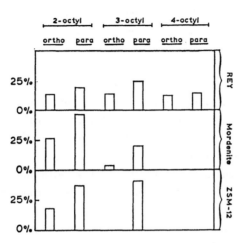

Fig. 11. Selectivity and p/(p+o)-ratio in the alkylation of phenol with 1-octanol (135).

The alkylation of aromatic amines over zeolites includes the reaction of aniline with methanol and with olefins. In principle, reaction can take place at the N-containing group forming N-alkylated compounds or at the nucleus forming C-alkylated compounds. The methylation of aniline, for example, yields toluidines, N-methylaniline and N,N-dimethylaniline. All are useful intermediates for dyestuffs, agrochemicals and drugs as well as for the organic synthesis.

As to the aniline/methanol reaction it should be mentioned that alumina and MgO are highly selective catalysts in providing N,N-dimethylaniline and N-methylaniline, respectively. Zeolites could play an important role in C-alkylation.

Upon methylating aniline over zeolites H-Y and H-ZSM-5 generally mixtures of N- and C-methylated products are observed (see e.g. ref. 136). Ione et al. demonstrated (137) that the product composition obtained over ZSM-5 type catalysts is strongly affected by the reaction temperature. With increasing temperature from 275-450 °C the content of N-alkylated products decreases from 100 to 0.3% and the C-methylated products increase.

Within the C-alkylate the regioselectivity depends on temperature, WHSV and on zeolite chemical composition. Using H-ZSM-5 the p-isomer of the toluidines is favoured in the temperature range between 275 and 350 °C and its content being about 80%. Under these conditions no m-toluidine is found. Below 275 °C the o-isomer and above 425 °C the m-isomer is favoured. At 285 °C the o/p-ratio increases with increasing WHSV indicating the p-isomer to be a secondary product. At 450 °C the thermodynamic equilibrium of the isomers is formed on H-ZSM-5. By modification (cf. Section 2.2) of the zeolite diffusional effects become operational and the p-isomer predominates again.

Burgoyne et al. have studied large pore H-zeolites, amorphous silica-alumina and alumina as catalysts in the alkylation of anilines with propene and isobutene (138). Reactions were performed either cofeeding the reactants over the solid catalyst in a fixed-bed reactor or by reacting the reagents and catalyst in a stirred autoclave.

The temperature is again an important variable: with increasing temperature N-alkylated, o-alkylated and p-alkylated anilines can become the major product when alkylating aniline with isobutene.

The catalyst activity order observed is H-Y > silica-alumina > alumina. Also dealuminated H-mordenite (139) and SAPO-37 (140) proved active catalysts.

Under proper reaction conditions high yields of o-alkylated products may be obtained. The proposed mechanism (138) involves concerted cycloaddition of alkene to protonated arylamine.

In the prime commercial process for arylamine ethylation homogeneous aluminum anilide catalysis is applied. This process is highly selective for ortho-ethylation but the rate falls off sharply with higher alkenes.

The t-butylation of toluene-2,6-diamine over H-Y shows shape-selectivity as it leads (138) essentially to the mono-t-butyl compound whereas on silica-alumina di-t-butylation is observed (reaction 19).

Silica—alumina	conv. 57%	76	:	15
Zeolite H—Y	conv. 70%	82	:	2

Another interesting reaction (138) is the conversion (at 120 °C) of aniline and butadiene over H-Y (reaction 20). The different attachment of the butenyl group at the 2- and 4-position is noteworthy.

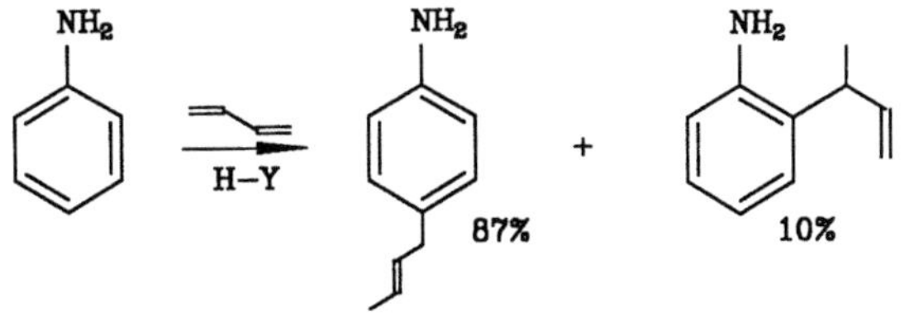

(20)

5.3. Alkylation of heteroarenes

Only a few references are dealing with zeolite-catalyzed alkylation of heteroaromatics perhaps because these systems are apparently intrinsically more complex.

Thiophene can be alkylated with methanol over H-ZSM-5 yielding, at 450 °C, a product containing 10% of 2-methylthiophene, 10% of 3-methylthiophene and 9% of dimethylthiophenes (141, 142).

The alkylation of pyridine over faujasite type catalysts (143) with methanol takes place primary at the aromatic nucleus. In subsequent reactions, cf. eqn. 21, the picolines formed can either undergo further ring alkylation or side chain alkylation to yield ethyl- or vinylpyridines. When HY, LiY, SrY or BaY are employed, ring methylation occurs almost exclusively to give picolines and lutidines (Table 6), with the alkaline earth doped Y-zeolites leading to higher yields than HY and LiY.

When the two latter zeolites are applied the β-position (3-position) is preferentially attacked whereas the use of Sr- and Ba-Y leads to preferential α- and γ-methylation. Isomers of lutidine (dimethylpyridine) are formed as by-products.

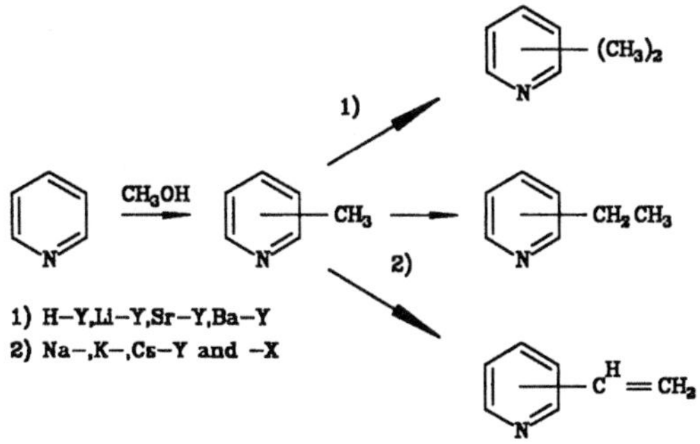

1) H-Y,Li-Y,Sr-Y,Ba-Y
2) Na-,K-,Cs-Y and -X

(21)

Table 6. Product spectrum of the reaction of pyridine and methanol.

Catalysts	H-Y	Li-Y	Sr-Y	Ba-Y	Na-Y[a]	Cs-Y[a]
conversion (%)	32	31	62	63	65	82
yield (%):						
pyridines	18	13	31	35	13	15
lutidines	6	6	22	22	3	2
ethylpyridines	1	2	/	/	13	27
vinylpyridines	/	/	/	/	5	7

Reaction conditions: 400 $^{\circ}$C, LHSV = 1.3 h^{-1}, molar ratio methanol/pyridine = 8.
[a] 450 $^{\circ}$C.

On the other hand, when pyridine is treated with methanol in the presence of X- and Y-zeolites ion-exchanged with alkali metals (not Li), side chain alkylation prevails as a consecutive reaction (Table 6). The principal products are ethylpyridine and vinylpyridine as well as isomers of picoline and lutidine.

5.4. Acylation of arenes

Present industrial acylation of alkylaromatics generally involves acid chlorides as the acylating reactants together with stoichiometric amounts of metal chloride ($AlCl_3$, $TiCl_4$, $FeCl_3$) "catalysts". In this way a substantial waste problem arises. Other acylating agents include acid anhydrides which exhibit similar disadvantages due to a high catalyst demand. Moreover, the combination metal chloride catalyst and hydrochloric acid formed in the reaction gives rise during work-up to highly corrosive media.

For the above reasons a direct and clean route involving the free acids as acylating agents would be most attractive. Zeolite catalysts show promise in this respect.

The direct acylation of toluene and p-xylene by straight chain aliphatic acids (C_2-C_{22}) using zeolite NaCe-Y (70% exchange) as the catalyst was reported (144) by Chiche et al. A stirred batch autoclave is used, and an excess of aromatic hydrocarbon is applied.

$$\text{toluene} + \text{RCOOH} \xrightarrow[150\,^{\circ}\text{C}]{\text{NaCe-Y}} \text{p-acyltoluene} + H_2O \qquad (22)$$

The acylation of toluene proceeds with increasing yield as the chain length of the acylating agent increases, and reaches a maximum of 96% with dodecanoic acid. Perhaps here the best balance of adsorption exists. It is to be expected - in view of the hydrophilic nature of the zeolite used - that the water formed will remain mainly in the zeolite pores and deactivate the acidic sites. This will limit the turnover number.

Also acid clays, particularly Al^{III}-exchanged montmorillonite, were found (145) to catalyze the direct acylation of toluene.

As to the regioselectivity of the toluene acylation in all NaCeY-catalyzed reactions more than 93% of the p-isomer is obtained. This is only partly due to the shape selectivity of the zeolite since acylation on the Al-montmorillonite or with the homogeneous catalyst $AlCl_3$ also leads to a preponderance of the p-isomer (see Table 7). However, the zeolite is superior in this respect.

Table 7. Comparison of regioselectivity among clay, zeolite and $AlCl_3$ in the acylation of toluene at 130 $^{\circ}$C (ref. 145).

Acylating agent	Isomeric composition (%) over NaCe-Y			Al-mont.			$AlCl_3$		
	o	m	p	o	m	p	o	m	p
octanoic acid	3	3	94	5	9	86	4	16	80
dodecanoic acid	3	3	94	5	11	84	4	15	81
palmitic acid	0.5	1	98.5	5	10	85	4	12	84

The relatively low reactivity of short chain carboxylic acids in the liquid phase over NaCe-Y is in line with a report (146) on the gas phase acylation of toluene and ethylbenzene by acetic acid over H-ZSM-5. At 250 $^\circ$C low conversions - with high regioselectivity - are observed.

In the acylation of toluene with octanoic acid in the presence of various cation-exchanged Y-type zeolites. Geneste et al. found (147) that the most efficient catalysts are the rare earth-exchanged zeolites. The following order of activity is observed for 70% exchanged NaY zeolites:

Cr^{III}, Zr^{IV} < Hg^{II}, Cu^{II}, Co^{II} << H^+ << Pr^{III}, La^{III}, Yb^{III}, Ce^{III}.

The rare earth-exchanged Y-zeolites are twice as active as H-Y. The activity of transition metal and alkaline earth exchanged Y-zeolites is only 2-5% of that of RE-Y.

As to the mechanism, acylium ions are expected to be the intermediates. So, Brönsted acidity and not the multivalent cation itself is supposed to exert the catalytic action.

The French group (148) also noticed a Hammett relationship in the acylation of substituted benzenes with octanoic acid over NaCe-Y which shows that the substituents govern the relative activity in the same way as in homogeneous solution.

Some patents deal with zeolite-catalyzed synthesis of anthraquinone, which compound is an important intermediate. Thus NaCe-Y is used (149) at 550 $^\circ$C to convert benzene and phthalic anhydride (molar ratio 25:1) towards anthraquinone (reaction 23). Using CO_2 as the carrier gas a conversion of 65% together with a selectivity of 92% is mentioned.

(23)

In another reported route (150) phthalic anhydride is applied as the sole reactant; thus decarboxylation and acylation are combined here. Temperature is 400-420 $^\circ$C and CO_2 is the carrier gas. Among several silicates tested the Zn-containing systems - $ZnCO_3$-on-silica and zeolite NaZnX - show the highest selectivity (up to 96%).

5.5. Acylation of phenol and phenol derivatives

Upon reacting phenol or a substituted phenol and a carboxylic acid in the proper temperature range over an acid catalyst ester formation followed by Fries rearrangement is to be expected (reaction 24). A set of reactions exist - in principle as equilibria - but catalyst and temperature determine (151) which part of the network is in operation (cf. 13).

$$(24)$$

Thus relatively high proportions of the ortho-isomer are observed in the direct acetylation of phenol over ZSM-5 type catalysts. On a silicalite (Si/Al = 480) at 300 °C with a residence time of 8 sec., phenol reacts with acetic acid to yield 2-hydroxyacetophenone with 47% selectivity at 67% phenol conversion (152). The para-isomer is only formed to an extent of 2% and 1% of 4-acetylphenyl acetate is present. The remainder of the reaction product consists largely of phenyl acetate (32%). In fact, on H-ZSM-5 (Si/Al = 24) at approximately 250 °C the ortho-isomer is formed with almost 99% selectivity at 15% phenol conversion (153).

The ortho-selectivity might stem from an intramolecular rearrangement towards the ortho-isomer with stabilization by chelation (Al), followed by a slow intermolecular ortho-para rearrangement. The results of direct acylation of phenol seem to be at variance with the results of the Fries rearrangement starting with PhOAc (cf. Section 4.6.4).

It may be noted that also sulfonic acid resins, e.g. Amberlyst 15, appear (151, 154) to be good catalysts for the direct acylation of phenols. With the present state of the art it cannot be decided beforehand which catalyst type will be best suited for a given phenol acylation.

A recent study (155) deals with the zeolite-catalyzed acylation of anisole (methoxybenzene) with phenylacetic and 3-phenylpropionic acid and with the two corresponding acid chlorides. The zeolites tested are a series of H(Na)-Y with 21, 50 and 100% exchange and Si/Al = 9-24, a zeolite H-beta (Si/Al = 14) and a H-ZSM-5 (Si/Al = 40). A slurry technique is applied using CCl_4 or toluene as the solvent at 78 and 110 °C, respectively.

The relatively high Si/Al ratio of the zeolites applied is supposed to have the advantage that reaction water will not strongly adsorb in the catalyst. On the other hand the system chosen by Corma et al. exhibits some complicating features such as residual water in the fresh catalyst and the possibility of intramolecular reaction leading to indanone as a major side-product in the case of phenylpropionic acid and its acid chloride.

Some further observations from this work are: (i) the activity order is H-Y, H-beta > H-ZSM-5, which might be due to limited accessibility of the latter zeolite at the mild temperatures applied, (ii) demethylation of anisole is absent, (iii) selective p-acylation takes place.

5.6. Acylation of heteroarenes

Almost all the known processes for the acylation of heteroaromatics such as thiophene, furan and pyrrole involve homogeneous Lewis acid catalysts. Recently it was reported (156) that it is possible to perform this acylation in the gas phase on zeolite catalysts with high selectivity according to eqn. 25.

$$X = O,S,NH \qquad (25)$$

The reaction of thiophene with acetic anhydride at 250 °C on a boron ZSM-5 zeolite leads to 2-acetylthiophene with 99% selectivity at 24% conversion. On a Ce-doped boron-ZSM-5 at 200 °C 2-acetylfuran is obtained with 99% selectivity and 23% conversion. On the other hand, in the case of pyrrole, which tends to polymerise, the acidity of the catalyst and also the temperature must be reduced in order to achieve a high selectivity. A boron zeolite doped with 0.2% by wt. Cs yields 2-acetylpyrrole with 98% selectivity and 41% conversion at 150 °C. As in the homogeneously catalysed reactions, the acylation on zeolites proceeds largely at the 2-position of the heteroaromatics so sterical differences, if any, between substitution at the 2- and 3-position are overruled by the electronic effect.

Direct C-acylations of imidazoles and pyrazoles by means of Friedel-Crafts reactions were unknown, and it was predicted that this reaction is not possible at all; one was therefore forced to adopt more complicated processes. Surprisingly, with the aid of zeolite catalysts at high temperature it has now become possible to carry out direct acylation (157). Thus upon passing a mixture of 2-methylimidazole and acetic acid or acetic anhydride over a boron-ZSM-5 zeolite at 400 $^{\circ}$C, a conversion of 63% and a selectivity of 85% for 2-methyl-4-acetylimidazole is observed.

In conclusion the use of zeolites in aromatic acylation seems to provide several interesting procedures with potential application. Noteworthy is the environmental friendliness of the new methods compared with the existing ones.

5.7. Aromatic nitration

Arenes, halogenoarenes and alkoxyarenes are nitrated with NO_2 and related N-oxides in the gas phase using zeolite catalysts. Particularly H-mordenite and H-ZSM-5 have been explored for this purpose.

For instance, using N_2O_4 as the nitrating agent for benzene at 200 $^{\circ}$C, nitrobenzene is formed on H-ZSM-5 with a selectivity of 98% at a conversion of 64% (158). Another patent (159) reports 93% selectivity upon nitrating benzene with NO_2 over H-mordenite at 150 $^{\circ}$C. It has been recommended to perform high temperature nitrations in the presence of some water and of oxygen to counteract NO_2 dissociation towards NO.

Toluene is nitrated at about 160 $^{\circ}$C over H-mordenite with a selectivity of 78% towards mononitrotoluenes at a conversion of 30%. The regioselectivity found does not differ much from that obtained in conventional liquid phase nitration.

Zeolite-catalyzed nitration of chlorobenzene has been carried out using NO_2 in the gas phase (159) as well as with aqueous HNO_3 in the liquid phase over the organophilic silicalite (160). Para/ortho ratios of 4-5 together with relatively low conversions are observed.

Present-day technical nitration of benzene, toluene and chlorobenzene involves reaction of the aromatic in the liquid phase with mixtures of HNO_3-H_2SO_4-H_2O. The nitration can be carried out batchwise or continuously. Zeolite-catalyzed gas phase nitration offers an advantageous alternative if the conversion can be increased while retaining a high degree of selectivity.

Finally the reader is referred to Section 3 for the use of zeolites in small scale nitrations using acyl nitrate reagents (47).

5.8. Aromatic halogenation

Chloro- and bromo-substituted arenes are prepared industrially in the presence of Lewis acids, especially $FeHal_3$. The use of zeolite catalysts in the gas or liquid phase could reduce corrosion and disposal problems. Systems carrying an activating substituent such as OH, OR, NH_2, do not require a catalyst. So reports on zeolite-catalyzed halogenation of such compounds, e.g. anisole, should be considered with caution.

The abilities of zeolites in - high temperature - isomerization and in separation of halobenzenes have already been mentioned in Section 4.6.2.

5.8.1. Chlorination of arenes

Upon chlorinating benzene in the gas phase at 175 °C over various zeolites (molar ratio $Cl_2:C_6H_6$ = 5) the composition of the reaction product was found (162) to depend strongly on the nature of the zeolite catalyst (see Fig. 12).

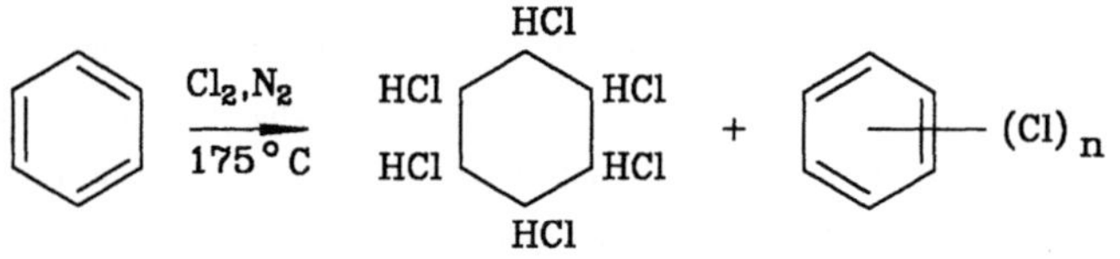

	Benzene conv. %	Add:Subst
Silicalite-1	97	96:4
NaH-ZSM-5	98	90:10
H-Mordenite	91	84:16
NaH-ZSM-11	86	72:28
KL	86	11:89
RE-Y	100	3:97
LA-LPV (13% alum.)	94	7:93

Fig. 12. Chlorination of benzene.

On the high silica zeolites *addition* and not substitution (Fig. 12) prevails; apparently a radical reaction is induced which - in view of the dimensions of the hexachlorocyclohexanes formed (0.8 nm) - mainly takes place at the outer surface and/or in the pore mouth openings. Recently free radicals were detected (ESR) upon adsorption of o-dichlorobenzene on zeolite H-beta (163).

Over Y and L zeolites the gas phase benzene chlorination mainly leads to substitution with an ionic mechanism involved.

Zeolites of the X-, Y- and L-type have been employed, particularly by Japanese workers, in the liquid phase chlorination of toluene (164, 165) of halobenzenes (166-168), and of halotoluene (169). Mild temperatures are applied and often p/o-ratios are somewhat higher than obtained with conventional $FeCl_3$-catalysis. At 100 °C in the liquid phase over Na-ZSM-5 chlorination proceeds (170) to the dichloro stage whereas over NaY also tri- and tetrachlorobenzenes are formed.

Further improvement of p-selectivity, e.g. in p-dichlorobenzene synthesis, has been explored by impregnation with salts (170) and/or by adding quaternary ammonium compounds (171) or lower aliphatic acids (172) to the liquid phase. The latter type of addition (e.g. chloroacetic acid) also improves (173) the 4,4'-selectivity in the chlorination of biphenyl over KL. The background of this effect is as yet unclear. A complicating factor is the fact that the acid additive can also exert a catalytic action in the halogenation.

Anisole chlorination proceeds uncatalyzed towards 2,4-dichloroanisole in CCl_4 at 25 °C. Upon adding NaCeY-62 as a catalyst subsequent chlorination towards 2,4,6-trichloro- and 2,3,4,6-tetrachloroanisole takes place (174).

Similarly a third chloro-substituent can be introduced in phenyl acetate (reaction 26).

$$(26)$$

The 2,4,5-trichlorophenyl acetate is found (174) to prevail whereas with $FeCl_3$-catalysis only the 2,4,6-isomer is observed.

Apparently shape selectivity is involved. It may be noted that the o,p-directing power of the $OCOCH_3$-group is less than that of OH or OCH_3. Furthermore competitive adsorption experiments showed 2,4,5-trichlorophenyl acetate to be stronger adsorbed onto a Y-zeolite than the 2,4,6-isomer; the better accommodation of the 2,4,5-isomer might be perceptible in the transition state of chlorination.

Finally, the ad- and desorption of 2,4,6-trichloro-anisole and -phenyl acetate in/from zeolite Y is noteworthy in view of the diameter of these compounds which is almost 0.9 nm. This underlines the margin in accessibility mentioned in Section 2.1.

As to scavenging of the HCl generated in the chlorination and probably damaging the zeolite, see the next section.

5.8.2. Bromination of arenes

A paper of one of the authors et al. (175) gives details on the liquid phase bromination of halobenzenes over various Y-zeolites. Ref. (175a) contains a study on the zeolite Y catalyzed bromination of toluene.

Using the bromination of bromobenzene in CCl_4 at 25 °C towards o- and p-dibromobenzene as a test reaction the following order of activity is found:

$$KY < NaY << CaX < RE-Y, CeY < HY, NaMgY, NaCaY$$

Activity as well as p/o selectivity are found to increase with increasing degree of cation exchange of NaY. Regioselectivity obtained in bromination of bromo-, chloro- and fluorobenzene is substantially higher than observed with $FeBr_3$-catalysis.

From adsorption isotherms (CCl_4, 25 °C) the following order of adsorption strength onto NaCaY-82 can be deduced (175):

$$HBr > o-diBr > p-diBr, \text{ mono-Br benzene} > Br_2$$

Hydrogen bromide, generated during the reaction is held responsible for the decay of the catalyst which manifests itself in reduced p/o-selectivity in the course of an experiment and upon re-use of a zeolite Y catalyst.

Attempted remedies by scavenging HBr include the addition of $NaHCO_3$ and KA (175), of KA (174) and of propylene oxide (176).

With the latter method a high (98:2) initial p/o-ratio in the bromination of toluene over NaY has been reported (176). The catalyst is deactivated completely however, after some 10% conversion.

The order of reactivity found in zeolite-Y-catalyzed liquid phase bromination is

$$\text{toluene} > \text{benzene} > \text{fluoro-} > \text{chloro-} > \text{bromobenzene}$$

which suggests an electrophilic attacking species. A proposed mechanism (Fig. 13) involves activation of bromine - which is already in interaction with the aromatic nucleus - by zeolitic Brönsted sites. Also bromine activation by cations, e.g. Na^I, may be envisaged, as evidenced by Raman spectroscopy (176a).

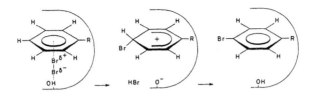

Fig. 13. Mechanism of zeolite-catalyzed bromination.

Reported p-dibromobenzene syntheses over zeolites at higher temperature include the bromination (177) of bromobenzene or benzene over Li-faujasite at 200 °C (p-selectivity 92%) and the oxidative bromination (178) using Br_2 and/or HBr/O_2 at 170 °C over CuY.

For a comparison of various reagents in the presence of solids in chlorination and bromination of aromatics we refer to the work of K. Smith (47, 178a).

5.8.3. Aromatic iodination

Direct iodination requires - due to the reductive power of HI - oxidative conditons. Zeolite-catalyzed oxyiodination of benzene (179), naphthalene (180) and several substituted arenes (181) has been described. Iodine, air and the aromatic substrate are fed in the gas phase at atmospheric pressure over the catalyst. Non-acidic faujasite, e.g. KX, sometimes impregnated with K-salts, and divalent metal exchanged pentasil zeolites, e.g. Ca-ZSM-5, are recommended as catalysts. Good conversions and high selectivities are reported.

5.9. Hydroxyalkylation of arenes

Hydroxyalkylation of aromatics is known to be a difficult reaction because the consecutive reaction towards diarylmethane systems usually is fast. Thus reaction of aniline with formaldehyde is reported (182) to yield bis(4-aminophenyl)methane with a selectivity of 85% at 40% conversion over Y-type zeolites. Similarly the reaction of toluene and trioxane over several molecular sieves, e.g. SAPO-11, of the AlPO family (183) leads to bis(4-methylphenyl)methane.

Two patents (184) claim successful hydroxymethylation and hydroxyethylation of phenol by reacting phenol at 150 °C with the cyclic aldehyde trimers trioxane and paraldehyde, respectively, over a H-Y type catalyst in the liquid phase. Remarkably high selectivities (up to 88%) for the p-hydroxy(m)ethyl-phenol are mentioned. Problems may arise due to consecutive reactions.

In another interesting approach (185) the aldehyde is reacted in the form of an acetal. Thus acetaldehyde dimethyl acetal and phenol (molar 1:10) react over H-ZSM-5 at 250 °C in the gas phase to give 2-(1-methoxyethyl) phenol in 59% selectivity (reaction 27).

(27)

6. OXIDATION REACTIONS

In the field of oxidation catalysis, zeolites also serve as supports for active components such as Pd, Cu, V, Fe, for performing oxidation and ammoxidation reactions in the presence of elementary oxygen, and influence the product spectrum by virtue of their shape selectivity.

The direct application of zeolites, especially Ti-containing materials, as catalysts for oxidation reactions using H_2O_2 as the oxygen source has been investigated intensively during the last seven years. However, only little is known about transoxidation reactions using organic peroxides such as t-butyl hydroperoxide in the presence of zeolites.

6.1. Oxidation reactions with oxygen

The selective partial oxidation of C_4 to C_{10}-hydrocarbons to maleic anhydride offers a commercially interesting alternative to the conventional processes by the oxidation of benzene, 2-butene or crotonaldehyde. Maleic anhydride is obtained with a selectivity of 80% at a conversion of 17% from n-butane at 525 °C using zeolites doped with V, P or alternatively Zn (186).

The oxidation of lower olefins to carbonyl compounds according to reaction (28) can be carried out in the gas phase on Pd/Cu-Y zeolites. It is essential to add water in order to ensure high selectivity and long catalyst service life. The rate of oxidation falls off as the chain length increases, as is known from the homogeneously catalyzed reaction using Pd and Cu salts (187, 188).

$$R-CH=CH_2 \quad + \quad 0.5\ O_2 \quad \longrightarrow \quad R-\overset{\displaystyle O}{\overset{\|}{C}}-CH_3 \qquad (28)$$

In the oxidative acetoxylation of propylene to give allyl acetate, Pd/Cu-erionite (~ 90% selectivity) and Pd/Cu-mordenite (~ 70% selectivity) are superior to Pd/Cu-faujasite (~ 40% selectivity). However, the activity and selectivity of conversion in this reaction depend on the alkali metal ion present in the zeolite before replacement by Pd and Cu (189).

The same Russian group reported (190) on the gas phase oxidation of 2-methylpyridine over PdCu-zeolites. Best results are obtained over PdCu-mordenite: at 375 °C selectivity to pyridine-2-carbaldehyde is 40%. On Pd- or Cu-containing zeolites deep oxidation is the main reaction. With both Pd and Cu present the activity of the catalyst is lower but service life and selectivity increase.

Zeolitic catalysts can also be used (191) for the oxidative dehydrogenation of olefinic, alkylaromatic and naphthenic hydrocarbons. The high temperature reaction is sensitive to the zeolite structure and Si/Al ratio as well as to the nature and concentration of the counter cations.

Table 8 gives results on the dehydrogenation of 2-methyl-2-butene at 500 °C over various Na-zeolites to yield isoprene. Faujasites give the best performance.

Table 8. Oxidative dehydrogenation of 2-methyl-2-butene over Na zeolites (T = 773 K, i-C_5H_{10}:air mole ratio = 1:3.8, space velocity = 3150 h^{-1}) (ref. 191).

Zeolite	Conversion, %			Selectivity to isoprene, %
	Total	to isoprene	to CO_2	
NaX	35.1	29.8	1.6	82
NaY	38.8	30.5	1.9	80
Na chabasite	23.2	17.7	3.2	76
Na erionite	24.4	18.5	3.4	76
Na mordenite	20.8	16.1	3.0	77

6.2. Ammoxidations

Already in 1966 it was shown (192) that benzonitrile can be obtained by reacting toluene with NH_3 at 540 °C in the presence of various X-zeolites. In this amination/dehydrogenation - in the absence of oxygen - NaZnX gives the highest initial activity. Rapid catalyst aging is observed, however.

$$CH_3 \text{(benzene ring)} + NH_3 + 1.5\ O_2 \xrightarrow{\text{CuH-ZSM-5}} CN \text{(benzene ring)} + 3H_2O \qquad (29)$$

When using CuH-ZSM-5 and related materials in the ammoxidation of toluene (reaction 29) high activity and selectivity is obtained (193, 194) at 350 °C. Benzene is the major side product, deep oxidation is < 10%. Water has a beneficial influence. Table 9 shows some results; also a CuH-ferrisilicate is an excellent catalyst. For a postulated mechanism, starting with Cu^{II} reduction to Cu^{I} by NH_3, see ref. (193).

Table 9. Ammoxidation of toluene[a].

Catalyst	Si/Al	Conversion %	Selectivity (wt %) to benzonitrile
Na-ZSM-5	17	7	3
H-ZSM-5	17	7	35
ZnH-ZSM-5	26	10	29
CuH-ZSM-5	17	70	86
CuCrH-ZSM-5	26	100	92
CuH-[Fe]-ZSM-5	45 (Si/Fe)	99	94

[a] Toluene:NH_3:H_2O:O_2 (as air) molar ratio: 1:2:6:4.7; WHSV toluene 0.17 h^{-1}; 350 °C; results after 4 h on stream.

Different types of vanadium-zeolite catalysts were tested in the ammoxidation of p- and m-xylene (195, 196). H-ZSM-5 and H-ZSM-11 impregnated with NH_4VO_3 proved to be stable in the reaction environment while NH_4Y exchanged with VO^{2+}

or impregnated with NH_4VO_3 collapsed. H-ZSM-5-V and H-ZSM-11-V are more active than V-silicalite. H-ZSM-11-V is more selective of low conversion than H-ZSM-5-V. At 350 °C the ammoxidation of p-xylene over H-ZSM-11-V yields 45% p-tolunitrile and 15% terephthalonitrile at 90% conversion. At higher temperature the yield of p-tolunitrile decreases and the yield of terephthalonitrile, CO, CO_2 benzene and benzonitrile increases. The ammoxidation of m-xylene leads under the same reaction conditions only to 20% m-tolunitrile and 8% isophthalonitrile. The catalytic behaviour of the V-containing H-ZSM-5 and H-ZSM-11 systems is similar to that of a conventional V_2O_5/TiO_2 catalyst.

Some recent reports deal with the high temperature ammoxidation of small saturated hydrocarbons.

Thus, at 500 °C a mixture of n-butane/NH_3/air (1:2:10 molar) is converted on Pt-ZSM-5 to achieve 16% selectivity of acrylonitrile and 12% of acetonitrile at 43% conversion of n-butane (197). Acetonitrile is produced with high yield by ammonolysis of propane at about 600 °C when a pentasil zeolite with Si/Al > 50 is employed as catalyst (198).

Upon incorporating vanadium in AlPO-5 an active catalyst is obtained (199) for the direct ammoxidation of propane towards acrylonitrile. At 500 °C selectivities up to 33% to acrylonitrile and 11% to acetonitrile are observed.

$$H_3CCH_2CH_3 \xrightarrow[O_2, NH_3]{VAPO-5} H_2C=CHCN + H_3CCN \tag{30}$$

The VAPO-5 molecular sieve is superior to a V-silicalite and to a conventional V-P-oxide catalyst.

6.3. Oxidation reactions with hydrogen peroxide

Chang and Hellring (200) recognized the use of the pentasil zeolites ZSM-5 and ZSM-11 for the oxygen-transfer from hydrogen peroxide onto hydrocarbons. The shape selective oxidation of phenol with 40% H_2O_2 at 80 °C over H-ZSM-5 (Si/Al 128) yields hydroquinone and catechol in a ratio 99:1. The product shape selectivity is responsible for the high yield of the p-isomer. Current liquid phase, homogeneous, e.g. perchloric acid, catalyzed technology for hydroquinone by the hydroxylation of phenol (as practiced by Rhone Poulenc) makes substantial quantities of coproduct, the o-isomer catechol (minimum catechol/hydroquinone ratio is about 1.5:1).

Recent work (201) showed also H-beta and H-USY to be active catalysts for phenol hydroxylation (reaction 31). The hydroquinone/catechol ratio is found to be 0.7 in the case of H-USY, 0.8 for H-beta, and 8.0 when using H-ZSM-5, confirming the p-selectivity of ZSM-5.

$$\text{(31)}$$

The discovery of the titanium-containing MFI-type zeolite TS-1 by Enichem workers has, in recent years, led to remarkable progress in the field of oxidations with H_2O_2 (cf. Chapter 13). First, the direct hydroxylation of benzene derivatives with H_2O_2 in the presence of TS-1 is possible (202). The oxidation of phenol to hydroquinone and catechol is already employed industrially on a large scale, 10,000 t/a, in Italy. The ratio of the p/o-isomers is about 1, the selectivity based on phenol is 92%, based on H_2O_2 80%. A comparison of the different catalysts for this reaction shows that the homogeneous catalysts are less efficient than TS-1. The zeolite-catalyzed route also appears to be more attractive than the p-di-isopropylbenzene approach.

For an excellent review on the phenol hydroxylation the reader is referred to Jacobs et al. (114).

Further examples of the use of the H_2O_2/TS-1 system in direct aromatic hydroxylation are the hydroxylation of toluene, ethylbenzene, anisole and the cresols. Deactivated substrates, e.g. benzonitrile or nitrobenzene, appear to be non-reactive. The tendency for the p-isomer to be formed preferentially and the miniminzation of undesired side reactions leading to multinuclear arenes and tar are indications of the shape selectivity of the titanium zeolite.

Temperature used is typically 80 °C . Solvents applied include methanol, t-butanol, water and acetone. Note that the acetone/H_2O_2 combination might be hazardous. Other important variables in the direct aromatic hydroxylation seem to be the Al content, the crystal size and the activation temperature of the catalyst (201).

The precise mechanism of action of the Ti-sites in TS-1 in the aromatic hydroxylation with H_2O_2 is as yet unsolved. Isolated lattice Ti, capable of adopting a coordination number in excess of four has been mentioned (204) as an important feature.

The application of the H_2O_2/TS-1 combination to primary and secondary alcohols offers a new route for the synthesis of aldehydes and ketones (207).

For example, benzyl alcohol is oxidized to benzaldehyde by means of H_2O_2 in a yield of approximately 90% (relative to H_2O_2) on TS-1 at 70-90 $^\circ$C. Obvious consecutive reactions when oxidizing primary alcohols over TS-1 are formation of acetals and further oxidation towards carboxylic acids.

Secondary alcohols are very selectively oxidized to ketones, also the H_2O_2 decomposition is negligible. In competitive oxidation over TS-1 (208) 2-butanol was 1.2 times faster than 2-pentanol, and more strikingly, 2-pentanol reacted 13 times faster than 3-pentanol. Cyclohexanol is oxidized very slowly. These facts point to inner-channel catalysis.

Liquid phase epoxidations of olefins and diolefins at temperatures between 0-80 $^\circ$C proceed with selectivities of 75-96% (based on olefin), with H_2O_2-conversion being usually quantitative (209-212). TS-1 is a highly effective catalyst for the oxidation of propylene to the industrially very important compound propylene oxide, using H_2O_2 as the oxidant (204, 209). In acetone at 40 $^\circ$C, propylene oxide is obtained with 93% selectivity at 94% H_2O_2-conversion. If the catalyst is silylated, it is possible to raise the selectivity to 98% while maintaining a comparable conversion (211). This synthetic route is environmentally friendly and offers a genuine alternative (204) to the technical epoxidation processes presently in operation such as the chlorohydrin and hydroperoxide methods. The principal by-product in this reaction is a glycol monomethyl ether (in methanol) or a glycol ketal (in acetone). The side-reaction leading to glycol monomethyl ether apparently involves addition of methanol to the epoxide and can be made the main reaction (209) if the process is carried out above 100 $^\circ$C (reaction 32).

$$R-CH-CH-R' \ + \ H_2O_2 \ + \ CH_3OH \ \longrightarrow \ R-\overset{\displaystyle H}{\underset{\displaystyle OH}{C}}-\overset{\displaystyle H}{\underset{\displaystyle OCH_3}{C}}-R' \ + \ H_2O \qquad (32)$$

R,R^1 = H, alkyl, aryl, alkylaryl, cycloalkyl

From competition experiments the following order of conversion rate in epoxidation over TS-1 is obtained for the butenes:

cis-2-butene > 1-butene > isobutene > trans-2-butene,

which order differs from the sequence in homogeneous electrophilic epoxidation which is iso- > cis-2 > trans-2 > 1-butene. The order for H_2O_2/TS-1 reflects diffusion as well as adsorption/reaction differences. The difference between cis- and corresponding trans-alkenes is so pronounced over TS-1 that the cis-alkene can be selectively epoxidized in cis/trans mixtures of an alkene.

The versatile Ti catalyst can also combine functions, as shown (210) by the epoxidation/isomerization of styrene via styrene oxide into phenylacetaldehyde (cf. Section 4.5.3).

Very recently two groups observed (213a,b) oxidation towards alcohols and ketones when subjecting saturated hydrocarbons to the H_2O_2/Ti-zeolite system.

6.4. Reactions of and with alkyl hydroperoxides

In Section 3.1 the selective adsorption of organic hydroperoxides onto zeolite NaX was mentioned. Zeolites containing protons or suitable redox cations are able to catalyze hydroperoxide rearrangement and decompositon, respectively. Thus, over NaCo-A zeolite, cyclohexyl hydroperoxide is converted at 30-160 °C into cyclohexanol and cyclohexanone with a selectivity of over 95% at 100% conversion (214); apparently a case of outer surface catalysis. For comparison Co on graphite gives a selectivity of only 82% at a peroxide conversion of 94%. The decomposition of cumene hydroperoxide to give phenol (96% selectivity) and acetone can be carried out over a mixed alumino/boro-zeolite at 40-60 °C (215).

The alkyl hydroperoxide decomposition on redox metal-ion exchanged zeolites may serve a synthetic purpose. Thus, 2,6-dialkylphenols can be selectively converted - reaction 33 - into 2,6-dialkylquinones with t-butyl hydroperoxide and NaCoX-72 at room temperature. t-Butyloxy and -peroxy radicals are formed inside the zeolite and attack the phenolic substrate. The highly selective oxidation of 2,6-di-t-butylphenol can be understood from the dimensions of the corresponding 4,4′-diphenoquinone, which show that this dimer - which is a major side-product under homogeneous conditions - is too large to be formed in the cages of zeolite X.

Brönsted acid sites will catalyze the ionic rearrangement of t-butyl hydroperoxide. Thus cyclohexanol and tBuOOH react over NaLaY-70 at 80 °C to give cyclohexyl formate as the main product (217). This compound is thought to originate from a zeolite-catalyzed rearrangement of tBuOOH to acetone and methanol, oxidation of methanol to formic acid, and electrophilic addition of formic acid to cyclohexene also catalytically formed from cyclohexanol.

6.5. Oxidation reactions with hydrogen peroxide and ammonia

Another noteworthy reaction (218) which can be carried out on titanium zeolites is the liquid phase conversion of cyclohexanone with ammonia and hydrogen peroxide to cyclohexanone oxime according to eqn. 34. This route avoids the co-production of ammonium sulphate associated with the classical reaction of cyclohexanone and the toxic hydroxylamine sulphate, and is, therefore, of industrial interest for environmental reasons.

$$\text{(cyclohexanone, } =O) \quad +NH_3 + H_2O_2 \longrightarrow \text{(cyclohexanone oxime, } =NOH) \quad + 2H_2O \qquad (34)$$

The ammoxidation reaction is carried out by dispersing the Ti-catalyst in an ammonia-cyclohexanone-water-organic liquid phase and by feeding the H_2O_2 to the well-stirred slurry. t-Butanol/water is a very efficient solvent, even if other solvents such as benzene and toluene give similar results. Preferred reaction temperature is 80 $^\circ$C. Table 10 compares TS-1 as a catalyst with some other materials.

Table 10. Ammoxidation[a] of cyclohexanone over various catalysts (218).

Catalyst	Ti %	$H_2O_2/C_6H_{10}O$ molar	$C_6H_{10}O$ conv.	Oxime select.	Yield on H_2O_2
none	-	1.07	53.7	0.6	0.3
SiO_2 amorphous	0	1.03	55.7	1.3	0.7
Silicalite	0	1.09	59.4	0.5	0.3
H-ZSM-5	0	1.08	53.9	0.9	0.4
TiO_2/SiO_2	1.5	1.04	49.3	9.3	4.4
TiO_2/SiO_2[b]	9.8	1.06	66.8	85.9	54.0
Ti-silicalite (TS-1)	1.5	1.05	99.9	98.2	93.2

[a] In water/tBuOH; catalyst 2 wt %; 80 $^\circ$C; NH_3:H_2O_2 molar 2.0; reaction time 1.5 h except for b) (5 h).

Though amorphous silica-supported Ti shows catalytic activity, TS-1 is the superior catalyst. The same group could demonstrate (219) that production of ketoximes in the liquid phase on titanium-silicalite is also possible starting from a secondary alcohol in the presence of O_2, H_2O_2 and NH_3. The intermediate ketone has not to be isolated, i.e. several reaction steps can be combined in one. For example, cyclohexanol at about 100 $^{\circ}$C is converted to cyclohexanone oxime with up to 90% yield.

The mechanism of the ketone ammoxidation is as yet not clarified. The sequence might contain an NH hydroxylation step as Ti-silicalite proves also to be a useful catalyst for the synthesis of N,N-dialkylhydroxylamines by reaction of dialkylamines with H_2O_2 (220).

$$R_2NH + H_2O_2 \longrightarrow R_2NOH + H_2O \qquad (35)$$

Diethylamine diluted with t-butanol reacts with H_2O_2 at 80 $^{\circ}$C over Ti-silicalite to form N,N-diethylhydroxylamine with 92% selectivity at 80% conversion of the amine. The H_2O_2-conversion is 99.8% and the selectivity based on H_2O_2 is 81%.

In conclusion, it can be stated that titanium-silicalite catalysts are very useful and promising tools for oxidation reactions with H_2O_2.

6.6. Oxidation with "ship in a bottle" systems and other enzyme mimics

The discovery of the titanium zeolite TS-1 and its successful employment as a selective oxidation catalyst were a milestone in the area of zeolite catalysis in the 1980's. Recent developments in the selective partial oxidation of non-activated alkanes on metal phthalocyanine (MPc) complexes trapped in faujasite supercages and on a FePd system in zeolite A are also very promising. The results point to potential applications of these inorganic mimics of, for instance, the natural Fe-containing monooxygenase enzyme cytochrome P 450 which hydroxylates organic substrates with high regioselectivity.

MPc complexes are synthesized (221, 222) within the zeolite framework by treating a metal-ion exchanged X or Y zeolite with molten o-dicyanobenzene which cyclotetramerises around the metal ion. The porphyrin-related iron phthalocyanine complex has a diameter of 1.6 nm whereas the zeolite supercage offers only a 1.3 nm diameter to host this molecule. This paradox is resolved by giving the complex a saddle distortion and allowing its arms to protrude through the supercage windows. Now, the complex fits into the zeolite.

These "ship in a bottle" complexes cannot leave the zeolite without destroying the framework. Zeolite catalysts of this type, in which the supercages function quasi as "reaction vessels" with molecular dimensions while imposing steric constraints.

The expected substrate selectivity of FePc-NaY-zeolite has been demonstrated by means of a competitive oxidation, in which a CH_2Cl_2-solution of cyclohexane and cyclododecane was oxidized at room temperature to the corresponding alcohols and ketones using iodosobenzene and air as oxidants (222-225). The rate of oxidation of cyclohexane is approximately twice that of cyclododecane (62:38). A further reduction of the pore diameter by substituting Rb^+ for Na^+ raises the selectivity for the smaller substrate molecule so as to yield a product ratio of 90:10. In contrast, when the homogeneous catalyst FePc is used, the two substances are oxidized at the same rate.

"Ship in a bottle" complexes also display stereoselectivity. With methylcyclohexane as substrate, the ratio of trans- to cis-4-methylcyclohexanol (hydroxylation at the 4-position) is approximately 2 for the zeolite catalyst, but only approximately 1.1 for the homogeneous FePc. Similarly in the oxidation of norbornane one finds an exo:endo norborneol ratio of approximately 6 on FePc-NaY-zeolite, but a ratio of approximately 9 on FePc. This altered preference of oxidation at one of the two diastereotopic hydrogens reflects the effect of the zeolite on the relative orientation of the substrate to the catalyst.

The "ship in a bottle" complexes can also be used for the partial oxidation of intermediates with air, e.g. of cumene (226). The oxidation of cumene in the presence of FePc-NaY-zeolite gives $PhCMe_2OOH$. PhCOMe and $PhCMe_2OH$ in a ratio similar to that obtained using the unsupported homogeneous FePc. However, the FePc-NaY shows an activity twenty times that of FePc.

Recently, the preparation of FePc in the super large pore molecular sieve VPI-5 and its use in the oxyfunctionalization of n-octane have been reported (226a). t-Butyl hydroperoxide is used as the oxidant. FePc-VPI-5 is less active than FePc-Y.

Another type of "ship in a bottle" system is formed by introducing the large but not bulky ligand N,N-ethylenebis(salicylideneamine), "salen", into a Y zeolite containing selected cations. Once the tetradentate salen has assembled its metal complex, it cannot desorb anymore.

692

M= Co, Mn

The complex can be accommodated in a supercage by involving 2 windows (227). The Co(salen) complex in Y zeolite has been studied (227) as a hemoglobin mimic; its oxygen binding differs considerably from the oxygen binding of the free complex in solution. In a recent study (228) a Mn(salen) complex encapsulated in zeolite Y serves as a catalyst for olefin epoxidation.

A completely inorganic mimic of Cytochrome P450 was prepared (229) by combining the ability of colloidal Pt or Pd metal to convert oxygen and hydrogen into H_2O_2 with the ability of Fe^{II} ions to use H_2O_2 for hydroxylating organic compounds. Such a bimetallic $Pd(0)Fe^{II}$-system inside zeolite A displayed substrate selectivity in the oxidation of octane and cyclohexane and an even more remarkable regioselectivity in the partial oxidation of octane. Fig. 14 gives some results (229).

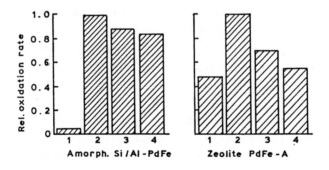

Fig. 14. Position of oxidation of n-octane with H_2/O_2 over two Fe/Pd-systems.

With respect to the substrate selectivity the ratio of oxidation products from octane and cyclohexane is about 45:55 on PdFe on amorphous aluminosilicate and increases to 100:1 on zeolite A. In these experiments the outer surface of zeolite A was poisoned with 2,2'-bipyridine.

Regarding the regioselectivity of the FePd-A system, the oxidation products of n-alkanes indicate a ratio primary C-H/secondary C-H oxidized of 0.67 which is very high. Apparently the zeolite matrix exerts some control on the substrate/active site interaction (cf. Section 3.4 regioselective chlorination).

The reader should note that this interesting system (i) has to be dissolved to release the oxidation products, (ii) uses H_2/O_2 which combination is not without its dangers.

Other examples of zeolite-encapsulated species with special properties include
- $Ru(bipy)_3^{2+}$ entrapped in zeolite Y and reported (230) to be an effective heterogeneous photosensitizer to generate singlet oxygen,
- discrete CdS cubes located and stabilized within the small sodalite units of Y zeolite (231) and representing a new well-defined semiconductor system.

The examples given in this paragraph demonstrate that the "ship in a bottle" complexes, the really inorganic mimic of cytochrome P450, the zeolite based photosensitizer and semiconductors as well as the recent developments of Pd-clusters in zeolites (122) are promising leads for the organic chemists in the future when sterically demanding supports for active sites are required.

The above approach could develop to a general means of immobilization of homgeneous catalysts.

6.7. Nitrous oxide as oxidants

Recently (232, 233) the use of N_2O has been reported in zeolite-catalyzed oxidation of benzene towards phenol. As shown in reaction 36, N_2O, which is conveniently prepared from ammonium nitrate, is a clean oxidant.

$$C_6H_6 + N_2O \longrightarrow C_6H_5OH + N_2 \qquad (36)$$

When using H-ZSM-5 at 400 °C 95% selectivity to phenol at 9.5% benzene conversion is obtained (233). Recently it was shown (233a) that iron impurities in H-ZSM-5 play an important role in the decomposition of N_2O while generating a very reactive form of surface oxygen.

7. ZEOLITE-CATALYZED CYCLIZATIONS

Some ring-forming reactions - e.g. epoxidation - and some ring transformations - e.g. the benzamine rearrangement - were mentioned already.

For the well-known methanol-to-gasoline (MTG) process, which is in fact a distribution of H and C (ratio 2:1) over aliphatic and aromatic hydrocarbons, the reader is referred to Chapter 15 and to ref. (234).

The analogous ethanol-to-gasoline conversion over H-ZSM-5 (235) proceeds also in the presence of excess of water which enables integrated conversion (236) of carbohydrates to hydrocarbons by consecutive fermentation and dehydration/aromatization.

In addition to the reaction of lower alcohols (and olefins) also light alkanes, e.g. propane, can be converted over ZSM-5 catalysts into aromatics; preferably a dehydrogenating component (Ga (237) or Zn (238)) is incorporated or added.

Cyclization/aromatization of hexane and higher n-alkanes is selectively attained (239) using Pt-loaded neutral L-zeolites. For instance n-hexane is converted over 0.6% Pt/K-L at 460 $^{\circ}$C and WHSV 2 h^{-1} in 80% selectivity to benzene. n-Octane yields over Pt/BaK-L at 460 $^{\circ}$C 30% aromatics of which 88% consists of ethylbenzene and o-xylene (and 12% m- and p-xylene). Mechanistic insight in this reaction is provided by Derouane et al. (206).

Some additional zeolite-catalyzed cyclization reactions will be dealt with in this section.

7.1. Carbocyclic ring formation

7.1.1. Diels-Alder cycloadditions

In paragraph 3.5 physical catalysis was shown to operate in a [4 + 2] cyclodimerization reaction over microporous solids. By adding a suitable cation (e.g. Cu^{I}) a chemo-catalytic component becomes active too.

Thus, highly selective cyclodimerization of butadiene to vinylcyclohexene over Cu^{I}Y (reaction 37) has been reported by Maxwell et al. (4a, 240). By contrast, in homogeneous catalysis, a mixture of cyclic dimers and trimers is normally obtained. The selectivity of the zeolite is attributed to reaction selectivity, since the Cu^{I} intermediate required to form vinylcyclohexene is less space-demanding than the intermediates necessary for the formation of the other oligomers. The stability of the Cu^{I}Y catalyst is found to be strongly dependent on its way of preparation.

$$(37)$$

Another example is the $Cu^{I}Y$ catalysis of the Diels-Alder reaction - 38 - of furan and α,β-unsaturated carbonyl compounds such as methyl vinyl ketone (241) at the required low temperature (0 °C). Essentially, the same *exo/endo* product ratio (2.5) is observed as found using Fe^{III}-doped clay (bentonite) as the catalyst (242). Here, the dienophile might well be activated by cation coordination to the carbonyl oxygen.

$$(38)$$

Another mechanism of activation by zeolites has been postulated in a study (243) on cyclodimerization of 1,3-cyclohexadiene and 2,4-dimethyl-1,3-penta-diene on NaX. Electron transfer to the zeolite is assumed to take place leading to a reactive cation radical.

These authors also observed zeolite (NaX) catalysis in the [2 + 2] cyclodimerization of the electron-rich styrene derivative 4-(1-propenyl)anisole (anethole) yielding a cyclobutane system. Here the question arises of the contributions of inner and outer zeolite surface.

Another interesting zeolite-catalyzed [2 + 2] cyclo-addition is reaction 39, the conversion of cyclopropene towards tricyclo[3.1.0.0]hexane (244). This cyclodimerization proceeds with high selectivity over NaA and KA (yields of 95 and 97%, respectively), whereas on zeolites with pore size larger than .4 nm,such as CaA, NaX and NaY polymerization takes place. Similarly 1- and 3-methylcyclopropene can be selectively cyclodimerized to the corresponding tricyclohexanes.

$$(39)$$

This new method of cyclodimerization of cyclopropenes is superior to known procedures of preparing tricyclohexanes. As to the mechanism the authors propose a stepwise ionic mechanism in which spatial restrictions imposed by the small pore zeolite prevent approach of a third cyclopropene molecule.

7.1.2. Cyclopropanation

Copper-exchanged X and Y zeolites are active catalysts for the decomposition of diazo compounds (245, 246) leading to carbenoid intermediates. Cyclopropanation of various olefins has been carried out (247) by the application of NaCuX as a catalyst for the decomposition of ethyl diazoacetate, see reaction 40.

$$H_5C_2OOCCHN_2 \quad + \quad \diagup\!\!=\!\!\diagdown \quad \xrightarrow{\text{Cu-X,-Y}} \quad \overset{\text{COOEt}}{\triangle} \quad + \quad N_2 \qquad (40)$$

Typically the ethyl diazoacetate is gradually added to a suspension of NaCuX in the olefin at 80 °C. The reaction is followed by monitoring the nitrogen evolution. Side products are the dimeric compounds diethyl fumarate and maleate and polymeric material assumed to be formed at the outer surface of the zeolite. Compared to conventional copper catalysts, the zeolite catalysts give rise to relatively low amounts of polymeric side-products. Important variables are the degree of copper exchange and the water content of the NaCuX. Some dienes were found to give mono-cyclopropane systems. For instance, 1,1-dichloro-4-methyl-1,3-pentadiene can be efficiently monocyclopropanated.

Some chiral complexes were constructed inside the zeolite, e.g. the neutral Cu(L-alanine)$_2$ and the cationic Cu(R-1,2-diaminopropane)$_2$. When used as a catalyst, these systems induce just a small asymmetric cyclopropanation of 1,1-dichloro-4-methyl-1,3-pentadiene.

7.1.3. Anthracene formation

When passing (248) benzyl alcohol over alumina, fluorided alumina, zeolite H-Y or heteropolyacids at temperatures of 300-500 °C anthracene is formed (reaction 41). Side products are toluene and benzaldehyde. The authors assume o-benzylbenzyl alcohol to be an intermediate. Zeolite H-Y gives the best performance. In a preparative example a solution of benzyl alcohol in benzene was passed over a bed of H-Y at 400 °C and WHSV .95 h^{-1} to give 64.5% (on benzyl alcohol) of anthracene.

$$\text{CH}_2\text{OH} \qquad\qquad (41)$$

7.2. Ring closure towards heterocyclic compounds

Heterocyclic compounds play an important role as intermediates and end-products in fine chemistry. For instance many bio-active molecules - pharmaca and agrochemicals - contain a heterocyclic ring. Ono et al. have demonstrated (249) the great potential of zeolite catalysis in heterocyclic-compound syntheses. Applications of zeolites in heterocyclic-ring (trans)formations include:
- cyclization reactions;
- heteroatom substitution (O → N);
- ring-size isomerization (5 → 6);
- aromatization reactions.
In this section some examples will be given on oxygen- and nitrogen-containing heterocycles. The reader will find additional information in some recent reviews (ref. 2a-e) by one of the present authors.

7.2.1. Oxygen-containing ring systems

Several examples exist of zeolite-catalyzed dehydration of diols to form five-membered ring heterocycles. For instance, trans-1,4-cyclohexanediol is transformed at 215 °C into 7-oxabicyclo[2.2.1]heptane (reaction 42) over a specially treated NaA zeolite (250). Selectivity is 71% at 98% conversion; apparently outer surface catalysis is involved.

$$\text{HO} \qquad \text{OH} \longrightarrow \qquad\qquad (42)$$

One of the present authors studied the cyclodehydration of 1,4-butanediol and 1,4-but-2-enediol towards tetrahydrofuran and dihydrofuran, respectively. Highly selective conversion of 1,4-butanediol is obtained (251) e.g. over a boron- or ironsilicate of the pentasil type at 200 or 300 °C. When passing 1,4-but-2-enediol over various boronsilicates (252) crotonaldehyde (2-butenal) showed up as a major side product.

698

Finally we mention the dehydration of 1,4-diketones over H-zeolites (253), leading to 2,5-disubstituted furans according to reaction 43. For example, 2,5-dimethylfuran is obtained with 98% yield from acetonylacetone (R = CH$_3$) over H-ZSM-5 (Si/Al 140) at 300 °C. The reaction may involve intramolecular acetalization of the mono-enolized diketone followed by dehydration.

Recently (253a) it was reported that on basic Na-ZSM-5 under identical conditions acetonylacetone enters into an intramolecular Claisen-condensation leading to 3-methyl-2-cyclopenten-1-one in 89% yield. In this way the acetonylacetone conversion is of diagnostic value for the nature of the active sites present in medium pore zeolites.

$$\underset{\text{RCCH}_2\text{CH}_2\text{CR}}{\overset{\text{O} \quad\ \text{O}}{\|\qquad\|}} \longrightarrow \quad R\text{—}\!\!\overset{}{\underset{\text{O}}{\bigcirc}}\!\!\text{—}R \qquad (43)$$

7.2.2. Nitrogen-containing ring systems

For a recent review on the use of zeolites as catalysts in the synthesis of N-containing compounds the reader is referred to ref. (254).

7.2.2.1. O/N replacement in cyclic compounds

Zeolites offer an interesting alternative to conventional acid non-shape-selective catalysts for the replacement of O by N in heterocyclic compounds.

Tetrahydrofuran (THF) and NH$_3$ (molar ratio 1:7) react at 350 °C over H-L to give pyrrolidine - reaction 44 - with a selectivity of 91% at 53% conversion and over H-Y with a selectivity of 82% at 61% conversion. Alumina and silica-alumina are also active in this reaction but the selectivity is poor. The alkali forms of zeolites L and Y are not active at all, indicating that Brönsted acid sites exert the catalytic action (255). A logical sequence would seem: oxygen protonation/nucleophilic substitution/hydroxyl protonation/intramolecular nucleophilic substitution. In view of the presence of (excess of) ammonia and pyrrolidine just a small part of the protons is available for catalysis.

$$\overset{}{\underset{\text{O}}{\bigcirc}} + \text{RNH}_2 \longrightarrow \overset{}{\underset{\underset{R}{|}}{\underset{N}{\bigcirc}}} + \text{H}_2\text{O} \qquad (44)$$

R= H, alkyl

Primary amines also enter into zeolite-catalyzed O/N substitution of THF. Thus THF reacts with propylamine ($R = C_3H_7$) over Al-doped H-Y at 360 °C to give 1-propylpyrrolidine with 75% selectivity at 61% conversion (249).

The corresponding six-membered ring, tetrahydropyran undergoes O/N replacement towards piperidine over dealuminated HY- and HL-zeolites which give a better performance than the parent H-zeolites (256). The alkali forms are inactive whether dealuminated or not, so Brönsted acidity seems to be required.

Five- and six-membered lactones are converted to the corresponding lactams (reaction 45) over various Y-zeolites.

$$(CH_2)_n \underset{}{\overset{O}{\big|}} \; + \; NH_3 \;\longrightarrow\; (CH_2)_n \underset{C=O}{\overset{NH}{\big|}} \; + \; H_2O \qquad (45)$$

For instance, γ-butyrolactone (n = 3) and NH_3 (molar ratio 1:5) react over CuY at 260 °C to give 2-pyrrolidinone with 80% selectivity at 31% conversion (257).

By contrast caprolactone (n = 5) and NH_3 are found to give 5-hexenenitrile as the main product upon reaction over pentasil type or HY zeolites (258) in contrast to the expected caprolactam. This may be related to the more difficult ring closure of the supposed acyclic intermediate compared to 5- and 6-membered ring formation. For a full discussion of the different behaviour of the lactones, see ref. (254).

7.2.2.2. Cyclocondensation with ammonia

Aldol condensations of aldehydes and ketones on zeolites have been extensively described. Such condensation reactions, which are preferably carried out in the gas phase, lead in the presence of NH_3 over acid catalysts to pyridine and alkylated pyridines (254). Various aldehyde/aldehyde and aldehyde/ketone combinations were reacted (259) over H-ZSM-5 (Si/Al 200-400).

In some cases high selectivities are obtained. A fine example is the synthesis of β-alkylpyridines (260) by reaction of acrolein, an alkanal and ammonia over pentasil zeolites (reaction 46).

$$\underset{H}{\overset{}{\diagup}}\!\!\diagdown_O \; + \; NH_3 \; + \; \underset{O}{\overset{R}{\diagdown}}\!\!\diagup\underset{H}{} \;\longrightarrow\; \underset{N}{\overset{R}{\bigcirc}} \; + \; 2H_2O \; + \; H_2 \qquad (46)$$

When passing a mixture of ammonia acrolein and butanal (molar ratio 3:1:1) over a HF-treated borosilicate at 400 °C and WHSV 3 h^{-1} one obtains β-ethylpyridine with 72% selectivity. For R = C_4H_9 and C_6H_{13} the selectivities are 78% and 90%, respectively. In all cases conversion is complete and catalyst lifetime > 48 h. The increasing selectivity with increasing chain length indicates shape selectivity; the zeolite might induce the long chain alkanals to adopt - on average - favourable positions with respect to the other reactants.

Another cyclo-condensation in which the position of the reactants in a final product is defined is reaction 47. When feeding a mixture of acetone, ^{13}C-labelled methanol, ammonia and water (molar ratio 2:1:4:13.7) to H-ZSM-5 (Si/Al 96) at 450 °C at WHSV (acetone) 0.3 h^{-1}, 2,6-lutidine is formed (13% selectivity) which is exclusively labelled at the 4-position (261).

$$+ 3H_2O + H_2 \qquad (47)$$

The reaction may involve methyl vinyl ketone as an intermediate, resulting from reaction of acetone and formaldehyde which latter compound may be formed by dehydrogenation of methanol. Under similar conditions dehydrogenation of methanol over silicalite is observed (262).

7.2.2.3. Oxidative pyridine synthesis

The reaction of ethanol with ammonia on zeolite catalysts leads to ethylamine (254). If, however, the reaction is carried out in the presence of oxygen, then pyridine is formed (263, 264). One of the present authors and coworkers recognized that H-boron zeolite with Si/B = 42, or Fe-containing ZSM-5 type catalysts are particularly suitable for this purpose. Thus, a mixture of ethanol, NH_3, H_2O and O_2 (molar ratio 3:1:6:9) reacts on H-boralite at 330 °C and WHSV - 0.17 h^{-1} to yield pyridine with 48% selectivity. The conversion is 24% and can be improved by increasing the number of boron atoms per unit cell, i.e. the number of acid sites, or by raising the temperature. At 360 °C the conversion is 81% but there is increased ethylene formation at the expense of pyridine. Further by-products include diethyl ether, acetaldehyde, ethylamine, picolines, acetonitrile and CO_2. When applying H-mordenite, HY or silica-alumina as the catalyst under similar conditions pyridine yields are very low and ethylene is the main product. The one-dimensional zeolite H-Nu-10 (Si/Al 45) turned out to be another pyridine-forming catalyst (265). A mechanism starting with partial oxidation of ethanol to acetaldehyde followed by aldolization, reaction with ammonia, cyclization and aromatization can be envisaged. An intriguing question is why pyridine is the main product and not methylpyridines (picolines) (265). It has been suggested in this connection that zeolite radical sites induce C_1-species formation.

7.2.3. Formation of bicyclic compounds

Zeolites give higher selectivities than existing industrial catalysts in the synthesis of diazabicyclo[2.2.2]octane (DABCO) from precursors such as N-hydroxyethyl- and N-aminoethylpiperazine. At 400 °C 10% conversion and 87% selectivity are obtained (266) over H-ZSM-5 (Si/Al 35-55). Conventional catalysts, e.g. alumina, are more active but less selective leading to difficultly separable byproducts. When using a high silica zeolite, the only byproduct is piperazine.

It is also possible (267) to use ethanolamine as the starting material for DABCO. At 400 °C and WHSV 10 h^{-1} a 64% yield of DABCO is claimed using a ZSM-5 catalyst (reaction 48).

$$3 \text{ HOCH}_2\text{CH}_2\text{NH}_2 \longrightarrow \text{N} \underset{}{\diagup} \text{N} \qquad (48)$$

The Fischer indole synthesis involves acid-catalyzed rearrangement of arylhydrazones. Ammonia is expelled. Hydrazones originating from nonsymmetric ketones afford two isomeric indoles. Recently it was found by Carlson et al. (268) that zeolites catalyze this transformation while exerting in some cases a profound effect on the regioselectivity. An example is given in reaction 49.

CH$_3$COOH (homog.)	100 : 0
H – Y	83 : 17
H – Mord.	7 : 93

The phenylhydrazone of 1-phenyl-2-butanone rearranges with homogeneous H$^+$-catalysis completely towards 2-ethyl-3-phenylindole. When applying zeolites (in refluxing isooctane or xylene) the other isomer is formed too and becomes the predominant product when using the parallel channel zeolite H-mordenite as the catalyst. Considering the shape of the two products this is perfectly understandable.

An interesting feature of the reaction is that H-zeolites also catalyze the formation of the phenylhydrazone. In this way a one-pot procedure is allowed starting from phenylhydrazine and 1-phenyl-2-butanone (or another suitable ketone).

The parent compound, indole, can be prepared by gas phase reaction of aniline and ethylene glycol over H-ZSM-5 (Si/Al 180) at 300 $^{\circ}$C in a H_2 atmosphere (269). The reaction is assumed to involve subsequent nucleophilic substitution, electrophilic substitution and dehydrogenation. The yield is moderate (54%) but again the procedure seems environment-friendly compared to the existing method.

Zeolites and clays have been postulated as prebiotic in the formation of biomolecules. In this connection, the observed formation (269a) of adenine and guanine from C_1 precursors in zeolite NaX is interesting.

Reaction 50 formulates the formation of adenine (1%) from ammonium cyanide. Note the C:N ratio of 1 in the product molecule.

$$NH_4CN \xrightarrow[\text{72h,120 }^{\circ}\text{C}]{\text{NaX}}$$

(50)

8. NUCLEOPHILIC SUBSTITUTION AND ADDITION

In nucleophilic substitution - a type of reaction frequently encountered in aliphatic chemistry - zeolites can play their role by activating the "leaving group" through protonation or metal ion coordination and by displaying shape selectivity e.g. limiting consecutive reactions.

In this paragraph some additional examples will be given of nucleophilic substitution together with some addition reactions in which a zeolite-activated olefin reacts with an O- or N-nucleophile.

8.1. Reactions of alcohols with ammonia

Methylamines, which are of considerable technical importance, are produced by the reaction of methanol with ammonia on acidic heterogeneous catalysts. In many cases, however, a product mixture is desired which differs from the equilibrium composition, in that it should contain as high as possible a fraction of mono- or dimethylamine (MMA or DMA) (270). Various zeolite catalysts have been employed in order to minimize the formation of trimethylamine (TMA). Mordenite displays a high DMA selectivity which surpasses the equilibrium level. The greatest increase in selectivity can be achieved by careful adjustment of the alkali content of the mordenite, at the expense, however, of the activity; for instance at 350 °C and a methanol conversion of < 80% the TMA content is < 20% and thus considerably below the equilibrium value of 62% (271). On small pore eight-ring zeolites such as H-Rho and H-ZK-5 at 325 °C the selectivity for DMA rises to over 60% at 90% methanol conversion (272-274). Recently it was found (275) that, surprisingly, when a Na-mordenite which has been treated with $SiCl_4$ is employed, the TMA content can be lowered below 0.5% - even at 100% methanol conversion (350 °C). The selectivity for DMA is 73%; however, the effective lifetime of the catalyst is only several hours. On the other hand the large pore HY-zeolite permits the formation of 96% TMA under comparable conditions. Although it is not yet possible to carry out the reaction on shape selective catalysts in such a way that each methylamine can be produced selectively, one can nevertheless prepare mixtures which are either almost totally free of TMA or are composed almost entirely of TMA. The selectivity for DMA can be markedly increased when the acid centres for the catalyst are, at least in part, neutralized by alkali metal ions, e.g. Na. Certainly, the restricted growth type shape selectivity of zeolites and the possibility of adjusting their acidity enable us to match the composition of the product mixture more closely to commercial requirements than in the case of the classical catalyst Al_2O_3 (276).

Shape selective C_2 to C_4 alcohol amination is described in a recent patent (277). Reactions are performed in the presence of hydrogen using cobalt- or nickel-exchanged mordenite or zeolite Y at 300 °C. Selectivity towards monoalkylamines is substantially higher on the zeolite catalysts than on a cobalt on silica catalyst. The mechanism of the metal in zeolite catalysis of this reaction is not elucidated as yet. It may be noted that the use of H-Y mainly yields olefins.

The amination of alcohols on zeolite catalysts is not restricted to monofunctional molecules. It has been shown, for instance, that nucleophilic substitution of the OH-group in ethanolamine with NH_3 yields ethylenediamine (278). A dealuminated (Si/Al = 7.2) rare earth or H-exchanged mordenite is used as the catalyst. With ethanolamine and ammonia in a molar ratio of 1:4 at 300 °C ethylenediamine is obtained with 84% selectivity at 15% conversion. As by-product aminoethylethanolamine is formed exclusively. This behaviour contrasts with that of many conventional catalysts such as Co, Ni, Cu/Cr, Pt on supports where selectivity to ethylenediamine is sacrificed in favour of conversion.

Another patent (279) describes the synthesis of polyethylenepolyamines by zeolite-catalyzed reaction of ethanolamine and ethylenediamine.

8.2. Addition of ammonia and amines to olefins

Another approach towards amines is the direct addition of ammonia or alkylamines to olefins. Various zeolites have been studied as catalysts: HY and rare earth Y (280-283, 285), H-offretite (280, 282, 286), H-clinoptilolite, H-chabazite-erionite and H-erionite (280, 282, 284), H-mordenite (280-283, 285) as well as pentasil zeolites (280, 286, 287, 288). The amination is favoured by low temperature, high pressure and high ammonia-to-olefin ratio. However, a minimum of reaction temperatures is required to activate the olefins and this treshold temperature depends on the olefin structure.

Ethylene is aminated with significant conversion of about 2%, at a minimum temperature of 320 $^{\circ}$C using HY, H-erionite and H-mordenite and the reaction conditions being 760 psig, GHSV = 1000 h^{-1} and a 4:1 molar NH$_3$ olefin ratio (280, 282). The conversion increases with temperature, i.e. from approximately 2% conversion at 320 $^{\circ}$C to about 12% conversion at 380 $^{\circ}$C over HY. In the presence of H-erionite 2.5% conversion is achieved at 320 $^{\circ}$C and 13.5% at 380 $^{\circ}$C. Mono- and diethylamines (wt ratio > 9:1) are the main products and the selectivity for these compounds is > 98%. At temperatures above 380 $^{\circ}$C the formation of nitriles and higher olefins is observed.

In the case the amination of propene the minimum temperature is 300 $^{\circ}$C to achieve 2% olefin conversion over HY. Because of steric constraints the mole ratio of mono- to diisprorylamine is 93 to 7, which is higher than in the case of ethylene amination.

Isobutene reacts with NH$_3$ over HY to t-butylamine (reaction 51) already at 220 $^{\circ}$C (281). At 300 $^{\circ}$C the equilibrium conversion (9%) is reached. Over this temperature range > 99% selectivities are obtained using a 2:1 ammonia/isobutene molar feed ratio. Oligomerization of isobutene is not significant below 300 $^{\circ}$C provided ammonia is present in excess.

$$\begin{array}{c} H_3C \\ \diagdown \\ C=CH_2 \; + \; NH_3 \; \rightleftharpoons \; H_3C-\overset{\displaystyle H_3C}{\underset{\displaystyle H_3C}{C}}-NH_2 \end{array} \qquad (51)$$

Although equilibrium conversions are low, the simplicity and high selectivity of this process, and the absence of inorganic coproducts, provide advantages

over the traditional HCN-based Ritter route to t-butylamine. Although ethylene
is effectively aminated by all zeolite catalysts and especially by small pore
zeolites such as H-erionite or H-clinoptilolite, propene shows only low
conversion and isobutene doesn't react at all over these small pore zeolites.
These results reflect shape selective restrictions. Concerning the reaction
mechanism it is found (280-282) that strongly acidic sites are necessary and
the amination occurs via protonated intermediates as evidenced by the relative
ease: isobutene > propene > ethene.

On amorphous silica-alumina low conversions and on non-acidic zeolites such
as NaY no aminations are observed.

Upon zeolite-catalyzed addition of a diamino-compound such as
ethylenediamine to isobutene only mono-alkylation is observed (289), even when
the olefin is present in large excess. Thus at 300 $^{\circ}$C and 300 bar N-t-butyl-
ethylenediamine is obtained in 20.5% yield in the presence of a borosilicate
pentasil zeolite.

The amination of dienes over zeolites is possible too (290) and suitable
dienes can give rise to N-containing ring systems. For example 2,5-dimethyl-
1,5-hexadiene - containing two active double bonds - reacts with NH_3 (reaction
52) over Ce-doped borosilicate at 300 $^{\circ}$C to give 2,2,5,5-tetramethylpyrrolidine
with 25% selectivity at 24% conversion. The noncyclic mono-addition product is
obtained with 31% selectivity.

8.3. Synthesis of anilines

The synthesis of anilines is of industrial interest. Existing methods of
preparation are not without disadvantages and a new clean zeolite-based
technology would be welcomed. Reported routes involve amination/dehydrogenation
of alicyclic systems and nucleophilic substitution of aromatic compounds with
NH_3.

The preparation by reaction of alicyclic alcohols like cyclohexanol, or
ketones like cyclohexanone with ammonia is carried out in the presence of a
crystalline ZSM-5 type silicate catalyst which contains a metal promotor having
dehydrogenation activity (291). Using NiHZSM-5 at 480 $^{\circ}$C and 200 psig only
16.2% aniline selectivity is observed. Dimeric compounds such as diphenylamine
and carbazole are preferentially formed.

706

Since amination and simultaneous dehydrogenation proceed with only moderate
yield, it is preferable to choose phenol as staring material for the synthesis
of aniline. Over H-ZSM-5 in the gas phase at 510 °C and 28 atm a phenol
conversion of 94% and an aniline selectivity of 91% (95, 292) are achieved.
Main side product is 2-methylpyridine, formed in a consecutive reaction (cf.
Section 4.5.6). Over Na-ZSM-5 at lower WHSV the selectivity to aniline raises
to 96.5%. The reaction is also catalyzed by Y-, X- and mordenite zeolites, but
these catalysts display a markedly shorter lifetime than H-ZSM-5, and in the
case of X- and Y-zeolites the conversion is lower (292, 293). By-products such
as diphenylamine and carbazole, which are obtained in the presence of non-
zeolitic silica-alumina according to the Halcon-Scientific Design process are
suppressed or eliminated by virtue of the shape selectivity of the ZSM-5
catalyst.

Ammonolysis of anisole on zeolites of the faujasite type also yields
aniline. Rearrangement towards cresols is a side reaction here. Again, the
lifetime of the catalysts is short. In a comparative study of the amination of
phenol and anisole over Y-zeolites (294) it is suggested that both protons and
cations play a role in the O → N substitution.

Several groups have reported (294-296) on the reaction of chlorobenzene and
ammonia in the presence of zeolite catalysts (reaction 53). Particularly
Cu-exchanged faujasites have been tested, which seems a logical choice. So far
conversions are modest and benzene shows up as a side product.

$$\langle\bigcirc\rangle\!-Cl \; + \; 2\;NH_3 \; \longrightarrow \; \langle\bigcirc\rangle\!-NH_2 \; + \; NH_4Cl \qquad\qquad (53)$$

On the basis of the foregoing results the only suitable route for the synthesis
of aniline is the nucleophilic substitution of phenol with ammonia in the
presence of H-ZSM-5. Substitution of conventional heterogeneous catalysts such
as non-zeolitic silica-alumina, mixtures of manganese and boron oxides and
alumina-titania by shape selective ZSM-5 catalysts has advantages with regard
to aniline selectivity. In comparison to the homogeneous, Lewis-acid-catalyzed
Halcon Process (297), the use of pentasil zeolites opens up an environmentally
and energetically more favourable route.

8.4. Nucleophilic addition to epoxides

The addition of polar compounds such as H_2O and NH_3 to epoxides on zeolite
catalysts yields the correponding diol or alcoholamine (2a). In the ring

opening of asymmetrical epoxides with amines, higher catalytic activity is achieved with weakly acidic or weakly basic (NaY, NaX) than with strongly acidic (HY) or strongly basic (CsY) zeolites (298). An interesting feature is the regioselectivity of this addition which turns out to be dependent on the type of zeolite and counterion.

$$R-CH \underset{O}{\overset{O}{\longrightarrow}} CH_2 \; + \; PhNH_2 \; \longrightarrow \; \underset{1}{RCHCH_2NHPh} \; + \; \underset{2}{RCHCH_2OH} \qquad (54)$$

with OH on the first product and NHPh on the second product.

For instance, during the reaction of 1,2-epoxyoctane with aniline in benzene at 80 °C according to reaction 54, R = C_6H_{13}, addition takes place at both the 1- and the 2-position, i.e. on the less or more hindered site of the epoxide ring. The ratio of the two isomers 1/2 is dependent upon the zeolite catalyst employed. On both NaX and NaY the two isomers are obtained in 90% yield, but on NaX the ratio 1/2 is 36 and on NaY 73. In other words, on NaY the content of isomer 1 is twice as high as on NaX.

In the reactin of styrene oxide (R = C_6H_5) Brönsted acid type Y zeolite (HY, CaY) give a 1/2 ratio of < .01 whereas the use of NaY and KY leads to 1/2 ratios of 0.16 and 8.3, respectively. The high preference for 2 formation when H^+-catalysis is applied may indicate some developing positive charge on the α-carbon which is stabilized by the phenyl substituent. In cation catalysis steric factors apparently dominate.

In the reaction of ammonia and ethylene oxide zeolites are able to restrict the products to mono- and diethanolamine.

Addition of an azido group to 2,3-epoxyalcohols leads to 2-azido-1,2-diols (299). If a CaY-zeolite loaded with sodium azide is used as reagent, a much higher regioselectivity is achieved than with conventional systems such as NaN_3/NH_4Cl in aqueous methanol, or when Al_2O_3 or SiO_2 loaded with NaN_3 are employed. Ring opening to yield the isomer in which the azido group is adjacent to the long carbon chain is favoured. In the case of 2,3-epoxyoctanol, the ratio is 94:6 at 85% yield on NaN_3/CaY, but only 66:34 at 65% yield on NaN_3/Al_2O_3 and 76:24 at 88% yield with NaN_3/NH_4Cl. The regioselectivity is explained by assuming bidentate coordination of the substrate to Ca^{II}.

These examples illustrate the property of zeolite catalysts to direct the regioselectivity in organic syntheses.

8.5. Addition of XH-compounds to unsaturates

Various XH-compounds (X = O, S, P) add - under zeolite catalysis - to the double bonds of alkenes to give saturated compounds. The addition of amines was already discussed in Section 8.2. Generally first step is proton addition followed by addition of the nucleophilic XH-compounds and transfer of a proton.

8.5.1. Addition of oxygen-compounds to alkenes

Numerous reports exist (1, 2) concerning the addition of hydroxyl-compounds to olefins. Generally zeolites do not offer distinct advantages over conventional Brönsted catalysts in these reactions. We will confine ourselves to some comments and examples.

Water addition. Alcohols are prepared by the acid-catalyzed hydration of olefins. When zeolites are used it is also necessary to accept low conversions (< 20%), in order to achieve high selectivities.

For the hydration of C_2 to C_4 olefins we refer to some comparative studies (300-302) in which several types of zeolites including the effect of the Si/Al ratio have been investigated. Ferrierite and pentasil type catalysts give the best performance.

Medium and large pore zeolites might be tuned to serve challenging hydrations in the terpene field, such as the isomerization/hydration of α-pinene towards (iso)borneol (303, 304).

Addition of alcohols. Ethers can be obtained by the acid-catalyzed addition of alcohols to olefins. This reaction is of commercial interest for the production of the octane booster, MTBE, from isobutene and methanol (reaction 55). When zeolite H-ZSM-5 (305) or Nu-10 (306) are used as catalysts, then conversions of approximately 35% and a MTBE selectivity of 95% are achieved. At approximately 100 °C and 35 bar a weakly acidic boron zeolite affords MTBE in 86% yield (307). Industrially, MTBE continues to be produced with the aid of acidic ion exchangers, since zeolite catalysts are at present less effective.

$$H_3C\diagdown$$
$$C = CH_2 + CH_3OH \rightleftharpoons H_3C - C - OCH_3 \qquad (55)$$
$$H_3C\diagup H_3C\diagup$$

Addition of carboxylic acids. The addition of acids (in most cases acetic acid) to olefins - reaction 56 - is described in several patents (308-310). Conversions of only 30% are achieved with H-ZSM-5 and H-ZSM-12 at 150-200 °C and 25 bar. In addition, mixtures of isomers are formed if asymmetrical olefins are used. Thus, 1-olefins afford the 2-carboxylates with a selectivity of 80% (310). The use of layer silicates in which replacement has been carried out with trivalent ions offers better results and zeolites. Yields of up to 90%, based on olefins, have been obtained (311, 312).

$$R-C^H=CH_2 + R'COOH \longrightarrow \underset{\underset{OCOR'}{|}}{RCH_2CH_2} + \underset{\underset{OCOR'}{|}}{RC^HCH_3} \qquad (56)$$

Zeolites, especially of the pentasil type, as well as phosphates, catalyze the intramolecular addition of carboxylate groups in unsaturated acids (313). Thus a mixture of 2-, 3- and 4-pentenoic acid esters reacts in the presence of water to yield 5-methylbutyrolactone. Using HY at 180 °C 93% selectivity to the lactone at 49% conversion of the esters is obtained.

8.5.2. Esterification and acetalization

The nucleophilic addition of alcohols to carbonyl compounds as aldehydes/ketones and carboxylic acids is the major reaction step in acetalization and esterification, respectively. As water is formed in these reactions, a zeolite catalyst should be relatively hydrophobic. Accordingly, a recent report (314) mentions the use of acid-leached mordenite (Si/Al 40-150) as the catalyst in the esterification of propionic acid and 1-butanol towards butyl propionate.

The equilibrium mixtures obtained in these reactions can be shifted towards the desired product by distilling of the water azeotropically or by selective water adsorption (cf. Section 3.6). A further example is the esterification of fatty acids with ethanol (315) with high yields in the presence of KA or NaA. Zeolites A can also serve as HCl or HBr scavengers when reacting alcohols with acid chlorides. An example is the regioselective benzoylation of 1,6-anhydroglucose in the presence of NaA (316).

Transesterification using zeolites, i.e. the formation of a new ester starting from another one, is described analogously to the synthesis of esters (317, 318). For example, yields of approximately 50% are obtained at 150 °C in the preparation of methyl acetate from acetic acid and methyl formate on zeolite H-ZSM-5.

Alcoholysis of esters, the other type of transesterification, is best performed with a basic catalyst (alkoxide). Zeolites can be of help by adsorbing the displaced alcohol. Thus dimethyl terephthalate and t-butanol are quantitatively converted into di-t-butyl terephthalate in the presence of NaA or CaA (319).

For the use of zeolites as catalysts in acetalization the reader is referred to ref. (50) and (320).

8.5.3. Addition of hydrogen sulfide

The acid-catalyzed reaction of olefins with H_2S furnishes thiols and thioethers (321-323). Non-doped X- and Y-zeolites catalyze the conversion of branched olefins into tertiary thiols in high yields (324, 325) but only give moderate yields in the case of linear olefins. Recourse is made in this case to RE-Y- and Cd-X-zeolites, which give conversions of > 99% and selectivities > 90% (321, 322). The reactions are carried out at 100-250 °C and under elevated pressure (up to 250 bar). Compared with conventional acidic catalysts, zeolites give better yields and service lives.

8.5.4. Addition of P-H and Si-H compounds

As discovered by one of the authors and his co-workers (326) zeolites catalyze the addition of phosphine and mono- and dialkylphosphines to olefins (reaction 57). Olefin conversion is much higher than in the case of amination, while maintaining high selectivity.

$$\bowtie \quad + \ H-P\big< \quad \longrightarrow \quad \underset{H}{\overset{}{\rightthreetimes}} \underset{P-}{\overset{}{\leftthreetimes}} \qquad (57)$$

At 100 °C and about 8 bar in a glass ampoule isobutene reacts with PH_3 (mol ratio 1:2) on a borosilicate pentasil zeolite to give t-butylphosphine with 98% selectivity at 42% conversion. When the temperature is raised to 200 °C the conversion increases but the selectivity falls to 85%. Under the same conditions at 100 °C the reaction of isobutene with methylphosphine yields 92% selectivity of methyl-t-butylphosphine at 41% conversion.

Also cyclic olefins and dienes have been successfully converted (326) with PH_3 into phosphines (see Table 11).

Table 11. Conversion of olefins or dienes (I) with phosphines (II).

educt I	educt II	I/II (molar)	temp. (°C)	product III	convers. of I (%)	selectivity of III (%)
C_2H_4	PH_3	1 : 1	200	$C_2H_5PH_2$	7	80
(propene)	"	1 : 1	200	>—PH_2	16	83
(isobutene)	"	1 : 2	200	+—PH_2	62	85
(cyclopentene)	"	1 : 2	100	◻—PH_2	10	85
(cyclohexene)	"	1 : 2	100	⬡—PH_2	12	81
(2,5-dimethyl-1,5-hexadiene)	"	1 : 2	100	⋀⋁—PH_2[1]	19	79
(cyclohexadiene)	"	1 : 2	100	⬡—PH_2[1]	20	75

[1] all isomers included

These high yields are by no means easily predictable, since the P-reactants, both educts and products, are thermally not very stable and highly reactive, especially towards oxygen.

Hydrosilation of olefinic compounds is a reaction which has found widespread application. Presently, homogeneous catalysts based upon transition metal and noble metal complexes are employed. Recently, hydrosilation on heterogeneous catalysts such as zeolites and phosphates has been reported (326). Using a Rh-doped pentasil zeolite a mixture of isobutene and trimethylsilane (mol ratio 1:1.5) is converted at 100 °C with 70% conversion to give t-butyltrimethylsilane with 92% selectivity. As in the case of the homogeneously catalyzed hydrosilation, the presence of a transition or noble metal is recommended to achieve a high efficiency of the reaction.

9. ZEOLITE-CATALYZED TWO- AND MULTI-STEP SYNTHESES

In the field of organic synthesis increasing attention is being paid to multifunctional catalysis, in which chemical conversion consisting of a number of individual reactions are brought about in the minimum number of steps (see e.g. ref. (327 and (2c,d)). Zeolite catalysts enable several catalytic steps to be combined in mutually optimized fashion. Examples which have already been mentioned include

712

- the reaction of caprolactone with ammonia involving addition, ring cleavage and dehydration to yield 5-hexenenitrile;
- the amination/rearrangement of phenol with ammonia leading to 2-methyl-pyridine;
- the condensation/imination/cyclization/dehydrogenation of acrolein, aldehyde and ammonia to yield 3-alkylpyridines;
- the dehydrogenation/condensation/imination/cyclization/dehydrogenation of ethanol and ammonia to yield pyridine;
- the condensation/rearrangement of phenylhydrazine and ketones towards 2,3-alkylated indoles;
- the isomerization/hydration of α-pinene towards isoborneol.

Some further examples of such two-step or multistep zeolite-catalyzed conversions will be considered in the following.

9.1. Dehydration and hydrogenation

Zeolites, particularly alumino- and borosilicate pentasil zeolites are suitable catalysts (328) for the dehydration of α-hydroxyketones to α,β-unsaturated ketones in high yield. If the zeolite is doped with a hydrogenating component such as Pd or Cu, then saturated unsymmetrical ketones can be obtained according to reaction 58.

$$
-\overset{|}{\underset{\underset{H}{|}}{C}}-\overset{|}{\underset{\underset{OH}{|}}{C}}-\overset{O}{\overset{\|}{C}}-CH_3 \longrightarrow \longrightarrow -\overset{|}{\underset{\underset{H}{|}}{C}}-\overset{|}{\underset{\underset{H}{|}}{C}}-\overset{O}{\overset{\|}{C}}-CH_3 \tag{58}
$$

Thus, using a Ce/Pd-doped borosilicate zeolite (Si/B 94, 2.3 wt % Ce, 0.5 wt % Pd) 3-hydroxy-2-methylbutan-2-one is converted quantitatively at 375 ^{0}C under H_2 to methyl isopropyl ketone. Under similar conditions non-zeolitic catalysts such as Pd (0.5 wt %) on alumina give relatively poor results.

9.2. Aldol condensation, dehydration and hydrogenation

Another route to α,β-unsaturated ketones is by means of the aldol condensation, which proceeds smoothly on acidic zeolite catalysts (329, 330). The aldol condensation of acetone on acidic zeolites yields mesityl oxide, isobutene, phorones, mesitylene and alkylphenols. In the presence of H-ZSM-5 (Si/Al = 34) acetone reacts at 200-250 ^{0}C in a sealed ampoule under autogeneous pressure to give mesityl oxide with 90% selectivity at 25% conversion (331).

Basic zeolites, such as NaX, CsNaX and NaKL, induce acetone oligomerization/dehydration through a carbanion mechanism which also leads to the dimeric mesityl oxide, the trimer phorone which compound can cyclize to isophorone and to higher oligomers. Martens et al. have shown (336) that zeolites give rise to smaller amounts of heavy products than alumina does.

Under hydrogenating conditions the aldol condensation yields saturated ketones (332-335). Thus acetone reacts in a H_2-stream at 180 °C on Pd-ZSM-5 (Si/Al - 24, 0.5 wt % Pd) according to reaction 59 to give methyl isobutyl ketone with 98% selectivity at 29% conversion (335).

$$2\ H_3C-\overset{\overset{\textstyle O}{\|}}{C}-CH_3 \longrightarrow \longrightarrow \quad \overset{\textstyle H_3C}{\underset{\textstyle H_3C}{>}}CH-CH_2-\overset{\overset{\textstyle O}{\|}}{C}-CH_3 \qquad (59)$$

Under identical reaction conditions but using Pd-Y-zeolite one achieves the same conversion but a selectivity of only 30%. In this case hydrogenation of the acetone to isopropanol occurs to a much greater extent.

Both the methods described above for the synthesis of unsymmetrically substituted saturated ketones are superior to the normal technical route involving the condensation/decarboxylation of two different organic acids, since the latter method has the disadvantage that symmetrically substituted ketones and CO_2 are unavoidably formed as by-products.

9.3. Hydroformylation and ketone formation

Rh^{III}-exchanged zeolites NaX and NaY are found (337) to catalyze the reaction of propene, CO and H_2 towards C_7 ketones according to reaction 60.

$$2\ H_3CC\overset{H}{=}CH_2 + CO + H_2 \longrightarrow \longrightarrow C_3H_7\overset{\overset{\textstyle O}{\|}}{C}C_3H_7 \qquad (60)$$

Typical reaction conditions are 150 °C, 1 atm, 1.25 s residence time over a fixed catalyst bed containing RhNaY (3.5 wt % Rh). Butyraldehyde and isobutyraldehyde are assumed to be intermediates. A mechanism for the second step is not given. The product ketones, 4-heptanone and 2-methyl-3-hexanone are formed in a ratio 2:1.

The most active catalyst for ketone synthesis is prepared by cation exchange of zeolite NaY with aqueous $RhCl_3$ in 0.2 M NaCl at 90 °C at pH 6. With

714

this catalyst all of the intermediate aldehydes are converted into ketones.

For a comparison of homogeneous and zeolite-supported Rh-catalysts in hydroformylation of various C_6-olefins the reader is referred to ref. (338).

9.4. Amidation and dehydration

Aliphatic dinitriles, such as adipodinitrile, n = 4, are prepared from dicarboxylic acids and ammonia (reaction 61) on a large scale using fluidized and fixed bed processes in the gas phase with heterogeneous catalysts. Amides would seem logical intermediates in this reaction. Selectivity-reducing cyclization to cyclopentanone and cyanocyclopentanonimine occurs in this reaction using conventional catalysts.

$$HOOC\ (CH_2)_n\ COOH\ \longrightarrow\ \longrightarrow\ NC\ (CH_2)_n\ CN \qquad (61)$$

This type of side reaction can be substantially reduced if a boron-zeolite of the ZSM-5 type is used (339) which is charged with 5.6 wt % Na and 7.6 wt % P, as the comparison with conventional catalysts based on SiO_2 shows (Table 12). The transition state selectivity of the zeolite catalysts, which does not allow cyclization, is clearly responsible for this. This example also shows the interrelationship between dehydration property and shape selectivity of the pentasil zeolites.

Table 12. Adipodinitrile out of adipic acid[a]

Products (mol %)	Catalyst	
	Na/P-B-zeolite[b]	Na/P-SiO$_2$
Adiponitrile	94.0%	83.4%
Cyanovaleric acid	0.7%	3.0%
Cyanovaleramide	2.8%	0.6%
useful products	97.5%	87.0%
Cyanocyclopentanonimine	0.5%	2.4%
Cyclopentanone	0.9%	4.3%

[a] Fluidized bed reactor, 400 °C, 200 g adipic acid/500 g catalyst.
[b] Si/B = 94.

9.5. Hydrolysis and hydrogenation

An interesting example of a single-stage zeolite-catalyzed method for a two-step reaction is the direct conversion of polysaccharides of the glucan type, especially starch, towards D-glucitol (sorbitol). In reaction 62 this is formulated for the amylose component of starch.

$$\text{amylose} \quad \longrightarrow \quad \text{glucose} \quad \longrightarrow \quad \text{sorbitol} \qquad (62)$$

The present industrial process requires purification of the intermediate glucose because the enzymatic hydrolysis does not reach completion.

A recently reported process (240) combines hydrolysis and hydrogenation by using Ru-loaded H-USY (3 wt % Ru) as a dual-function catalyst. The outer zeolite surface would seem to provide the Brönsted acidity required for the hydrolysis of the polymeric substrate. Surface roughness and crystal size are expected to be important factors. Perhaps also a homogeneous component contributes. The Ru hydrogenation component of the catalyst can exert its action at the inner as well as at the outer surface of the zeolite as the Y pore system is accessible to glucose.

Typical reaction conditions are: 180 $^{\circ}$C, batch autoclave, 5.5 MPa H_2, starch concentration 30 wt %, Ru/starch wt/wt 0.002. With this formulation a reaction time of 1 h suffices to obtain essentially quantitative conversion. The selectivity to sorbitol is > 95%. Just minor amounts of mannitol and pentitols are formed. The catalyst can be re-used many times.

Similar excellent results are also obtained (340) by combining a 5% Ru-on-carbon catalyst with an acidic zeolite catalyst (H-USY, H-mordenite or H-ZSM-5).

Altogether a fine example of zeolite catalysis demonstrating that zeolites also show promise in the conversion of polymeric materials.

FINAL REMARKS

The rapid development of zeolite catalysis in the synthesis of various types of organic compounds is reflected in the considerable body of recent literature including many patents and patent applications.

As mentioned in the introduction of this chapter the size of the molecules to be converted or synthesized seems limited by the pore dimensions of zeolites. However, a growing number of examples related to outer surface catalysis, or to selective scavenging of side products or impurities or to slow release of reagents illustrates that zeolites can be of much value also in the synthesis or conversion of large molecules or materials. Moreover, the recently discovered super-large pore molecular sieves will enlarge the spectrum of molecules having access to zeolites considerably.

Another limiting factor sometimes mentioned is the relatively rapid decay of zeolite catalysts due to coke formation. Here many examples exist - as shown in the foregoing - of stabilizing zeolites by adding components or by adapting the reaction mixture, e.g. by adding water or carbon dioxide. Also process conditions - fluidized bed with continuous regeneration - can be applied to overcome this trouble. Moreover and very importantly, zeolites can generally be completely reactivated by calcination.

Advantages - as outlined amongst others in Section 2.6 - are manifold and justify the expectation that a further exponential growth of the use of zeolites and related materials will take place. The present authors are convinced that zeolite catalysis and technology will be future cornerstones of a clean environmentally friendly organic chemicals industry.

REFERENCES

1   P.B. Venuto, ChemTech., 215 (1971); P.B. Venuto and P.S. Landis, Adv. Catal., 18 (1968) 259.
2   a. W.F. Hölderich, M. Hesse and F. Näumann, Angew. Chemie, 100 (1988) 232; b. W.F. Hölderich, in K. Tanabe, H. Hattori, T. Yamaguchi and T. Tanaka (eds.), Acid-Base Catalysis, Kodansha, Tokyo (1989), p. 1; c. W.F. Hölderich, Stud. Surf. Sci. Catal., 49 (1989) 69; d. W.F. Hölderich, Stud. Surf. Sci. Catal., 41 (1988) 83; e) ref. (254).
3   H. van Bekkum and H.W. Kouwenhoven, Recl. Trav. Chim. Pays-Bas, 108 (1989) 283; Stud. Surf. Sci. Catal., 41 (1988) 45.
4   a. I.E. Maxwell, Adv. Catal., 31 (1982) 1; b. J.M. Thomas and C.R. Theocharis, in Modern Synthetic Methods (ed. R. Scheffold), 5 (1989), p. 249. c. G. Perot and M. Guisnet, J. Mol. Catal., 61 (1990) 173.
5   U. Hammon, G.T. Kokotailo, L. Riekert and J.Q. Zhou, Zeolites, 8 (1988) 338.
6   J. Weitkamp, S. Ernst, H. Dauns and E. Gallei, Chem.-Ing.-Tech., 58 (1986) 623.
7   S.M. Csiscery, ACS Monograph, 171 (1976) 680; Zeolites, 4 (1984) 202.
8   W.O. Haag, R.M. Lago and P.B. Weisz, Faraday Disc., 72 (1982) 317.
9   N.Y. Chen, ACS Symposium Series, 368, 29 (1988) 468.
10 S.M. Csiscery, J. Catal., 108 (1987) 433.
11 P.A. Jacobs and J.A. Martens, Stud. Surf. Sci. Catal., 28 (1986) 23.
12 E.G. Derouane, J. Catal., 100 (1986) 541.
13 W.F. Hölderich, in NATO ASI Series B: Physics, D. Barthomeuf, E.G. Derouane and W. Hölderich (Eds.), 221, L1.
14 C. Herrmann, J. Haas and F. Fetting, Appl. Catal., 35 (1987) 299.
15 L.D. Rollmann and E.W. Valyocsik, US Patent 4.205.052 (1980), Mobil Oil Corp.
16 J.P. Verduyn, Eur. Pat. Appl. 219.354 (1986), Exxon Co.
17 J.H. Lunsford, Rev. Inorg. Chem., 9 (1987) 1.
18 E.F.T. Lee and L.V.C. Rees, Zeolites, 7 (1987) 446, 473.
19 A. Corma, V. Fornés, R.M. Martin-Aranda, H. Garcia and J. Primo, Appl. Catal., 59 (1990) 237.
20 P.E. Hathaway and M.E. Davis, J. Catal., 116 (1989) 263.
21 L.R.M. Martens, P.J. Grobet, W.J.M. Vermeiren and P.A. Jacobs, Stud. Surf. Sci. Catal., 28 (1986) 935; 31 (1987) 531.
21a D. Barthomeuf, G. Coudurier and J.C. Vedrine, Mater. Chem. Phys., 18 (1988) 553.
22 S.T. Homeyer and W.M.H. Sachtler, J. Catal., 118 (1989) 266; Stud. Surf. Sci. Catal., 49 (1989) 975, and references cited herein.
23 W.R. Hastings, C.J. Cameron, M.J. Thomas and M.C. Baird, Inorg. Chem., 27 (1988) 3024.
24 H.K. Beyer, I.M. Belenykaja, F. Hange, M. Tielen, P.J. Grobet and P.A. Jacobs, J. Chem. Soc. Far. Trans. I, 81 (1985) 2889; G.W. Skeels and D.W. Breck, in Proc. 6th Int. Zeolite Conference, Reno, Butterworth (1983) p. 87.
25 L.V.C. Rees and E.F.T. Lee, PCT, WO 88/01254 (1988).
26 T. Hibino, M. Niwa, Y. Murakami, M. Sano, S. Komai and T. Hanaichi, J. Phys. Chem., 93 (1988) 7847.
26a A. Thys, G. Peeters, E.F. Vansant and I. Verhaert, J. Chem. Soc. Faraday Trans. I, 79 (1983) 2821, 2835.
27 P.G. Rodewald, US Patent 4.477.583 (1984), Mobil Oil Corp.
28 G.H. Kuehl, US Patent 4.482.531 (1984), Mobil Oil Corp.
29 E.W. Valyocsik, US Patent 4.490.342 (1984), Mobil Oil Corp.
30 J.R. Anderson, K. Foger, T. Male, R.A. Rajadhyaksha and J.V. Sanders, J. Catal., 58 (1979) 114.
31 W.O. Haag and D.H. Olson, US Patent 4.117.026 (1978), Mobil Oil Corp.
32 J.A. Martens, M. Tielen, P.A. Jacobs and J. Weitkamp, Zeolites, 4 (1984) 98.

718

32a W.F. Hölderich, L. Riekert, M. Kotter and U. Hammon, Ger. Offen DE 3.231.498 (1984), BASF AG.

33 For an example see: Th.M. Wortel, W. van den Heuvel, N.A. de Munck and H. van Bekkum, Acta Phys. Chem. (Szeged), 24 (1978) 341.

34 For a review see: D.M. Ruthven, Chem. Eng. Progr., (2) (1988) 42.

35 D.R. Burfield, G.H. Gan and R.H. Smithers, J. Appl. Chem. Biotechnol., 28 (1978) 23. See for drying of other solvents by zeolites: D.R. Burfield et al., J. Org. Chem., 42 (1977) 3060; 43 (1978) 3966; 46 (1981) 629.

36 Y. Gao, R.M. Hansen, J.M. Klunder, H. Masamune and K.B. Sharpless, J. Am. Chem. Soc., 109 (1987) 5765.

37 K. Mikami, M. Terada and T. Nakai, J. Am. Chem. Soc., 111 (1989) 1940.

38 Th.M. Wortel and H. van Bekkum, J. Org. Chem., 45 (1980) 4763.

39 For a review on bulk separations see: R.V. Jasra and S.G.T. Bhat, Separ. Sc. and Technol., 23 (1988) 945.

40 M.E. Davis, C. Montes, P.E. Hathaway, J.P. Arhancet, D.L. Hasha and J.M. Garces, J. Am. Chem. Soc., 111 (1989) 3919.

41 J.D. Shermann and C.C. Chao, in Proc. 7th Int. Zeolite Conference, Kodansha (1986), p. 1025.

42 M.C. Heslinga, H. Pluim and H. van Bekkum, in "Zeolites for the nineties", 8th Int. Zeolite Conference (1989), p. 347.

43 T.P.J. Izod, Eur. Pat. Appl. 13.451 (1980), Union Carbide Corp.

44 M. Floor, A.P.G. Kieboom and H. van Bekkum, Recl. Trav. Chim. Pays-Bas, 108 (1989) 128.

45 C.A.A. van Boeckel, T. Beetz, A.C. Kock-van Dalen and H. van Bekkum, Recl. Trav. Chim. Pays-Bas, 106 (1987) 596.

46 M. Onaka and Y. Izumi, Chem. Lett., (1984) 2007.

47 K. Smith, Bull. Soc. Chim. France, (1989) 272.

48 N.J. Turro, J.R. Fehlner, D.. Hessler, K.M. Welsh, W. Ruderman, D. Fernberg and A.M. Braun, J. Org. Chem., 53 (1988) 3731.

49 R.M. Dessau, J. Chem. Soc. Chem. Commun., (1986) 1167.

50 Th.M. Wortel, W.H. Esser, G. van Minnen-Pathuis, R. Taal, D.P. Roelofsen and H. van Bekkum, Recl. Trav. Chim. Pays-Bas, 96 (1977) 44.

51 R. Huang, Q. Chen and Z. Yu, C.N. 85.102.395, Chem. Abstr., 108 (1988) 133817.

52 D.P. Roelofsen and H. van Bekkum, Recl. Trav. Chim. Pays-Bas, 91 (1972) 605.

53 M.P. de Nijs and A.P.G. Kieboom, Preprints Eurocarb V, Prague 1989, paper C-30.

54 H. Pines, "The Chemistry of Catalytic Hydrocarbon Conversions", Academic Press (1981) 123.

55 K. Tanabe, M. Misono, Y. Ono and H. Hattori, Stud. Surf. Sci. Catal., 51 (1989) 215.

56 E.g. L.R. Martens, P.J. Grobet, W.J. Vermeiren and P.A. Jacobs, Stud. Surf. Sci. Catal., 28 (1986) 935.

57 W. Hölderich, A. Lindner, W.D. Mross, M. Schwarzmann, K. Volkammer and U. Wagner, Eur. Pat. 129.899 (1985), BASF AG.

58 A. Lindner, U. Wagner, K. Volkamer, W. Hölderich, W. Hochstein and W. Immel, Eur. Pat. 129.000 (1985), BASF AG.

59 W. Hölderich, A. Lindner, W.D. Mross and M. Strohmeyer, DBP 3.427.979 (1986), BASF AG.

60 D.L. Sikkenga, US Pat. 4.503.282 (1985), Standard Oil Co.

61 R. Fischer, W. Hölderich, W.D. Mross and H.M. Weitz, Eur. Pat. 0.167.021 (1986), BASF AG.

62 H.W. Schneider and R. Kummer, DEP 3.317.163 (1984), BASF AG.

63 W. Hölderich, R. Fischer, K.D. Malsch and H. Lendle, DEP 3.521.380 (1986), BASF AG.

64 W. Hölderich, W. Richter, H. Lendle and K.D. Malsch, DEP 3.521.381 (1986), BASF AG.

65 W. Hölderich, H. Aichinger, F. Näumann and R. Fischer, DEP 3.638.011 (1988), BASF AG.

66   W. Hölderich, R. Fischer, W.D. Mross and H.M. Weitz, Eur. Pat. 0.132.737 (1985), BASF AG.
67   M. Stoecker, P. Hemmersbach, H. Raeder and J.K. Grepstad, Appl. Catal., 25 (1986) 223.
68   H.W. Kouwenhoven, Adv. Chem. Ser., 121 (1973) 529.
69   J. Weitkamp, W. Gerhard and P.A. Jacobs, Acta Phys. Chem. (Szeged), (1985) 261.
70   K. Honna, M. Sugimoto, N. Shimizu and K. Kurisaki, Chem. Lett., (1986) 315.
71   G.C. Lau and W.F. Maier, Langmuir, 3 (1987) 164.
72   M. Nomura and Y. Fujihara, Nippon Nogei Kagaku Kaishi, 57 (1983) 1227; CA, 101 (1984) 7440f.
73   W. Hölderich, F. Merger, W.D. Mrosz and R. Fischer, Eur. Pat. 0.162.387 (1985), BASF AG.
74   K.G. Ione, L.A. Vostrikova and V.M. Mastikin, J. Mol. Catal., 31 (1985) 355.
75   H.C. Linstid and G.S. Koermer, US Pat. 4.537.995 (1985), Celanese Corp.
76   Q. Wang and Z. Chen, Kexue Tongbao, 29 (1984) 1130.
77   A. Molnar, S. Bucsi and H. Bartok, Stud. Surf. Sci. Catal., 41 (1982) 203.
78   W. Hölderich, N. Götz, L. Hupfer, R. Kropp, H. Theobald and B. Wolf, DOS 3.546.372 (1987), BASF AG.
79   C. Neri and F. Buonomo, Eur. Pat. 100.117 (1983), Enichem Anic SpA.
80   C. Neri and F. Buonomo, Eur. Pat. 102.097 (1984), Enichem Anic SpA.
81   M. Orisaku and K. Sano, Jap. Pat. 61.112.040 (1986), Mitsubishi Gas Chem.
82   W. Hölderich, N. Götz, L. Hupfer and H. Lermer, DEP 3.632.529 (1988), BASF AG.
83   E.g. Y. Fujihare, N. Nomura and K. Igawa, Jap. Pat. 62.19.549 and 62.114.926 (1987), Toyo Soda Mfg.
84   J. Kaminska, M.A. Schwegler and H. van Bekkum, to be published.
85   W. Hölderich, F. Merger and R. Fischer, Eur. Pat. 199.210 (1986), BASF AG.
86   Z. Ore, Z. Chen and O. Jiang, Kexue Tongbao, 32 (1987) 462.
87   W. Hölderich, R. Fischer and W. Mesch, DOS 3.636.430 (1987), BASF AG.
88   H.G. Schmidt, Eur. Pat. 75.680 (1986), Dynamit Nobel AG.
89   W.K. Bell and C.D. Chang, Eur. Pat. 056.698 (1985), Mobil Oil Corp.
90   P.S. Landis and P.B. Venuto, J. Catal., 5 (1966) 245.
91   H. Sato, N. Ishii, K. Hirose and S. Nakamura, Stud. Surf. Sci. Catal., 28 (1986) 755.
92   H. Sato, K. Hirose, N. Ihii and Y. Umada, Eur. Pat. 234.088 (1987), Sumitomo Chem. Co.
93   H. Sato, K. Hirose, M. Kitamura, H. Tojima and N. Ishii, Eur. Pat. 236.092 (1987), Sumitomo Chem. Co.
94   K.D. Olson, Eur. Pat. 251.168 (1988), UCC.
95   C.D. Chang and P.D. Perkins, Zeolites, 3 (1983) 298.
96   H. LeBlanc, L. Puppe and K. Wedemeyer, DE 3.332.687 (1985), Bayer AG.
97   V. Solinas, R. Monaci, B. Marongiu and L. Forni, Appl. Catal., 9 (1984) 109.
98   M. Neuber, H.G. Karge and J. Weitkamp, Catalysis Today, 3 (1988) 11.
99   F.J. Weigert, J. Org. Chem., 51 (1986) 2653.
100   F.J. Weigert, J. Org. Chem., 52 (1987) 3296.
101   K. Eichler, E. Leupold, H.J. Arpe and H. Baltes, DE 3.420.707 (1985), Hoechst AG.
102   R.H. Hardy and H. Burton, J. Catal., 111 (1988) 146.
103   K. Iwayama, Y. Magatani and K. Tada, Jap. Pat. 6.393.739 (1988), Toray Ind.
104   K. Tada, Y. Imada and K. Iwayama, Jap. Pat. 6.393.738 (1988), Toray Ind.
105   K. Iwayama, Y. Magatani and K. Tada, Eur. Pat. Appl. 0.278.729 (1988); Jap. Pat. 01.254.633 (1989), Toray Ind.

720

106   K. Eichler, H.J. Arpe, H. Baltes and E. Leupold, Eur. Pat. Appl. 164.045
       (1985), Hoechst AG.
107   H. Litterer and E.J. Leupold, DBP 3.537.288 (1987), Hoechst AG.
108   U. Dettmeyer, K. Eichler, K. Kühlein, E.J. Leupold and H. Litterer,
       Angew. Chem., 99 (1987) 470; Angew. Chem. Int. Ed., 26 (1987) 468.
109   Y. Pouilloux, N.S. Gnep, P. Magnoux and G. Perot, J. Mol. Catal., 40
       (1987) 231.
110   C.S. Cundy, R. Higgins, S.A.M. Kibby, B.M. Lowe and R.M. Paton,
       Tetrahedron Lett., 30 (1989) 2281.
110a   Y. Pouilloux, J.P. Bodibo, I. Neves, M. Gubelmann, G. Perot and M.
       Guisnet, Preprints 2nd Symposium Heterogeneous Heterogeneous Catalysis
       and Fine Chemicals, Poitiers, 1990, C-145. To appear in Stud. Surf. Sci.
       Catal.
111   F.G. Dwyer, in W.R. Moser (ed.), Chemical Industries, Vol. 5, Marcel
       Dekker (1981) p. 29.
112   J. Weitkamp, Acta Phys. Chem., 31 (1985) 271.
113   W.F. Hölderich and E. Gallei, Ger. Chem. Eng., 8 (1985) 337; Chem.-Ing.-
       Tech., 56 (1984) 908.
114   R.F. Parton, J.M. Jacobs, D.R. Huybrechts and P.A. Jacobs, Stud. Surf.
       Sci. Catal., 46 (1989) 163.
115   L.B. Young, US Pat. 3.962.364 (1976), Mobil Oil Corp.
116   H.A. Bouncer and I.A. Cody, Eur. Pat. 160.144 and 160.145 (1985), Exxon
       Co.
117   M.W. Anderson and J. Klinowski, J. Am. Chem. Soc., 112 (1990) 10.
118   D. Fraenkel and M. Levy, J. Catal., 118 (1989) 10.
119   W.W. Keading, G.C. Barile and M.M. Wu, Catal. Rev.-Sci. Eng., 26 (1984)
       597.
120   W.W. Keading, L.B. Young and C.C. Chu, J. Catal., 89 (1984) 267.
121   K.H. Chandavar, S.G. Hedge, S.B. Kulkarni, P. Ratnasamy, G. Chitlangia,
       A. Singh and A.V. Deo, in Proc. 6th Int. Zeolite Conference, Reno,
       Butterworth (1984) p. 325.
122   K. Eichler and E. Leupold, DBP 3.334.084 (1985), Hoechst AG.
123   R.J. Sampson, Ch.B. Hanson and J.P. Candlin, Eur. Pat. 202.752 (1986),
       ICI.
124   O. Takahashi, Jap. Pat. 62.29.536 (1987), Idemitsu Kosan Co.
125   D. Fraenkel, M. Cherniavsky, B. Ittah and J. Levy, J. Catal., 101 (1986)
       273.
126   M. Neuber and J. Weitkamp, in "Zeolites for the nineties", 8th Int.
       Zeolite Conference (1989), p. 425.
127   J. Weitkamp, M. Neuber, W. Höltmann and G. Collin, DEOS 37.23.104 (1989),
       Rütgerswerke AG.
128   E.G. Derouane, J. Catal., 100 (1986) 545.
129   K. Tanaguchi, M. Tanaka, K. Takahata, N. Sakamoto, T. Takai, Y. Kurano
       and M. Ishibashi, Eur. Pat. Appl. 288.582 (1988), Mitsui Petrochem. Ind.
130   G.S. Lee, J.J. Maj, S.C. Rocke and J.M. Garcès, Catalysis Letters, 2
       (1989) 243.
131   R.F. Parton, J.M. Jacobs, H. van Ooteghem and P.A. Jacobs, Stud. Surf.
       Sci. Catal., 46 (1989) 211.
132   M. Marczewski, J.P. Bodibo, G. Perot and M. Guisnet, J. Mol. Catal., 50
       (1989) 211.
133   M. Marczewski, G. Perot and M. Guisnet, Stud. Surf. Sci. Catal., 41
       (1988) 273.
134   A. Corma, H. Garcia and J. Primo, J. Chem. Research, (1988) 40.
135   L.B. Young, Eur. Pat. 29.333 (1981), Mobil Oil Corp.
136   P.Y. Chen, M.C. Chen, H.Y. Chu, N.S. Chang and T.K. Chuang, Stud. Surf.
       Sci. Catal., 28 (1986) 739.
137   K.G. Ione and O.V. Kikhtyanin, Stud. Surf. Sci. Catal., 49 (1989) 1073.
138   W.F. Burgoyne, D.D. Dixon and J.P. Casey, ChemTech, (1989) 690, and
       references cited therein. W.F. Burgoyne and D.D. Dale, Appl. Catal., 62
       (1990) 161 and 63 (1990) 117.

721

139 R. Agrawal, S. Auvil and M. Deeba, Eur. Pat. 240.018 (1987), Air Prod. Chem. Inc.
140 R. Pierantozzi, Eur. Pat. 245.797 (1987), Air Prod. Chem. Inc.
141 K. Eichler and E. Leupold, Eur. Pat. 175.969 (1986), Hoechst AG.
142 V. Solinas, R. Monaci, G. Longu and L. Forni, Acta Phys. Chem., 31 (1985) 291.
143 H. Kashiwagi, Y. Fujiki and S. Enomoto, Chem. Pharm. Bull., 30 (1982) 404, 2575.
144 B. Chiche, A. Finiels, C. Gauthier, P. Geneste, J. Graille and D. Pioch, J. Org. Chem., 51 (1986) 2128.
145 B. Chiche, A. Finiels, C. Gauthier, P. Geneste, J. Graille and D. Pioch, J. Mol. Catal., 42 (1987) 229.
146 B.B.G. Gupta, Eur. Pat. 239.383 (1987), Celanese Corp.
147 C. Gauthier, B. Chiche, A. Finiels and P. Geneste, J. Mol. Catal., 50 (1989) 219.
148 B. Chiche, A. Finiels, C. Gauthier and P. Geneste, Appl. Catal., 30 (1987) 365.
149 Jap. Pat. 81.142.233 (1981), Mitsui Toatsu Chemicals Inc.
150 G. Friedhoven, O. Immel and H.-H. Schwarz, DBP 2.633.458 (1978), Bayer AG.
151 A.J. Hoefnagel and H. van Bekkum, unpublished results.
152 I. Nicolau and A. Aguilo, US Pat. 4.652.683 (1987), Celanese Corp.
153 B.B.G. Gupta, US Pat. 4.668.826 (1987), Celanese Corp.
154 H. Urano and H. Kikuchi, Jap. Pat. 63.264.543 (1988), Toyo Gosei Kogyo Co. Ltd.
155 A. Corma, M.J. Climent, H. Garcia and J. Primo, Appl. Catal., 49 (1989) 109.
156 W.F. Hölderich, H. Lermer and M. Schwarzmann, DOS 3.618.964 (1987), BASF AG.
157 H. Lermer, W.F. Hölderich, T. Dockner and H. Koehler, DEOS 3.724.035 (1987), BASF AG.
158 Jap. Pat. 58.157.748 (1983), Sumitomo Chem. Co.
159 I. Schumacher and K. Wang, US Pat. 4.426.543 (1984), Monsanto Co.
160 I. Schumacher and K. Wang, Eur. Pat. 078.245 (1985), Monsanto Co.
161 M. Furuya and H. Nakajima, Jap. Pat. 01.09.960 (1989), Technol. Res. Assoc.
162 T. Huizinga, J.J.F. Scholten, Th.M. Wortel and H. van Bekkum, Tetrahedron Lett., (1980) 3809.
163 J. Pardillos, D. Brunel, B. Coq, P. Massiani, L.C. de Ménorval and F. Figueras, J. Am. Chem. Soc., 112 (1990) 1313.
164 Y. Higuchi and T. Suzuki, Eur. Pat. 112.722 (1984), Ihara Chem. Ind.
165 K. Sekizawa, T. Hironaka and Y. Tsutsumi, Eur. Pat. 171.256 (1986), Toyo Soda Mfg.
166 T. Miyake, K. Sekizawa, T. Hironaka and Y. Tsutsumi, US Pat. 4.861.929 (1989), Tosoh Corp.
167 K. Sekizawa, T. Miyake, T. Hironaka and Y. Tsutsumi, US Pat. 4.849.560 (1989), Toyo Soda Mfg.
168 T. Suzuki and Y. Higuchi, US Pat. 4.822.933 (1989), Ihara Chem. Ind.
169 Y. Higuchi and T. Suzuki, Eur. Pat. 118.851 (1984), Ihara Chem. Ind.
170 T. Miyake, K. Sekizawa, T. Hironaka, M. Nakano, S. Fujii and Y. Tsutsumi, Stud. Surf. Sci. Catal., 28 (1986) 747.
171 T. Hironaka, K. Sekizawa and Y. Tsutsumi, DEOS 32.20.391 (1987), Toyo Soda Mfg.
172 T. Suzuki and C. Komatsu, Eur. Pat. 154.236 (1985), Ihara Chem. Ind.
173 A. Botta, H.J. Buysch and L. Puppe, DEOS 38.09.258 (1989), Bayer AG.
174 J.C. Oudejans, A.C. Kock-van Dalen and H. van Bekkum, to be published.
175 Th.M. Wortel, D. Oudijn, C.J. Vleugel, D.P. Roelofsen and H. van Bekkum, J. Catal., 60 (1979) 110.
175a F. de la Vega and Y. Sasson, Zeolites, 9 (1989) 418.
176 F. de la Vega and Y. Sasson, J. Chem. Soc., Chem. Commun., (1989) 653;

722

176a R.P. Cooney and P. Tsai, J. Raman Spectrosc., 8 (1979) 195, 236.
177 S. Kai, Jap. Pat. 60.224.644 (1985), Asahi Chem. Ind.
178 S. Kai, Jap. Pat. 60.224.645 (1985), Asahi Chem. Ind.
178a K. Smith, Preprints 2nd Symposium Heterogeneous Catalysis and Fine Chemicals, Poitiers, 1990. To appear in Stud. Surf. Sci. Catal.
179 G. Paparatto and M. Saetti, Eur. Pat. 181.790 and 183.579 (1986), Montedipe SpA.
180 G.C. Tustin and M. Rule, US Pat. 4.778.939 (1988), Eastman Kodak Co.
181 M. Rule, G.C. Tustin, D.W. Lane and T.H. Larkins, US Pat. 4.778.940 (1988), Eastman Kodak Co.
182 Y. Kiso, T. Takai and T. Hayashi, Eur. Pat. 329.367 (1989), Mitsui Petrochem. Ind.
183 K.D. Olson, Eur. Pat. 251.169 (1987), Union Carbide Corp.
184 T. Kiyora, Jap. Pat. 63.307.834 and Jap. Pat. 63.307.835 (1988), Mitsui Toatsu Chem. Inc.
185 B.B.G. Gupta, US Pat. 4.694.111 (1987), Celanese Corp.
186 E.L. Moorehead, US Pat. 4.604.371 (1986), Union Oil Co.
187 H. Arai, T. Yamashiro, T. Kubo and H. Tominaga, Bull. J. Pet. Inst., 18 (1976) 39.
188 Kh.M. Minachev, N.Y. Usachev, A.P. Rodin, V.P. Katalin and Y.I. Isakov, Izv. Akad. Nauk SSSR Ser. Khim, (1981) 267, 724; (1982) 132.
189 Kh.M. Minachev, O.M. Nefedov, V.V. Kharlamov, S.Y. Panov and S.F. Politanskii, Izv. Akad. Nauk SSSR Ser. Khim., (1981) 1490.
190 L. Ya Leitio, K.J. Rubina, M.V. Shimanskaya, V.V. Kharlamov and Kh.M. Minachev, React. Kinet. Catal. Lett., 29 (1985) 175.
191 D.B. Tagiev and Kh.M. Minachev, Russ. Chem. Rev., 50 (1981) 1009 and Stud. Surf. Sci. Catal., 28 (1986) 981.
192 D.G. Jones and P.S. Landis, US Pat. 3.231.600 (1966), Mobil Oil Corp. and ref. (1).
193 J.C. Oudejans, F.J. van der Gaag and H. van Bekkum, Proc. 6th Int. Zeolite Conf. Reno, Butterworth, 1984, p. 536.
194 F.J. van der Gaag, Thesis Delft University of Technology, 1987.
195 F. Cavani, F. Trifiro, P. Jiru, K. Habersberger and Z. Tvaruzkova, Zeolites, 8 (1988) 12.
196 F. Trifiro and P. Jiru, Catal. Today, 3 (1988) 519.
197 Th.G. Attig, K.J. Kuruc and R.K. Grasselli, US Pat. 4.736.054 (1988), Standard Oil, Ohio.
198 S.S. Shepelev and K.G. Ione, SU Pat. 1.321.722 (1987), Inst. Catal. Novosibirsk.
199 A. Miyamoto, Y. Iwamoto, H. Matsuda and T. Inui, Stud. Surf. Sci. Catal., 49 (1989) 1233.
200 C.D. Chang and S.D. Hellring, US Pat. 4.578.521 (1986), Mobil Oil Corp.
201 C. Ferrini and H.W. Kouwenhoven, Stud. Surf. Sci. Catal., 55 (1990) 53.
202 A. Esposito, M. Taramasso, C. Neri ande F. Buonomo, Br. Pat. 2.116.974 (1985), Anic SpA.
203 G. Belussi, M. Clerici, F. Buonomo, U. Romano, A. Esposito and B. Notari, Eur. Pat. 200.260 (1986), Enichem.
204 B. Notari, Stud. Surf. Sci. Catal., 37 (1987) 413.
205 M. Constantini, J.M. Popa and M. Gubelmann, Eur. Pat. 314.582 (1989), Rhone Poulenc Chimie.
206 E.G. Derouane and D.J. Vanderveken, Appl. Catal., 45 (1988) L15.
207 A. Esposito, C. Neri and F. Buonomo, US Pat. 4.480.135 (1984), Anic SpA.
208 U. Romano, A. Esposito, F. Maspero, C. Neri and M.G. Clerici, Stud. Surf. Sci. Catal., 55 (1990) 33.
209 C. Neri, B. Anfossi, A. Esposito and F. Buonomo, Eur. Pat. 100.118 and 100.119 (1984), Anic SpA.
210 C. Neri and F. Buonomo, Eur. Pat. 102.097 (1984), Anic SpA.
211 M. Clerici and U. Romano, Eur. Pat. 230.949 (1987), Enichem.
212 F. Maspero and U. Romano, Eur. Pat. 190.609 (1986), Enichem.

213a D.R.C. Huybrechts, L. de Bruycker and P.A. Jacobs, Nature (London), 345 (1990) 240.

213b T. Tatsumi, M. Nakamura, S. Negishi and H. Tominaga, J. Chem. Soc. Commun., (1990) 476; M. Nakamura, S. Negishi, K. Yuasa, T. Tatsumi and H. Tominaga, Shokubai, 32 (1990) 99.

214 J. Hartig, A. Stoessel, G. Hermann and L. Marosi, DOS 3.222.144 (1983), BASF AG.

215 U. Romano, M.G. Clerici, G. Bellussi and F. Buonomo, Eur. Pat. 203.632 (1986), Enichem.

216 J.C. Oudejans and H. van Bekkum, J. Mol. Catal., 12 (1981) 149.

217 M. Floor, A.P.G. Kieboom and H. van Bekkum, Recl. Trav. Chim. Pays-Bas, 107 (1988) 362.

218 P. Roffia, G. Leofanti, A. Cesana, M. Mantegazza, M. Padovan, G. Petrini, S. Tonti and P. Gervasutti, Stud. Surf. Sci. Catal., 55 (1990) 43.

219 P. Roffia, G. Paparatto, A. Cesana and G. Tauszik, Eur. Pat. 301.486 (1989), Montedipe SpA.

220 S. Tonti, P.Roffia, A. Cesana, M.A. Mantegazza and M. Padovan, Eur. Pat. 314.147 (1989), Montedipe SpA.

221 V.Y. Zakharov and B.V. Romanovsky, Vestn. Mosk. Univ. Khim., 20 (1979) 94; B.V. Romanovsky, Proc. 8th Int. Congr. Catal., Verlag Chemie (1984) (4) 657. B.V. Romanovsky, Acta Phys. Chem., 31 (1985) 215.

222 N. Herron, G.D. Stucky and C.A. Tolman, J. Chem. Soc. Chem. Commun., (1986) 1521.

223 C.A. Tolman and N. Herron, Catal. Today, 3 (1988) 235.

224 C.A. Tolman, J.D. Douliner, M.J. Nappa and N. Herron, in C.L. Hill (Ed.,), Act. Funct. Alk. Wiley, 1989, 303.

225 N. Herron, ACS Symp. Series, 392 (1989) 141.

226 T.V. Korolkova and B.V. Romanovsky, Neftekhimiya, 26 (1986) 546.

226a R.F. Parton, L. Uytterhoeven and P.A. Jacobs, Preprints 2nd Symposium Heterogeneous Catalysis and Fine Chemicals, Poitiers, 1990, C-93. To appear in Stud. Surf. Sci. Catal.

227 N. Herron, ChemTech, (1989) 542; Inorg. Chem., 25 (1986) 4714.

228 C. Bowers and P.B. Dutta, J. Catal., 122 (1990) 271.

229 N. Herron and C.A. Tolman, J. Am. Chem. Soc., 109 (1987) 2837.

230 F.L. Pettit and M.A. Fox, J. Phys. Chem., 90 (1986) 1353.

231 N. Herron, Y. Wang, M.M. Eddy, G.D. Stucky, D.E. Cox, K. Moller and T. Bein, J. Am. Chem. Soc., 111 (1989) 530.

232 Y. Ono, K. Tohmori, S. Suzuki, K. Nakashiro and E. Suzuki, Stud. Surf. Sci. Catal., 41 (1988) 75.

233 M. Gubelman and J.P. Tirel, Eur. Pat. Appl. 341.165 (1989), Rhone Poulenc Chimie.

233a G.I. Panov, V.I. Sobolev and A.S. Kharitonov, J. Mol. Catal., 61 (1990) 85.

234 C.D. Chang and A.J. Silvestri, ChemTech, (1987) 624 and references cited therein.

235 See e.g. a. J.C. Oudejans, P.F. van den Oosterkamp and H. van Bekkum, Appl. Catal., 3 (1982) 109; b. G.A. Aldridge, X.E. Veryklos and R. Mutharasan, Ind. Eng. Chem. Process Des. Dev., 23 (1984) 733.

236 P.A. de Boks, T.F. Huber, W.J. van Nes, P.F. van den Oosterkamp, H. van Bekikum, G.C. van Eybergen, J.H. van den Hende, N.W.F. Kossen, C.I. Mooring, J.C. Oudejans, A.C. Snaterse and J.A. Wesselingh, Biotechnology Lett., 4 (1982) 447.

237 a. H. Kitagawa, Y. Sendoda and Y. Ono, J. Catal., 101 (1986) 12; b. N.S. Gnep, J.Y. Doyemet and M. Guisnet, Stud. Surf. Sci. Catal., 46 (1989) 153; c. L. Petit, J.P. Boursonville and F. Raatz, Stud. Surf. Sci. Catal., 49 (1989) 1163; d. T. Inui, Y. Ishihara, K. Kamachi and H. Matsuda, ibid., 1183; e. P. Meriaudeau, G. Sapaly and C. Nacchae, ibid., 1423; f. C.R. Bayense and J.H.C. van Hooff, Rec. Res. Reports, 8th Int. Zeol. Conf. Amsterdam, 1989, p. 181.

238 T. Mole, J.R. Anderson and G. Greer, Appl. Catal., 17 (1985) 141.

724

239 a. J.R. Bernard, in L.V.C. Rees (ed.), Proc. 5th Int. Zeol. Conf., Naples, Heyden, London, 1980, p. 686; b. T.R. Hughes, W.C. Buss, P.W. Tamm and R.L. Jacobson, in Proc. 7th Int. Zeol. Conf., Tokyo, Kodansha-Elsevier, 1986, p. 725.

240 I.E. Maxwell, R.S. Downing, J.J. de Boer and S.A. van Langen, J. Catal., 61 (1980) 485, 493.

241 J. Ipaktschi, Z. Naturforsch., 416 (1986) 496.

242 P.Laszlo and J. Lucchetti, Tetrahedron Lett., (1984) 4387.

243 S. Ghosh and N.L. Bauld, J. Catal., 95 (1985) 300.

244 A.J. Schipperijn and J. Lukas, Recl. Trav. Chim. Pays-Bas, 92 (1973) 572.

245 J.C. Oudejans, J. Kaminska and H. van Bekkum, Recl. Trav. Chim. Pays-Bas, 102 (1983) 537.

246 M. Onaka, H. Kita and Y. Izumi, Chem. Lett., (1985) 1895.

247 J.C. Oudejans, J. Kaminska, A.C. Kock-van Dalen and H. van Bekkum, Recl. Trav. Chim. Pays-Bas, 105 (1986) 421.

248 K. Ganesan and C.N. Pillai, J. Catal., 119 (1990) 8.

249 For a review, see Y. Ono, Heterocycles, 16 (1981) 1755.

250 D.J. Sikkema, P. Hoogland, J. Bik and P. Lam, Polymer, 27 (1986) 1441.

251 W.F. Hölderich and M. Schwarzmann, Eur. Pat. 309.911 (1989), BASF AG.

252 W.F. Hölderich and M. Schwarzmann, Eur. Pat. 309.906 (1989), BASF AG.

253 R.M. Dessau, Zeolites, 10 (1990) 205; US Pat. 4.658.045 (1987), Mobil Oil Corp.

254 W.F. Hölderich, Stud. Surf. Sci. Catal., 46 (1989) 193.

255 Y. Ono, K. Hatada, K. Fujita, A. Halgeri and T. Klii, J. Catal., 41 (1976) 322.

256 Y. Ono, A. Halgeri, M. Kanedo, K. Hatada, ACS Symp. Ser., 40 (1977) 596.

257 K. Hatada and Y. Ono, Bull. Chem. Soc. Japan, 50 (1977) 2517.

258 W.F. Hölderich and M. Schwarzmann, Eur. Pat. 267.438 (1988), BASF AG.

259 D. Feitler, W. Schimming and H. Wetstein, US Pat. 4.675.410 (1987), Nepera Inc.

260 W.F. Hölderich, N. Götz and G. Fouquet, Eur. Pat. 263.464 (1988), BASF AG.

261 F.J. van der Gaag, R.J.O. Adriaansens, H. van Bekkum and P.C. van Geem, Stud. Surf. Sci. Catal., 52 (1989) 283.

262 Y. Matsumura, K. Hashimoto and S. Yoshida, J. Catal., 100 (1986) 392.

263 F.J. van der Gaag, F. Louter, J.C. Oudejans and H. van Bekkum, Appl. Catal., 26 (1986) 191.

264 F.J. van der Gaag, F. Louter and H. van Bekkum, Proc. 7th Int. Zeol. Conf., Tokyo, Kodansha, Elsevier (1986), p. 763.

265 R.A. le Fêbre, Thesis Delft University of Technology, 1989.

266 R.A. Budnik and M.R. Sandner, Eur. Pat. 158.139 (1985), Union Carbide Corp.

267 H. Sato and M. Tsuzuki, Int. Appl. WO 8.703.592 (1987), Idemitsu Kosan Co.

268 M.P. Prochazka, L. Eklund and R. Carlson, Acta Chem. Scand., in the press; M.P. Prochazka, Thesis Umea University, Sweden, 1990.

269 H. Sato and M. Tsuzuki, Jap. Pat. 63.196.561 (1988), Idemitsu Kosan Co.

269a F. Seel, M. von Blon and A. Dessaur, Z. Naturforsch., 37b (1982) 820.

270 F.J. Weigert, J. Catal., 103 (1987) 20.

271 Y. Ashina, T. Fujita, M. Fukatsu, K. Niwa and J. Yagi, Proc. 7th Int. Zeol. Conf., Tokyo, 1986, Kodansha, p. 779.

272 R.D. Shannon, M. Keane, L. Abrams, R.H. Staley, T.E. Gier and G.C. Sonnichsen, J. Catal., 113 (1988) 367; 114 (1988) 8; 115 (1989) 79.

273 H.E. Bergna, M. Keane, D.H. Ralston, G.C. Sonnichsen, L. Abrams and R.D. Shannon, J. Catal., 115 (1989) 148.

274 H.E. Bergna, D.R. Corbin and G.C. Sonnichsen, US Pat. 4.683.334 (1987) and 4.752.596 (1988), DuPont.

275 K. Segawa, A. Sugiyama, M. Sakaguchi and K. Sakurai, in K. Tanabe et al., Eds., Acid-Base Catalysis, Kodansha, 1989.

276 I. Mochida, A. Yasutake, H. Fujitsu and K. Takeshita, J. Catal., 82 (1983) 313.
277 M. Deeba, Eur. Pat. Appl. 311.900 (1989), Air Prod. Chem. Inc.
278 M. Deeba, M.E. Ford and T.A. Johnson, Eur. Pat. 252.424 (1988), Air Prod. Chem. Inc.
279 T.A. Johnson, M.E. Ford and M. Deeba, DEOS 3.543.228 (1986), Air Prod. Chem. Inc.
280 M. Deeba, M.E. Ford and T.A. Johnson, Stud. Surf. Sci. Catal., 38 (1987) 221.
281 M. Deeba and M.E. Ford, J. Org. Chem., 53 (1988) 4594.
282 M. Deeba, M.E. Ford and T.A. Johnson, J. Chem. Soc. Chem Commun., 8 (1987) 562.
283 J. Peterson and H.S. Fales, US Pat. 4307250 (1981); Eur. Pat. 39918 (1984), Air Prod. Chem. Inc.
284 M.E. Deeba, W.J. Ambs, Eur. Pat. 77016 (1985), Air Prod. Chem. Inc.
285 J. Peterson and H.S. Fales, US Pat. 4375002 (1983), Air Prod. Chem. Inc.
286 W.J. Ambs, M.E. Deeba and J.F. White, Eur. Pat. 101921 (1986), Air Prod. Chem. Inc.
287 V. Taglieber, W. Hölderich, R. Kummer, W.D. Mross and E. Saladin, Eur. Pat. 133938 (1985) and Eur. Pat. 132736 (1985), BASF AG.
288 W. Hölderich, V. Taglieber, H. Pohl, R. Kummer and K.G. Baur, DBP 36342747 (1987), BASF AG.
289 M. Hesse, W. Hölderich and M. Schwarzmann, Eur. Pat. 296.495 (1988), BASF AG.
290 W. Hölderich, M. Hesse and M. Schwarzmann, Eur. Pat. 297.446 (1989), BASF AG.
291 C.D. Chang and W.H. Lang, US Pat. 4.434.299 (1984), Mobil Oil Corp.
292 C.D. Chang and W.H. Lang, Eur. Pat. 062.542 (1982); C.D. Chang and P.D. Perkins, Eur. Pat. 082.613 (1986), Mobil Oil Corp.
293 K.K. Pitale and R.A. Rajadhyaksha, Curr. Sci., 54 (1985) 447.
294 M.G. Warawdekar and R.A. Rajadhyaksha, Zeolites, 7 (1987) 574.
295 D.G. Jones and P.S. Landis, US Pat. 3.231.600 (1966).
296 K. Hatada, Y. Ono and K. Tominaga, in Adv. Chem. Ser. ACS, 121 (1973) 501.
297 M. Gans, Hydrocarbon Proc., 55 (1976) 145.
298 M. Onaka, M. Kawai and Y. Izumi, Chem. Lett., (1985) 779.
299 M. Onaka, K. Sugita and Y. Izumi, Chem. Lett., (1986) 1327; and in K. Tanabe et al., Eds., Acid-Base Catalysis, Kodansha, 1989, p. 33.
300 E. Kikuchi, T. Matsuda, K. Shimomura, K. Kawahara and Y. Morita, Stud. Surf. Sci. Catal., 28 (1986) 771.
301 M. Iwamoto, M. Tajima and S. Kagawa, J. Catal., 101 (1986) 195.
302 K. Eguchi, T. Tokiai and H. Arai, Appl. Catal., 34 (1987) 275.
303 M. Nomura and Y. Fujihara, Kinki Daigaku Kokagubu Kenkyu Hokoku, 19 (1985) 1.
304 Q. Chen, Zh. Yu and Zh. Liu, CN 85.102.398 (1986), Guangzhu Chem. Int.
305 P. Chu and G.H. Kuehl, US Pat. 4.605.787 (1986), Mobil Oil Corp.
306 J.A. Daniels and A. Stewart, Eur. Pat. 055.045 (1986), ICI.
307 M.R. Klotz, US Pat. 4.584.415 (1986), Amoco.
308 L.B. Young, US Pat. 4.365.084 (1982), Mobil Oil Corp.
309 Jap. Pat. 56.092.833 (1979), Mitsubishi Gas Chem. Ind.
310 S. Haruhito, DBP 3.149.979 (1982), Idemitsu Kosan Co.
311 R. Gregory, D.J. Westlake and J.J. Harry, Eur. Pat. 045.618 (1986), BP Co.
312 J.A. Ballantine, M. Davies, E.M. O'Neil, I. Patel, J.H. Purnell, M. Ravanakorn and K.J. Williams, J. Mol. Catal., 26 (19840 57.
313 F. Näumann, W. Hölderich and M. Schwarzmann, DPB 2.802.980 (1988) BASF AG.
314 R. Huang, Q. Chen and Z. Yu, CN 85.102.395 (1986).
315 J. Wang, Y. Li and Y. Zhou, Riyong Huaxue Gongye, 2 (1988) 52.
316 T.B. Grindley and R. Thangarasa, Carbohydr. Res. 172 (1988) 311.

726

317 L.A. Hamilton, US Pat. 3.328.439 (1967), Mobil Oil Corp.
318 H. Eckhardt, K. Halbritter, W. Hölderich and W.D. Mross, DBP 3.506.632 (1985), BASF AG.
319 D.P. Roelofsen, J.W.M. de Graaf, J.A. Hagendoorn, H.M. Verschoor and H. van Bekkum, Recl. Trav. Chim. Pays-Bas, 89 (1970) 193.
320 A. Corma, M.J. Climent, H. Garcia and J. Primo, Appl. Catal., 59 (1990) 333.
321 H.E. Fried, Eur. Pat. 122.654 (1984), Shell Int. Res.
322 D. Kallo, G. Onyestyak and J. Papp, in Proc. 6th Int. Zeol. Conf., Reno, Butterworth, 1984, p. 444.
323 G. Onyestyak, D. Kallo, J. Papp and J.E. Detrekoy, Hung. Pat. 29.972 (1984).
324 G.T. Kerr and P.S. Landis, US Pat. 3.312.751 (1967), Mobil Oil Corp.
325 B. Buchholz, US Pat. 4.102.931 (1978); E.J. Dziera and B. Buchholz, Eur. Pat. 082.289 (1983), Pennwalt Corp.
326 W.F. Hölderich, M. Hesse and E. Sattler, in Proc. 9th Int. Congr. Catal. Calgary, 1988, Vol. 1, p. 316.
327 W.F. Hölderich, Stud. Surf. Sci. Catal., 41 (1988) 83.
328 W.F. Hölderich, K. Schneider and L. Hupfer, DPB 3.632.530 (1988), BASF AG.
329 G.P. Hagen, US Pat. 4.433.174 (1984), Standard Oil Co.
330 Y.I. Isakov and Kh.M. Minachev, Russ. Chem. Rev., 51 (1982) 1188.
331 C.J. Plank, E.J. Rosinski and G.T. Kerr, US Pat. 4.011.278 (1977), Mobil Oil Corp.
332 Y. Isakov, N.Y. Usachev, T. Isakova and K. Minachev, Izv. Akad. Nauk SSR Ser. Khim., (1985) 1965 and (1986) 299.
333 Y. Isakov, K. Minachev and T. Isakova, Dokl. Akad. Nauk SSR, (1985) 284.
334 G.I. Golodets, N.V. Pavlenko, L.F. Korzhova, K.M. Vaisberg and Y.I. Churkin, Kinet. Katal., 25 (1984) 1015.
335 T.J. Huang and W.O. Haag, US Pat. 4.339.606 (1982) and Eur. Pat. 112.821 (1986), Mobil Oil Corp.
336 L.R.M. Martens, W.J. Vermeiren, D.R. Huybrechts, P.J. Grobet and P.A. Jacobs, in Proc. 9th Int. Congr. Catal., Calgary, 1988, Vol. 1, p. 420.
337 E. Rode, M.E. Davis and B.E. Hanson, J. Chem. Soc. Chem. Commun., 1985, 716.
338 M.E. Davis, P.M. Butler, J.A. Rossin and B.E. Hanson, J. Mol. Catal., 31 (1985) 385.
339 W. Hölderich, H. Lendle, P. Magnussen, H. Leitner, J.H. Mangegold and W. Leitenberger, Eur. Pat. 196.554 (1986), BASF AG.
340 P.A. Jacobs and H. Hinnekens, Eur. Pat. Appl. 329.923 (1989), Synfina-Olefina S.A.

Appendix I

Zeolite Structures

This appendix[*] presents a number of structural characteristics of some important zeolites and of $AlPO_4$-5. The zeolite structure types are listed in alphabetical order according to their IUPAC mnemonic codes.

The Figures labeled **(a)** show a schematic representation of the channel system pictured as it is oriented within a crystal of typical morphology. The framework structures are labeled **(b)**; T-O-T links in these Figures are represented as straight lines.

Some of the framework structures have been reproduced from W.M. Meier and D.H. Olson, *Atlas of Zeolite Structure Types*, IZA Structure Commission, 1987. Crystallographic and other data on a large number of other zeolite structures not listed here can also be found in the *Atlas*.

Contents:

AFI
FAU
FER
LTA
LTL
MFI
MOR

[*] composed by M.S. Rigutto

AFI

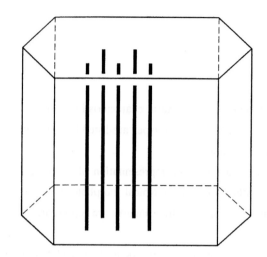

(a)

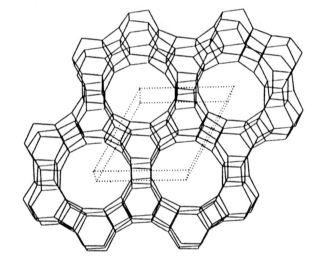

(b)

IUPAC name:	AFI (AlPO$_4$-5)
Composition:	[Al$_{12}$P$_{12}$O$_{48}$] • (C$_3$H$_7$)$_4$NOH • x H$_2$O
Important structural isotypes:	SAPO-5; SSZ-24 (all-silica)
Channel system:	1-D; straight circular 12-ring 7.3 Å channels

FAU

(a)

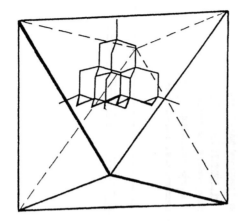

(b)

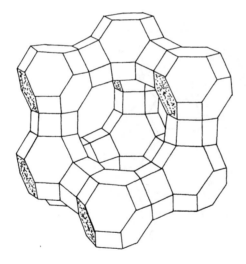

IUPAC name:	FAU (Faujasite)
Composition:	$Na_{58}[Al_{58}Si_{134}O_{348}]$ • 240 H_2O (NaY)
Important structural isotypes:	X (1 ≤ Si/Al ≤ 1.5); Y (Si/Al ≥ 2.5); SAPO-37
Channel system:	3-D; circular 12-ring 7.4 Å windows connecting ~ spherical 11.8 Å cavities (supercages)

730

FER

(a)

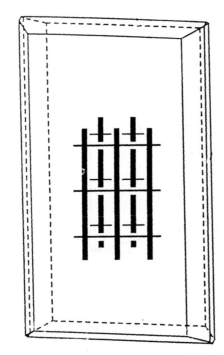

(b)

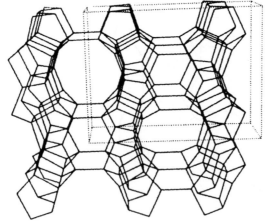

IUPAC name:	FER (Ferrierite)
Composition:	$Na_2Mg_2[Al_6Si_{30}O_{72}] \cdot 18\ H_2O$
Important structural isotypes:	high silica types (e.g. ZSM-35, NU-23; $5 \leq Si/Al \leq 25$)
Channel system:	2-D; straight 10-ring 4.2×5.4 Å channels connected by straight 8-ring 4.8×3.5 Å channels

LTA

(a)

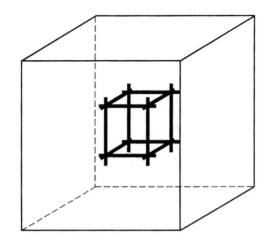

(b)

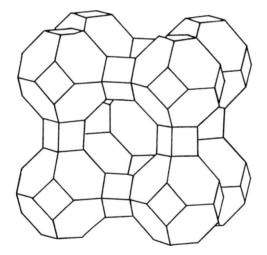

IUPAC name: LTA (Linde Type A)

Composition: $Na_{12}[Al_{12}Si_{12}O_{48}] \cdot 27\ H_2O$

Important structural isotypes: SAPO-42; high silica types ($1 \leq Si/Al \leq 3$)

Channel system: 3-D; circular 8-ring 4.1 Å windows connecting ~ spherical 11.4 Å cavities

LTL

(a)

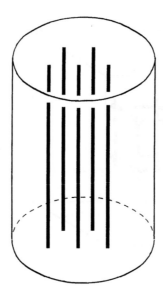

(b)

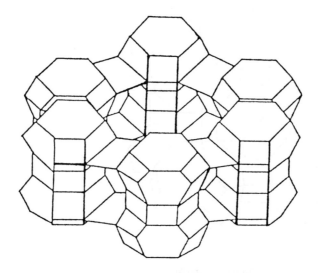

IUPAC name:	LTL (Linde Type L)
Composition:	$K_6Na_3[Al_9Si_{27}O_{72}] \cdot 21\ H_2O$
Channel system:	1-D; straight circular 12-ring 7.1 Å channels

MFI

(a)

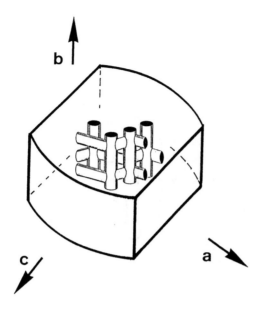

(b)

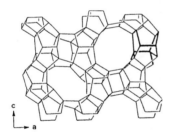 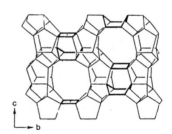

IUPAC name:	MFI (ZSM-5)
Composition:	$Na_n[Si_{96-n}Al_nO_{192}] \cdot 16\ H_2O$ (n ≤ 8)
Important structural isotypes:	Silicalite-1 (Si/Al = ∞); Boralite; TS-1; (Si,Ge)-MFI
Channel system:	3-D; straight 10-ring 5.2×5.7 Å channels connected by sinusoidal 5.3×5.6 Å channels; intersection cavities ~ 9 Å

MOR

(a)

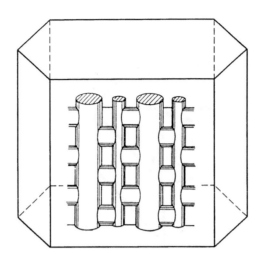

(b)

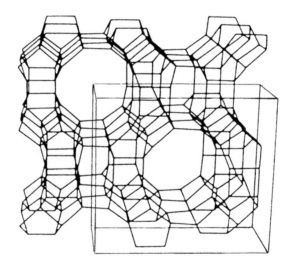

IUPAC name:	MOR (Mordenite)
Composition:	$Na_8[Al_8Si_{40}O_{96}] \cdot 24\ H_2O$
Important structural isotypes:	high silica types ($5 \leq Si/Al \leq 20$)
Channel system:	2-D; straight 12-ring 7.0×6.5 Å channels connected by short alternating 8-ring channels (~3 Å)

Appendix II

Pore sizes and structurally isotypic species

This appendix - compiled by H.E. van Dam from W.M. Meier and D.H. Olson, Atlas of Zeolite Structure Types, IZA Structure Commission, 1987 - presents a list of zeolite and $AlPO_4$ pore sizes and geometries, together with the names of structurally isotypic species. The structures are listed according to the largest free diameter of the largest channel.

STRUCTURE TYPE (IUPAC)	CHANNEL FREE DIAMETERS		CHANNEL SYSTEM DIMENSIONALITY	ISOTYPIC FRAMEWORK STRUCTURES
--	$\underline{12}$		1	VPI-5, MCM-9
--	$\underline{8.7}$·7.9		1	$AlPO_4$-8
HEU	$\underline{7.6}$·3.0	4.7·2.6	2	Heulandite
	4.7·2.6	4.6·3.3	2	Clinoptilolite
				LZ-219
--	$\underline{7.5}$·5.7	6.5·5.6	3	Beta
FAU	$\underline{7.4}$		3	Linde X
				Linde Y
				SAPO-37
				CSZ-3
				LZ-210
MAZ	$\underline{7.4}$		1	Mazzite types
	5.6·3.4		1	LZ-202
				Omega
				ZSM-4
AFI	$\underline{7.3}$		1	$AlPO_4$-5
				SAPO-5
LTL	$\underline{7.1}$		1	Linde L
				Gallosilicate L
				Perlialite
				LZ-212

GME	<u>7.0</u>	3.9·3.6	3.9·3.6	3	Gmelinite
MOR	<u>7.0</u>·6.5	5.7·2.6		2	Mordenite types Ca-Q Zeolon LZ-211
PAR	<u>6.9</u>·3.5			1	Partheite
OFF	<u>6.7</u>	4.9·3.6	4.9·3.6	3	Offretite Linde T LZ-217 TMA-O
AFS	<u>6.3</u>	4.0	4.0	3	MAPSO-46
AEL	<u>6.3</u>·3.9			1	$AlPO_4$-11 $SAPO$-11
APD	<u>6.3</u>·2.1	5.8·1.3		2	$AlPO_4$-D
AFY	<u>6.1</u>	4.3·4.0	4.3·4.0	3	CoAPO-50
STI	<u>6.1</u>·4.9	5.6·2.7		2	Stilbite Stellerite Barrerite
ATF	<u>6.1</u>·2.3	4.1·3.7		2	$AlPO_4$-25 $AlPO_4$-21
CAN	<u>5.9</u>			1	Cancrinite types ECR-5 Tiptopite
MTW	<u>5.9</u>·5.5			1	ZSM-12 CZH-5 Nu-13 Theta-3 TPZ-12
EUO	<u>5.7</u>·4.1 (with large side pockets)			1	EU-1 TPZ-3 ZSM-50
APC	<u>5.7</u>·2.9	3.7·3.4		2	$AlPO_4$-C $AlPO_4$-H3
MFI	<u>5.6</u>·5.3	5.5·5.1		2	ZSM-5 Silicalite
EPI	<u>5.6</u>·3.4		5.2·3.7	2	Epistilbite
TON	<u>5.5</u>·4.4			1	Theta-1 ISI-1 KZ-2 NU-10 ZSM-22

MEL	<u>5.4</u>·5.3			3	ZSM-11 Boralite D Silicalite 2
FER	<u>5.4</u>·4.2	4.8·3.5		2	Ferrierite types FU-9 ISI-9 NU-23 Sr-D ZSM-35
LAU	<u>5.3</u>·4.0			1	Laumontite Leonhardite
DAC	<u>5.3</u>·3.4	4.8·3.7		2	Dachiardite Svetlozarite
MTT	<u>5.2</u>·4.5			1	ZSM-23 EU-13 ISI-4 KZ-1
EAB	<u>5.1</u>·3.7			2	TMA-E
ERI	<u>5.1</u>·3.6			3	Erionite $AlPO_4$-17 LZ-220 Linde-T
MER	<u>5.1</u>·3.7 3.3·3.3	3.6·2.7 3.6·2.7	3.5·3.1 3.5·3.1	3 3	Merlinoite types K-M Linde W
YUG	<u>5.0</u>·3.1	3.6·2.8		2	Yugawaralite
BRE	<u>5.0</u>·2.3	4.1·2.8		2	Brewsterite
WEN	<u>4.9</u>·2.6	2.7·2.2		2	Wenkite
GIS	<u>4.8</u>·2.8	4.5·3.1		3	Gismondine types Amicite Garronite Gobbinsite MAPSO-43 Na-P1 Na-P2
LEV	<u>4.8</u>·3.6			2	Levyne SAPO-35 ZK-20
GOO	<u>4.7</u>·2.9	4.1·2.7	4.0·2.8	3	Goosecreekite
ATT	<u>4.6</u>·4.2	3.8		2	$AlPO_4$-12-TAMU $AlPO_4$-33

DDR	<u>4.4</u>·3.6	4.4·3.6		2	Deca-dodecasil 3R Sigma-1
LOV	<u>4.4</u>·3.2	3.7·3.6		2	Lovdarite
CHI	<u>4.3</u>·3.9			1	Chiavennite
PHI	<u>4.3</u>·3.0	3.6	3.3·3.2	3	Phillipsite Harmotome Wellsite ZK-19
ROG	<u>4.2</u>			1	Roggianite
ANA	<u>4.2</u>·1.6			1	Analcime types Pollucite types Leucite types Hsianghualite Kehoeite Viseite Wairakite $AlPO_4$-24 Na-B Ca-D
LTA	<u>4.1</u>			3	Linde A Alpha Gallogermanate A LZ-215 N-A SAPO-42 ZK-4 ZK-21 ZK-22
THO	<u>4.0</u>·2.2	3.9·2.3	variable	3	Thomsonite
KFI	<u>3.9</u> <u>3.9</u>			3 3	ZK-5 types P Q
NAT	<u>3.9</u>·2.6	<u>3.9</u>·2.6	variable	3	Natrolite types Gonnardite Mesolite Scolecite
PAU	<u>3.8</u> <u>3.8</u>			3 3	Paulingite
CHA	<u>3.8</u>			2	Chabazite CoAPO-44 CoAPO-47 LZ-218 Linde D Linde R MeAPO-47

					MeAPSO-47
					SAPO-34
					Wilherdersonite
					ZK-14
					ZYT-6
EDI	3.8·2.8	3.8·2.8	variable	3	Edingtonite
					K-F
					Linde F
ABW	3.8·3.4			1	Li-A
					$CsAlSiO_4$
					Gallosilicate ABW
					$RbAlSiO_4$
BIK	3.7·2.8			1	Bikitaite types
RHO	3.6			3	Rho
	3.6			3	Pahasapaite
					LZ-214

Impressions of the Summer School on Zeolites, 1989, Zeist, The Netherlands

Professor Peter Jacobs, President of the IZA, convincingly explaining catalysis by zeolites

Dr. Edith Flanigen and Dr. Gunther Engelhardt, caught by the photographer, in a moment of relaxation

742

Dr. Rosemary Szostak, recommending controlled zeolite modification

Lecturers (and authors) facing the camera lens. From left to right: G. Engelhardt, M.F.M. Post, R.P. Townsend, R. Szostak, I.E. Maxwell, H. van Koningsveld, D.P. de Bruyn, H.W. Kouwenhoven, P.A. Jacobs, J.C. Jansen, H.G. Karge, S.T. Wilson, J.H.C. van Hooff, E.M. Flanigen, R.A. van Santen, R.A. Schoonheydt, H. van Bekkum

Keyword index

STUDIES IN SURFACE SCIENCE AND CATALYSIS

Advisory Editors: B. Delmon, Université Catholique de Louvain, Louvain-la-Neuve, Belgium
J.T. Yates, University of Pittsburgh, Pittsburgh, PA, U.S.A.

752

Printed and bound by CPI Group (UK) Ltd, Croydon, CR0 4YY

03/10/2024

01040328-0014